Control Systems Engineering Exam Reference Manual

A Practical Study Guide

Fourth Edition
2022 Release

Control Systems Engineering Exam Reference Manual

A Practical Study Guide

Fourth Edition
2022 Release

For the NCEES Professional Engineering (PE) Licensing Examination

Bryon Lewis, PE (CSE), CAP, CCST III

Controls engineering encompasses a broad range of industries: power, paper and pulp, pharmaceuticals, manufacturing, water treatment, and chemical plants. Although this fourth edition addresses many different applications used by all these industries, it focuses on petrochemical applications.

The National Council of Examiners for Engineering and Surveying (NCEES) Professional Engineer (PE)/Control Systems Engineer (CSE) exam tends to be concentrated toward chemical and pharmaceutical plant design applications of code and control systems. The purpose of this book is to introduce new engineers to the depth of knowledge they will need to tackle the NCEES PE/CSE exam.

ISA™

Notice

The information presented in this publication is for the general education of the reader. Because neither the author nor the publisher has any control over the use of the information by the reader, both the author and the publisher disclaim any and all liability of any kind arising out of such use. The reader is expected to exercise sound professional judgment in using any of the information presented in a particular application.

Additionally, neither the author nor the publisher has investigated or considered the effect of any patents on the ability of the reader to use any of the information in a particular application. The reader is responsible for reviewing any possible patents that may affect any particular use of the information presented.

Any references to commercial products in the work are cited as examples only. Neither the author nor the publisher endorses any referenced commercial product. Any trademarks or tradenames referenced belong to the respective owner of the mark or name. Neither the author nor the publisher makes any representation regarding the availability of any referenced commercial product at any time. The manufacturer's instructions on the use of any commercial product must be followed at all times, even if in conflict with the information in this publication.

Printed in the United States of America.

ISBN 13: 978-1-64331-060-2 (print)

ISBN 13: ePub: 978-1-64331-061-9 (ePub)

ISBN 13: Kindle: 978-1-64331-062-6 (Kindle)

ISA
67 T. W. Alexander Drive
P.O. Box 12277
Research Triangle Park, NC 27709

Library of Congress Cataloging-in-Publication Data in process

SKY10161083_070126

Contents

Introduction to This Study Guide . . . . **xv**

Chapter 1 Welcome to Control Systems Engineering

Licensing as Professional Engineer/Control Systems Engineer 1
Why Become a Professional Engineer? 2
This Is the Fourth Edition of This Study Manual 3
Notes for Reading This Manual 3
Overview of Recommended Flowchart of Study for the CSE 5

Chapter 2 Exam General Information **7**

State Licensing Requirements 7
Eligibility 7
Exam Schedule 8
Description of Exam 8
Exam Content – The NCEES 2019 Specifications 8
- I. Measurement – 20 Questions 9
- II. Control Systems – 20 Questions 9
- III. Final Control Elements – 16 Questions 11
- IV. Signals, Transmission, and Networking – 12 Questions 12
- V. Safety Systems – 12 Questions 13

Exam Scoring 13

Chapter 3 Reference Materials for the Exam **15**

Recommended Books and Materials to Take to the Exam 15
Books and Materials for Testing 16
Books for Additional Study 16
Courses for Additional Study 16
Free Video Review Course for the (CSE) PE Exam 18

Chapter 4 Measurement Standards and Terminology **19**

Overview of Process Measurement, Control, and Calibration 19
Process Signal and Calibration Terminology 20
Definition of the Range of an Instrument 20
Definition of the Span of an Instrument 20
Definition of the Use of Zero in Instrumentation 21
- Live Zero 21
- Elevated Zero 21
- Suppressed Zero 21

Illustrations of Range and Span Terminology 22
Illustrations of Measured Variable, Measured Signal, Range, and Span . . . 22
Standard Abbreviations Used Throughout the Book and Their Definitions 22

Chapter 5 Fluid Mechanics in Process Control 25
Relationship of Pressure and Flow 25
Applications of Velocity and Pressure in Pipe Systems 29
Sizing a Pump Head Using the Specific Gravity of the Pumped Fluid. . . . 34
Summary of Fluid Mechanics for Process Control 36

Chapter 6 Temperature Measurement and Calibration 37
Temperature Measurement Devices and Calibration 37
Thermocouple – Worked Examples
(How to Read the Thermocouple Tables) 39
Resistance Temperature Detector. 39
RTD – Worked Examples. 40
Installing RTDs and Thermocouples in a Process Stream 43
Typical RTD and Thermocouple Applications. 43
Thermocouple Wake Frequency Calculations 44

Chapter 7 Pressure Measurement and Calibration 49
Pressure Measurement and Head Pressure 49
Applying Pressure Measurement and Signals – Worked Examples. . . . 50
Differential Pressure and Meter Calibration 50
Pressure Change in a Pipe for a Given Flow Rate 51
Pressure Change across the Flow Element for a Given Flow Rate 52
Pressure Calibration of Transmitter. 53

Chapter 8 Level Measurement and Calibration 55
Applying Level Measurement and Calibration – Worked Examples . . . 55
Level Displacer (Buoyancy). 58
Bubbler Level Measurement. 60
Density Measurement 61
Interface Level Measurement 63
Radar and Ultrasonic Level Measurement. 66
Capacitance Level Measurement 68
Radiometric (Gamma) Level Measurement 69
Level Gauging System in a Tank Farm 70
Calculating the Volume in Tanks. 70

Chapter 9 Flow Measurement and Calibration. 73
Applying Flow Measurement Devices 73
Turndown Ratio in a Flowmeter 73
ISA Standard Flow Meter Symbols 74
Flow Meter Applications Chart 74
Pressure Tappings (Impulse Line Taps) 75
Orifice Tap Dimensions and Impulse Line Connections 75
Various Types of Flowmeters 77
Applying Bernoulli's Principle for Flow Control. 77
Types of Head-Pressure Based Meters 78
Differential Head Meter Calculations. 80
The Beta Ratio. 84
Pipe Size Is Important 85
Standard Flow Measurement Equations. 86
Spink Flow Measurement Equation. 86

The Basic Spink Equation Derived . . . 87
The Basic Spink Equation for Liquid . . . 88
The Basic Spink Equation for Gas and Vapor . . . 89
The Basic Spink Equation for Steam . . . 89
ISO 5167 – Flow Measurement Equation . . . 91
Sizing Orifice Type Devices for Flow Measurement – Worked Examples . . . 95
Mass Flow Measurement and Control . . . 98
Applying Mass Flow Measurement with an Orifice – Worked Example . . . 101
Turbine Meter Applications . . . 103
Turbine Flowmeter – Worked Example . . . 106
Vortex Flowmeter Applications . . . 108

Chapter 10 Weight Measurement and Calibration . . . 111
Weight Measurement Devices and Calibration . . . 111
Load Cells . . . 111
Load Cells for (Flow, Level, Force) Applications in Process . . . 112

Chapter 11 Process Analyzers . . . 113
Electrical Conductivity and pH Correction . . . 113
How Are pH and Electrical Conductivity Measured? . . . 113
Control of pH Values in Processes . . . 114
Control of Conductivity . . . 114
Common Plant Analyzers . . . 115
Combustion and Analyzers . . . 117
Examples of Process Analyzers . . . 119
Select the Appropriate Analyzer and Configuration . . . 119

Chapter 12 Sensors for Fire and Rotating Equipment . . . 121
Sensors for Fire Protection and Flame Monitoring . . . 121
Sensors for Rotating Equipment Monitoring . . . 123

Chapter 13 Process Control Valves and Actuators . . . 127
Process Control Valves . . . 127
Considerations When Sizing a Control Valve . . . 128
ISA Standard Valve Symbols . . . 129
ISA Standard Pressure-Regulating Valve Symbols . . . 130
Valve Actuators . . . 130
ISA Standard Actuator Symbols . . . 131
Limit Switches on a Valve – ISA Standard Symbol . . . 131
Calculating the Size of the Actuator . . . 132
Split Ranging Control Valves . . . 134
Valve Positioner Applications . . . 135
Control Valve Application Comparison Chart . . . 138
Understanding Flow with Valve Characteristics . . . 138
Gain and Rangeability (Turndown Ratio in Valves) . . . 143
Proper Control Valve Sizing . . . 144
Oversized Valves Present Problems . . . 146
Summary of Control Valve Characteristics . . . 147
Control Valve Sizing . . . 148

The Valve Sizing Equations. 148
Sizing Valves for Liquid – Worked Example 150
Liquid. 150
Sizing Valves for Gas – Worked Example. 151
Gas 152
Sizing Valves for Vapor and Steam – Worked Example 154
Steam – Worked Example 154
Sizing Valves for Two-Phase Flow – Worked Example. 157
Two-Phase Flow 158
Flowing Quantity (Turndown Ratio of a Valve). 159
Flashing 160
Joule-Thomson Effect – Auto-Refrigeration in Valves. 160
Choked Flow. 160

Chapter 14 Pressure-Relief Valves and Rupture Disks 161
Pressure-Relief Valves and Pressure Safety Valves 161
Environmental Protection Agency Regulations. 162
Pilot-Operated Safety Valve 164
Bellow or Balanced Bellow and Diaphragm Valves 164
Venting Atmospheric and Low-Pressure Storage Tanks 166
API Standards for Pressure-Relieving Systems 168
CFR Standards for Pressure Relief Required by Federal Law 169
ISA Pressure-Relief Valve and Rupture Disc Symbols 172
Sizing Equations for Relief Valves and Rupture Disks. 173
Sizing Rupture Disks – Worked Examples. 175
Sizing Pressure-Relief Valves – Worked Examples 178

Chapter 15 Review of Feedback Control Fundamentals 185
Compare Open-Loop Control to Closed-Loop Control 185
Closed-Loop Example – A Mathematical Analysis. 187

Chapter 16 Review of Frequency Response Fundamentals 191
Electrical Application – A First-Order System. 191
Bode Plot of First-Order System. 192
Calculate the Data for the Bode Plot 193
Creating a Bode Plot – First-Order System Using Frequency 195
Hydraulic Application – A First-Order System 196

Chapter 17 Control Theory and Controller Tuning 199
Degrees of Freedom in Process Control Systems. 199
Controllers and Control Strategies (Models/Modes) 201
Process Loop Gain 204
Control Valve Linearization 205
Signal Filtering in Process Control 206
Applying Signal Filters 207
Example of Filter Time Selection 209
DCS/PLC Sample and Scan Time Consideration 210
Tuning of Process Controllers. 211
Closed-Loop Tuning of the Controller 212
Open-Loop Tuning of the Controller. 213
Advanced Tuning Methods for Controllers 216
The Integral Criteria Method 216

Process Characteristics from the Transfer Function 217
Poles, Zeros, and Dampening from the Transfer Function 218
Block Diagram Algebra 221
Example of Block Diagram Algebra Reduction 222
Nyquist Stability Criterion 222
Routh Stability Criterion 223
Check for Stability Using Routh (Example) 226
Feedforward and the Lead/Lag Compensator with
Heat Exchangers (Example) 228

Chapter 18 Communications and Control Networks 231
Overview of Corporate and Plant Networks 231
Open System Interconnect and TCP/IP Network Layer Model. 232
The Typical Network Model 234
Overview of Industrial Networks 236
Summary – Automation and Process Control Networks 251
Plant Facility Monitoring and Control Systems (FMCS). 251
Networked Intelligent and Smart Devices 252

Chapter 19 Digital Logic in Control Systems 255
Overview of Digital Logic. 255
Digital Logic Gate Symbols. 255
Digital Logic Gate Truth Tables 256
ISA Binary Logic. 256
Relay Ladder Logic. 257
Standard Relay Ladder Logic Symbols 258
Sealing Circuits. 259
Control System Architectures. 259
DCS Plant-Wide Control System Architecture – Networked. 260
PLC Control System Architecture 260
Programmable Logic Controller versus
Process Automation Controller 261
PLC Programming Languages. 262
LD or RLL 263
Structured Text 263
Functional Block Diagram 263
Sequential Function Chart 264

Chapter 20 Motor Control and Logic Functions. 265
Plant Electrical System. 265
Motor Control Center (MCC) 265
Typical MCC Design 266
How to Control a Motor 267
The Basic National Electrical Manufacturers Association
Stop-Start Station 268
NEMA and IEC Terminal Designations 269
Coil Lettering and Relay Socket Numbers
(NEMA and IEC Numbers). 269
Standard Symbols. 271
NEMA and IEC Comparisons 273
Stop-Start Station Control Circuit Schematic 273
Starter Control Circuit Schematic 274

Relay Ladder Logic and Function Blocks 274
RLL and its Boolean Functions 274
Putting RLL into the PLC 275
Safety Logic in the PLC 276
The PLC Logic for Valve and Alarm Monitoring 278

Chapter 21 Analog Circuits in Control Systems 281
Overview of Analog Signals 281
Typical Analog Loop Wiring Diagram 281
Constant Current Loops and Ohm's Law 283
Using Current to Transmit Transducer Data 284
Designing a Current Loop System 286
Adding More Transducers and Instruments 288
A Typical Current Loop Repeater 290
Sinking and Sourcing Devices 292

Chapter 22 Overview of Motion Controller Applications 295
Motion Control Systems 295
Stepper Motors 296
Servomotor Systems 297
Electrohydraulic Servo System 298
Soft-Starter Applications 299
Variable Frequency Drive 300
Volts to Hertz Ratio Relationship 304
PID Control with VFD or DC Drive 306

Chapter 23 Electrical Systems and Power Quality 307
Filtering Power and Harmonics 307
Proper Grounding Procedures 309

Chapter 24 Emergency Standby Systems 313
Uninterruptible Power Supply 313
Backup Generator 314
Building Monitoring and Controls System Implementation 314

Chapter 25 Hydraulics and Pneumatics 317
Fluid Power Systems 317
Hydraulic Systems 317
Pneumatic Systems 318

Chapter 26 Fluid Power Schematic Symbols 321

Chapter 27 ISA Standards for Documentation 329
ISA Instrument or Function Symbol—ANSI/ISA-5.1-2009 329
ISA Line-Type Symbols—ANSI/ISA-5.1-2009 330
Standard Line Types—ANSI/ISA-5.1-2009 330
ISA Identification Letters 331
ISA P&ID Identification (Controllers and Readouts) 332
ISA P&ID Identification (Transmitters, Switches, and Alarms) 333
ISA P&ID Identification (Compute, Relay, and Elements) 334
Piping and Equipment Symbols 335
Standard Piping and Instrumentation Diagram 335

ISA Standard P&ID as Might Be Included on the CSE Exam. 336
A More Complex P&ID as Might Be Included on the CSE Exam . 338
ISA Simplified P&ID as Might Be Included on the CSE Exam. . . . 339
ISA Standard Piping Flow Diagram or Mechanical Flow Diagram . . . 339
PFD or MFD as Might Be Included on the CSE Exam 340
Block Flow Diagram (BFD) 342
ISA Standard Loop Diagram 342
Instrument Location and Elevation Plan Drawing 345
Instrument Index Sheet 346
DCS or PLC I/O List (a List of Inputs and Outputs with Tags and Calibration Data). 347
ISA Standard HMI Graphical Display Symbols and Designations 348
NFPA 79 Colors for Graphical Displays (Industrial Machinery) 348
Battery Limits of the Plant 349

Chapter 28 Overview of Safety Instrumented Systems
Overview of Process Safety and Shutdown 351
SISs 351
Complying with IEC 61511/ISA-84 351
ISA and OSHA Letter Defining the Requirements for Implementation of SIS 352
Initiating Events of Safety Instrumented Systems. 354
The Difference Between BPCSs and SISs 354
IEC 61508 Mandatory and Guidelines. 355
Safety Instrumented Function and SIL. 356
Designing an SIS System. 357
SFF 359
PFD and PFH 360
SIL Capability and Safety System 361
SIF. 361
Voting or Polling of the System 363
Voting Probabilities 364
The SIS Calculations 364
Failure Models – The Bathtub Curve. 365
Reliability Laws 366
SIL and Availability 367
Metrics Used in the Reliability Engineering Field Involving SIS 368
SIS Calculations – Worked Example 371
SIS and SIL – Worked Examples. 373
Sample SIS Problems That Might be on the CSE Exam 373

Chapter 29 Overview of NEC/NFPA and Other Codes. 377
List of NFPA Codes (Be Familiar with These Codes) 377
NFPA 70 – NEC. 378
Voltage Drop Calculations 378
Voltage Drop Calculations – Worked Examples 380
NEC Article 500 Explosion-Proof Installations 381
Class I Hazardous Location NEC Article 501. 382
Class II Hazardous Location NEC Article 502. 384
Class III Hazardous Location NEC Article 503 385
Use of IEC Zone Classifications 386
Designation of NEC/CEC Classification. 387

Summary of the Various Hazardous (Classified) Locations 388
Hazardous Location Wiring Methods. 388
Purged and Pressurized Systems. 389
NEMA Electrical Enclosures Types and Uses 390
Nonhazardous Location NEMA Enclosure Types 390
NFPA 70E Standard for Electrical Safety 394
NFPA 77 Static Electricity 396
NFPA 780 Lightning Protection (Formerly NFPA 78) 399
NFPA 780 and NFPA 70 (NEC). 399
Summary of Lightning Protection Components 401
NFPA 79 Industrial Machinery 403
NFPA 496 Purged and Pressurized Systems 404
Type X Equipment (NFPA 496, Section 5-4.4). 406
Type Y Equipment (NFPA 496, Section 5-4.5). 407
Type Z Equipment (NFPA 496, Section 5-4.5). 407
Examples of Purged and Pressurized Systems 407
40 CFR and the Environmental Protection Agency –
Leak Detection and Repair 408

Chapter 30 Equations for Pumping, Piping, and Sizing Motors............ 411
Find Pipe Diameter with Velocity of Flow Known 411
Find Flow Velocity with Pipe Diameter Known 411
Change acfm to scfm with Standard Temperature and
Pressure Correction 411
Change scfm to acfm with Standard Temperature and
Pressure Correction 411
Find the Reynolds Number for the Flow 411
Calculate the Piping Head Losses to Size a Control Valve........... 412
Calculating the Motor Horsepower of Process Pumps.......... 413
Calculating the Motor Horsepower of Hydraulic Fluid Pumps .. 413
Calculating the Motor Horsepower of Fans...................414
Calculating the Motor Horsepower of Conveyors...............414
Piping Absolute Roughness Values 415

Chapter 31 Calculating Volume in Tanks 417
Cylindrical Tanks Upright 417
Cylindrical Tanks on Side.. 418
Spherical Tanks... 418
Bullet Tanks.. 419

Chapter 32 Exam Sample Questions...................................421
Sample Questions.. 421
Answers to Exam Sample Questions............................... 427
Explanations and Proofs of Exam Sample Questions 428

Chapter 33 Guide to the Fisher Control Valve Handbook.................439
Important Sections to Review.................................... 439
Important Pages to Possibly Tab................................. 439

Appendix and Data Tables 443

References...510

This Reference Manual Covers Most Subject Content for the PE/CSE Exam

This book thoroughly covers most subject content currently listed in the NCEES PE/CSE exam specifications for the PE in the discipline of control systems engineering as of 2019.

I have compiled numerous college textbook subjects into this one comprehensive guide. The International Society of Automation (ISA) offers books on most of these subjects that cover this information in extensive detail. The textbook subjects covered in this manual are:

- Fluid mechanics
- Fundamentals of instrumentation
- Process analyzers
- Control valves
- Pressure-relief valves
- Process control theory
- Digital fundamentals
- Motor controls
- Industrial electronics
- Electrical system design for industrial plants
- Fluid power systems
- Fundamentals of industrial networking
- ISA documentation standards
- Safety instrumented systems
- State and federal codes for process plant installations

Plan Your Study Time

If your only job is engineering and designing control systems, you should be able to adequately prepare for the PE/CSE exam in 8 weeks.

If you are very familiar with control systems, you should be able to adequately prepare for the PE/CSE exam in 12 weeks.

If you are new to control systems engineering and only have 4 years of experience, you should plan to spend 12–18 weeks studying this review material to be thoroughly prepared for the PE/CSE exam, as many of the subjects in the exam are not taught or covered in a college 4-year degree.

You should try to complete at least one module a week. You may want to combine some of the smaller modules, such as the ones on instrumentation, and study them at the same time. For modules on important subjects, such as safety instrumented systems (SISs) and documentation, you should plan on dedicating a significant amount of time. These subjects encompass a large percentage of the time spent on questions when taking the CSE exam.

Introduction to This Study Guide

This manual helps prepare the Professional Engineer (PE) candidate for the NCEES Practice of Engineering exam in the PE discipline option of control systems engineering (CSE). The CSE exam covers a broad range of subjects from the electrical, mechanical, and chemical engineering disciplines. This exam is not on systems theory, but on sound judgment of the application of process control systems and applicable codes. Basic process control systems (BPCSs) and safety instrumented systems (SISs) are presented in detail. Experience in engineering or designing process control systems is a necessity to pass this discipline of the NCEES Principles and Practice of Engineering exam.

Studying this manual and other reference materials recommended within it should adequately prepare the experienced engineer or designer to take the CSE exam. This manual presents many practical problems that may be presented on the CSE exam, with explanations and worked solutions. State and federal codes needed for the exam are reviewed, and standard documentation and design practices are demonstrated for the design of real-world plant control systems.

Most state licensing boards in the United States recognize the CSE exam; however, some states do not offer it. Check with your state licensing board to determine if it offers the CSE exam. If you live in a states that does not offer the CSE exam, you may choose to pursue licensing in another discipline (e.g., electrical, mechanical, or chemical engineering). You may also try to arrange to take the CSE exam in a neighboring state. More details about the exam content, application process, and study materials are presented later in this manual.

About the Author

Professional Engineer (PE)—Control Systems Engineering
Certified Manufacturing Engineer (CMfgE)—Integration and Control
Certified Automation Professional (CAP)
Certified Control System Technician Level III (CCST)
Certified Electronics Technician (CET)—Industrial Electronics
State of Texas Master Electrician

email: bryon.lewis@integrated.cc
profile: http://www.linkedin.com/in/bryonlewis

Bryon Lewis is a licensed PE in control systems engineering (CSE). He is a senior member of ISA and has held senior membership with SME. Lewis has over 30 years of experience in electrical, mechanical, instrumentation, and control systems. He holds letters of recommendation from Belcan Engineering, S & B Engineers and Constructors, Enron Corporation, and Lee College. Lewis' experience in diversified engineering and competitive projects is as follows:

- 12 years of engineering and design experience using AutoCAD 9 through 2014
- 16 years of field experience including start-up and troubleshooting and calibration of instruments
- Projects consist of compressor stations, and petrochemical and food process plants
- Turbine and compressor control systems, material handling systems, and burner management systems
- Development of P&IDs, MFDs, electrical power distribution systems, and control diagrams
- Engineering and implementation of Foxboro I/A and Honeywell DCS systems and security
- Network support including servers, workstations, routers, switches, and cabling
- Allen-Bradley automation and programmable logic controller (PLC) programming for the Allen-Bradley family of processors

Lewis has participated in projects for clients such as Shell Oil, Exxon, Diamond Shamrock, Eli Lilly Pharmaceuticals, Procter and Gamble (fault analysis), JVC America (solvent recovery project), Keebler Corporation, Mission Foods, Enron Transportation and Storage. He also worked at the Comanche Peak Steam Nuclear Station in 1987 and the powerhouse addition and computer grounding at the Johnson Space Center in 1985. If there are any questions, please contact Bryon Lewis at his email address.

1 Welcome to Control Systems Engineering

Licensing as Professional Engineer/Control Systems Engineer

A Professional Engineer (PE) license must be obtained to perform engineering work for the public and private sectors in the United States and most countries in the world. To protect the health, safety, and welfare of the public, the first engineering licensure law was enacted in 1907 in Wyoming. Now every state regulates the practice of engineering to ensure public safety by granting only Professional Engineers the authority to sign and seal engineering plans and offer their services to the public. *Individuals who do not have a PE license cannot use the title of* engineer *to advertise for engineering work.*

A control systems engineer (CSE) takes on responsibilities beyond those of most other disciplines of professional engineering. If a pump quits working, there is no water. If an electrical panelboard fails, there is no power. In plant control systems, a failure can mean absolute disaster. Without proper attention to these failures, the plant can explode, resulting in fatalities. The failure of systems can mean the loss of hundreds of thousands of dollars, and loss of product and production can cost millions of dollars. There may also be class action and environmental lawsuits into the billions of dollars.

This is why I have taken a complete plant design approach to show the vast exposure and experience needed to be a CSE. Just like the saying in the *Spiderman* movie, "With great power comes great responsibility." The CSE's job cannot be taken lightly. People's lives depend on CSEs knowing what they are doing and getting it right the first time. You cannot guess at control systems engineering. *You must know. Being a PE does not just mean answering a minimum number of questions on an 8-hour exam.*

The CSE cannot just say "the bottle is in place, now fill it." The CSE must ask questions such as:

1. Is the bottle in place?
2. Is the valve open?
3. Is fluid available to fill the bottle in the tank?
4. Is the pump running?
5. Is the fluid flowing?
6. Did the bottle fill?

7. Did the valve close?
8. Did the fluid stop flowing?
9. Did the pump stop?
10. Did something fail?

The CSE must be ready to handle abnormal conditions and upsets at any time. This will be a major part of the programming and a large part of the instrumentation, with increasing concern for safety and compliance with government regulations now requiring safety instrumented systems (SISs) to be installed.

Explosions can occur in petrochemical and other similar hazardous plants, even though the electrical and process systems are designed to be explosion-proof per National Fire Protection Association (NFPA), American National Standards Institute (ANSI)/International Society of Automation (ISA), American Petroleum Institute (API), Occupational Safety and Health Administration (OSHA), International Electrotechnical Commission (IEC), and other codes.

Why Become a Professional Engineer?

Being licensed as a PE is an important distinction that can enhance one's career options. Many engineering jobs require that a person have a PE license to work as an engineering consultant or senior engineer, testify as an expert witness, conduct patent work, work in public safety, or advertise to provide engineering services. Although you may never need to be registered for "legal" reasons, you may find that you must be a PE to be eligible for engineering management positions.

On average, a PE makes significantly more money than an unlicensed engineer. Even if your first job does not require a PE license, you may need a license later in your career. In today's economic environment, it pays to be in a position to move to new jobs and compete with others who have a PE license or are on a professional engineering track. It is also highly unlikely that a job requiring a PE license will be outsourced overseas.

The following excerpt is from the NCEES website.

What makes a PE different from an engineer?

- Only a licensed engineer may prepare, sign and seal, and submit engineering plans and drawings to a public authority for approval, or seal engineering work for public and private clients.
- PEs shoulder the responsibility for not only their work, but also for the lives affected by that work and must hold themselves to high ethical standards of practice.
- Licensure for a consulting engineer or a private practitioner is not something that is merely desirable; it is a legal requirement for those who are in responsible charge of work, be they principals or employees.

- Licensure for engineers in government has become increasingly significant. In many federal, state, and municipal agencies, certain governmental engineering positions, particularly those considered higher level and responsible positions, must be filled by licensed professional engineers.
- Many states require that individuals teaching engineering must also be licensed. Exemptions to state laws are under attack, and in the future, those in education, as well as industry and government, may need to be licensed to practice. Also, licensure helps educators prepare students for their future in engineering.

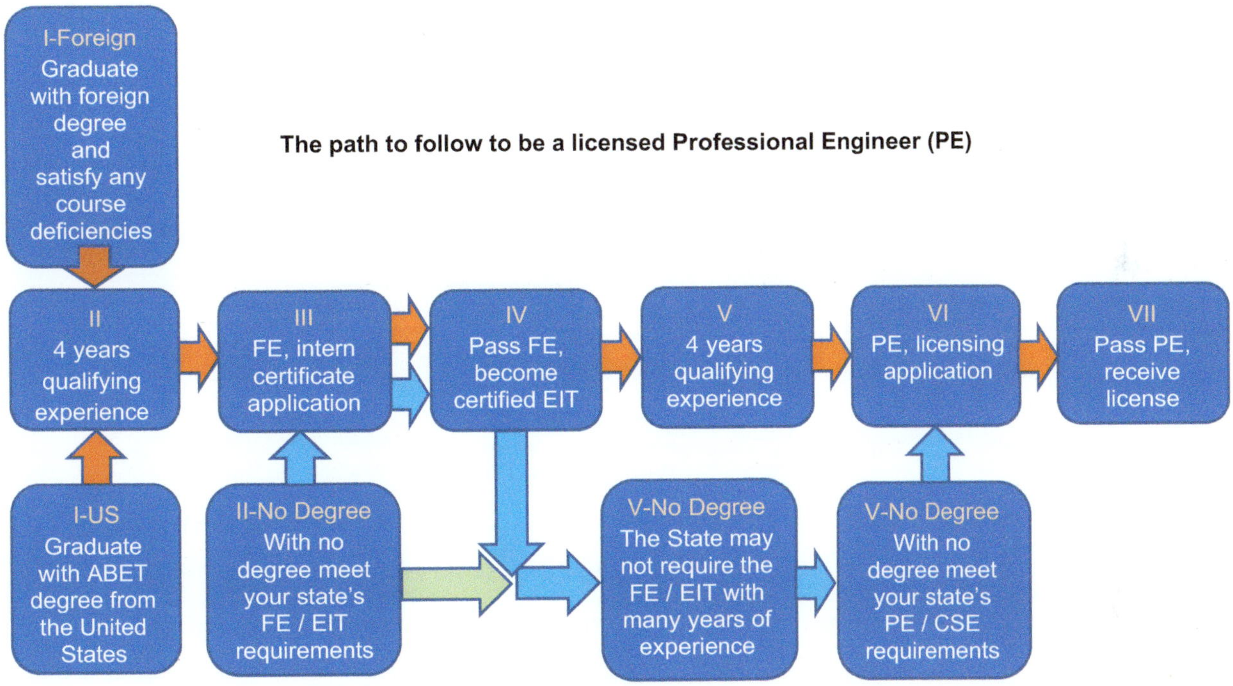

This Is the Fourth Edition of This Study Manual

This new edition is based on the NCEES 2019 NCEES PE/CSE exam specification. It has been streamlined to be more focused on those specifications, while including new and expanded chapters on numerous control systems subjects.

I have taken a systematic approach to explaining how to design a complete plantwide control system for multiple processes, a task that PE/CSEs may encounter throughout their career.

Notes for Reading This Manual

The following box indicates that the material or example problems may be on the CSE exam. Major topics of study are in a box with a gray background.

Sample Problem: The problem will be stated in this colored area with any parameters needed to solve the problem.

The solution will be in this colored area.

Major topics of study will appear in highlighted text of this color. This color will break up the topic.

Important information for the CSE exam will appear in highlighted text of this color.

Recommended Flowchart of Study for the CSE Exam

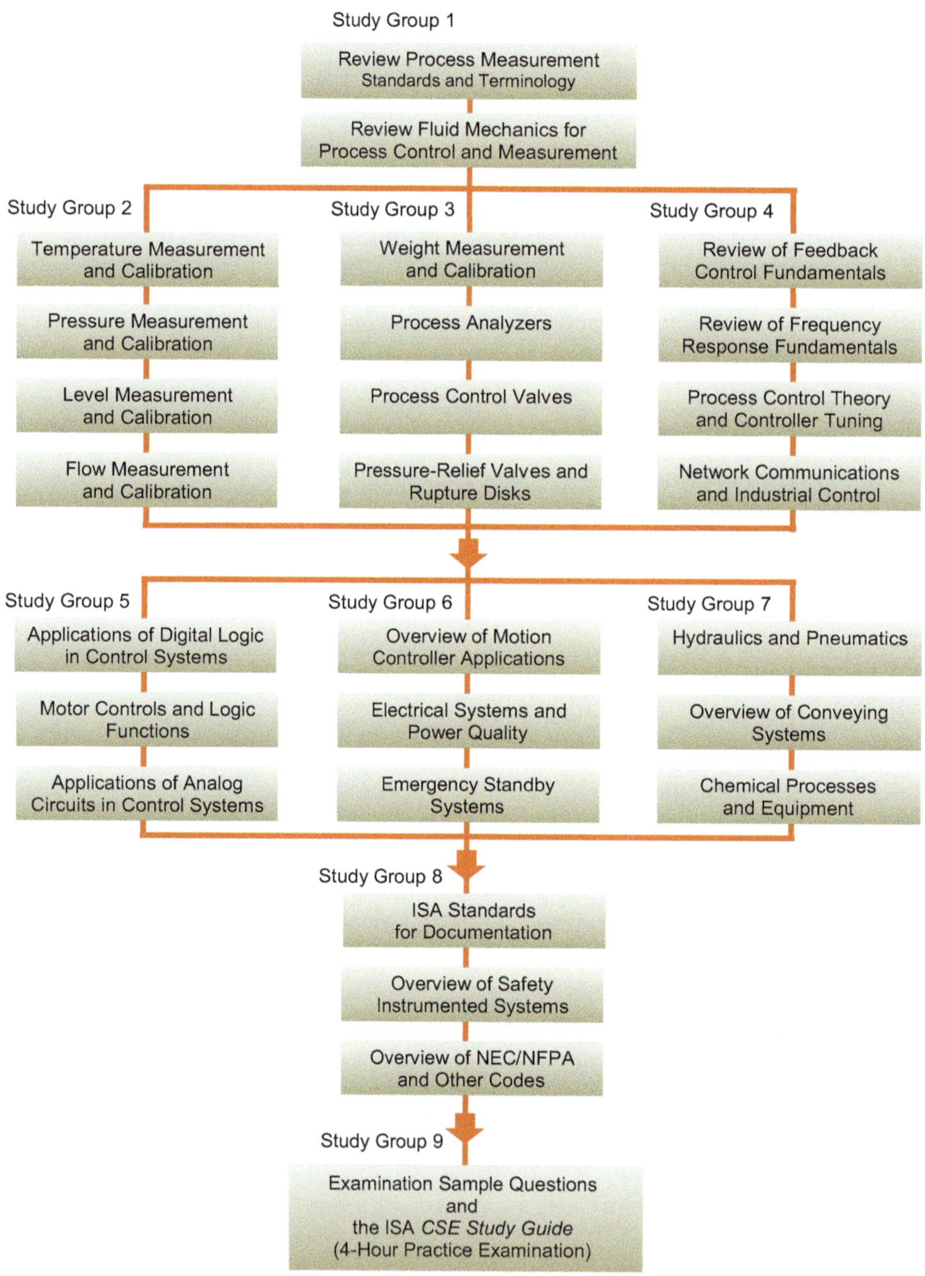

Overview of Recommended Flowchart of Study for the CSE

Note: *The recommended flow of study using this manual is as follows:*

Study Group 1

Definitions of *span, range, zero, suppressed zero, elevated zero,* and *lower-* and *upper-range values*

Pressure, head, and the relationship between DP (pressure) and flow

Differential pressure in pipes and head-type flow elements

Pump head and work done by the total energy in fluids

Study Group 2

Temperature of liquids, gases, and vapors

Level of liquids

Density and interface level

Flow rates of liquids, gases, and vapors

Mass rate of liquids, gases, and vapors

Study Group 3

Level of solids and liquids

Mass rate of solids and liquids

Molecular composition

Flow control of liquids, gases, and vapors

Pressure control and regulation

American Society of Mechanical Engineers (ASME) VIII and NFPA 30 codes for equipment and piping protection from overpressure

Study Group 4

Laplace transforms and block algebra reduction

Process control modes and loop applications

Process system response and tuning of controllers

Networks and communications

Process control through networking

Study Group 5

Digital functions and truth tables

Relay Ladder Logic and motor control applications with National Electrical Manufacturers Association (NEMA) and IEC standards

Constant current and constant voltage control-loop applications

Analog loop signals and impedance

Study Group 6

Stepper motor and servo systems (electric and hydraulic)

Speed control and variable frequency drive operations

Pulse wave modulation

Harmonics and electrical noise

Power systems for critical loads

Study Group 7

Control with hydraulics and pneumatics

Study overview of basics of conveyors

Study overview of basics of process systems

Study Group 8

ISA standard piping and instrumentation diagram (P&ID), mechanical flow diagram (MFD), block flow diagram (BFD), loop sheets (electrical and pneumatic) and human-machine interface (HMI)

Basic process control system (BPCSs) and safety instrumented systems (SISs)

Designing a safety instrumented system using safety integrity level (SIL) and safety instrumented function (SIF) values

SIS calculations for the CSE exam

OSHA, NFPA, National Electric Code (NEC), Code of Federal Regulations (CFR) and US Environmental Protection Agency (EPA) codes

Study Group 9

ISA *CSE Study Guide* (containing a 4-hour sample test), available from ISA

ISA PE/CSE exam review course

LearnControlSystems.com online and hands-on supplement course (putting it all together review)

2 Exam General Information

State Licensing Requirements

Licensing of engineers is intended to protect the public health, safety, and welfare. State licensing boards have established requirements to be met by applicants for licenses that will, in their judgment, achieve this objective.

Licensing requirements vary somewhat from state to state but have some common features. In all states, candidates with a 4-year engineering degree from an Accreditation Board for Engineering and Technology (ABET)/Engineering Accreditation Commission (EAC) accredited program and 4 years of acceptable experience can be licensed if they pass the Fundamentals of Engineering (FE) exam and the Principles and Practice of Engineering exam in a specific discipline. References must be supplied to document the duration and nature of the applicant's work experience.

Eligibility

Some state licensing boards will accept candidates with engineering technology degrees, related-science (such as physics or chemistry) degrees, or no degree, with indication of an increasing amount of work experience. Some states will allow waivers of one or both exams (FE and PE) for applicants with many years (6–20) of experience. Additional procedures are available for special cases, such as applicants with degrees or licenses from other countries. Most states have abandoned the no-degree statute and will only accept as minimal, an accredited associate degree.

Note: Recipients of waivers may encounter difficulty in becoming licensed by "reciprocity" or "comity" in another state where waivers are not available. Therefore, applicants are advised to complete an ABET accredited degree and to take and pass the FE/Engineer in Training (EIT) exam. Some states require a minimum of 4 years of experience after passing the FE/EIT exam before allowing a candidate to sit for the PE (principles and practices) exam. Some states will not accept experience incurred before passing the FE/EIT exam as qualifying experience.

It is necessary to contact your licensing board for the up-to-date requirements of your state. Phone numbers and addresses can be obtained by calling the information operator in your state capital, or by checking the Internet at www.ncees.org or nspe.org.

Exam Schedule

The CSE exam is offered once per year, on the last weekend in October (typically on Friday). Application deadlines vary from state to state, but typically are about 3 or 4 months ahead of the exam date.

Requirements and fees vary among state jurisdictions. Enough time must be allotted to complete the application process and assemble the required data. PE references may take a month or more to be returned. The state board needs time to verify professional work history, references, and academic transcripts or other proof of the applicant's engineering education.

After accepting an applicant to take one of the exams, the state licensing board will notify him or her where and when to appear for the exam. The board will also describe any unique state requirements, such as allowed calculator models or limits on the number of reference books taken to the exam site.

Description of Exam

Exam Format

The NCEES Principles and Practice of Engineering exam (commonly called the *PE exam*) in control systems engineering (CSE) is an 9.5-hour exam administered as computer-based testing (CBT) with a 50-minute break. The test contains 85 questions in a multiple-choice format.

Each question has a correct or "best" answer. Questions are independent, so an answer to one question has no bearing on the following questions.

All the questions are compulsory; applicants should try to answer all the questions. If a question is omitted or the answer is incorrect, a score of zero will be given for that question. There is no penalty for guessing.

Exam Content – The NCEES Specifications

The subject areas of the CSE exam are described by the exam specification and are given in five areas. ISA supports CSE licensing and the exam for professional engineering. ISA is responsible for the content and questions in the NCEES exam. Refer to the ISA website (http://www.isa.org) for the latest information concerning the CSE exam.

For a copy of the latest PE/CSE exam format and content, visit NCEES at: http://www.ncees.org.

The following is an overview of the categories and content that might be included the exam. The NCEES website has the latest information on the exam, as the format and specifications change over the years.

I. Measurement – 17 to 27 Questions

- Sensors
  - Sensor technologies applicable to general measurement (e.g., flow, pressure, level, temperature, counters, and motion)
  - Sensor technologies applicable to general analytical instruments and sampling systems (e.g., pH, ORP, density, O_2, conductivity, effects of sampling systems, and GC)
  - Grounding, shielding, segregation, and alternating current (AC) coupling
  - Sensor technologies applicable to fire and gas detection
  - Sensor technologies applicable to machinery monitoring and protection (e.g., vibration, bearing temperature, lube oil pressures, thrust, and speed)
  - Sensor characteristics (e.g., rangeability, accuracy and precision, temperature effects, response times, reliability, repeatability, maintenance, and calibration)
  - Sensor selection (e.g., plugging service, process severity, environmental effects and constraints, and costs)
  - Material compatibility
  - Installation details (e.g., process, pneumatic, electrical, location, maintenance, and calibration)
- Flow, Level, and Pressure Calculations
  - Flow (e.g., element sizing, pressure-temperature compensation, and mass/volume)
  - Level
  - Pressure drop
- General Calculations
  - Unit conversions
  - Velocity
  - Square root extraction and interpolation
  - Variables involved in wake frequency calculations (e.g., thermowell length/ diameter, velocity, natural frequency, and wake frequency)

II. Control Systems – 17 to 27 Questions

- Drawings
  - Drawings (e.g., process flow diagrams, piping and instrumentation diagrams—P&IDs, loop diagrams, Ladder Diagrams, logic drawings, cause-and-effects drawings, electrical drawings, schematics, and wiring diagrams)

- Theory
  - Basic control of processes (e.g., pumps, compression, combustion, evaporation, distillation, hydraulics, reaction, dehydration, heat exchangers, crystallization, filtration, refrigeration, and fluidization)
  - Process dynamics (e.g., loop response, pressure-volume-temperature relationships, and simulations)
  - Basic control (e.g., regulatory control, feedback, feedforward, cascade, ratio, proportional-integral-derivative—PID control, split-range control, and gap control)
  - Discrete control (e.g., relay logic, Boolean algebra, and aliasing)
  - Sequential control (e.g., batch, assembly, conveying, computer numerical control—CNC, state machine, and Sequential Function Chart—SFC)
- Implementation
  - Human-machine interface (HMI) (e.g., graphics, alarm management, trending, historical data, and operator panels)
  - Equipment layout (e.g., human factors engineering, physical control room arrangement, and panel layout)
  - Limited variability programming languages for distributed control systems (DCSs) and programmable logic controllers (PLCs) (e.g., IEC 61131-3 languages/Ladder Diagrams, Function Block Diagrams—FBDs, SFCs, Structured Text—ST, and Instruction List—IL)
  - System design comparisons and compatibilities (e.g., advantages and disadvantages of system architecture, distributed architecture, remote input/output—I/O, buses, and wireless)
  - Installation requirements (e.g., shielding, constructability, I/O termination, environmental, heat load calculations, power load requirements, purging, lighting, and maintainability)
  - System testing (e.g., factory acceptance test—FAT, integrated system test, and site acceptance test—SAT)
  - Commissioning (e.g., performance tuning and loop checkout)
  - Performance evaluation (e.g., troubleshooting, and root cause failure analysis and correction)
- Security of Industrial Automation and Control Systems
  - Security (e.g., physical, cyber, network, firewalls, routers, switches, protocols, hubs, segregation, and access controls)
  - Security life cycle (e.g., assessment, controls, audit, and management of change—MOC)
  - Requirements for a security management system

- Security risk assessment and system design
- Product development and requirements
- Verification of security levels (e.g., SL 1, SL 2, SL 3, and SL 4)

III. Final Control Elements – 14 to 23 Questions

- Valves
  - Types (e.g., globe, ball, and butterfly)
  - Trim characteristics (e.g., linear, low noise, equal percentage, and seat leakage class)
  - Calculation (e.g., sizing, split range, noise, actuator, response time, pressure drop, and air/gas consumption)
  - Selection of motive power and failure mode (e.g., hydraulic, pneumatic, electric, and spring)
  - Applications of fluid dynamics (e.g., cavitation, flashing, choked flow, Joule-Thompson effects, and two-phase)
  - Material selection based on process characteristics (e.g., erosion, corrosion, plug, extreme pressure, temperature, and material compatibility)
  - Accessories (e.g., limit switches, solenoid valves, positioners, transducers, air regulators, servo amp, boosters, and quick exhaust)
  - Environmental constraints (e.g., fugitive emissions, packing, special sealing, and fire rating)
  - Installation practices (e.g., vertical, horizontal, bypasses, location, and flow direction)
- Pressure-Relieving Devices
  - Pressure-relieving valve types (e.g., conventional spring, balanced bellows, and pilot-operated)
  - Pressure-relieving valve characteristics (e.g., modulating and pop action)
  - Pressure-relieving valve calculations (e.g., sizing considering inlet pressure drop, back pressure, and multiple valves)
  - Material selection based on process characteristics
  - Pressure-relieving valve installation practices (e.g., linking valves, sparing the valves, accessibility for testing, car sealing inlet valves, piping installation, and combination devices)
  - Rupture discs and buckling pin valves (e.g., types, characteristics, application, and calculations)

- Motor Controls
  - Types (e.g., motor starters and variable-speed drives)
  - Applications (e.g., speed control, soft starters, and motor-operated valve actuators)
  - Calculations (e.g., sizing, tuning, and location)
  - Accessories (e.g., encoders, positioners, relays, and limit switches)
- Other Final Control Elements
  - Motion (e.g., damper controls, types, orientation, actuators, servos, and encoders)
  - Solenoid valves (e.g., types and sizing)
  - On-off devices/relays (e.g., types, applications, and energize and de-energize to trip)
  - Self-regulating devices (e.g., types, sizing, pressure, temperature, and level and flow regulators)

IV. Signals, Transmission, and Networking – 11 to 18 Questions

- Signals
  - Pneumatic, electronic, optical, hydraulic, digital, analog, buses, wireless, and thermocouple
  - Transducers (e.g., analog to digital—A/D, digital to analog—D/A, current to pneumatic —I/P, current to current—I/I, splitters, and filters)
  - Hazardous area classification and instrument installation techniques (e.g., intrinsically safe—IS barriers, cabinet purges, and nonincendive)
  - Grounding, shielding, segregation, and electromagnetic interference
  - Basic signal circuit design (e.g., two-wire, four-wire, isolated outputs, loop powering, and buses)
  - Circuit calculations (voltage, current, impedance, and power)
  - Unit conversion calculations
- Transmission
  - Different communication systems architecture and protocols (e.g., fiber optics, coaxial cable, wireless, paired conductors, buses, Transmission Control Protocol/Internet protocol—TCP/IP, and Open Platform Communications—OPC)
  - Distance considerations versus transmission medium (e.g., data rates and sample rates)
- Networking
  - Routers, bridges, switches, firewalls, gateways, network loading, error checking, bandwidth, crosstalk, parity, and hubs

V.Safety Systems – 11 to 19 Questions

- Documentation
  - Basic documentation required (e.g., process hazards analysis, safety requirements specification—SRS, logic diagrams/narratives, test procedures, SIL selection report, SIL verification report, safety life-cycle plan)
- Theory
  - Reliability and availability (e.g., bathtub curve, failure rates types, voting, proof test intervals, and common cause and diversity)
  - SIL selection (e.g., safety layer matrix, risk graph, and Layer of Protection Analysis—LOPA)
- Implementation
  - Safety system design (e.g., SRS, I/O assignments, redundancy, segregation, logic design, and failure direction)
  - SIL verification calculations (e.g., failure rates types, voting, proof test intervals, and common cause and diversity)
  - Installation, commissioning, and validation (e.g., methods, procedures, and test records)
- Safety Life-Cycle Management
  - Modifications (e.g., management of change, scope of change, impact of change, and documentation)
  - Operations and maintenance (e.g., methods, procedures, test records, partial stroke testing, demand tracking, bypass and override management, failure analysis, and validation of design assumptions)

Exam Scoring

NCEES exams are scored independently. There are no prespecified percentages of candidates that must pass or fail.

Assisted by a testing consultant, a panel of licensed CSEs uses recognized psychometric procedures to determine a passing score corresponding to the knowledge level needed for minimally competent practice in the discipline.

The passing score is expressed as the number of questions out of 80 that must be answered correctly. The method used for pass-point determination assures that the passing score is adjusted for variations in the level of exam difficulty and that the standard is consistent from year to year.

When an exam is introduced or when its specifications change, a committee of subject matter experts works with experienced psychometricians (testing experts with a background in statistics) to determine the level of performance that corresponds with minimal competence in that discipline. This becomes the passing score. NCEES does not publish passing scores because they change with each administration. NCEES scores each exam with no predetermined percentage of examinees that should pass or fail. All exams are scored the same way. First-time takers and repeat takers are graded to the same standard.

3 Reference Materials for the Exam

Recommended Books and Materials to Take to the Exam

I have included a review of all subject material that is in the NCEES PE/CSE exam specifications and almost any data you may need to look up for questions on the exam.

The list of recommended books and materials for testing is provided to help you pass the CSE exam. Use the materials with which you are most comfortable. A substitution with the same material and information may be used.

The list of recommended books and materials for additional study can be helpful in the review of subjects and preparation for the exam. See the International Society of Automation (ISA) website (www.isa.org) for more books that may give you knowledge and deeper insight into various subjects in instrumentation and control systems.

Remember to keep the review simple. The test is not focused on control systems theory studies, but rather on general functional design. Again, keep your studies simple and practical; control systems theory will encompass only about 3% of the exam.

National Council of Examiners for Engineering and Surveying

NCEES
advancing licensure for engineers and surveyors

Nonprofit organization

The National Council of Examiners for Engineering and Surveying is a national nonprofit organization composed of engineering and land surveying licensing boards representing all US states and territories.

Founded: 1920

NCEES on Wikipedia

NCEES on LinkedIn

Click on any link above to visit the site.

Books and Materials for Testing

- NCEES-approved calculator (Have a spare with new batteries installed.) I recommend the TI-36X Solar (any light). Practice with the calculator you will be using. (See www.ncees.org for a current list of approved calculators.)
- ANSI/ISA-5.1-2009, *Instrumentation Symbols and Identification*
- ISA-5.2-1976 (R1992), *Binary Logic Diagrams for Process Operations*
- ISA-5.3-1983, *Graphic Symbols for Distributed Control/Shared Display Instrumentation, Logic, and Computer Systems*
- ISA-5.4-1991, *Instrument Loop Diagrams*
- Fisher or Masoneilan *Control Valve Sizing Handbook/Catalog* (Most valve-sizing data needed for the CSE exam is in these reference manuals)
- A safety relief valve book (could be useful)

Books for Additional Study

- *The CSE Study Guide* from ISA (I highly recommend purchasing this book which contains a 4-hour review exam.)
- *Instrumentation for Process Measurement and Control*, 3rd ed., CRC Press, LLC, by Norman A. Anderson (The Foxboro Company)
- *Basic and Advanced Regulatory Control: Systems Design and Application*, 3rd ed., ISA, by Harold Wade
- *Measurement and Control Basics*, ISA, by Thomas A. Hughes
- *Process Control: A Practical Approach*, Wiley, by Myke King
- *Crosby® Pressure Relief Valve Engineering Handbook*
- *Pentair Pressure Relief Valve Engineering Handbook*
- *Programmable Controllers: Theory and Implementation*, Industrial Text Co., by L. A. Bryan and E. A. Bryan
- Visit the ISA webpage on preparing for the CSE exam (https://www.isa.org/isa-certification/cse-licensure-preparation/) for a list of recommended study material.

Courses for Additional Study

ISA Control Systems Engineer (CSE) PE Exam Review

- ISA offers an instructor-led Control Systems Engineer (CSE) PE exam review course at various locations across the nation.

The ISA CSE information page:
http://www.isa.org/isa-certification/cse-licensure-preparation

This course is typically taught by Gerald Wilbanks, PE He is a registered professional engineer in four states, a member of NSPE and ASQ, and an international former president (1995) of ISA. Gerald is a graduate of Mississippi State University with a BS in electrical engineering and was recognized as the Engineer of the Year in 1991 by the Engineering Council of Birmingham.

Gerald Wilbanks, PE

He is a Distinguished Engineering Fellow of MSU and a Life Fellow member of ISA. He has served as an instructor for many courses, seminars, and other educational sessions for ISA and in his own business.

See the ISA website (www.isa.org) for more books and training materials on advanced and basic subjects. ISA offers webinars with instructor-led training in many aspects and topics of process control and networking, as well as the popular topic of networking security. ISA also offers several online study courses in instrumentation and process control for people needing an introduction to the fundamentals of instrumentation. Many of the training videos and ISA/ANSI standards are free for ISA members.

ISA Control Systems Certifications

- **Certified Automation Professional (CAP)**

  ISA CAP certification provides a nonbiased, third-party, objective assessment and confirmation of an automation professional's skills. The CAP certification exam is focused on direction, definition, design, development/application, deployment, documentation, and support of systems, software, and equipment used in control systems, manufacturing information systems, systems integration, and operational consulting. The CAP certification exam reflects the documented knowledge, skills, and abilities needed for competent job performance.

- **Certified Control System Technician (CCST)**

  ISA's Certified Control Systems Technician (CCST) program provides a nonbiased, third-party, objective assessment and confirmation of a technician's skills. CCSTs are skilled in pneumatic, mechanical, and electronic instrumentation. They understand process control loops and process control systems, including computer-based systems.

  ISA's three levels of CCST certification require differing degrees of technical experience, education, and training, with CCST Level III requiring the most. To help you demonstrate your specific level of experience and expertise, the program allows you to apply for and test at the level for which you qualify.

Free Video Review Course for the (CSE) PE Exam

The author offers online study courses in CSE as a supplement to the ISA (CSE) PE exam review course. They include streaming videos of instructor-led training. These courses demonstrate real-world applications of calculations and the tuning and response of process systems. The videos include typical plant control systems and instrumentation seen every day in a large chemical plant or manufacturing environment.

Topics include instrumentation calculations and calibration procedures, industrial networking configuration and troubleshooting, motor controls, programmable logic controller (PLC) programming, electrical installations and codes, instrumentation and electrical safety grounding, applications of fluid mechanics, and applied thermodynamics for process control and measurement.

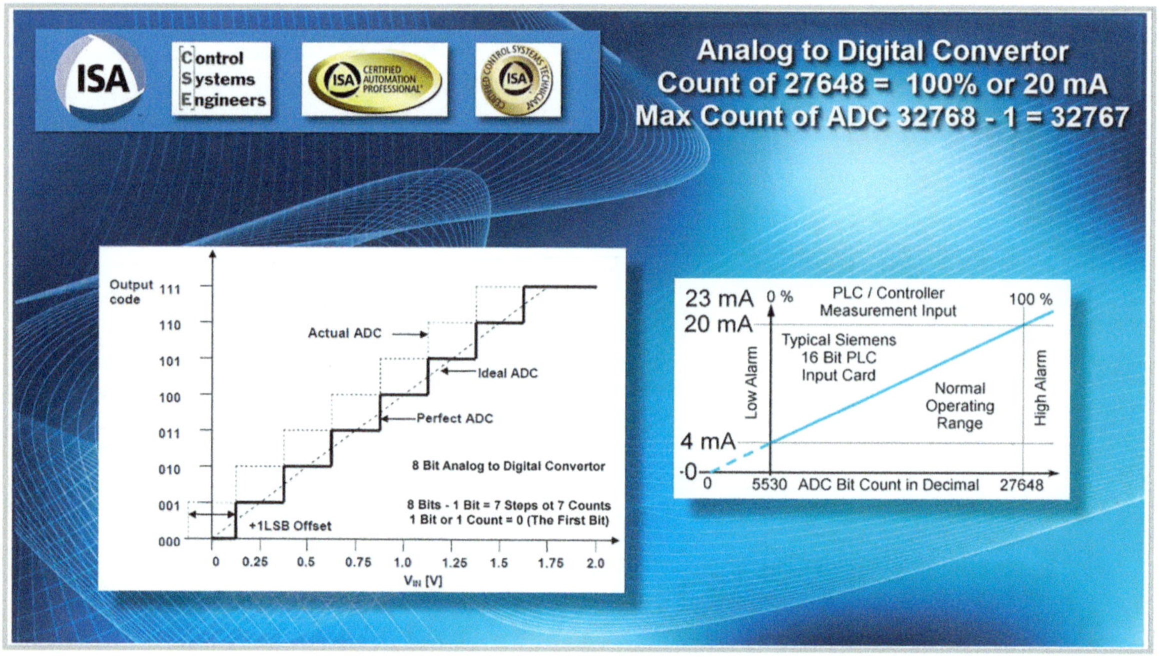

Free Video Review Course

The author's free 1080p video course can be found on YouTube. Search for "ISA PE Exam" or "Bryon Lewis PE" on YouTube or your smart TV YouTube app. Click on the "ISA Control Systems Exams" playlist.

4 Measurement Standards and Terminology

Overview of Process Measurement, Control, and Calibration

The process control industry covers a wide variety of applications: petrochemical, pharmaceutical, pulp and paper, food processing, material handling, and even commercial applications. Experience designing process control systems is almost a necessity to pass the CSE PE exam.

Process control in a plant can include discrete logic, such as relay logic or a programmable logic controller (PLC); analog control, such as single-loop control or a distributed control system (DCS); and pneumatic, hydraulic, and electrical systems. The CSE must be versatile and have a broad range of understanding of the engineering sciences. The CSE is typically referred to as *instrumentation and electrical* (I&E), though the CSE must have in-depth knowledge of mechanical and process systems.

The CSE exam encompasses a broad range of subjects to ensure minimum competency. This book will review the foundations of process control and demonstrate the breadth and width of the CSE exam. It will then review the basic process control elements and their theory of operation, and then apply the elements to real-world applications. Next, this book reviews the calculations for sizing the elements, as well as the applicable laws, standards, and codes governing the installation of a process control system.

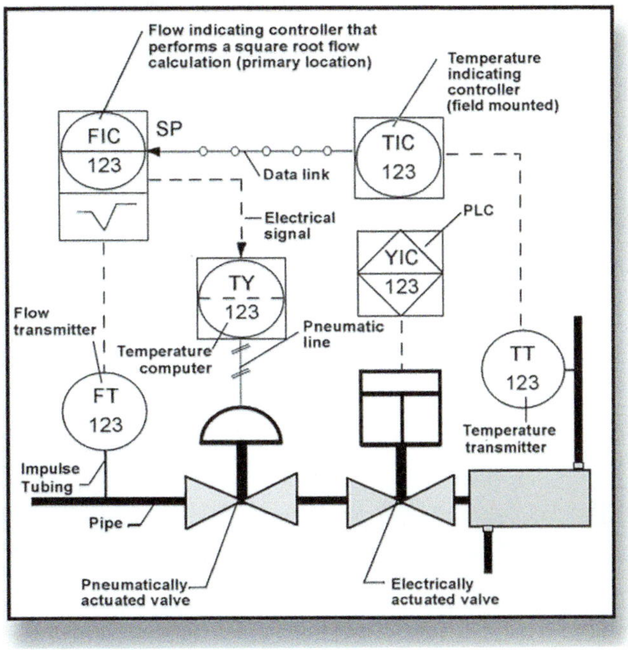

Process Signal and Calibration Terminology

The most important terms in process measurement and calibration are *range, span, zero, accuracy,* and *repeatability.* Let us start by first defining *span, range, lower range value* (LRV), *upper range value* (URV), *zero, elevated zero,* and *suppressed zero.*

Definition of the Range of an Instrument

The range is the region in which a quantity can be measured, received, or transmitted, by an element, controller, or final control device. The range can usually be adjusted and is expressed by stating the lower and upper range values.

Note 1: For example:

	Full Range	Adjusted Range	LRV	URV
a)	0 to 150°F	None	0°F	150°F
b)	–20 to 200°F	–10 to +180°F	–10°F	+180°F
c)	20 to 150°C	50 to 100°C	50°C	100°C

Note 2: Unless otherwise modified, input range is implied.

Note 3: The following compound terms are used with suitable modifications in the units: *measured variable range, measured signal range, indicating scale range, chart scale range,* etc. See Tables 4-1 and 4-2.

Note 4: For multirange devices, this definition applies to the particular range that the device is set to measure.

- **LRV** – The lowest value of the measured variable that a device is adjusted to measure
- **URV** – The highest value of the measured variable that a device is adjusted to measure

Note 5: The following compound terms are used with suitable modifications to the units: *measured variable lower range-limit, measured signal lower range-limit,* etc. See Tables 4-1 and 4-2.

Note 6: The following compound terms are used with suitable modifications to the units: *measured variable upper range-limit, measured signal upper range-limit,* etc. See Tables 4-1 and 4-2.

Definition of the Span of an Instrument

The span is the algebraic difference between the URVs and LRVs.

Note 1: For example:

Range:	0	to	150°F	Span 150°F
Range:	–10	to	+180°F	Span 190°F
Range:	50	to	100°C	Span 50°C

Note 2: The following compound terms are used with suitable modifications to the units: measured variable range, measured signal range, and so on.

Note 3: For multirange devices, this definition applies to the particular range that the device is set to measure. See Tables 4-1 and 4-2.

Definition of the Use of Zero in Instrumentation

Live Zero

The LRV is said to be set to zero, as a reference point, whether it is at zero or not. This LRV can be 0%; –40°F; 4 mA; 1V, or 3 psi. All LRVs are an example of the ZERO (live zero), in process control signals or elements.

Elevated Zero

The LRV of the range is below the value of zero. The LRV of the range must be raised to live zero for the instrument to function properly. The output signal of the measured value will always be 0%–100%. If the LRV of the range is too low, the instrument may not be able to reach 100% output.

Note: For example: Input signal = (–100 in H_2O to 25 in H_2O)
Output signal = (4 mA–20 mA)

The output signal may only reach 12 mA for 25 in H_2O (100%) input, due to limitations in the electronics or pneumatics. Therefore, the elevate jumper must be set in the transmitter or an elevation kit must be installed in a pneumatic transmitter. See Table 4-1.

Suppressed Zero

The LRV of the span is above the value of zero. The LRV of the range must be lowered to live zero for the instrument to function properly. The output signal of the measured value will always be 0%–100%. If the LRV of the range is too high, the instrument may not be able to reach 0% output.

Note: For example: Input signal = (50 in H_2O to 200 in H_2O)
Output signal = (4 mA to 20 mA)

The output signal may only reach 6 mA for 50 in H_2O (0%) input, due to limitations in the electronics or pneumatics. Therefore, the suppress jumper must be set in the transmitter or a suppression kit must be installed in a pneumatic transmitter. See Table 4-1.

Illustrations of Range and Span Terminology

Table 4-1. Examples of range and span terminology.

Typical Ranges		Name	Range	LRV	URV	Span	Supplementary Data
0	+100	–	0 to 100	0	+100	100	–
20	+100	Suppressed Zero Range	20 to +100	20	+100	80	Suppression Ratio = 0.25
–25	+100	Elevated Zero Range	–125 to +100	–25	+100	125	–
–100	0	Elevated Zero Range	–100 to 0	–100	0	100	–
–100	–20	Elevated Zero Range	–100 to –20	–100	–20	80	–

Illustrations of Measured Variable, Measured Signal, Range, and Span

Table 4-2. Examples of measured variables, measured signal ranges, and spans.

Typical Ranges	Type of Range	Range	LRV	URV	Span
Thermocouple 0 to 2000°F TYPE K T/C	Measured Variable	0 to 2,000°F	0°F	2000°F	2000°F
–0.68 to +44.91 mV	Measured Signal	–0.68 to +44.91 mV	–0.68 mV	+44.91 mV	45.59 mV
Flowmeter 0 to 10,000 lb/h	Measured Variable	0 to 10,000 lb/h	0 lb/h	10,000 lb/h	10,000 lb/h
0 to 100 in H_2O	Measured Signal	0 to 100 in H_2O	0 in H_2O	100 in H_2O	100 in H_2O
0 to 10 • 1000 = lb/h	Scale and/or Chart	0 to 10,000 lb/h	0 lb/h	10,000 lb/h	10,000 lb/h
4 to 20 mA	Measured Signal	4 to 20 mA	4 mA	20 mA	16 mA
1 to 5 V	Measured Signal	1 to 5 V	1 V	5 V	4 V

Standard Abbreviations Used Throughout the Book and Their Definitions

A = The area of a pipe or vessel.

β = The ratio of orifice diameter to exact interior piping diameter in meters.

d = The diameter of the orifice opening.

D = The inside diameter of a pipe.

DP = The differential pressure across the orifice or valve. The symbol for this term varies by organization and industry. Other symbols include: Δp, ΔP, dP, and dp.

G_f = The specific gravity of a material (liquid, solid, gas, or vapor) flowing in pipes or in the vessel of a process at the time of normal operation.

h = The head, which is often listed in this book as the height of the water column in inches.

H = The height in feet or meters.

p_f = The pressure exerted on or by a material (liquid, solid, gas, or vapor) flowing in pipes or in the vessel of a process at the time of normal operation.

p_1 = The upstream pressure before the orifice or valve (entry).

p_2 = The downstream pressure after the orifice or valve (exit).

ρ = The density of a material such as liquid, solid, gas, or vapor.

Q = The quantity of flow, such as gallons per minute (gpm) or standard cubic feet per hour (ft^3/h).

SG = Specific gravity. The symbol for this term varies by organization and industry. Other symbols include: s.g., S.G., and sg.

t = The time in seconds.

T_f = The temperature of a material (liquid, solid, gas, or vapor) flowing in pipes or in the vessel of a process at the time of normal operation.

v = The velocity, which may also be shown in some cases as: V.

V = The volume of a pipe or vessel.

wc = Water column. A column of water or H_2O. The symbol for this term varies by organization and industry. Other symbols include: W.C., w.c., and WC.

γ_f = The specific weight of steam flowing in pipes or in the vessel of a process at the time of normal operation.

DCS = Distributed control system.

PLC = Programmable logic controller.

5 Fluid Mechanics in Process Control

Relationship of Pressure and Flow

In a pipe, the static pressure distributed across the pipe is even during no flow or deadheaded with pump. The pressure at both ends of the pipe is the same because the total energy in the system is at equilibrium. As the fluid flows, it is accelerated through the pipe. There is a pressure drop across the pipe. The static pressure is a measurement of the potential energy in the fluid. It is changed to the form of kinetic energy and is used up in the form of heat and vibration doing work on the pipe to overcome the friction of the pipe.

$$Q = A \bullet v$$

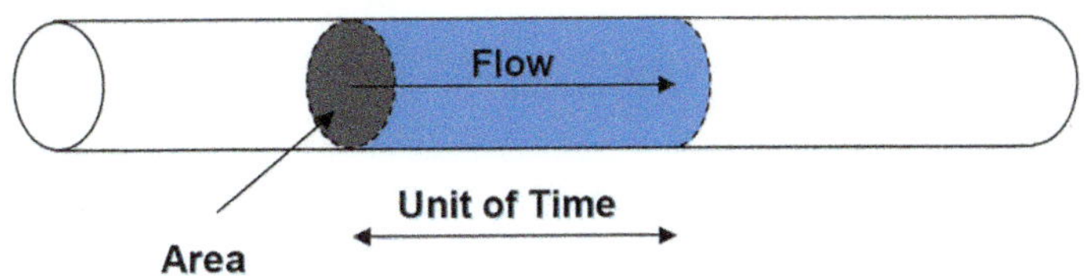

The higher the flow rate, the greater the pressure drop across the pipe. The work done to transfer the fluid through the pipe at higher flow rates becomes greater. Therefore, the pressure drop across the pipe increases as the velocity of the fluid increases through the pipe. The static pressure (available pressure) at the end of the pipe will be lower than the supply or pump pressure at the start of the pipe because work is being done on the pipe. The pump head energy is used up doing work on the pipe.

The DP measurement across the flow element acts a bit differently. Flow is measured in the units of *DP* (differential pressure). There is a pressure drop across the orifice element, and there will be additional pressure drop across the element as the flow rate (the fluid's velocity) increases. This is the same thing that is happening in the pipe, because more work is being done on the element as the velocity increases. However, remember that the pressure on the downstream side of the flow element drops as the velocity increases. How does the pressure for the flow measurement increase? It does not; there is an increase in *DP*, not in the static pressure.

We are measuring the DP across the element, and this is an inferred measurement of flow rate. Flow rate equals the velocity (distance per time) multiplied by the area of the pipe. We achieve the measurement of velocity using differential pressure. The difference between the upstream pressure and the downstream pressure across the element is a measurement of the difference in height in two different water columns. This difference in height is a direct proportional measurement of the velocity of the fluid flowing through the pipe.

The pump endows potential energy into the fluid and accelerates the fluid upward into a measurable column of water. The water column is typically measured in feet of head pressure but can be measured in pounds per square inch (psi). The water is constantly "falling" down the pipe toward the other end of the pipe, and the pump must constantly accelerate the water upward against the pull of gravity to keep the water column up in the air. The potential energy endowed into the water column turns into kinetic energy as the water column falls.

The kinetic energy is used to overcome the resistance of the pipe and the work done on the pipe as the fluid flows to the other end. If there is energy left over in the fluid, it is again transformed into potential energy at the other end of the pipe, as an available pressure at the end of the pipe. This potential energy left over can now fall through a pipe, device, or piece of equipment and do work, finally resting at a state of equilibrium. At this point, all the energy endowed into the fluid by the pump will be used up.

Note: The image at the right shows that the pump must develop enough head to raise the fluid to the pipe's top elevation plus enough head to overcome the friction loss of the piping (suction and discharge). We will also need to add head for any differential pressure across the valve and the orifice or head type meter.

The velocity of the fluid is measured as the fluid falls: $V^2 = 2gH$, where H is the height in feet (the head). The volumetric flow rate can then be an inferred measurement of the height of the water column. By knowing the size of the pipe, the coefficient of the orifice, and the properties of the fluid, we can accurately measure the volumetric flow rate of the fluid.

As the fluid flows through the opening of the orifice restriction, kinetic energy is transformed into potential energy in the form of a difference of water column on each side of the restriction orifice element. The height of the water column is the "scaled" velocity of the fluid through the pipe. Remember that the slower the fluid travels, the less work it has to do. The fluid has to accelerate through the small opening in the orifice to maintain the same mass flow rate through the pipe. Mass in must equal mass out.

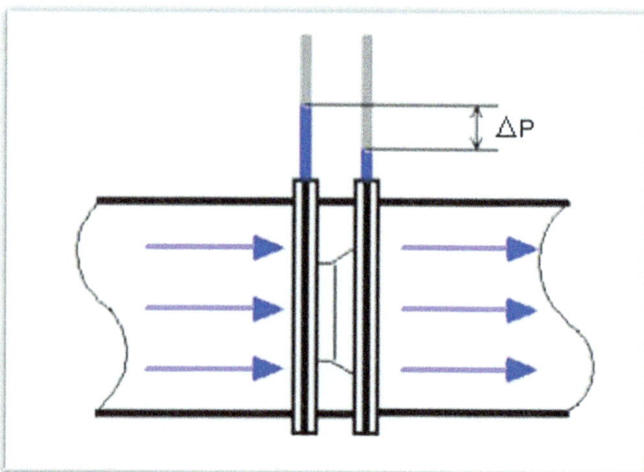

Energy is lost doing work on the orifice plate, and the pressure drops on the exit side of the orifice. This can be seen in the profile of the vena contracta of the fluid flowing and the DP across the orifice element. As the fluid exits the small opening into the much larger area of the pipe, the fluid decelerates and a portion of the kinetic energy endowed into the fluid by the pump is transformed back into potential energy. This potential energy can be seen in the form of a water column, of varying height, on the entry and exit sides of the orifice.

If the pipe were blocked at the exit end, the water would squirt out the taps on both sides of the orifice and the two water columns of equal height would become obvious. Again, as the fluid starts to accelerate through the pipe and through the orifice, the fluid's potential energy tends to change back into kinetic energy to do work. This means the water columns start to fall on both side of the orifice. The exit side will fall even more than the entrance side, because work is done on the orifice restriction element as the flow rate increases. The difference in height the column falls on the exit side compared to the upstream column is its scaled velocity of the flow rate. The higher the fluid's velocity, the more work is done on the orifice and the pressure drops even more on the exit side of the orifice. This gives a greater DP across the orifice. Note that as the pressure drops in the pipe due to increased velocity, the DP at the measurement meter becomes greater. This is because the total system pressure (total hydraulic head) is decreasing by doing work on the pipe, and the potential energy (pressure head) is being transformed back to kinetic energy (velocity head) to do the work.

The lower the fluid's velocity through the orifice, the higher the pressure on the exit side of the orifice. This means there is less difference between the pressure on the high side (entry side) water column and the low side pressure (exit side) water column. Therefore, there is less measured DP across the orifice when the fluid decelerates, even though the pressure increased on the exit side of the orifice and everywhere in the pipe system.

Note that as the fluid flow approaches a stop, the two water columns are almost even in height. The DP becomes almost nothing. The static pressure on the exit side of the orifice, which represents the potential energy in the fluid, becomes greater. The pipe system will try to reach equilibrium or uniform distribution of static pressure across the pipe system as the work across the pipe becomes less and less. The kinetic energy will change back into potential energy.

Remember that the total energy in the system equals the kinetic + potential + work done. As the fluid starts to accelerate down the pipe once again, the exit side water column starts to drop in height. The potential energy (pressure head) is once again being transformed back into kinetic energy (velocity head), to do work across the element and pipe. The distance in height the exit side water column falls compared to the height of the entry side water column is the *scaled* velocity of the flowing fluid.

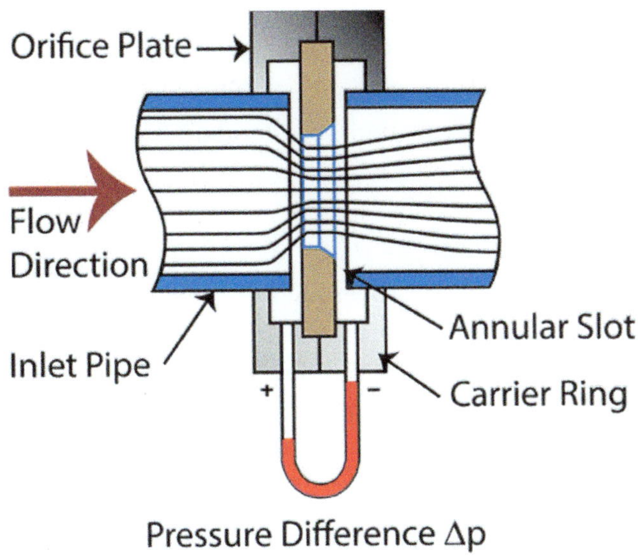

Because we know the fluid's specific gravity (SG), we can now calculate the fluid's height as if it were a column of water. Remember ($F = m \bullet a$) and weight is a measure of the force exerted by the pull of gravity. Pressure equals (density • height) and force equals (pressure • area); therefore, the pressure measurement is a representation of the fluid's height.

Two different Stack 231 in^3 of water on top of each other, to form a tall column of water, with a base of 1 in^2. The column of water will be 231 in tall. Divide the height of the column of water, 231 in, by the weight of 1 gal of water, 8.334 lb at 60°F. The result will be 27.718 or 27.7 in of water column per pound of water, over 1 in^2 of area. *Therefore, 27.7 in H_2O equals 1 psi of head pressure, or 2.31 ft of H_2O equals 1 psi of head pressure.*

Therefore, the column of water can be measured in psi, not just *head pressure* as a height of inches of water in the measurement meter. Knowing just the height of the column, we can determine the pressure it can exert and the inferred amount work it can do. A column of fluid with a lesser weight or density compared to water has an SG less than 1.

SG is the ratio of the density or weight of a fluid compared to the density or weight of water. The denser the fluid is, the more mass it has; therefore, the more force it exerts due to the acceleration of gravity ($F = m \bullet a$). So, a fluid with an SG less than 1 cannot exert as much force as water because it has less mass. Thus, a column of fluid with an SG less than 1 exerts less pressure on a measurement meter than a column of water does. Therefore, we divide the pressure head by the SG to give it a "gain" of force equal to that exerted by water, the industrial standard of measurement.

The previous demonstration shows that a column of fluid with an SG less than 1 must be taller than a column of water to exert the same pressure on the measurement meter. A fluid, such as solvent, may have an SG of 0.7874. We use the industrial standard of water to calibrate the meter. So, to measure the height of the column of solvent in the standard of calibration with water, the column of solvent must be taller than a column of water to exert the same force on a weight scale. It would seem that the taller column of solvent would be falling faster than the velocity we need to measure, and it is. It has less mass; therefore, it must be accelerated faster than the column of water to develop more force on impact. This force at impact will be the same force generated by the column of water falling from a lower height, and the pressure on the measurement element will be the same. There is an equivalent force and an equivalent pressure on the meter, for the height columns of fluid.

In level measurement, the column of water used to calibrate the meter will be less than the column of solvent being measured. The water must fall from a lower height to exert the same pressure as the taller column of solvent. So, if we have an SG of 0.7874 for the solvent, the column of water will be 0.7874 times the height of the solvent column or 78.74% of the intended height measurement. This will produce a 78.74 in column of water (100 in $H_2O \bullet 0.7874$ SG = 78.74 in H_2O). The solvent column will be 100 in tall but will appear to be only 78.74 in of water to the measurement meter. Zero to 100% output will equal 0 to 100 in of solvent.

The height of solvent needed to produce a pressure equal to that of 100 cm of water is shown to the right.

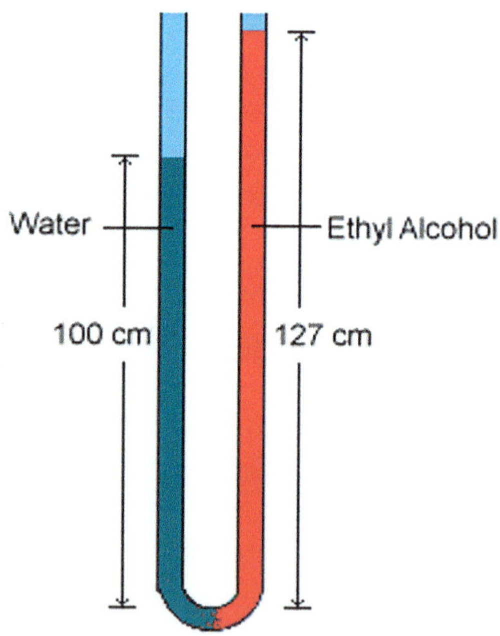

The solvent column is taller than the column of water, 100 cm/(SG = 0.7874). So, the column of solvent equals 100 cm/0.7874 = 127 cm. Both columns produce the same pressure at the bottom of the "U" tube.

The same thing is happening in the flowmeter. The solvent is less dense than water and exerts less differential pressure on the meter for the same flow rate as water. Ten gallons a minute of water traveling down a pipe or conveyor weighs 10 • 8.33 lb = 83.3 lb. Ten gallons a minute of solvent traveling down a pipe or conveyor weighs 10 • 8.33 lb • 0.7874 SG = 65.59 lb. The pressure the solvent exerts on the scale is less for the same volumetric flow rate. Again, the flowmeter will be calibrated in water with a lower measure of water column applied to the meter to read the desired flow rate of solvent.

Applications of Velocity and Pressure in Pipe Systems

This section gives a quick overview of how we use fluid mechanics in process control and then discusses how fluid mechanics are used in detail. It also explains how to use the formulas in the following chapters to provide safe and accurate control of process plants. *These topics may be on the CSE exam.*

Part One

Find the liquid flow velocity in a pipe for a given flow rate.

$$v\,(\text{ft/s}) = \frac{\text{gpm} \bullet 0.4085}{ID^2\,(\text{in})}$$

This type of question might be on the exam. There are 120 gpm flowing in a 2-inch schedule 40 pipe. What is the fluid velocity in the pipe? The pipe inside diameter (ID) can be found in the Appendix in Table A15, "Pipe Standard Dimensions and Data."

$$v\,(\text{ft/s}) = \frac{120 \bullet 0.4085}{2.067^2\,(\text{in})} = 11.473\,(\text{ft/s})$$

Find the pipe size required for a given flow rate and flow velocity.

$$ID\,(\text{in}) = \sqrt{\frac{\text{gpm} \bullet 0.4085}{v\,(\text{ft/s})}}$$

This type of question might be on the exam. There are 400 gpm flowing in a schedule 40 pipe at 10.2 ft/s. What is the pipe size required for this flow rate?

$$ID\,(\text{in}) = \sqrt{\frac{400 \bullet 0.4085}{10.2\,(\text{ft/s})}} = 4.002\,(\text{in})$$

Using the equation above, the pipe size is 4 in.

Find the change in velocity for a change in the pipe size in a simple piping system.

$$v\text{ outlet (ft/s)} = \left(\frac{\textit{pipe ID}\text{ inlet (in)}}{\textit{pipe ID}\text{ outlet (in)}}\right)^2 \bullet v\text{ inlet (ft/s)}$$

This type of question might be on the exam. There is a 4-inch pipe with a fluid velocity of 10 ft/s. If the pipe is reduced to 2 in, what is the fluid velocity in the 2-inch pipe?

$$v\text{ outlet (ft/s)} = \left(\frac{4\text{ in (inlet)}}{2\text{ in (outlet)}}\right)^2 \bullet 10\text{ (ft/s) } v\text{ inlet} = 40\text{ (ft/s)}$$

Part Two

Let us apply some basic formulas to calculate to the pressure drop in pipes.

Find the change in static pressure for a change in the flow rate of the pipe.

For a change in the static pressure anywhere in the piping system for a change in flow rate:

$p_1F_1^2 = p_2F_2^2 \qquad p_2 = \left(\frac{F_1}{F_2}\right)^2 p_1$ These formulas can be used for simple problems on the CSE exam.

This type of question might be on the exam. There is a 2-inch pipe with 100 gpm flowing, and the pressure gauge in the pipe reads 30 psig (pounds per square inch gauge). If the flow rate is reduced to 60 gpm, what will the pressure gauge read in psig?

$$p_2 = \left(\frac{100\text{ gpm}_1}{60\text{ gpm}_2}\right)^2 \bullet 30\text{ psig} = 83.33\text{ psig}$$

This is practical for a pressure meter to measure the available static pressure at any point with a given flow rate, but it does not tell the loss of differential pressure across the piping system or a flow element.

Find the change in differential pressure for a change in flow rate across an orifice plate.

$$h_1F_2^2 = h_2F_1^2 \qquad h_2 = \left(\frac{F_2}{F_1}\right)^2 h_1$$

This type of question might be on the exam. There is a flow transmitter measuring 100 gpm with 240 in of differential head pressure across the orifice. If the flow rate is reduced to 50 gpm, how many inches of differential head pressure will the transmitter read?

$$h_2 = \left(\frac{50\text{ gpm}_2}{100\text{ gpm}_1}\right)^2 \bullet 240\text{ in } H_2O_{\text{head1}} = 60\text{ in } H_2O_{\text{head2}}$$

Part Three

Find the change in differential pressure across a simple piping system.

There are 100 gpm of water flowing through 100 ft of 2-inch schedule 40 pipe (ID = 2.067 in) at 60°F ($cS_T = 1.22$). The pump is producing 100 ft of water or 43.32 psi. When the pump is running at full speed and the pipe is blocked by a valve at the exit end of the pipe, the pressure of 100 ft of head is distributed evenly throughout the pipe.

We crack open the valve until the water is flowing at 100 gpm. We will calculate the head drop (delta pressure drop) across the pipe.

First determine the velocity of the fluid:

$$v\,(\text{ft/s}) = \frac{\text{gpm} \bullet 0.4085}{ID^2\,(\text{in})}$$

$$9.56\,(\text{ft/s}) = \frac{100 \bullet 0.4085}{2.067^2\,(\text{in})}$$

Calculate the Reynolds number for the pipe:

$$R_e = \frac{3160 \bullet \textit{flow rate}\,(\text{gpm}) \bullet \text{SG}}{\textit{pipe ID}\,(\text{in}) \bullet \textit{viscosity}\,(\text{cSt})}$$

Note: For liquids in turbulent flow (most applications).

$$R_e = 125{,}310 = \frac{3160 \bullet 100 \bullet 1}{2.067\text{ in}\,(\textit{pipe ID}) \bullet 1.22\,(\text{cSt})}$$

Note: These equations apply in most situations in which the Reynolds number for pipe with liquids in turbulent flow is calculated.

Find the friction factor (*f*) using the Moody equation (same friction factor as the Moody chart):

Find the friction factor:

<u>Friction factor for Darcy–Weisbach equation</u>

Note: $e = 0.00015$ for steel pipes

$$f = 0.0055 + 0.0055\left[20{,}000\left(\frac{e \bullet 12\,(\text{in})}{\textit{pipe ID}\,(\text{in})}\right) + \frac{10^6}{\text{Re}}\right]^{\frac{1}{3}}$$

$$0.0217 = 0.0055 + 0.0055\left[20{,}000\left(\frac{0.00015 \bullet 12\,(\text{in})}{2.067\,(\text{in})}\right) + \frac{10^6}{125{,}310}\right]^{\frac{1}{3}}$$

Find the head loss across the pipe using the Darcy–Weisbach equation:

$$h_L = f \bullet \left(\frac{Length\,(\text{ft}) \bullet 12}{pipe\ ID\,(\text{in})}\right) \bullet \frac{v^2(\text{ft/s}^2)}{2 \bullet g}$$

$$17.99\ \text{ft} = 0.0217\left(\frac{100\,(\text{ft}) \bullet 12\,(\text{in})}{2.067\,(\text{in})}\right) \bullet \frac{9.56^2(\text{ft/s}^2)}{64}$$

where

h_L = piping head loss in feet of water
f = moody friction factor of the pipe per foot
v = velocity of the fluid in ft/s
g = acceleration of gravity

There is a head loss (pressure drop) across the pipe of 17.99 ft of water (or 7.8 psi) at 100 gpm. This leaves 82 ft of head (100 ft – 18 ft = 82 ft) or 35.52 psi at the end of the pipe to do work across a control valve or to overcome a pressure in a vessel.

It is recommended that an additional 15%–40% increase in pump head be added to the required system pump pressure for normal pumping through the piping system, minus the required head to overcome any vessel pressure (pressurized tank, vessel, or column). We must add only the 10%–40% extra head to the pump head that is needed to overcome the friction loss of the pipe and to do the foot-pounds of work to accelerate the fluid through the valve.

Important Note: You cannot size the pump for just the pressure drop across the piping system because of friction loss and flow rate. The valve will not work. There must be extra available head pressure across the valve or the valve will not function.

The minimum DP across the valve should be 15%:

17.99 ft • 0.15 = 2.7 ft of head or 1.17 psi for the valve sizing calculation.

There will be 1.17 psi available across the valve if there is a 15% increase in the pump head for the piping system.

The recommended DP across the valve should be 40%:

17.99 ft • 0.40 = 7.196 ft of head or 3.12 psi for the valve sizing calculation.

There will be 3.12 psi available across the valve if there is a 40% increase in the pump head for the piping system.

Now calculate the head loss at 50 gpm.

First, calculate the velocity of the fluid:

$$4.78\,(\text{ft/s}) = \frac{50\,(\text{gpm}) \bullet 0.4085}{2.067^2\,(\text{in})}$$

Find the Reynolds number for the pipe:

$$R_e = 62{,}655 = \frac{3160 \bullet 50\,(\text{gpm}) \bullet 1}{2.067 \text{ in } \mathit{pipe\ ID} \bullet 1.22\,(\text{cST})}$$

Find the friction factor with the Moody equation:

$$0.0232 = 0.0055 + 0.0055\left[20{,}000\left(\frac{0.00015 \bullet 12\text{ in}}{2.067\text{ in}}\right) + \frac{10^6}{62{,}655}\right]^{\frac{1}{3}}$$

Find the head loss across the pipe:

$$4.8\text{ ft} = 0.0232\left(\frac{100\text{ ft} \bullet 12\text{ in}}{2.067\text{ in}}\right) \bullet \frac{4.78^2\,(\text{ft/s})}{64}$$

There is a head loss (pressure drop) across the pipe of 4.8 ft of water (or 2.08 psi) at 50 gpm. This leaves 95.2 ft of head (100 ft – 4.8 ft = 95.2 ft) or 41.21 psi at the end of the pipe to do work across a control valve or to overcome a pressure in a vessel.

Note: The psi drop across the control valve increases as the flow slows down and the valve absorbs the remaining pressure left in the system across the control valve. The difference in the system pressure is the pump head minus the head loss across the piping system and minus any head needed to overcome entry into a pressurized vessel. Similar to an electrical circuit, $I \bullet R = E$, the valve has more resistance to flow as it closes down, so the pressure drop across the valve increases to maintain the flow rate. So, even though the control valve is trying to slow down (restrict) the flow rate of the fluid, the fluid will try to maintain its flow rate as the valve absorbs the extra pressure in the system. (The energy is converted to heat and sound in vibrations.) The control valve controls the flow by burning up the extra energy head (dissipating energy) in the fluid as it flows through the piping system. We will discuss this in much more detail in the section on control valves.

$$1\text{ gpm} = 1\,C_V \bullet \sqrt{1\,DP_{psig}}$$

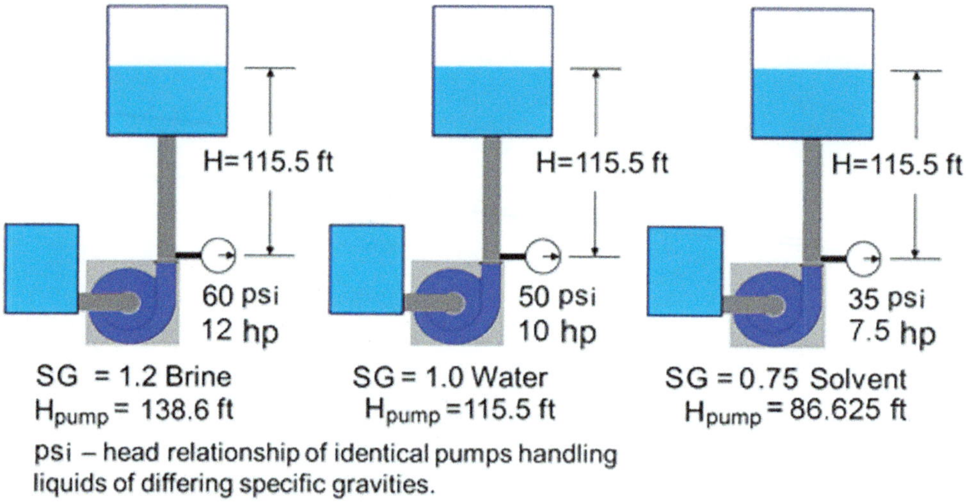

Figure 5-1.

Sizing a Pump Head Using the Specific Gravity of the Pumped Fluid

As with flow and level measurements, the work that is needed to raise the fluid (give it lift) is proportional to the weight of that fluid and the size of the motor. The work done by the pump motor is in foot-pounds. If the fluid weighs less than water (it has a lower SG), the pump can raise the fluid higher; and if the fluid weighs more than water (has a higher SG), the pump cannot raise the fluid as high.

As shown in Figure 5-1, the pump can produce less head or pressure to lift a light solvent with an SG of 0.75 to 115.5 ft. It only needs 86.625 ft of head in water, and it needs less horsepower (7.5 hp). With brine, you will need more head in water, 138.6 ft, and more horsepower (12 hp) to lift the brine to an elevation of only 115.5 ft of lift.

By dividing the head pressure in feet of water (H_{pump}) by the SG of the process fluid, the head pressure in feet of the process fluid is obtained. The feet of process fluid or head pressure

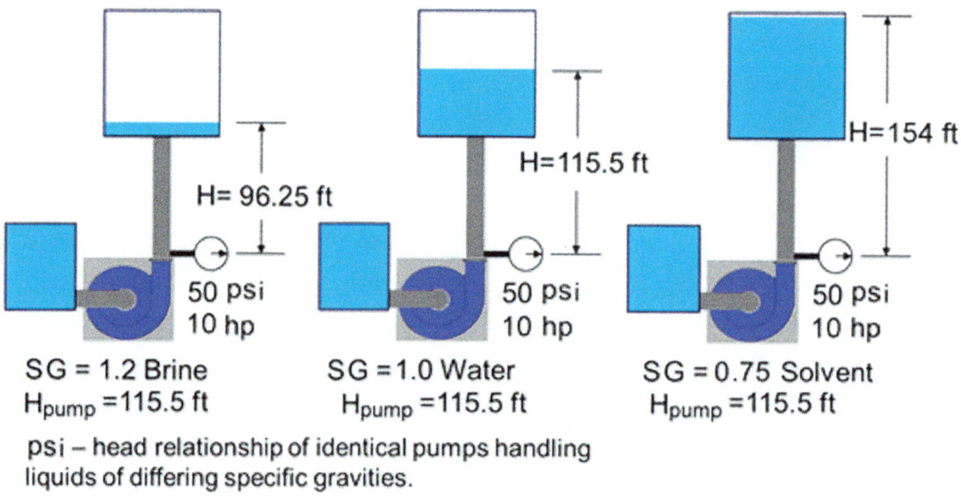

Figure 5-2.

is actual lift in feet of height. The pump is always lifting the process fluid, doing work on the process fluid to store potential energy in the process fluid. The height produces the hydrostatic pressure in the fluid and causes the process fluid to do work on the piping system when it changes to kinetic energy.

As shown in Figure 5-2, the pump producing a constant head of 115.5 ft of water, at a constant horsepower, will lift a light solvent much higher than that of water or brine. The solvent weighs less and requires less work, while the brine is much heavier and requires more work; therefore, you will get less head or lift from your pump for the reference to head in water.

All pumps are purchased with the specification for head versus flow in feet of water only. For fluids other than water, you must correct the head and flow rate for the SG and viscosity of the fluid.

Find the change in pressure after the pump in a simple piping system.

This type of question might be on the exam. A liquid with an SG of 0.8 flows to a pump at 100 psig. The viscosity of the liquid is 3 cP, the molecular weight is 94, and the temperature is 15°C. The pump supplies 100 ft of head. Piping losses are insignificant. What will the pressure (psig) at the inlet to the control valve most nearly be?

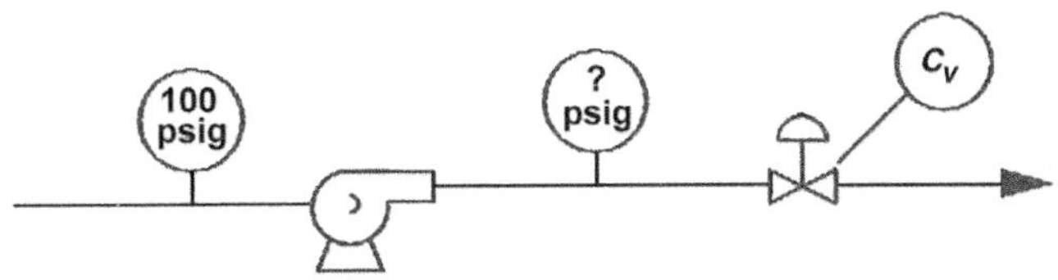

We are not concerned with the centipoise (viscosity) or molecular weight, only the temperature that can change the SG of the process fluid. The pump is rated in head feet of water or pressure (This must be understood.) The pump motor is rated in foot-pounds of work and is rated with water flowing through it. Because the process fluid has an SG of 0.8, which is less than water with an SG of 1, the pump will produce more pressure with a process fluid than if it were water.

We divide the pump head value of 100 ft by the SG. The pump outlet adds to any pressure already in the line at the pump inlet, so we will add our pump head value of (100 ft • SG = 0.8) to the 100 psig line pressure to get the increased line pressure at the valve.

$$100 \text{ ft of process fluid} \bullet 0.8 \text{ SG} = 80 \text{ ft of water}$$

Convert the feet of process fluid to psi:

$$\frac{80 \text{ ft of water}}{\left(\dfrac{2.31 \text{ ft}}{\text{psi}}\right)} = 34.632 \text{ psi}$$

Add the line pressure and the pump pressure together to get the psig at the valve:

$$100\ \text{psi}_{\text{line}} + 34.632\ \text{psi}_{\text{pump}} = 135\ \text{psi}_{\text{valve}}$$

Summary of Fluid Mechanics for Process Control

The DP across the orifice decreases as the velocity of the fluid decreases. The pressure on the exit side of the orifice increases as the fluid's velocity decreases and the pressure drop across the pipe decreases (less work is being done).

The velocity being measured is a "scaled" velocity. It is scaled by the orifice size; the beta factor or the Spink factor, or flow coefficient; the pipe ID and the SG of the fluid. The velocity is proportional to the "square root of (2gH)." The fluid's velocity through the pipe may be much different than the measured differential height of the two water columns that are being measured to obtain the fluid's velocity.

Depending on the orifice size and the beta factor of, for example, 0.3 for a given flow rate, the DP may be 1,000 in of water column differential across a small orifice opening. The fluid must do much more work to get through the high resistance of the small opening. The Δp could be only 100 in water column differential for a much larger beta ratio of, for example, 0.7. The larger opening has less resistance, and therefore much less work is being done to flow through it. Therefore, less potential energy has to change to kinetic energy, and the height of the water column on the exit side of the orifice is much higher than with a beta ratio of, for example, 0.3. Therefore, there is less DP across the larger orifice for the same flow rate that has been "scaled" to calculate the volumetric flow rate.

Remember that the pump will have to produce enough head pressure to provide energy for the work to be done on the piping system and any valves or head-type meters at the maximum flow rate. The valve cannot just be sized for some selected differential pressure; it must be sized for the pump head pressure not being used to do work on the rest of the piping system. The valve differential pressure must be checked at maximum flow, normal flow, and minimum flow rates to ensure the valve will control the flow rate properly.

Next we will discuss process measurement in detail, including the application of the fluid mechanics we just reviewed. These basic principles work for level, flow, orifice sizing, valve sizing, pump sizing, and pipe sizing, as well as understanding the basics of process operations.

6 Temperature Measurement and Calibration

Temperature Measurement Devices and Calibration

In the process industry, temperature measurements are typically made with thermocouples, resistance temperature detectors (RTDs), and industrial thermometers. Industrial thermometers are typically liquid (class I), vapor (class II), or gas (class III).

Standard Thermocouple Configurations

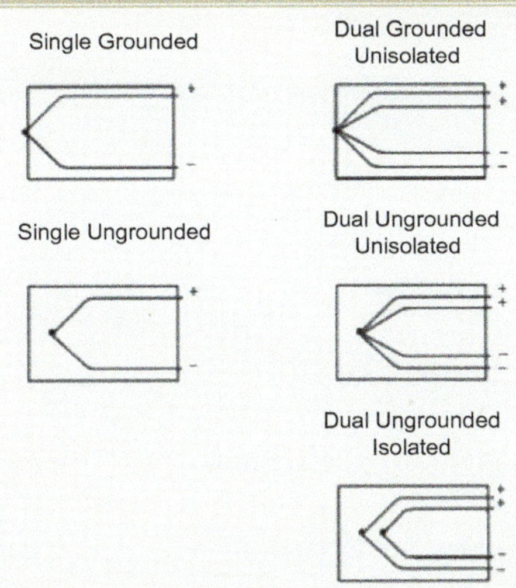

There are five major types of thermocouple configurations used in plants. They are shown to the left.

The first two thermocouples are welded or grounded, as shown, to the outside metal protective sheathing.

The last three thermocouples are ungrounded and should never touch the metal protective sheathing; otherwise, they are shorted to ground.

Thermocouple terminal junction blocks should be made of the same material as the thermocouple wire that is being connected to the terminal. This will prevent additional thermocouple junction points from being introduced into the temperature signal. Some companies use standard terminal strips, but they can cause an error in the signal.

Cold Junction Compensation in Thermocouple Temperature Measurements

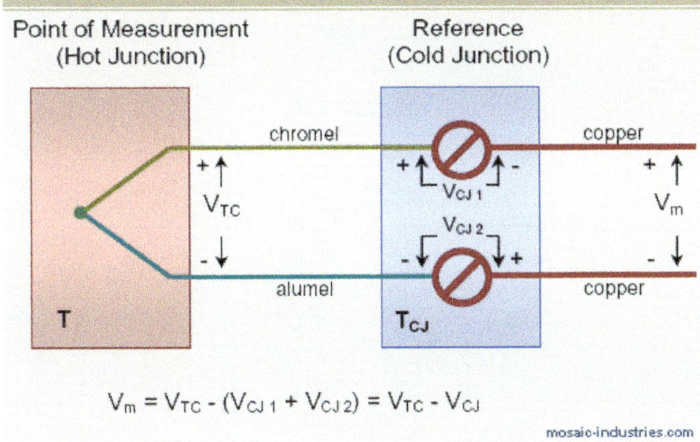

Thermocouple junctions are formed when connecting to the wire ends together or to a terminal block, so the total potential measured depends on the temperatures of those junctions as well as the junction at the measurement end of the thermocouple. Thermocouple voltages are relative values that must be measured with respect to a junction at a known temperature, called the *cold junction* or *reference junction*. Temperature should be measured with the cold junction at 0°C or 32°F.

The connection diagram to the left shows a thermocouple used to measure the temperature at its hot junction. The cold junction (CJ) is connected to the wires that connect to the meter leads.

Thermocouple Extension Wiring

Thermocouples *should* be extended with thermocouple extension wire and thermocouple termination blocks, but can be extended with standard copper wire and standard terminal blocks if some error is acceptable. With all thermocouples, one side is positive (the color: yellow, white, purple, etc.) and the other side is negative (always red, except in some extension wires).

Thermocouple millivolt tables for the examination can be found in Table A1, "Thermocouple Table" (Type J), Table A2, "Thermocouple Table" (Type K), Table A3, "Thermocouple Table" (Type E), and Table A4, "Thermocouple Table" (Type T), in the Appendix, "Data Tables."

Thermocouple linearity chart.

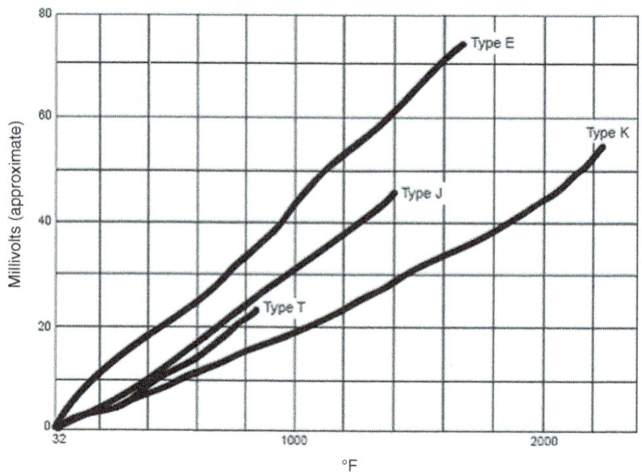

Thermocouple makeup material and color code.

Thermocouple Type	Thermocouple Material	Range for Calibration °F	Useful Range °F	Thermocouple Colors
E	Chromel (+) Constantan (−)	−300 to 1830	200 to 1650	Purple Wire Jacket Purple (+) Red (−)
J	Iron (+) Constantan (−)	−320 to 1400	200 to 1400 (300 to 800)	Black Wire Jacket Black (+) Red (−)
K	Chromel (+) Alumel (−)	−310 to 2500	200 to 2300	Yellow Wire Jacket Yellow (+) Red (−)
R	Platinum 13% Rhodium (+) Platinum (−)	0 to 3100	1600 to 2640	Green Wire Jacket Black (+) Red (−)
S	Platinum 10% Rhodium (+) Platinum (−)	0 to 3200	1800 to 2640	Green Wire Jacket Black (+) Red (−)
T	Copper (+) Constantan (−)	−300 to 750	−310 to 660	Blue Wire Jacket Blue (+) Red (−)

Thermocouple – Worked Examples (How to Read the Thermocouple Tables)

Sample Problem: What is the millivolt (mV) output of a Type J thermocouple at 218°F and referenced to a 32°F electronic ice bath?

Find the nearest temperature in Table A1, "Thermocouple Table" (Type J), in the Appendix of this guide.

The nearest temperature in the first column is 210. Look at the column headers at the bottom of the chart. Find the column header labeled 8. Follow the column up to the row with the 210 value. Where they meet is a total of 210°F + 8°F = 218°F.

Read the value of mV. The answer is 5.45 mV.

Sample Problem: What is the millivolt (mV) output of a type K thermocouple at 672°F from the data given? Assume the thermocouple is linear.

Given:
670°F = 14.479 mV
672°F = mV
680°F = 14.713 mV

Interpolate the mV value for the desired temperature as follows:

Interpolation:

$$\textit{output}\ \mathrm{mV} = \left[\left(\frac{\deg\ \textit{desired} - \deg\ \textit{lower value}}{\deg\ \textit{upper value} - \deg\ \textit{lower value}}\right)(\mathrm{mV}\ \textit{upper value} - \mathrm{mV}\ \textit{lower value})\right] + \mathrm{mV}\ \textit{lower values}$$

Therefore, the output (mV) for 672°F:

$$14.526 = \left[\left(\frac{672-670}{680-670}\right)(14.713-14.479)\right] + 14.479$$

The output at 672°F is 14.526 mV.

This can be verified in Table A2, "Thermocouple Table" (Type K), in the Appendix.

Resistance Temperature Detector

The process control industry also uses RTDs for many applications, for example, when precise temperature measurement is needed, such as mass flow measurements or critical temperature measurements of motor bearings.

RTDs typically come in 10 Ω copper and 100 Ω platinum elements. Their resistance is typically very linear over the scale.

Resistance values for the examination can be found in Table A5, "Platinum 100 Ohm RTD Table in Ohms," in the Appendix.

Typical wiring configurations and uses of RTDs.

Two-Wire RTD	Three-Wire RTD	Four-Wire RTD

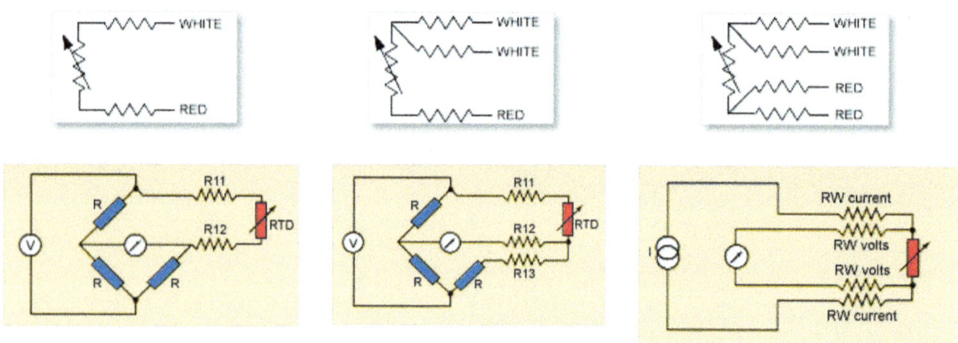

Good for close applications. At the transmitter.	Good for further distance applications. Remote from the transmitter.	Best application and usually uses 20 mA driving current and a voltage measurement.

RTD – Worked Examples

Sample Problem: An RTD is made of platinum; it has a resistance of 100 Ω at a temperature of 32°F and a resistance change (alpha) of 0.2178 Ω per °F. What is the resistance of the RTD at a temperature of 240°F?

Find the difference in the temperature first: 240°F − 32°F = 208°F

Now, find the resistance for the differential temperature:
208°F • 0.2178 Ω/°F = 45.3 Ω

Now, add the change in resistance to the resistance at 32°F:
100 Ω + 45.3 = 145.3 Ω

Refer to Table-A5, "Platinum 100 Ohm RTD Table in Ohms," in the Appendix for the resistance value at 32°F.

Sample Problem: In the bridge circuit below, if R_1 and R_2 are 200 Ω and the RTD is at 60°F, what resistance should R_3 measure to balance the circuit and give the meter a reading of 0 V? The RTD is platinum and measures 100 Ω at 32°F with an alpha of 0.2178 Ω per degree Fahrenheit.

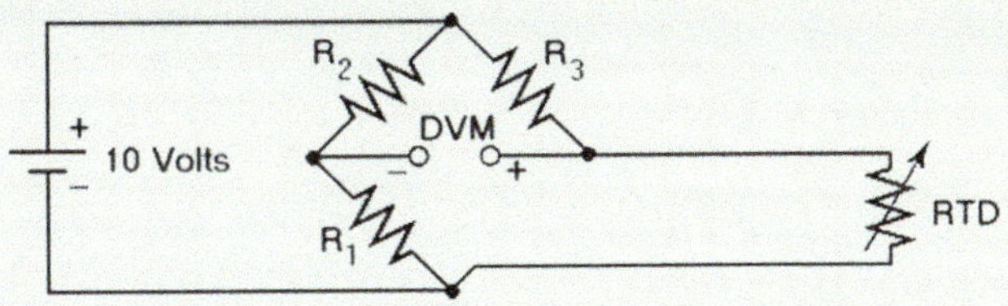

Find the difference in the temperature first: 60°F – 32°F = 28°F

Now find the resistance for the differential temperature:

28°F • 0.2178 Ω/°F = 6.0984 Ω

Now we the change in resistance to the resistance at 32°F:

100 Ω + 6.0984 = 106.0984 Ω

The resistor R3 must be 106 Ω to balance the bridge and give 0 V at the meter.

Sample Problem: In the bridge circuit above, R_1 and R_2 are 200 Ω. R_3 is 150 Ω. The excite voltage to the bridge is 10 V. If the meter is reading 0.4 V (the positive is on the right side and the negative is on the left side), what is the temperature at the RTD?

Find the voltage on the left side of the bridge. This is the voltage we will add to the meter voltage on the right side. We will use the voltage divider theorem to find the voltage across R_1.

$$V_{R1} = \frac{R_1}{R_1 + R_2}(10\text{ V}) = \frac{200}{200+200}(10\text{ V}) = 5\text{ V}$$

This means the voltage across the RTD is 5.0 V + 0.4 V = 5.4 V.

We will now use the voltage divider theorem to find the resistance of RTD.

$$V_{RTD} = \frac{R_{RTD}}{R_{RTD} + R_{R3}}(10\text{ V});\ 5.4\text{ V} = \frac{R_{RTD}}{R_{RTD}+150}(10\text{ V})$$

Solving for R_{RTD}:

$$5.4 = \left(\frac{R_{RTD}}{R_{RTD}+150}\right)10$$

$$\frac{5.4}{10} = \left(\frac{R_{RTD}}{R_{RTD}+150}\right)\frac{10}{10}$$

$$0.54(R_{RTD}+150)=\left(\frac{R_{RTD}}{R_{RTD}+150}\right)(R_{RTD}+150)$$

$$0.54(R_{RTD}+150)=R_{RTD}$$

$$0.54R_{RTD}+0.54(150)=R_{RTD}$$

$$0.54R_{RTD}+81=R_{RTD}$$

$$0.54R_{RTD}-0.54R_{RTD}+81=R_{RTD}-0.54R_{RTD}$$

$$81=R_{RTD}-0.54R_{RTD}$$

$$81=(1-0.54)R_{RTD}$$

$$81=(0.46)R_{RTD}$$

$$\frac{81}{0.46}=\frac{(0.46)R_{RTD}}{0.46}$$

$$176.087=R_{RTD}$$

We can prove that the value of 176.087 Ω for the RTD is correct by plugging the value into the voltage divider formula to find the 5.4 V at the meter.

$$V_{RTD}=\frac{176.087}{176.087+150}(10\text{ V})=5.4\text{ V}$$

We have the ohms of the RTD, now we can find the temperature.

$$100\ \Omega=32°\text{F}$$

Subtract the difference in ohms: 176.087 − 100 = 76.087 Ω.

Divide 76.087 Ω by the alpha 0.2178 Ω per °F.

$$°\text{F}=\frac{76.087\ \Omega}{\left(\frac{0.2178\ \Omega}{°\text{F}}\right)}=349.34°F$$

Add the 32°F bias for 100 Ω to the 349.34°F bias for 76.087 Ω:

$$349.34°\text{F}+32.00°\text{F}=381.34°\text{F}$$

Installing RTDs and Thermocouples in a Process Stream

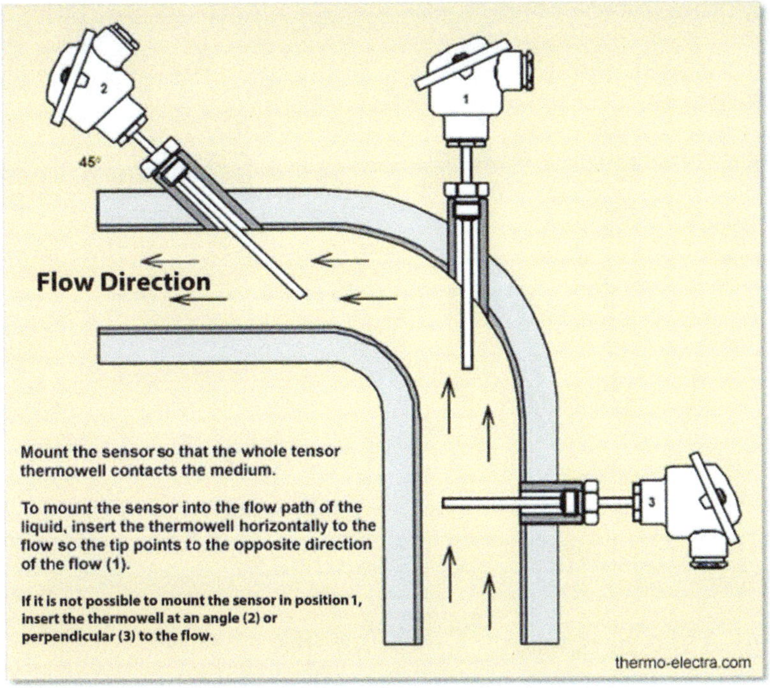

Typical RTD and Thermocouple Applications

A typical industrial thermocouple with a transmitter.

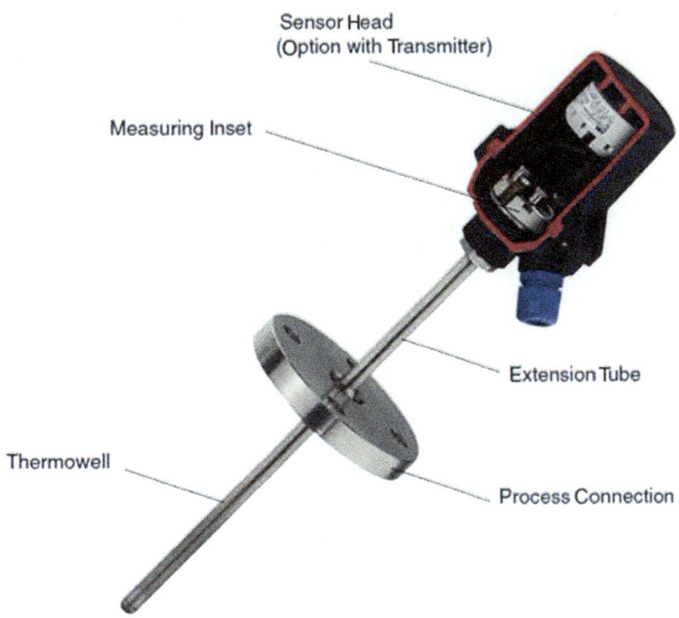

Industrial RTD or thermocouple with head. Straight and tapered thermowells are shown threaded for mounting in tanks and pipes.

Various industrial thermometers.

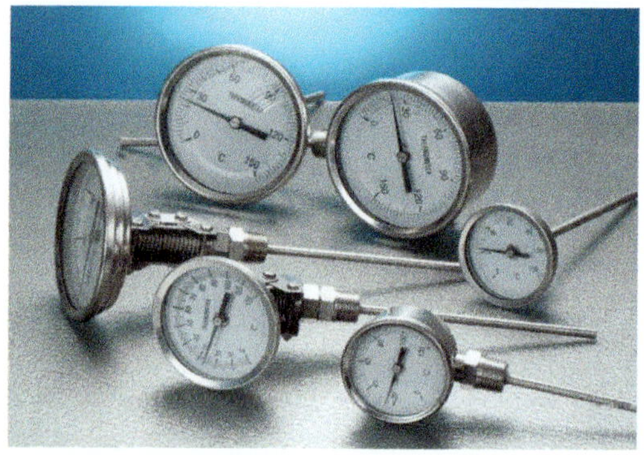

Thermocouple Wake Frequency Calculations

What Is a Thermowell Wake Frequency Calculation?

A wake frequency calculation is used to determine or prove the required dimensions and suitability of a thermowell based on the process conditions. The fluid will form a wake, known as a *Von Karman trail*. The wake has a specific frequency, which is a function of the diameter of the thermowell and the fluid velocity. If the wake frequency coincides with the natural frequency of the thermowell, the thermowell will vibrate, which leads to destruction. See Figure 6-1.

Figure 6-1.

Thermowells

Thermowells are principally used with thermocouples, RTDs, and bimetal thermometers in applications where it is necessary to measure temperature at high pressure or in hostile environments. They are also used for isolation, so a sensor can be replaced without having to shut down the process. Thermowells are machined from solid barstock (see Figure 6-2). Safe working pressures depend on the thermowell material, operating temperature, and the velocity of the flowing medium.

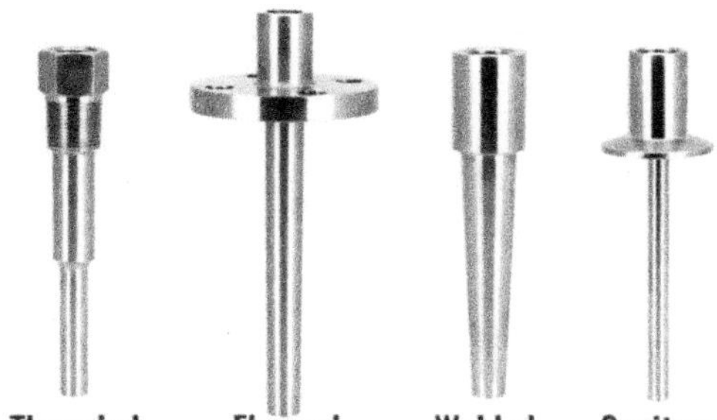

Figure 6-2.

Thermowell Vibration

When fluid flows past a thermowell, the change in fluid momentum creates a turbulent wake behind the well. Vortices form in this wake and shed from alternate sides of the well. The vortex shedding frequency or wake frequency (fw) is linear with flow velocity and inversely proportional to the thermowell tip diameter.

These shedding vortices impose on the thermowell a periodic force comprised of two components:

1. A lift force, normal to the direction of flow, oscillating at the wake frequency
2. A smaller drag force, parallel to flow, oscillating at twice the wake frequency

These vortex-induced forces, which cause thermowell vibration, are normally small, with the magnitude of the vibrations generally negligible. However, as the wake frequency approaches the "natural frequency" (ω_n) of the thermowell (within 20%), it can shift and lock in to the natural frequency. When fw $= \omega_n$, the thermowell goes into resonance, and vibrating forces increase rapidly. The resultant vibrations can cause mechanical failure of the well (see Figure 6-3). The in-line vibration is a result of the drag of the fluid. The transverse vibration is a result of the vortex shedding at a natural frequency.

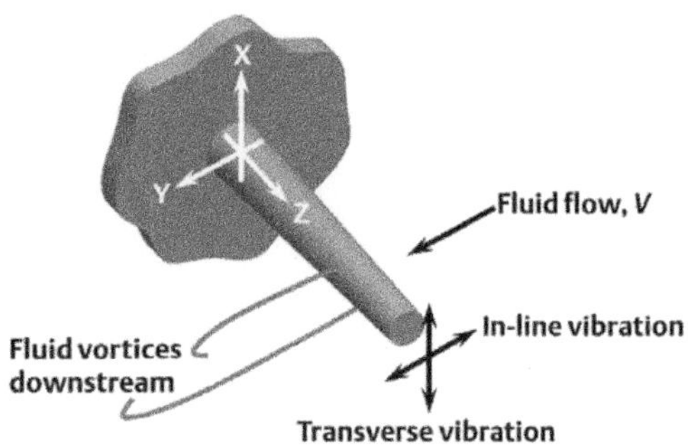

Figure 6-3.

The Murdock calculations and companion ASME PTC19.3 TW-1974 consider only the oscillating lift force as the cause of thermowell vibration. The ratio of wake to natural frequency is restricted to a maximum of 0.8 to eliminate the possibility of resonance.

Although the oscillating drag force is small, it can force the thermowell into resonance at lower velocities because it occurs at twice the wake frequency. For high-density fluids (liquids and high-pressure steam), the Murdock analysis is not adequate. When the oscillating drag component is included, the velocity rating can be reduced by up to 50%.

The calculations included herein are modified (ASME PTC19.3TW-2016) to include in-line resonance due to the oscillating drag force, correction for the magnification ratio, and use of the actual natural frequency of the well rather than the estimated value.

The results of these calculations should only be used as a guide in the selection of the correct thermowell. Other variables, such as corrosion, should also be evaluated and influence the decision.

Thermowell Velocity Calculations

When fluid flows past a thermowell, low-pressure vortices are formed in the wake downstream of the well. These vortices shed from alternate sides of the well, and the resulting differential pressure produces two periodic forces on the thermowell:

1. An oscillating lift force, transverse to the fluid flow at shedding frequency (fs)
2. An oscillating drag force, in-line with the fluid flow at frequency 2fs

Vortex shedding can occur at frequencies from 50–1500 Hz. The vortex shedding frequency (Strouhal frequency) increases linearly with fluid velocity, but the forces increase with the square of the velocity. When the Strouhal frequency approaches the natural frequency of the thermowell, it can lock in to the natural frequency causing resonance, with greatly magnified forces. To prevent lock-in, the natural frequency of the thermowell must be higher than either the in-line or the transverse resonance condition. Operation through the in-line resonance is acceptable only if the cyclic stresses at the resonance condition are acceptably small.

The fluid velocity at which resonance occurs is referred to as a *velocity critical*. There are two velocity criticals for each natural frequency of the thermowell: one describing the transverse and the other describing the in-line response. Because in-line force fluctuates at twice the frequency of the transverse force, the corresponding velocity critical is approximately one-half that required for transverse resonance. If the natural frequency of the thermowell overlaps either the fs or 2fs, a large resonant buildup in vibration amplitude can occur. The major cause of thermowell failure is fatigue due to resonance. A high enough level of damping may cause the thermowell to operate at the in-line or even the transverse resonance frequencies.

In addition to frequency limits, the stresses within the thermowell and forces applied are critical to evaluating the suitability of a thermowell for a specific process application. The four quantitative criteria to be evaluated are:

1. **Frequency limit** – Resonant frequency (*fn*) of the thermowell must be sufficiently high so that destructive oscillations are not excited by the fluid flow. The steady-state (s-s) fluid velocity should meet one of the following conditions:

$$fs(s\text{-}s) < 0.4 \bullet fn \quad \text{OR} \quad 0.6 \bullet fn < fs(s\text{-}s) < 0.8 \bullet fn$$

2. **Static stress limit** – Steady-state stresses are the result of hydrostatic fluid pressure and nonoscillating drag forces on the thermowell and are calculated at the location of maximum stress. If the thermowell is partially shielded or has a reduced tip, the calculation must be performed with those considerations. The maximum steady-state stress on the thermowell at design velocity must not exceed the allowable stress as determined by the Von Mises Criteria.

   Note: The exposed length of the thermowell that is not anchored is a consideration.

3. **Dynamic stress limit** – Dynamic stresses are a result of the periodic drag forces that cause in-line oscillations and the periodic lift forces that cause transverse oscillations. If the thermowell is intended to operate above the in-line velocity critical, there are cyclic stresses at the in-line resonance to consider as it passes through that point on the way to the design velocity. The maximum dynamic stress must not exceed the allowable fatigue stress limit. The magnification factors are calculated and applied to the cyclical stress equations, then the cyclic drag and lift forces are calculated at the design velocity. The maximum combined lift and drag stress must not exceed the fatigue stress limit

4. **Hydrostatic pressure limit** – The external pressure must not exceed the minimum pressure rating of the thermowell tip, shank, or flange (or threads) at the operating temperature.

Thermowell Standards

ASME PTC19.3-1974 has been used for many years to design most thermowells. This standard was applicable to tapered profiles only, and did not account for the stress caused by in-line resonance due to the oscillating drag force.

ASME PTC19.3TW-2010 was a new standard released in July 2010 to replace the 1974 standard. It uses more advanced methods for evaluating the suitability of a thermowell for a specific application, and is applicable to tapered, straight, and reduced tip profiles.

ASME PTC19.3TW-2016 is the latest standard and should be used for all thermocouple wake calculations.

Thermowell wake frequency calculation variables.

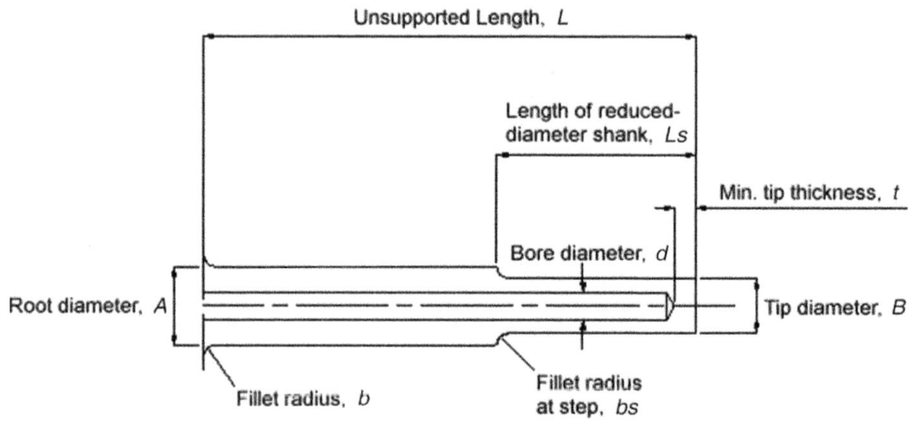

Variable Description	Variable Per ASME PTC 19.3 TW-2016
Root diameter	*A*
Tip diameter	*B*
Bore diameter	*d*
Insertion length	*L*
Minimum tip thickness	*t*
Bore ratio	*d/B*
Taper ratio	*B/A*
Cross-sectional ratio	*L/B*
Length ratio	*Ls/L*
Fillet radius	*b*
Fillet radius as step	*bs*

7 Pressure Measurement and Calibration

Pressure Measurement and Head Pressure

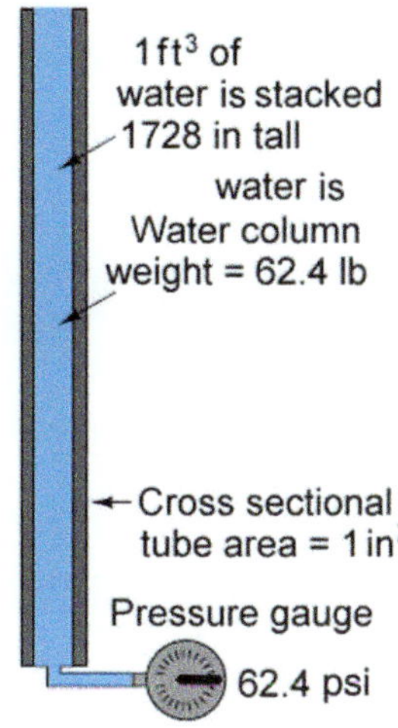

Pressure is typically measured in two different forms, pounds per square inch (psi) or head pressure. Head pressure is measured in inches or feet of water column (H_2O).

Head pressure is the measure of the potential energy or hydrostatic pressure in the system. Head pressure is independent of the tank's height or area. The transmitter measures head pressure. The transmitter measures the height from which the fluid is falling. The distance the fluid falls indicates the force generated ($F = ma$). The density of the fluid must be known to calibrate a pressure transmitter for a process, to obtain the fluid mass. The calibration process uses specific gravity (SG), the ratio of a known density of a fluid divided by the density of water (H_2O).

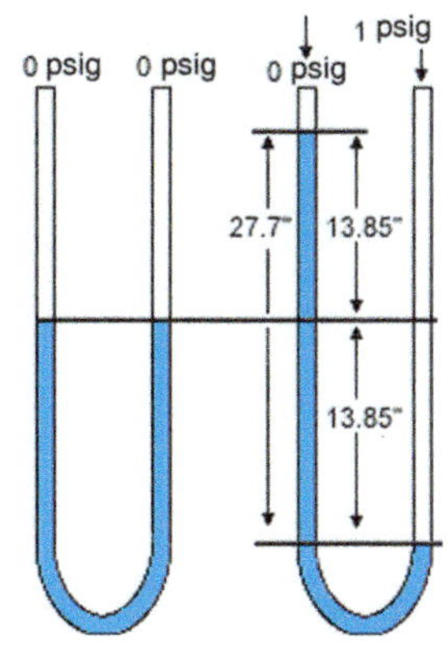

To illustrate these facts, we will start with 1 gal of water. The gallon of water equals 231 in^3 and weighs approximately 8.334 lb at 60°F. Pressure is measured in psi. Only 1 in^2 of area is needed to calculate the height of the water and the force it is exerting. Remember that force divided by area equals pressure.

Divide the height of the column of water, 231 in, by the weight of 1 gal of water, 8.334 lb. The result will be 27.718 or 27.7 in of water per pound of water, over a 1 in^2 area. *Therefore, 27.7 in of H_2O equals 1 psi of head pressure, or 2.31 ft of H_2O equals 1 psi of head pressure.*

The SG of the fluid to be measured multiplied by the height of the tank in inches provides an equivalent value in inches of water. The transmitter can now be calibrated in inches of water, regardless of the fluid. If the tank's fluid has an SG equal to 0.8 and a height of 100 in, then the height in inches of H_2O will be:

$$(100 \text{ in of fluid} \bullet 0.8 \text{ SG} = 80 \text{ in of } H_2O)$$

Pressure transmitters are purchased in different sizes of measurement. They are in ranges of inches H_2O, psig (the *g* stands for *gauge pressure*), or psia (the *a* stands for *absolute pressure*). When the symbol psid (the *d* stands for *differential pressure*) is called for, a standard psig transmitter is used. Most industrial pressure transmitters are differential pressure transmitters. They act on differential forces applied to each side of the transmitter. The force is produced by the pressure in the system multiplied by the area of the diaphragm.

Applying Pressure Measurement and Signals – Worked Examples

Differential Pressure and Meter Calibration

Differential pressure or differential head pressure is used to calibrate transmitters for pressure, level, flow, and density measurements. The transmitter has a high side, marked with an *H*, and a low side, marked with an *L*. The low side will typically go to atmospheric pressure or to a fixed height wet-leg measurement. The high side will typically go to the tank, where the varying height of fluid is to be measured. When calibrating an instrument, remember: The low side is the negative scale, below zero, and the high side is the positive scale, above zero. The transmitter's sensor element is static in position or elevation, and therefore the transmitter itself is always equal to zero elevation. This will be discussed in detail in Chapter 8 on level measurement.

Transmitters can be purchased in ranges of 25 in H_2O, 250 in H_2O, 1000 in H_2O, 300 psi, and 2000 psi.

The formula for calibration is:
(high side inches • SG) – (low side inches • SG) = lower or upper range value

Note: The equation provides the lower range value (LRV) when the tank is empty or the pressure is at the minimum level and the upper range value (URV) when it is full or the pressure is at the maximum level.

Sample Problem: A pressure gauge is reading 25 psig. It is attached to a tank filled with a fluid. The bottom of the tank is 65 ft above the ground. The pressure gauge is 5 ft above the ground. The fluid has a specific gravity of 0.7. What is the level of the fluid in the tank?

First, convert the psig measurement to feet of head measurement.

25 psig • 2.31 ft per psi = 57.75 ft of H_2O

Next, find the elevation of the bottom of the tank in relation to the elevation of the pressure gauge. Tank bottom in feet – pressure gauge elevation in feet = the height in feet to the bottom of the tank

65 ft – 5 ft = 60 ft of head to the bottom of the tank

Note: Head is always measured in the standard of inches or feet of water column (wc).

Multiply the head between the bottom of the tank and the pressure gauge times the SG to get the head equal to H_2O.

60 ft of fluid • 0.7 SG = 42 ft H_2O to bottom of tank from the pressure gauge

Next, subtract the height from the pressure gauge to the bottom of the tank in feet of H_2O from the total height of fluid in feet of in H_2O above the pressure gauge to find the height of the fluid in the tank in H_2O.

(57.75 ft of H_2O total head) – (42 ft of H_2O below the tank) = feet of fluid in H_2O in the tank
(57.75 ft total) – (42 ft to bottom tank from the pressure gauge) = 15.75 ft in H_2O in the tank

Next, convert height in feet of H_2O to height of fluid with an SG of 0.7.

15.75 ft of H_2O/0.7 SG = 22.5 ft of total height of the fluid column in the tank

Pressure Change in a Pipe for a Given Flow Rate

On the CSE exam, you will be asked to correlate signals and measurements using flow, pressure and the output in 4–20 mA signals. A change in flow in a pipe will cause a change in the head pressure across the pipe and measurement element. If the flow decreases in the pipe, the pressure in the pipe will increase at any point along the pipe.

If the flow rate increases, the pressure in the piping system decreases. If the flow rate decreases, the pressure in the piping system increases. This is because the total head of the system remains constant due to the head pressure developed by the pump. The total energy head being endowed into the pump and piping system remains constant. This is the case with a pump at a constant speed; two pressure gauges, one at each end of the pipe; and a hand valve at the end of the pipe.

$$h_1F_1^2 = h_2F_2^2 \qquad h_1\left(\frac{F_1}{F_2}\right)^2 = h_2$$

Sample Problem: There is a flow rate of 300 gallons per minute (gpm) in a piping system. There is a pressure gauge reading 100 psi somewhere in the piping system. If the flow rate is decreased to 240 gpm, what is the new pressure gauge reading (in psi) in the piping system?

Find the new pressure at the point of the gauge in the piping system for a flow rate of 240 gpm.

$$h_2 = h_1\left(\frac{F_1}{F_2}\right)^2 = 100\left(\frac{300}{240}\right)^2 = 156.25 \text{ psi}$$

Pressure Change across the Flow Element for a Given Flow Rate

If the flow in the pipe increases, the head pressure on the outlet of the measurement element will decrease. This correlation can be demonstrated by the following equations for differential head pressure (DP) across the orifice element (a fixed resistor) or a smaller section of pipe (venturi or Dall tube).

$$h_1F_2^2 = h_2F_1^2 \qquad h_1\left(\frac{F_2}{F_1}\right)^2 = h_2$$

Sample Problem: a) A flow of 250 gpm has a head pressure measurement of 309 in of H_2O. If the flow is decreased to 150 gpm, what is the new head pressure (DP) in H_2O for the measurement element?

b) What would be the new output to the programmable logic controller or distributed control system, in an mA signal, if the transmitter was calibrated in 0–400 in of H_2O? The signal is calibrated for 4–20 mA. The square root is extracted to linearize the flow signal.

a) Find the new head pressure for 150 gpm.

$$h_2 = h_1\left(\frac{F_2}{F_1}\right)^2;\ 309\left(\frac{150}{250}\right)^2 = 111.24 \text{ in } H_2O$$

b) Find the mA output.

The output signal is the square root of the ratio of change in head pressure (new measurement) to the full-scale calibrated range of the transmitter. First find the percentage of head pressure in the scale of 0–400 in H_2O.

$$\%\ head = \frac{111.24}{400} = 0.2781$$

The output is a 4–20 mA current signal. The span is 16 mA (20 mA – bias of 4 mA).

Because the flow rate is a squared function, we must first extract the square root of the percentage measurement to find the percentage of output signal.

$$\text{output mA} = \sqrt{0.2781} \bullet 16 \text{ mA} + 4 \text{ mA bias} = 12.44 \text{ mA}$$

Pressure Calibration of Transmitter

Sample Problem: The pressure in a pipe filled with natural gas is to be measured. The maximum pressure is measured as 462 ft of head. It is to be displayed in units of psig. What is the calibration of the transmitter to display this pressure in 0%–100% psig on the display? The minimum pressure measurement will be 0 ft of head.

Find the psig for the given maximum head pressure.
psig = feet head/2.31 psig per foot of head

Maximum measurement in psig:
200 psig = 462/2.31

Next, find the calibration range to order the transmitter.

The formula for calibration is:
(high side psi) – (low side psi) = lower or upper range value

Note: The formula provides the lower range value when minimum pressure or level and upper range value when maximum pressure or level.

URV = 200 – 0 = 200 psi
LRV = 0 – 0 = 0 psi

The transmitter will be calibrated as:
0–200 psig

Note: When measuring pressure in a pipe or vessel, it is just feet of water (H_2O) whether it be gas, oil, water, or gasoline. It is the molecules impinging against the side walls that produce the pressure. We do not use the specific gravity when measuring pipe wall static pressure.

8 Level Measurement and Calibration

Applying Level Measurement and Calibration – Worked Examples

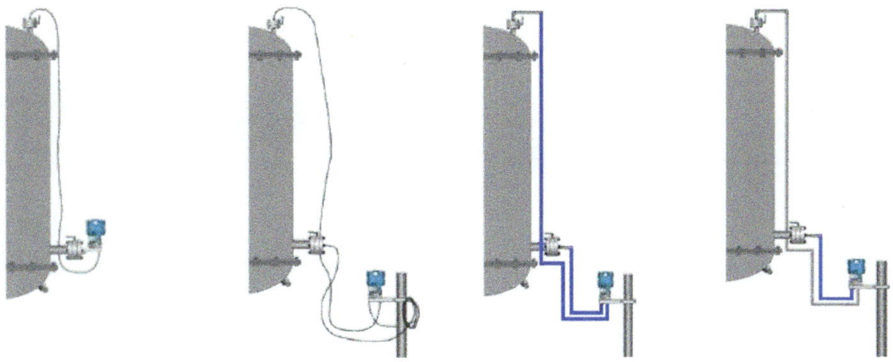

The Calibration Procedure

The level in a vessel or tank can be measured by many methods: differential pressure, displacement of volume, bubbler tube, capacitance, sonar, radar, and weight, to name a few. This book will focus on differential pressure, displacement of volume, and bubbler tube for the exam.

Remember: (high side inches • SG) − (low side inches • SG) = lower or upper range value

Example 1 – Open Tank

The low side of the transmitter is open to atmosphere. Atmospheric pressure is pushing on the low side. The high side of the transmitter is connected to the tank; it also has atmospheric pressure pushing on it. The atmospheric pressures on each side of the transmitter cancel out. In the example, the first line of math will be the lower range value (LRV) and the second line of math will be the upper range value (URV). The tank has 100 in of fluid with a specific gravity (SG) of 1.0. The calibrated *range* of the instrument will be 0–100 in of water or H_2O.

The *span* of the transmitter is: 100 in • 1.0 = 100 in

Example 2 – Open Tank, Suppress the Zero

The low side of the transmitter is open to the atmosphere. Atmospheric pressure is pushing on the low side. The high side of the transmitter is connected to the tank; it also has atmospheric pressure pushing on it. The atmospheric pressures on each side of the transmitter cancel out. In the example, the first line of math will be the LRV and the second line of math will be the URV.

The tank has a 100-inch level, and the tube dropping down below the tank adds 20 in of fluid height with an SG of 1.0. The calibrated range of the instrument will be 20–120 in of water or H_2O. Remember that the minimum measurement cannot be lower than the fixed tube height of 20 in. Suppress the zero with the hard wire jumper or set the variable in the transmitter and make 20 in a live zero for the instrument. In pneumatic instruments, a suppression kit must be installed.

The *span* of the transmitter is: 100 in • 1.0 = 100 in

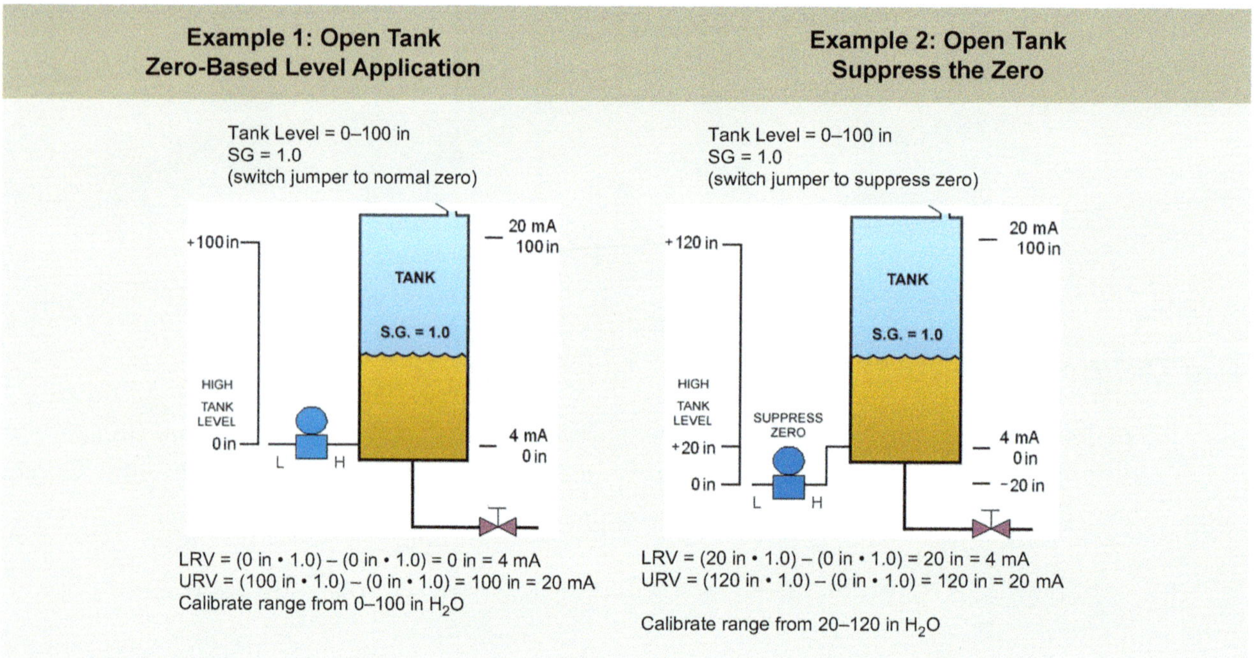

Example 3 – Closed Tank, Elevate the Zero

The low side is connected to the top of the closed tank. The high side is connected to the bottom of the closed tank. The tank's pressure does not matter because the pressures in the low and high side lines cancel each other out. Because the tank is pressurized, a "wet leg" or "reference leg" must be used. This is the piping going from the low side of the transmitter to the top of the tank. It typically will be filled with some other type of product, such as glycol or silicon. This prevents moisture from accumulating in the line.

If moisture accumulates in the line, it will give an error in the transmitter reading. The wet leg has 100 in of fluid with an SG of 1.1. In the example, the first line of math will be the LRV and the second line of math will be the URV. The tank has 100 in of fluid with an SG of 1.0. The calibrated *range* of the instrument will be –110 to –10 in of water or H_2O. Elevate the zero in the transmitter with the hard wire jumper or set the variable in the transmitter and make –110 in a live zero for the instrument. In pneumatic instruments, a suppression kit must be installed.

The *span* of the transmitter is: 100 in • 1.0 = 100 in

Example 4 – Closed Tank, Elevate the Zero (Transmitter below Tank)

The low side is connected to the top of the closed tank. The high side is connected to the bottom of the closed tank. The tank's pressure does not matter because the pressures in the low and high lines cancel each other out. The wet leg has 120 in of fluid with an SG of 1.1. The first line of math will be the LRV, and the second line of math will be the URV. The tank has 100 in of fluid, and the tube dropping down below the tank adds 20 in of fluid height with an SG of 0.8. The calibrated *range* of the instrument will be –116 to –36 in of water or H_2O. Remember that the minimum measurement cannot be lower than 20 in on the high side, due to the fixed 20-inch height of the tube dropping below the tank. **Elevate the zero** and make –116 in a live zero.

The *span* of the transmitter is: 100 in • 0.8 = 80 in

Remember: (high side inches • SG) – (low side inches • SG) = lower or upper range value

Note: The formula provides the LRV when the tank is empty and URV when it is full.

Example 3: Closed Tank
Elevate the Zero

Tank Level = 0–100 in
SG = 1.0, Wet Leg: SG = 1.1
Height = 100 in
(switch jumper to elevate zero)

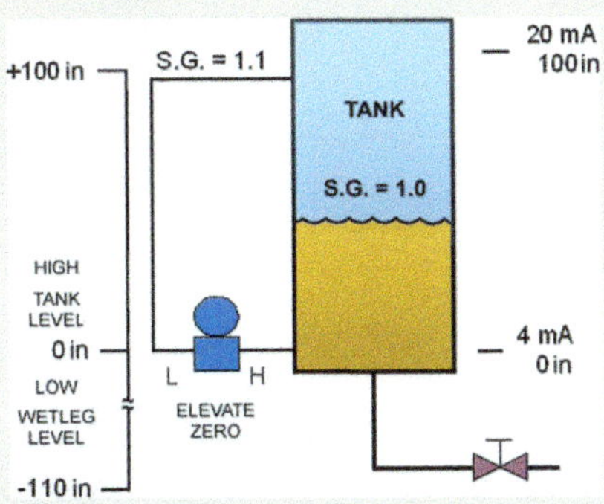

LRV = (0 in • 1.0) – (100 in • 1.1) = –110 in = 4 mA
URV = (100 in • 1.0) – (100 in • 1.1) = –10 in = 20 mA
Calibrate range from –110 in to –10 in H_2O

Example 4: Closed Tank
Elevate the Zero (transmitter below tank)

Tank Level = 0–100 in
SG = 0.8, Wet Leg: SG = 1.1
Height = 120 in
(switch jumper to elevate zero)

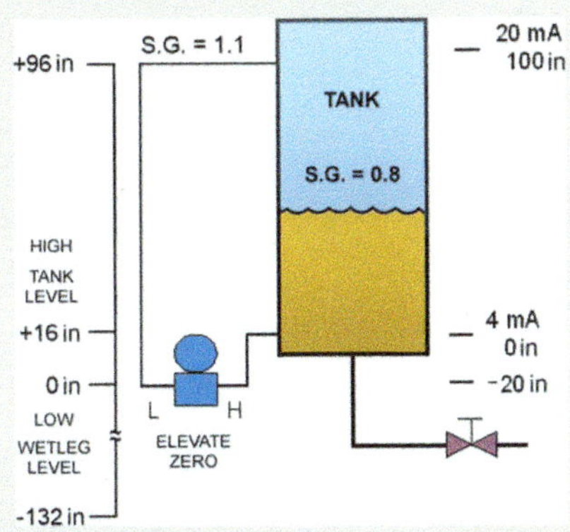

LRV = (20 in • 0.8) – (120 in • 1.1) = –116 in = 4 mA
URV = (120 in • 0.8) – (120 in • 1.1) = –36 in = 20 mA
Calibrate range from –116 in to –36 in H_2O

Transmitters with seals for density and level applications.

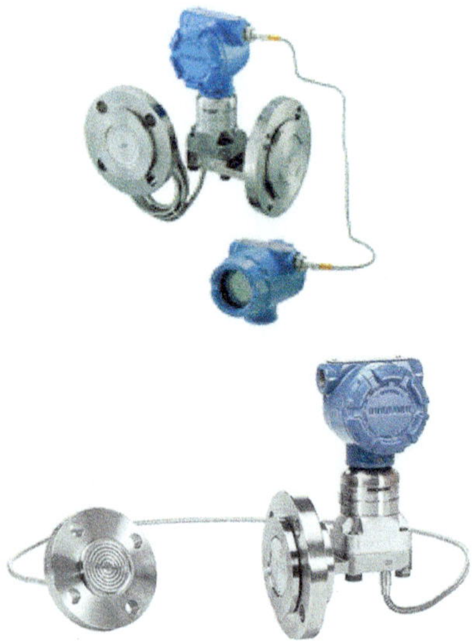

Rosemount suggested mounting with wet/dry leg to prevent freezing.

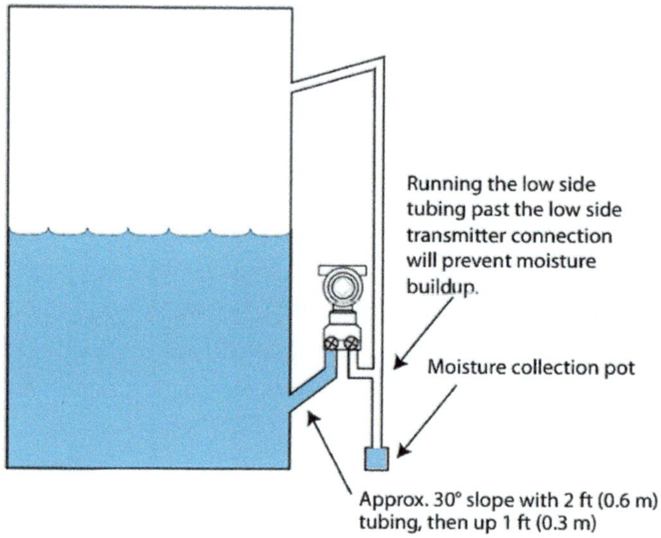

Level Displacer (Buoyancy)

The displacer tube for liquid level measurement is based on Archimedes' principle that the buoyancy force exerted on a sealed body immersed in a liquid is equal to the weight of the liquid displaced.

There are two types of displacer transmitters in common use today: torque tube and spring operated.

$$f = \frac{V_{df}}{231}(8.338)G_f$$

where

f = buoyancy force in pound-force
V_{df} = total volume of displaced process fluid in cubic inches
Ls = the submerged length of the displacer in process fluid
231 = cubic inches in 1 gallon of water
8.338 = weight of 1 gallon of water in pounds
G_f = specific gravity of displaced process fluid

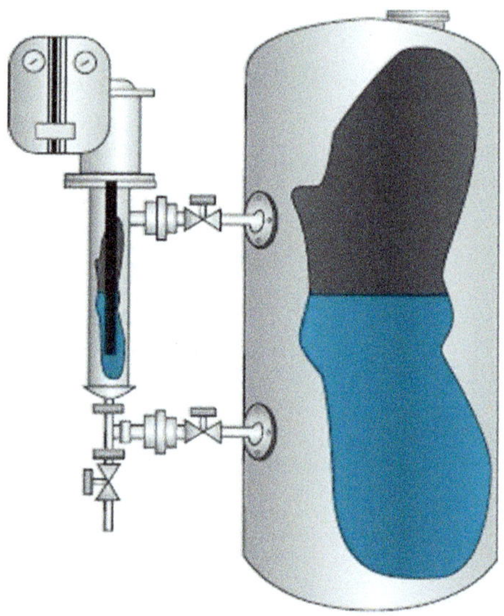

Sample Problem: a) What is the force upward on a 30 in displacer if the diameter of the displacer is 4 in and the displacer is submerged 10 in into a fluid with an SG of 0.72?

b) What is the mA output and percentage output of the process signal?

a) Find the displaced volume.

$$V_{df} = \left(\frac{\pi \bullet D^2}{4}\right) \bullet L_s = \left(\frac{\pi \bullet 4_{in}^2}{4}\right) \bullet L_s = \left(\frac{\pi \bullet 16\ \text{in}^2}{4}\right) \bullet 10\ \text{in} = 125.66\ \text{in}^3$$

Find the displacement force upward.

$$f = \frac{V_{df}}{231}(8.33)G_f = \frac{125.66}{231}(8.338)(0.72) = 3.266\ \text{lbf}$$

b) Find the displacement force upward for the total 30 in submerged.

$$V_{df} = \left(\frac{\pi \bullet D^2}{4}\right) \bullet L_s = \left(\frac{\pi \bullet 16}{4}\right) \bullet 30 = 376.99\ \text{in}^3$$

$$f = \frac{V_{df}}{231}(8.338)G_f = \frac{376.99}{231}(8.33)(0.72) = 9.798\ \text{lbf}$$

Find the percentage output and mA:

$$\%\ \text{output} = \frac{3.26}{9.79} = 0.333 \bullet 100 = 33.3\%\ \text{output}$$

$$(0.333 \cdot 16\ \text{mA}) + 4\ \text{mA} = 9.328\ \text{mA output}$$

Various types of displacement measuring devices and transmitters.

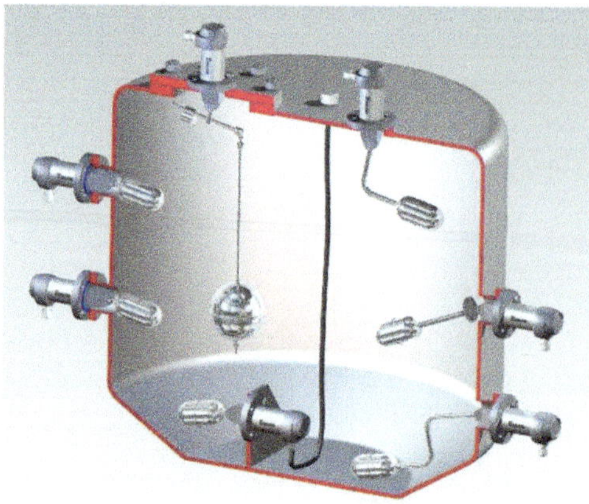

Bubbler Level Measurement

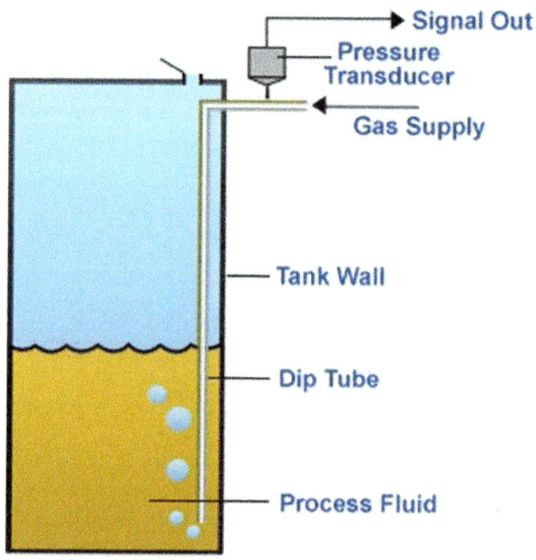

The bubbler tube, or dip tube, measures the level of the process fluid by measuring the back pressure on the bottom of the tube. This back pressure is the force exerted from the weight of the fluid in the tank against the tube opening. The tube will have to build up enough pressure for the gas to escape through the process fluid above the opening. The dip tube will have a static back pressure equal to the height or head of the process fluid above the bottom of the opening, as the bubbles escape the dip tube.

This simple level measurement has a dip tube installed with the open end close to the bottom of the process vessel. The lowest level that can be measured is from the bottom of the tank to the bottom of the dip tube. If the bottom of the dip tube is 2 in above the bottom of the tank, the minimum level that can be measured is 2 in. The maximum height that can be measured is only limited to the air supply pressure minus the minimum measurable level.

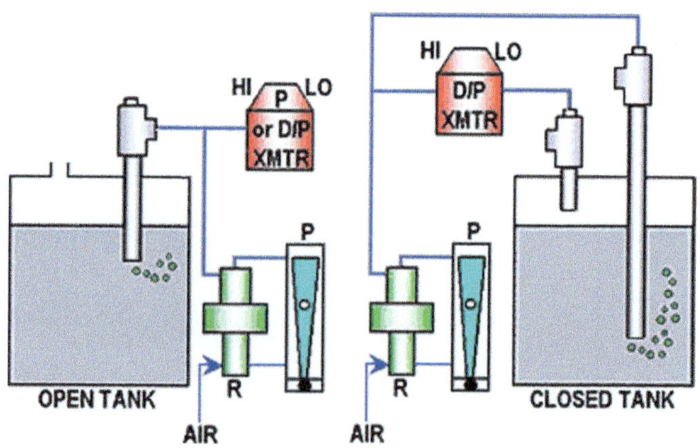

A flow of gas, usually air or nitrogen, is passed through a regulator to reduce the pressure. Then the flow of the gas will be controlled and monitored by passing through a rotameter (flowmeter). It then makes its way down the dip tube and the resultant back pressure, due to the hydraulic head of the process fluid, and forces back on the pressure transmitter. The pressure in the bubbler tube or dip tube equals the head pressure of level of the fluid in the vessel, and a proportional signal is sent to the programmable logic controller (PLC) or distributed control system (DCS). With a transmitter standard level calibration in inches of water, the signal out will vary proportionally with the change in level of the process fluid.

Sample Problem: a) What is the head pressure measurement of a bubbler tube submerged 24 in into a fluid with an SG of 0.85?

b) What is the percentage output and mA output at 24 in as in problem (a) if the transmitter is calibrated for a tube 100 in long and the transmitter is calibrated from 0–85 in H_2O (100 in • 0.85 SG = 85 in H_2O)?

a) Find the head pressure of the process fluid.

$$h = L_{DipTube}G_f = 24 \bullet 0.85 = 20.4 \text{ in } H_2O$$

(The water exerts a force of only 20.4 in H_2O against the bottom of the tube.)

b) Find percentage and mA output.

The transmitter is calibrated for 0–85 in H_2O, which equals 0%–100%

$$\% = \frac{20.4}{85} = 0.24 \bullet 100\% = 24\% \text{ output}$$

The transmitter output is a 4–20 mA current signal. The span is 16 mA (20 mA – bias of 4 mA) (0.24 • 16 mA) + 4 mA (bias) = 7.84 mA output, which equals 24% of the input measurement scale into the control room.

The control room computer (DCS or PLC) is scaling the input signal to a value of 0–100 in for the tank level. The 24% signal reads as 24 in in the tank for the control room.

Density Measurement

Head pressure and volume displacement can be used to measure density. By using a differential head pressure transmitter, calibrated in inches of water, you can connect the high and low lines to the tank at a fixed distance of separation, such as 10 in. Both taps of the density transmitter must be completely submerged below the top of fluid whose density is being measured. The height measured in inches of water divided by 10 in (in our example) is the SG of the unknown fluid. Example: The density transmitter is measuring 7 in H_2O, the SG = 0.7 (7 in/10 in = 0.7). See Figure 8-1.

With the SG known from the density transmitter and a second level transmitter calibrated in inches of H_2O, the tank level can be found. The level measurement can be divided by the SG measurement from the density transmitter, to show the true height of the process fluid in the tank.

Fine Powders and Sticky Fluids

- Paint, latex, animal fat, and soybean oil
- Sawdust, carbon black, titanium tetrachloride, salt, and grain
- Oils or grease in tanks

Capacitance Level Measurement

Commercial capacitance level transmitters are proven devices and were first introduced in the 1950s. They are also extremely versatile in that they can measure the continuous level and point level (a predetermined measurement point) of liquids, slurries, and liquid–liquid interface, as well as the point level of solids. Capacitance technology for level devices has also become known as *reactance, admittance,* or *radio frequency* (RF) technology.

The capacitance calculation for empty and full is important because a minimum change of capacitance of about 10 pF is needed for measurement. Last but not least, foam can be tricky with capacitance probes. If the foam is conductive, the capacitance probe will see the liquid and the foam as the complete level. Capacitance transmitters and switches can come with SIL 2 and 3 ratings.

Capacitance level does not work well in tanks with solids, in noncylindrical tanks, or with liquids of varying compositions and varying dielectric content.

Can Be Used For:

- Liquids level measurement
- Powered and granular solids
- Liquid metals at very high temperature
- Liquefied gases at very low temperature
- Corrosive material
- Very high-pressure industrial processes

Advantages of Capacitance Level Measurement:

- Inexpensive
- Reliable
- Versatile
- Minimal maintenance
- Contains no moving parts
- Good range of measurement

- Rugged
- Simple to use
- Easy to clean
- High-temperature and high-pressure application

Radiometric (Gamma) Level Measurement

Similar to radar devices, gamma level devices use electromagnetic radiation emitters and receivers to measure the level. Gamma devices can be used for liquids and solids in tanks. Gamma devices use electromagnetic radiation at a different part of the electromagnetic spectrum. They use gamma rays, which have much higher frequency and therefore smaller wavelengths than microwaves.

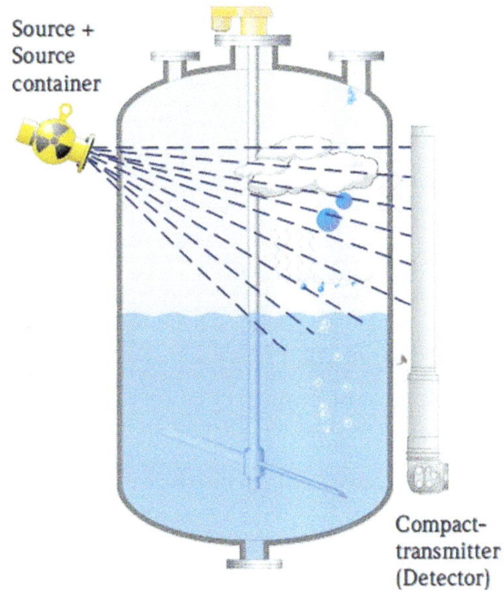

A source of gamma radiation, usually Cesium 137 or Cobalt 60, depending on the application, is placed in a lead source container. The container can be closed (emitting no radiation) or open (emitting gamma radiation). A detector, capable of measuring the amount of radiation from the source, is installed on the other side of the tank. If the tank is empty, the detector receives most of the available gamma radiation. If the tanks starts to fill with liquid or solid, as the level increases, the media will attenuate (absorb) some of the available gamma radiation. When the tank is full, the detector receives little radiation compared to the empty tank scenario.

This is an excellent level transmitter for difficult level measurements, such as for catalyst levels in tanks that are in series with other tanks or when the piping is in the way.

Gamma devices can also be used to measure the thickness, not just the level, of materials. Gamma devices can also be used as irradiators. Irradiators are devices or facilities that expose products to radiation to sterilize them, such as spices and some foods, milk containers, and hospital supplies.

Gamma level devices have been proven to be safe and reliable if safety procedures and regulations are followed. The safety of personnel is the number-one priority, and the amount of radiation over time that an employee can receive is well known and documented. All of this must be taken into account when purchasing gamma level devices. However, used safely, some of the most critical level measurements can be made with a gamma device.

Level Gauging System in a Tank Farm

Tank gauging is the measurement of liquids in large storage tanks with the purpose of quantifying the volume and mass of the product in the tanks.

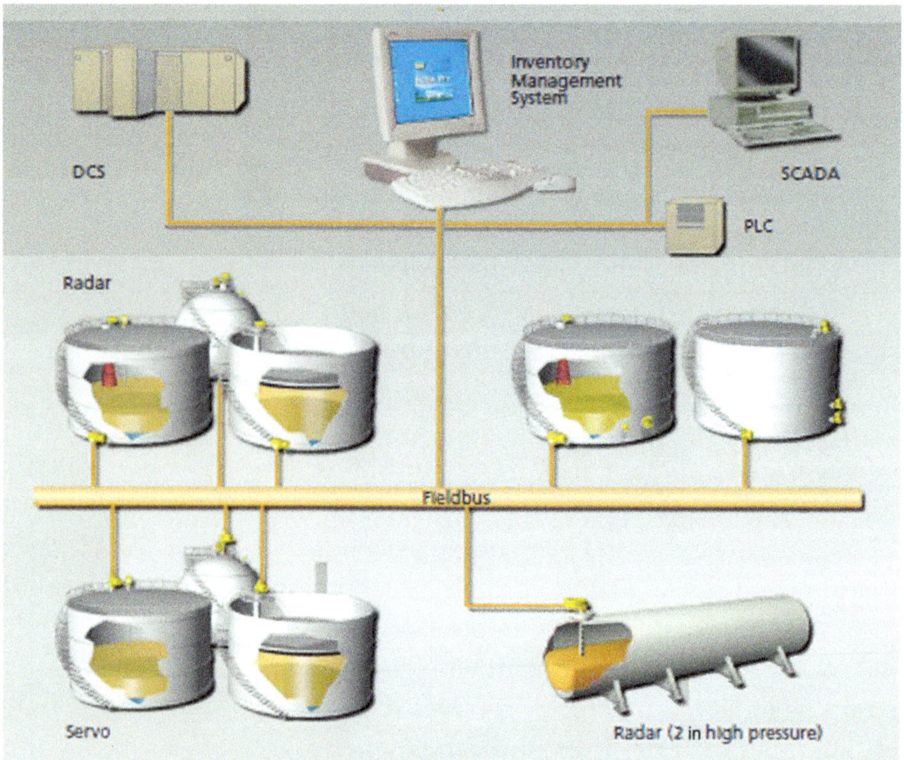

The oil and gas industry generally uses static volumetric assessments of the tank content. This involves level, temperature, and pressure measurements. There are different ways of measuring the liquid level and other properties of the liquid. The measurement method depends on the type of tank, the type of liquid, and the way the tank is used.

Storage tanks can contain large volumes of liquid product, representing a significant value. The accuracy performance of a tank gauging system is of high importance when assessing the tank content at any given time.

Tank gauging is used on large storage tanks in refineries, fuel depots, pipelines, airports, and storage terminals. Storage tanks usually come in four basic designs: cylindrical fixed roof tanks, cylindrical floating roof tanks, and pressurized tanks of either spherical or horizontal cylinder design. Tank gauges are available for all these tank types.

Calculating the Volume in Tanks

With a head pressure measurement, the height of the liquid in a tank can be measured. This is simple with standard cylindrical tanks but much more difficult with irregularly shaped tanks.

Calculating the volume in tanks will probably not be on the CSE exam, but the formulas to calculate the volume in these tanks is derived from calculus and included in the Appendix. The formulas show how to calculate the volume of spherical tanks and bullet tanks, so the volume can be calculated in the PLC or DCS. See Chapter 31, "Calculating Volume in Tanks," for the volume formulas.

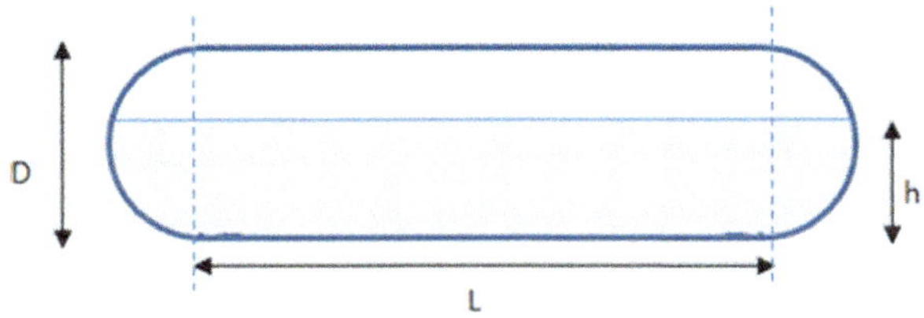

The tank ends can be flat (so the tank is just a horizontal cylinder). Tanks can come with different heads (end caps). They can be dished (ASME F&D, or flanged and dished), 2:1 elliptical, or hemispherical.

Tank Volume Calculation

Horizontal Cylinder

$$\left[\left(\frac{D}{2}\right)^2 \cos^{-1}\left(\frac{\frac{D}{2}-h}{\frac{D}{2}}\right) - \left(\frac{D}{2}-h\right)\left(Dh-h^2\right)^{0.5}\right] L$$

Head Volume Calculation

ASME F&D: $0.215483h^2\,(1.5D - h)$

Elliptical head: $\frac{\pi}{6}h^2(1.5D-h)$

Hemispherical head: $\frac{\pi}{3}h^2(1.5D-h)$

The liquid volumes in a horizontal cylinder, and ASME F&D, 2:1 elliptical, and hemispherical heads, are determined using these equations. The $\cos^{-1}$ (arccos or arcos) function must return radians, *not* degrees. In the Appendix, the volume for the tank section plus both heads combines into one formula. These formulas can be modified using the formulas defined previously for more accuracy with different heads (end caps).

The total volume of liquid in the tank is simply the liquid volume in the cylinder plus two times the liquid volume in the heads. (Hint: Multiply tank diameter, D, by the percentage-level signal to get h (the height shown on the human-machine interface or display) and then calculate the total tank volume with the math formula in the Appendix.)

9 Flow Measurement and Calibration

Applying Flow Measurement Devices

Like level measurement, flow measurement is head pressure and zero elevation-based. Head pressure is the measure of the endowed potential energy (hydrostatic pressure) in the system. The transmitter measurement is from how high the fluid falls, it is velocity squared. The velocity is squared because the fluid is constantly being accelerated through the pipe, as potential energy is endowed into the fluid by the pump head pressure.

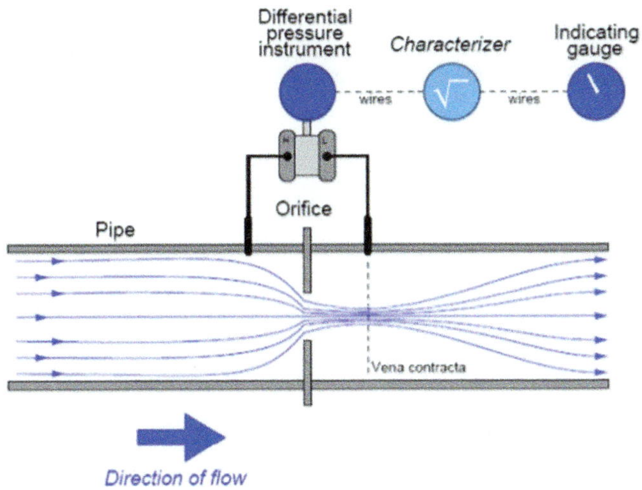

Head pressure is lost across an orifice element because energy loss is the product of energy flow multiplied by the resistance though which it flows (see figure at right).

Sizing of the orifice will be discussed in detail in the subsection titled "Orifice Tap Dimensions and Impulse Line Connections later in this chapter. You should familiarize yourself with the different types of flow-meters, their applications, and their ISA symbols. The ISA piping and instrumentation diagram (P&ID) symbols are shown in the ISA Standard Flowmeter Symbols section below.

Refer to ANSI/ISA-5.1-2009, *Instrumentation Symbols and Identification*, for more symbols and standards.

Turndown Ratio in a Flowmeter

The turndown ratio of a flowmeter is its ability to measure with acceptable accuracy the ratio of the maximum flow rate measurement to the minimum flow rate measurement. This is also known as the rangeability of the flowmeter. Turndown ratio is important when choosing a flowmeter technology for a specific application. If a gas flow to be measured will have a maximum measured flow rate of 1,000,000 scfm (standard cubic feet per minute) and a minimum measured flow rate of 100,000 scfm, the meter must have a minimum turndown ratio of 10:1 (1,000,000/100,000). For example, if the meter had an advertised turndown ratio of 20:1 and a maximum flow rate measurement of 2,000,000 scfm, then the minimum measurable flow rate would be 100,000 scfm.

The turndown ratio of each type of meter is limited by the constraints of the manufacturing process and materials used, as well as practical application considerations. For example, orifice meters create a pressure drop in the measured fluid proportional to the square of the velocity.

ISA Standard Flow Meter Symbols

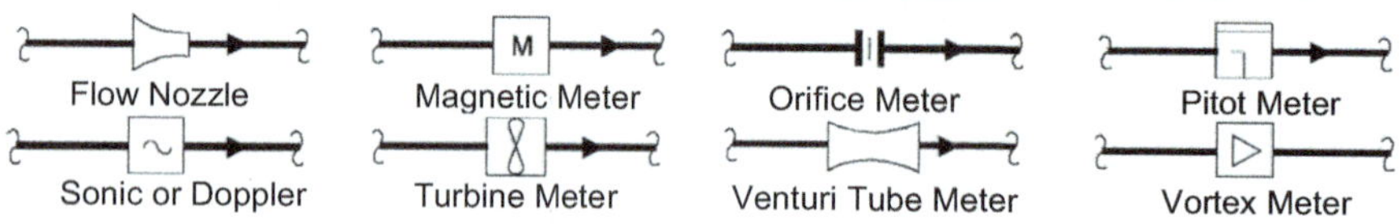

Flow Meter Applications Chart

Sensor	Rangeability	Accuracy	Advantages	Disadvantages
Orifice	3.5:1	2%–4% of full span	–Low cost –Extensive industrial practice	–High pressure loss –Plugging with slurries
Venturi	3.5:1	1% of full span	–Lower pressure loss than orifice –Slurries do not plug	–High cost –Line under 15 cm
Flow nozzle	3.5:1	2% full span	–Good for slurry service –Intermediate pressure loss	–Higher cost than orifice plate –Limited pipe sizes
Elbow meter	3:1	5%–10% of full span	–Low pressure loss	–Very poor accuracy
Annubar	3:1	0.5%–1.5% of full span	–Low pressure loss –Large pipe diameters	–Poor performance with dirty or sticky fluids
Turbine	20:1	0.25% of measurement	–Wide rangeability –Good accuracy	–High cost –Strainer needed, especially for slurries
Vortex shedding	10:1	1% of measurement	–Wide rangeability –Insensitive to variations in density, temperature, pressure, and viscosity	–Expensive
Positive displacement	10:1 or greater	0.5% of measurement	–High rangeability –Good accuracy	–High pressure drop –Damaged by flow surge or solids
Coriolis mass flow	100:1	0.05%–0.15% of measurement	–Very good accuracy	–Expensive

Pressure Tappings (Impulse Line Taps)

There are three standard positions for pressure tappings (also called *taps*), commonly named as follows:

1. *Corner taps* placed immediately upstream and downstream of the plate; convenient when the plate is provided with an orifice carrier incorporating tappings
2. *D* and *D/2 taps*, or *radius taps* or *vena contracta taps*, placed one pipe diameter upstream and half a pipe diameter downstream of the plate
3. *Flange taps* placed 25.4 mm (1 in) upstream and downstream of the plate, normally within specialized pipe flanges.

These types are covered by ISO 5167 and other major standards. Other types include:

- *2½D* and *8D taps* or *recovery taps* placed 2.5 pipe diameters upstream and 8 diameters downstream, at which point the measured differential is equal to the unrecoverable pressure loss caused by the orifice
- *Vena contracta tappings* placed one pipe diameter upstream and at a position of 0.3 to 0.9 diameters downstream, depending on the orifice type and size relative to the pipe, in the plane of minimum fluid pressure

The measured differential pressure differs for each combination, so the coefficient of discharge used in flow calculations depends partly on the tapping positions.

The simplest installations use single tappings upstream and downstream, but in some circumstances these may be unreliable; they might be blocked by solids or gas bubbles, or the flow profile might be uneven so that the pressures at the tappings are higher or lower than the average in those planes. In these situations, multiple tappings can be used, arranged circumferentially around the pipe and joined by a piezometer ring, or in the case of corner taps, annular slots running completely around the internal circumference of the orifice carrier.

Orifice Tap Dimensions and Impulse Line Connections

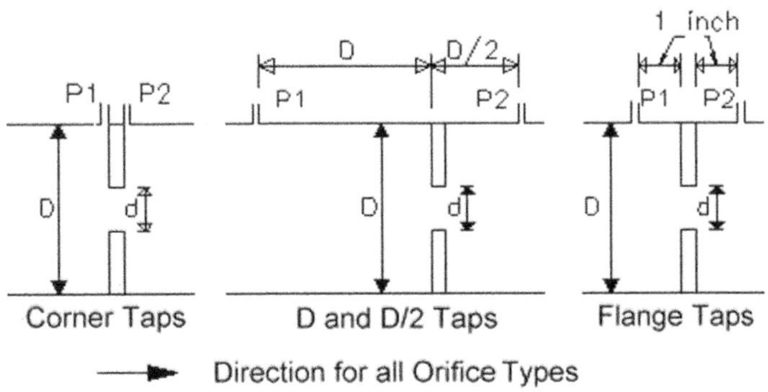

Flowmeter and pressure meter impulse line connections.

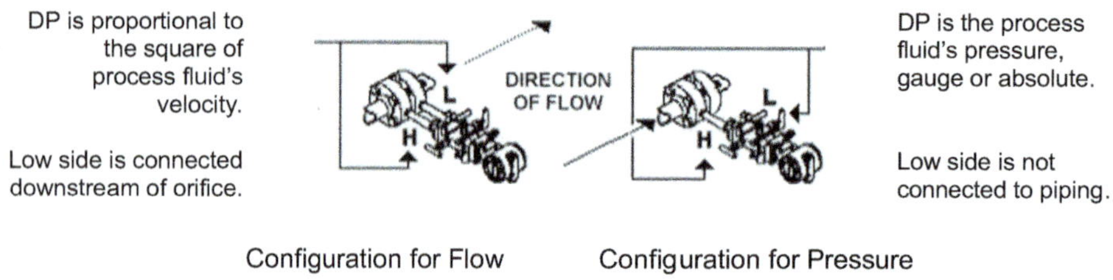

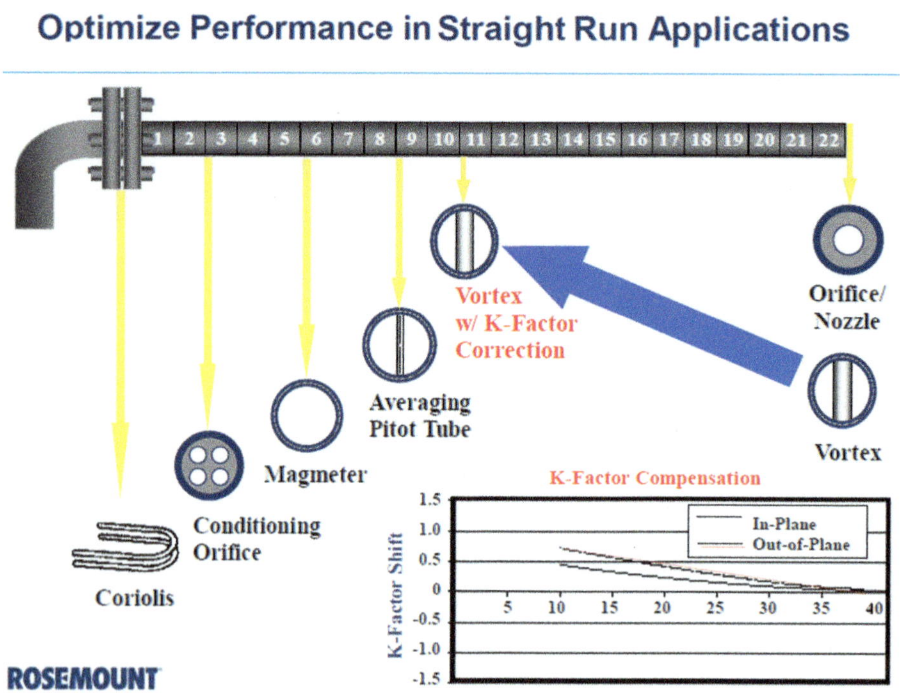

Be sure to verify distance information with your chosen vendor for product applications.

Flowmeter impulse line connections.

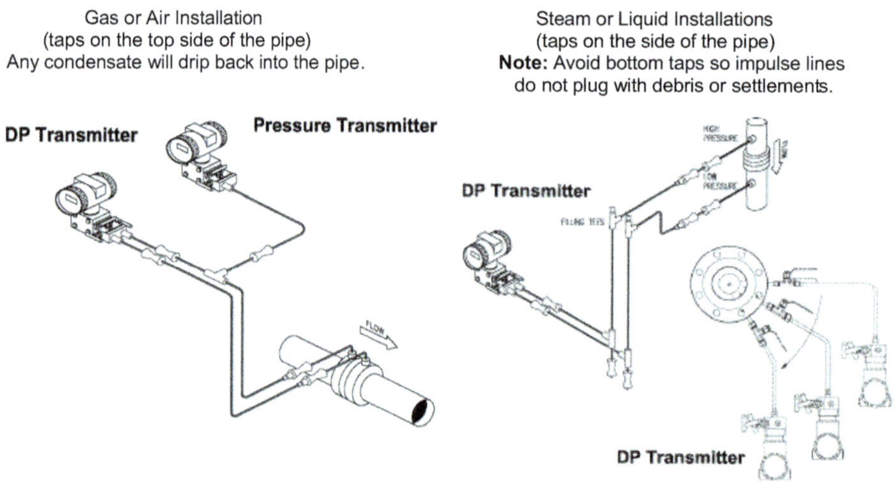

Various types of flowmeters

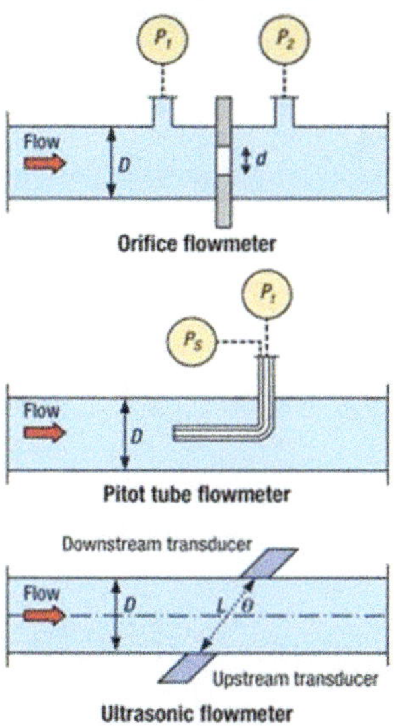

Applying Bernoulli's Principle for Flow Control

The process control industry covers a wide variety of applications of elements and final correction devices. The CSE exam encompasses a broad range of valve applications and sizing for different head or pressure devices, possibly an orifice meter, a turbine meter, a pressure-relief valve, or a safety rupture disk. This manual will cover the basics for the CSE exam.

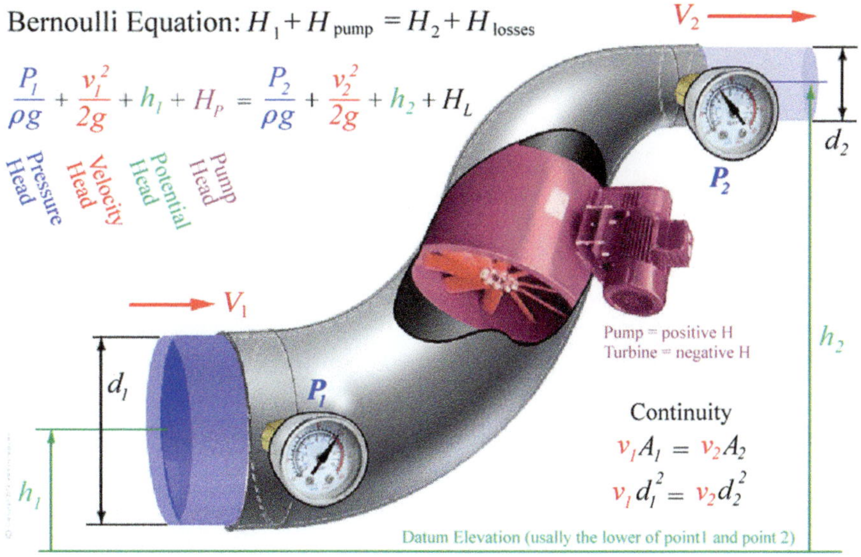

Let us start by reviewing the energy equation and its applicable laws governing flow control for processes. The relationship of the energy equation and flow rate through a pipe or head measurement device is shown by:

$$\frac{P_1}{\rho g} + \frac{V_1^2}{2g} + h_1 = \frac{P_2}{\rho g} + \frac{V_2^2}{2g} + h_2 \qquad A_1 V_1 = A_2 V_2$$

These equations are very useful on the CSE exam

For change in pressure anywhere in the piping system:

$$P_1F_1^2 = P_2F_2^2 \qquad P_2 = \left(\frac{F_1}{F_2}\right)^2 P_1$$

For change in head pressure across the flow measurement element:

$$h_1F_2^2 = h_2F_1^2 \qquad h_2 = \left(\frac{F_2}{F_1}\right)^2 h_1$$

For calculating the Reynolds number of a fluid to determine laminar or turbulent flow characteristics:

$$Re = \frac{3160 \bullet \textit{flow rate}\,(\text{gpm}) \bullet \text{SG}}{\textit{pipe ID}\,(\text{in}) \bullet \textit{viscosity}\,(\text{cSt})}\text{; for liquids} \qquad Re = \frac{6.316 \bullet \textit{flow rate}\,(\text{lb/h})}{\textit{pipe ID}\,(\text{in}) \bullet \textit{viscosity}\,(\text{cSt})}\text{; for vapors}$$

$$Re = \frac{7740 \bullet \textit{velocity}\,(\text{ft/s}) \bullet \textit{pipe ID}\,(\text{in})}{\textit{viscosity}\,(\text{cSt})}\text{; for liquids} \qquad Re = 1000\frac{v\,(\text{m/s})\,D\,(\text{mm})}{\upsilon\,(\text{cSt})}\text{; metric SI units}$$

Types of Head-Pressure Based Meters

There are several types of flowmeters that rely on Bernoulli's principle, either by measuring the differential pressure (DP) within a constriction or by measuring the difference between static and stagnation (impact) pressures to derive the fluid velocity.

Venturi Meter

A venturi meter constricts the flow in some fashion, and pressure sensors measure the differential pressure before and within the constriction. This method is widely used to measure flow rate in the transmission of gas through pipelines. The coefficient of discharge of C in venturi meters ranges from 0.93 to 0.97. The first large-scale venturi meter to measure liquid flows was developed by Clemens Herschel, who first used it to measure small and large flows of water and wastewater at the end of the nineteenth century.

Orifice Plate

An orifice plate is a plate with a hole through it, placed in the flow. It constricts the flow, and measuring the pressure differential across the constriction gives the flow rate. It is basically a crude form of Venturi meter but with higher energy losses. There are three basic types of orifice plates: concentric, eccentric, and segmental.

Dall Tube

The Dall tube is a shortened version of a venturi meter, with a lower pressure drop than an orifice plate. As with these flowmeters, the flow rate in a Dall tube is determined by measuring the pressure drop caused by restriction in the conduit. The pressure differential is typically measured using diaphragm pressure transducers with a digital readout. Because these meters have significantly lower permanent pressure losses than orifice meters, Dall tubes are widely used for measuring the flow rate of large pipe systems. The differential pressure produced by a Dall tube is higher than that of the venturi tube or nozzle, and all have the same throat diameters.

Pitot-Static Tube

A pitot-static tube is a pressure measuring instrument used to measure fluid flow velocity by differential pressure. DP is the difference of the static pressure and stagnation (impact) pressure. Bernoulli's equation is used to calculate the dynamic pressure, which is the kinetic energy of the fluid $V^2 = 2gh/SG$ (the difference of the height of the water column of the fluid) and therefore the fluid's velocity. Pitot tubes are often used to measure the air speed of an aircraft.

Multihole Pressure Probe

Multihole pressure probes (also called *impact probes*) extend the theory of the pitot tube to more than one dimension. A typical impact probe has three or more holes (depending on the type of probe) on the measuring tip arranged in a specific pattern. More holes cause the instrument to measure the direction of the flow velocity in addition to its magnitude (after appropriate calibration). Three holes arranged in a line cause the pressure probes to measure the velocity vector in two dimensions. Introduction of more holes (e.g., five holes arranged in a "plus" formation) enables measurement of the three-dimensional velocity vector.

Cone Meters

Cone meters are a newer differential pressure metering device. While working with the same basic principles as venturi- and orifice-type DP meters, cone meters do not require the same upstream and downstream piping. The cone acts as a conditioning device as well as a DP producer. Upstream requirements are between 0–5 pipe diameters compared to upstream diameters of up to 22 pipe diameters for an orifice plate or venturi. Because cone meters are generally of welded construction, it is recommended they always be calibrated prior to service. The inevitable effects of heat from welding can cause distortions and other effects that prevent tabular data on discharge coefficients with respect to line size, beta ratio, and operating Reynolds numbers from being collected and published. Calibrated cone meters have an uncertainty up to ±0.5%. Uncalibrated cone meters have an uncertainty of ±5.0%.

Annubar Meters

An Annubar primary element is an averaging pitot tube, such as a single point pitot tube used to measure the flow of gas, steam, or liquid in a pipe. An Annubar primary element or Annubar averaging pitot tube provides better accuracy than a single point pitot tube.

The pitot tube measures the difference between the static pressure and the flowing pressure of the media in the pipe. The volumetric flow is calculated from that difference using Bernoulli's principle and taking into account the pipe inside diameter.

The biggest difference between an Annubar averaging pitot tube and a pitot tube is that an Annubar averaging pitot tube takes multiple samples across a section of a pipe or duct. In this way, it averages the differential pressures encountered, accounting for variations in flow across the section. A pitot tube will give a similar reading if the tip is located at a point in the pipe cross section where the flowing velocity is close to the average velocity.

The T-shape cross section of the Rosemount Annubar primary element allows for increased signal strength and reduced signal noise compared to other averaging pitot tube shapes.

Note: Pitot tubes and Annubar tubes must be supported in high-velocity fluids because of vortex sheading. High frequencies can weaken the tube at the welds, and it can snap off inside the pipe. See the manufacturer's instructions for calculations and mounting procedures for installation in high-velocity fluids.

Differential Head Meter Calculations

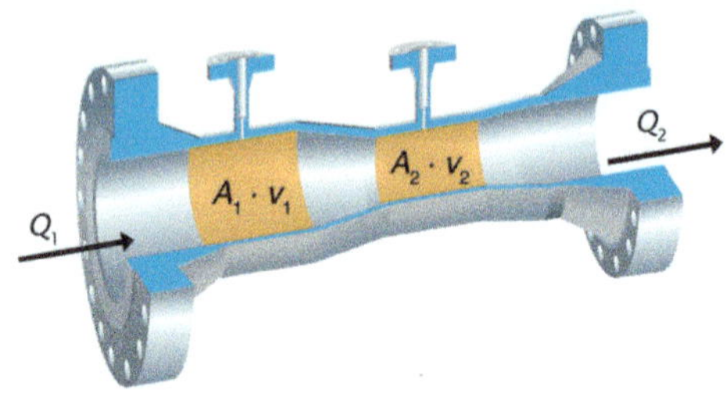

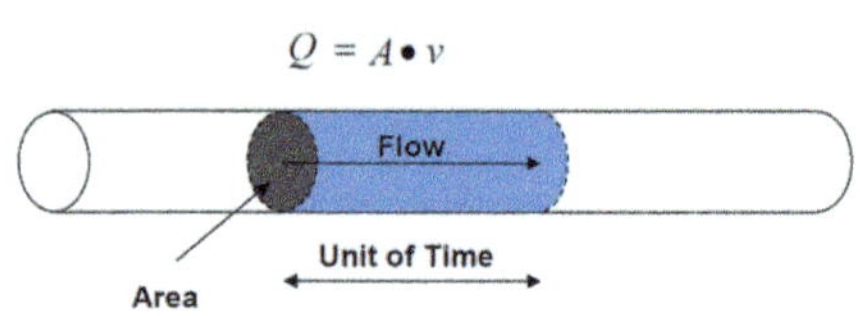

Classic Fluid Mechanics Model

The equation for flow through an orifice is easy to understand. The units are somewhat awkward, but they easily can be converted into terms that make them useful.

$Q = A \bullet V$
$Q =$ flow in cubic feet per second (ft^3/s)
$A =$ area of the orifice in square feet (ft^2)
$V =$ velocity of the liquid in feet per second (ft/s)

Experience shows that the actual flow is quite different than calculated flow because of the different shapes of the various orifices. The following diagrams illustrate some of these popular shapes. Each has been assigned a *k* value.

k value flow coefficients.

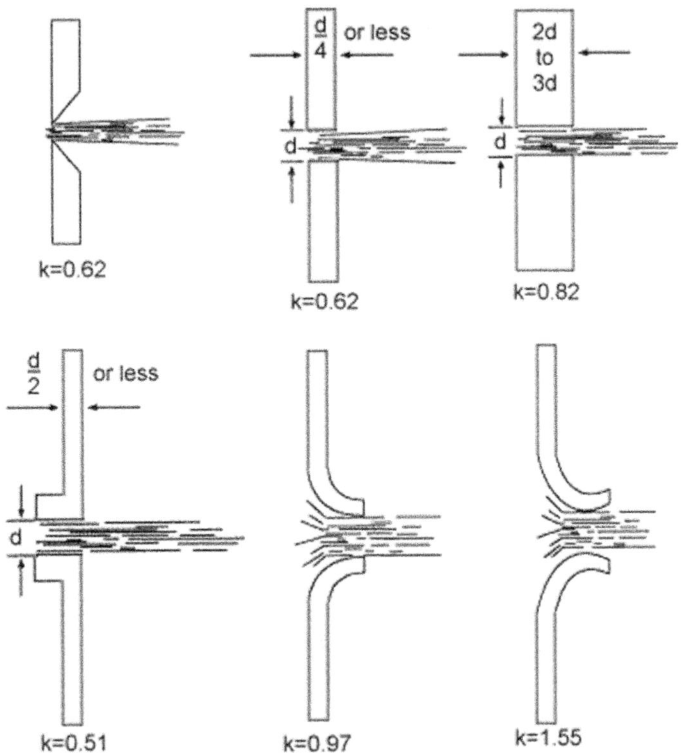

After entering the *k* value into our equation, the new equation becomes:

$$Q = AVk$$

To make the equation easier to handle, we can express the velocity, *V*, as a function of head in feet of water, *H*:

Note: Water column can be expressed as wc or H_2O.

$$\frac{V^2}{2g} = H \quad \text{or} \quad V = \sqrt{2gH} \quad \text{or} \quad V = \sqrt{64.34} \bullet \sqrt{H} = 8.02 \bullet \sqrt{H}$$

where

$g = 32.2$ ft/s^2
H = head across the orifice in feet. Use the differential head across the orifice.

Remember to convert pressure to head, using these formulas:

$$\frac{27.718 \text{ in}}{1 \text{ psi}} \bullet \frac{1 \text{ ft}}{12 \text{ in}} = \frac{2.31 \text{ ft}}{1 \text{ psi}}$$

It would also make sense to convert some of the terms in our equation to terms that are more convenient to use. As an example:

Q can be converted from cubic feet per second to gallons per minute:

$$1 \text{ ft}^3/\text{s} = 448.831 \text{ gpm}$$

$$\frac{1728 \text{ in}^3}{1 \text{ ft}^3} \bullet \frac{1 \text{ gal}}{231 \text{ in}^3} = \frac{7.4805 \text{ gal}}{1 \text{ ft}^3} \bullet \frac{60 \text{ s}}{1 \text{ min}} = \frac{448.831 \text{ gpm}}{1 \text{ ft}^3}$$

The area in square feet (A) can be converted to square inches:

$$1 \text{ ft}^2 = 144 \text{ in}^2$$

Putting all of this together gives us a new formula:

$$Q_{\text{gpm}} = \frac{A \bullet 8.02 \bullet 448.831 \bullet k \bullet \sqrt{H}}{144} = 25.0 \bullet A \bullet k \bullet \sqrt{H}$$

Let us plug in some numbers and calculate a flow through a typical orifice.

where

$H = 20$ ft
$A = 0.049$ in^2
$k = 0.62$ unitless

$$Q_{\text{gpm}} = 0.049 \bullet 25.0 \bullet 0.62 \bullet \sqrt{20}$$
$$Q = 3.397 \text{ gpm}$$

If we want to solve for the orifice area:

$$A = \frac{Q_{\text{gpm}}}{25.0 \bullet k \bullet \sqrt{H}} = \frac{3.397}{25.0 \bullet 0.62 \bullet \sqrt{20}} = 0.049 \text{ in}^2$$

Remember:

27.718 in = 1 psi at 60 °F, and head meters are typically in inches of water head, so we will change the feet into inches:

Note: Lowercase h is now head in inches.

$$Q_{\text{gpm}} = A \bullet 25.0 \bullet k \bullet \sqrt{H} = A \bullet 25.0 \bullet k \bullet \sqrt{H \bullet 12} = A \bullet 25.0 \bullet k \bullet 0.288675 \bullet \sqrt{h}$$
$$Q_{\text{gpm}} = A \bullet 7.21688 \bullet k \bullet \sqrt{h}$$

Let us plug in some numbers and calculate a flow through a typical orifice.

where

$h = 240$ in (20 ft)

$A = 0.049$ in^2

$k = 0.62$

$$Q_{\text{gpm}} = 0.049 \bullet 7.21688 \bullet 0.62 \bullet \sqrt{240}$$
$$Q_{\text{gpm}} = 3.396 \text{ gpm}$$

We made the formula more user-friendly by substituting in some conversions. Now we can make our calculations in gallons per minute and measure the area of the orifice hole in square inches, but the flow rate formula would be much better if we could measure the orifice diameter rather than the orifice area. This is a much more practical way to size an orifice. Use the following equation:

$$A = \frac{\pi d^2}{4} \quad \text{or} \quad d = \sqrt{\frac{4A}{\pi}}$$

Inserting the 0.049 in^2 we calculated from the prior formula, we get:

$$d = \sqrt{\frac{4 \bullet 0.049}{\pi}} = 0.250 \text{ in} \quad \text{or} \quad \tfrac{1}{4} \text{ in}$$

This exercise shows how the formulas we use to obtain flow rates are derived. We will rewrite the flow and orifice diameter formulas again, and maybe this time they will be simpler to use. We will start with the flow rate formula and then correct for the orifice formula:

$$A = \frac{\pi d^2}{4} \quad \text{or} \quad d = \sqrt{\frac{4A}{\pi}} \quad \text{so} \quad A = 0.7854d^2$$

Substituting the previous formula in the flow equation for A, we get:

$$Q_{\text{gpm}} = A \bullet 7.21688 \bullet k \bullet \sqrt{h} \quad \text{or} \quad \frac{\pi}{4} d^2 \bullet 7.21688 \bullet k \bullet \sqrt{h}$$

Ignoring the rounding errors through the previous equations, we get:

$$Q_{\text{gpm}} = 5.667 \bullet d^2 \bullet k \bullet \sqrt{h}$$

This equation is similar to the Spink equations we use for the CSE exam problems.

The formula for calculating the orifice diameter becomes:

$$d^2 = \frac{Q_{\text{gpm}}}{5.667 \bullet k \bullet \sqrt{h}} \quad \text{or...} \quad d = \sqrt{\frac{Q_{\text{gpm}}}{5.667 \bullet k \bullet \sqrt{h}}}$$

Determine if the formula still works. Here are the numbers:

- $d = 0.250$ or 1/4 in
- $k = 0.620$
- $Q =$ required flow rate of 3.4 gpm
- $h = 240$ in

We will begin by solving for flow (Q):

$$Q_{\text{gpm}} = 5.667 \bullet 0.25^2 \bullet 0.62 \bullet \sqrt{240} = 3.402 \text{ gpm}$$

Now we will try it to find the orifice size:

$$d = \sqrt{\frac{Q_{\text{gpm}}}{5.667 \bullet k \bullet \sqrt{h}}} = \sqrt{\frac{3.402}{54.44}} = 0.25 \text{ in or } 1/4 \text{ in}$$

The formula has proven correct; there is less than 1/1000 of an inch and 6/100 of a gallon error. All of the above numbers were generated assuming that you were moving water through the orifice at 60 °F. If you are making calculations for a liquid other than water or at a temperature other than 60 °F, you will have to factor in the SG of that liquid compared to water.

The Beta Ratio

The ratio of d/D has a special name. It is called the *beta ratio* (β) *of the orifice plate* or *head measuring element.*

where

$d =$ the orifice hole diameter or head restriction element throat diameter
$D =$ the exact internal diameter of the pipe carrying the fluid (related to pipe schedule and material)

Note: We will be using β repeatedly in our studies of flow and pressure-relief devices.

Some might say if d/D is greater than 30% (0.3), a modifier formula should be used. The following industrial standard formulas we will use to solve problems on the CSE exam demonstrate that the modifier formula gives results that are much greater than the actual flow rate.

The modifier (M) looks like this:

$$M = \sqrt{\frac{1}{1 - \left(\frac{d}{D}\right)^4}}$$

When you are using the modifier, the formulas look like this:

$$Q_{\text{gpm}} = 5.667 \bullet d^2 \bullet k \bullet \sqrt{h} \bullet M$$

$$d = \sqrt{\frac{Q_{\text{gpm}}}{5.667 \bullet k \bullet \sqrt{h} \bullet M}}$$

Now we will see what happens when a 0.250 in (1/4 in) orifice is put into a smaller cross section 0.5 in (1/2 in) pipe, assuming the other numbers stay the same:

$$\left(\frac{d}{D}\right)^4 = \left(\frac{0.250}{0.500}\right)^4 = 0.0625$$

$$M = \sqrt{\frac{1}{1-0.0625}} = 1.0328$$

This means that you would have to multiply by 1.0328, so the 3.402 gpm in the calculation in the previous section would become 3.514 gpm.

Pipe Size Is Important

The internal diameter of real pipe does not equal the nominal pipe size. Example: 1/2-inch, schedule 40 pipe has an internal diameter of 0.622 in; and 8-inch, schedule 40 pipe has an internal diameter of 7.981 in.

Let us recalculate the flow equation with the corrected size:

$$\left(\frac{d}{D}\right)^4 = \left(\frac{0.250}{0.622}\right)^4 = 0.0261$$

$$M = \sqrt{\frac{1}{1-0.0261}} = 1.0133$$

This means that you would have to multiply by 1.0133, so the 3.402 gpm result from the standard calculation would become 3.447 gpm.

That is a difference of $3.514 - 3.447 = 0.067$ gpm. This equals 4.02 gal error in 1 hour (0.067 gpm $\cdot$ 60 min $= 4.02$ gpm). A typical work year is 8760 hours. That equates to an error of 35,215.2 gal in 1 year. In a much larger flow rate, these inaccurate measurements can add up to millions of dollars in losses or overcharges.

We will address handling these errors in the following standard equations used for real-world industrial flow measurements. These include the Spink equation, taught in the ISA CSE review course, and the ISO 5167 equations used by most companies and software.

Other factors that affect the accuracy of the measurement include the following:

- The placement of the measurement instrument's line taps. (They are called *impulse lines.*)
- Solids buildup inside the piping. Calcium in water applications and coke in hot oil applications are typical. (The interior pipe diameter varies.) Higher temperature usually hastens the solids buildup.
- The SG of the fluid being measured, which can also change with temperature.
- The shape and design of the orifice plate and any defects, such as nicks in the orifice hole of the plate. (Nicks can cause up to 1% error in measurement.)
- The shape and design of restriction elements other than just an orifice plate.

Standard Flow Measurement Equations

There are many variations of flow equations. Books dedicated solely to how to take flow measurements can be from 500 to 1000 pages long. There are commercially available software packages for calculating flow measurements, such as Flowel (sold by ISA and Emerson). This software can cost in the area of $3,000. Most calculations do not need to be this accurate. The Spink equation is the author's favorite flow equation, but this manual lists others used by the top manufacturers of flow measurement equipment.

The most popular standard equations for flow calculation are:

- Spink
- ISO 5167
- AGA3 (now API-MPMS-14.3)
- AGA5/7/8
- AGA9
- *API Manual for Petroleum Measurement*

See the attachments with this manual for details on how to use these calculations for flow measurement.

Spink Flow Measurement Equation

Note: We will probably use only the Spink equations for problems on the CSE exam.

The basic Spink equation for liquid flow through an orifice plate is:

Volume Flow	Mass Flow	**Note:**
$Q = 5.667SD^2\sqrt{\frac{h}{G_f}}$	$Q_m = 5.667 \bullet \rho \bullet SD^2\sqrt{\frac{h}{G_f}}$	$\rho = 8.3378\frac{\text{lb}}{\text{gal}}$ @ 60°F for water

The book *Principles and Practice of Flow Meter Engineering* by L. K. Spink, first published in 1930, is generally recognized as the first, and for many years the only, definitive "body of knowledge" pertaining to industrial flow measurement. The book has undergone nine revisions, and the last edition was printed in 1978—21 years after Spink's death.

Another authoritative book on the subject of flow is the *Flow Measurement Engineering Handbook* by Richard Miller. This is an in-depth book almost 2 in thick that can be found at most libraries or obtained through interlibrary loan. This book is also used by many colleges.

The Basic Spink Equation Derived

Let us review the math that derives this volumetric flow equation.

Note: We will probably use only the Spink equations for problems on the CSE exam.

where

V = velocity
Q = quantity of flow
A = area of orifice
H = height in feet
h = height in inches
G_f = specific gravity at flowing condition
g = gravity

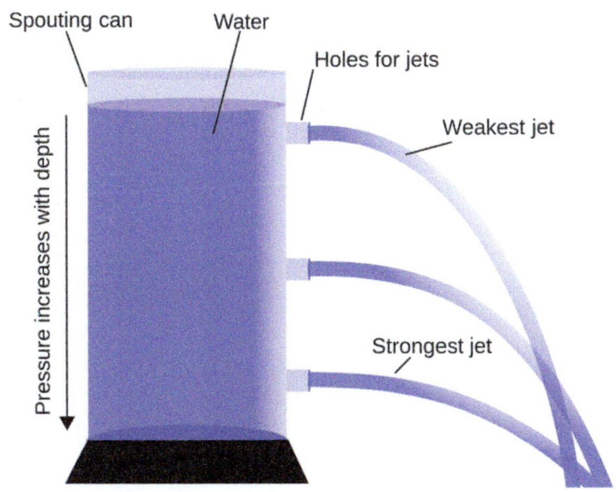

$$V^2 = 2gH$$

$$V = \sqrt{2gH}$$

$$Q = AV$$

$$Q = A\sqrt{2gH}$$

$$H = \frac{h}{12G_f}$$ **Note:** h is in inches; put it in feet.

$$Q = A\sqrt{2g\frac{h}{12G_f}}$$

$$Q\,(\text{gpm}) = [\textit{time scaling} \bullet \textit{volume scaling}] \bullet \left\{ \sqrt{\frac{2g}{1}} \bullet \frac{A}{144} \bullet \sqrt{\frac{h}{12G_f}} \right\}$$ **Note:** Scale inches to feet.

$$Q\,(\text{gpm}) = \left[\frac{60\,\text{s}}{1\,\text{min}} \bullet \frac{1728\,\text{in}^3}{231\,\text{in}^3} \right] \bullet \sqrt{\frac{2(g)}{12}} \bullet \frac{\pi d^2}{4 \cdot 144} \bullet \sqrt{\frac{h}{G_f}}$$

$$Q\,(\text{gpm}) = \left[\frac{60\,\text{s}}{1\,\text{min}} \bullet \frac{1728\,\text{in}^3}{231\,\text{in}^3} \right] \bullet \sqrt{\frac{64.34\,\text{ft}}{12\,\text{in s}^2}} \bullet \frac{\pi}{4 \cdot 144\,\text{in}^2} \bullet (d\,\text{in})^2 \bullet \sqrt{\frac{h\,\text{in}}{G_f}}$$

$$Q\,(\text{gpm}) = \frac{60\,\text{s}}{\text{min}} \bullet \frac{7.4805\,\text{gal}}{\text{ft}^3} \bullet \frac{2.3155\,\text{ft}}{\text{in} \bullet \text{s}} \bullet \left[\frac{0.00545}{\text{in}^2} \bullet (d\,\text{in})^2 \right] \text{ft}^2 \bullet \sqrt{\frac{h\,\text{in}}{G_f}}$$

Ignoring the rounding errors accumulated, we derive:

$$Q\,(\text{gpm}) = \frac{60}{\text{min}} \bullet 7.4805\,\text{gal} \bullet 2.3155 \bullet 0.00545 \bullet d^2 \bullet \sqrt{\frac{h}{G_f}} = 5.667 \bullet d^2 \bullet \sqrt{\frac{h}{G_f}} \bullet \frac{\text{gal}}{\text{min}}$$

$$Q\,(\text{gpm}) = 5.667 \bullet d^2 \bullet \sqrt{\frac{h}{G_f}}$$

Add the factors for coefficients of friction, viscosity, convergence, and divergence.

$$Q\,(\text{gpm}) = 5.667 \bullet Kd^2 \bullet \sqrt{\frac{h}{G_f}}$$

Note: S = the Spink factor used for sizing orifice flow measurements.

Because the K (the constant) and d (orifice diameter) are unknowns:

$$\text{Note that} \left(\frac{D^2}{D^2} \right) = 1 \text{ and } S = K \left(\frac{d^2}{D^2} \right)$$

We will cancel the orifice diameter (d^2) with S and we are left with S times the pipe internal diameter squared, (D^2):

$$Q\,(\text{gpm}) = 5.667 \bullet Kd^2 \bullet \frac{D^2}{D^2} \bullet \sqrt{\frac{h}{G_f}} = 5.667 \bullet \frac{Kd^2}{D^2} \bullet D^2 \bullet \sqrt{\frac{h}{G_f}} = 5.667SD^2 \sqrt{\frac{h}{G_f}}$$

The Basic Spink Equation for Liquid

The basic equation for liquid through an orifice or restriction type device is:

$$Q\,(\text{gpm}) = 5.667SD^2 \sqrt{\frac{h}{G_f}}$$

Using the sizing equation and the Spink sizing factor table, we can accurately size the orifice diameter and the dimensions for the orifice taps, pipe taps, nozzle and venturi, low-loss tube, and Dall (flow) tube for flow measurement.

The Basic Spink Equation for Gas and Vapor

The basic equation for gas or vapor through an orifice or restriction type device is:

$$Q\,(\text{scfh}) = 218.4SD^2 \frac{T_{abs}}{P_{abs}} \sqrt{\frac{hP_f}{T_f G_f}}$$ **Note:** scfh = standard cubic feet per hour

If the conditions are standard 60°F and 14.7 psia (pounds per square inch absolute), then the formula can be reduced to:

$$Q\,(\text{scfh}) = 7{,}727SD^2 \sqrt{\frac{hP_f}{T_f G_f}};$$ *only* at 60°F and 14.7 psia conditions.

The Basic Spink Equation for Steam

$$W\,(\text{lb/h}) = 359SD^2\sqrt{h\gamma_f}$$

where for all Spink formulas:

$$G_f = \text{SG for a gas: } \frac{\text{(molecular weight of gas)}}{28.97\text{ (molecular weight of air)}}$$

$$G_f = \text{SG for a fluid: } \frac{\text{(liquid weight of fluid)}}{\text{(liquid weight of water)}}$$

h = head in inches

P_{abs} = reference pressure (psia)

P_f = fluid operating pressure (psia)

T_{abs} = reference temperature (temperature absolute in Rankin, $°F + 460$)

T_f = fluid operating temperature (temperature absolute in Rankin, $°F + 460$)

γ_f = specific weight of the steam or vapor in pounds per cubic foot (operating condition)

Let us compare the Spink equation to the basic fluid mechanics equation we studied earlier. Here are the numbers:

- $d = 0.250$ or 1/4 in (orifice diameter)
- $D = 0.622$ in (1/2 in, schedule 40 pipe internal diameter)
- $h = 240$ in

This makes calculations much simpler. The size of ε depends on the pressure relation and the isentropic coefficient K; that is, the relation between a relative change in pressure and the corresponding relative change in density. (**Note:** This is a different K than the heat ratio factor used previously.) K is a property that is different for different media and varies also with the pressure and temperature of the medium. For many gases, there are no published data for K. The ISO standard recommends using C_P/C_v.

$\varepsilon = 1-\left(0.351+0.256 \bullet \beta^4 + 0.93 \bullet \beta^8\right) \bullet \left[1-\left(\frac{p_2}{p_1}\right)^{\frac{1}{K}}\right]$	Expansibility factor
Note: ε = for incompressible fluids.	

The Discharge Coefficient

The variable or coefficient C determines the relation between the real flow through a "primary device" and the theoretical possible flow. For a given primary device, this dimensionless parameter is determined using an incompressible fluid. With fixed geometry, C depends only on the actual Reynolds number (Re).

In a way, C can be regarded as a "calibration constant" for a primary device. Generally, in different installations that are geometrically equivalent, sense the same flow conditions, and are characterized by the same Reynolds number, render the same value for C .

Remember: Re varies with pressure, temperature, viscosity, and flow; therefore, so does C.

The ISO 5167 Equation Explained

$C = 0.5961+0.0261\bullet\beta^2-0.216\bullet\beta^8+0.000521\bullet\left(\frac{10^6\bullet\beta}{Re_D}\right)^{0.7} + (0.0188+0.0063\bullet A)\bullet\beta^{3.5}\bullet\left(\frac{10^6\bullet\beta}{Re_D}\right)^{0.3}$	Data on pipe orifice plate medium
$+(0.043+0.080\bullet e^{-10L1}-0.123\bullet e^{-7L1})\bullet(1-0.11\bullet A)\bullet\frac{\beta^4}{1-\beta^4}$	Data on upstream pressure tappings
$-0.031\bullet\left(M_2-0.8\bullet M_2^{1.1}\right)\bullet\beta^{1.3}$	Data on downstream pressure tappings
$\beta=\frac{d}{D}$ Relation between the orifice and pipe diameter	Beta ratio
$A=\left(\frac{19000\bullet\beta}{Re_D}\right)^{0.8}$ A function of β and Re_D	Reader-Harris Gallagher equation

$M_2 = \left(\frac{2 \cdot L_2}{1-\beta}\right)$	M_2 function of β and L_2	Reader-Harris Gallagher equation

Example: 40°C, 3 m/s flow

$$Re = \frac{DV\rho_1}{\mu} \quad Re = \frac{1.968 \text{ in/(12 in/ft)} \cdot 9.842 \text{ ft/s} \cdot 1.92 \text{ slugs/ft}^3}{1.35 \cdot 10^{-5} \text{ lb} \times \text{s/ft}^2} = 229{,}559$$

English units

Example: 40°C, 3 m/s flow

$$Re = \frac{DV\rho_1}{\mu} \quad Re = \frac{0.050 \text{ m} \cdot 3 \text{ m/s} \cdot 992.2 \text{ kg/m}^3}{0.656 \cdot 10^{-3} \text{ N} \times \text{s/m}^2} = 226{,}900$$

Metric units

Only the three following pairs of values for L_1 and L_2 are valid:

$L_1 = L_2 = 0$	Corner tappings
$L_1 = L_2 = \frac{0.0254}{D}$	Flange tappings
$L_1 = 1$ $L_2 = 0.47$	D and D/2 tappings

Let us compare the ISO 5167 equation to the Spink equation and to basic fluid mechanics. Here are the numbers:

- $d = 0.250$ in $= 0.00635$ m
- $D = 0.622$ in $= 0.0157988$ m
- $V = 3.5899$ ft/s $= 1.0942$ m/s
- operating temperature = 60°F = 20°C
- orifice taps are flange tappings
- $\beta = 4.02$
- $\Delta h = 240$ in $= 8.66426$ psi $= 6.096$ m H_2O
- $\Delta_p = \Delta h = 59737.9693$ Pa
- $\mu = 1.005$ N • s/m²
- $\rho_1 = 998.992$ kg/m³

Solve for the Reynolds Number (Re)

$$Re = \frac{DV\rho_1}{\mu} = \frac{0.00635 \text{ m} \cdot 1.0942 \text{ m/s} \cdot 998.992 \text{ kg/m}^3}{1.005 \cdot 10^{-3} \text{ N} \times \text{s/m}^2} = 6906.633$$

Solve for the Coefficient (C)

First we will solve for the needed variables:

$$A=\left(\frac{19000\bullet\beta}{\mathrm{Re_D}}\right)^{0.8}=\left(\frac{19000\bullet 0.402}{6906.633}\right)^{0.8}=1.0839$$

$$L_1=L_2=\frac{0.0254}{D}=\frac{0.0254}{0.0157988}=1.6077$$

$$M_2=\left(\frac{2\bullet L_2}{1-\beta}\right)=\left(\frac{2\bullet 1.6077}{1-0.402}\right)=5.3769$$

Next, we will solve for C in the three parts as shown previously, $C_1+C_2+C_3=C$.

Data on pipe orifice plate medium:

$$C_1=0.5961+0.0261\bullet\beta^2-0.216\bullet\beta^8+0.000521\bullet\left(\frac{10^6\bullet\beta}{\mathrm{Re_D}}\right)^{0.7}$$
$$+(0.0188+0.0063\bullet A)\bullet\beta^{3.5}\bullet\left(\frac{10^6\bullet\beta}{\mathrm{Re_D}}\right)^{0.3}$$

$$C_1=0.5961+0.0261\bullet(0.402)^2-0.216\bullet(0.402)^8+0.000521\bullet\left(\frac{10^6\bullet(0.402)}{6906.633}\right)^{0.7}$$
$$+(0.0188+0.0063\bullet 1.0839)\bullet(0.402)^{3.5}\bullet\left(\frac{10^6\bullet(0.402)}{6906.633}\right)^{0.3}=0.6127$$

Data on upstream pressure tappings:

$$C_2=\left(0.043+0.080\bullet e^{-10L1}-0.123\bullet e^{-7L1}\right)\bullet\left(1-0.11\bullet A\right)\bullet\frac{\beta^4}{1-\beta^4}$$

$$C_2=\left(0.043+0.080\bullet e^{10(1.6077)}-0.123\bullet e^{-7(1.6077)}\right)\bullet\left(1-0.11\bullet 1.0839\right)\bullet\frac{0.402^4}{1-0.402^4}=0.001016$$

Data on downstream pressure tappings:

$$C_3=-0.031\bullet\left(M_2-0.8\bullet M_2^{1.1}\right)\bullet\beta^{1.3}=-0.031\bullet\left(5.3769-0.8\bullet 5.3769^{1.1}\right)\bullet 0.402^{1.3}=-0.002725$$

Sum solutions for C as $C_1+C_2+C_3=C$

$$C=0.6127+0.001016+(-0.002725)=0.61099\ or\ 0.611$$

Solve for Mass Flow Rate

Note: This is the same K as used in the fluid mechanics examples.

$$q_m = K \bullet \varepsilon \bullet A \bullet \sqrt{2g \bullet m \bullet \mathrm{DP}} = \frac{C}{\sqrt{1-\beta^4}} \bullet \frac{\pi}{4} d^2 \bullet \sqrt{2\rho_1 \mathrm{DP}}$$

$$q_m = \frac{0.611}{\sqrt{1-0.402^4}} \bullet \frac{\pi}{4} 0.00635^2 \bullet \sqrt{2(998.992)(59{,}737.9693)} = 0.2142$$

Solve for Volumetric Flow Rate

$$q_v = \frac{q_m}{\rho_1}$$

$$q_v = \frac{0.2142}{998.992} = 0.0002144 \text{ m}^3$$

$$\text{gpm} = \frac{0.0002144 \text{ m}^3}{1 \text{ s}} \bullet \frac{60 \text{ s}}{1 \text{ min}} \bullet \frac{61023.7 \text{ in}^3}{1 \text{ m}^3} \bullet \frac{1 \text{ gal}}{231 \text{ in}^3} = 3.39856 \text{ or } 3.399 \text{ gpm}$$

Equation Comparison Summary

Fluid 60°F **0.5-inch, Schedule 40 Pipe, Beta = 0.4**		
Spink Basic Equation (uncompensated)	3.356 gpm	orifice = 0.25 in, head = 240 in H_2O
ISO 5167 Equation	3.399 gpm	orifice = 0.25 in, head = 240 in H_2O

Fluid 60°F **8-inch, Schedule 40 Pipe, Beta = 0.5**		
Spink Basic Equation	565.997 gpm	orifice = 3.9905 in, head = 100 in H_2O
ISO 5167 Equation	564.216 gpm	orifice = 3.9905 in, head = 100 in H_2O

There is a 0.315% flow rate disagreement between the Spink and ISO 5167 equations on the larger flow rate and a 1.27% disagreement on the smaller flow rate.

Sizing Orifice Type Devices for Flow Measurement – Worked Examples

Note: Table 9-1, "The Spink Factor" (S) (orifice sizing factor) will be used to size the orifice devices.

Liquid Sample Problem: Gasoline is carried in a 3-inch, schedule 40 pipe (ID = 3.068). A concentric sharp-edged orifice plate, with corner taps, is used to measure the flow. If the beta ratio (d^2/D^2) is 0.500, maximum flow rate is 100 gpm, and SG is $G_f = 0.75$, what is the differential head and span of the flowmeter transmitter?

$$Q\,(\text{gpm}) = 5.667SD^2\sqrt{\frac{h}{G_f}}$$

From Table 9-1, the Spink factor: beta = 0.500, $S = 0.1568$

$$100\text{ gpm} = 5.667(0.1568)(3.068)^2\sqrt{\frac{h}{0.75}}$$

$$\frac{100\text{ gpm}}{5.667(0.1568)(3.068)^2} = \sqrt{\frac{h}{0.75}}$$

$$\left(\frac{100\text{ gpm}}{8.3639}\right)^2 = \left(\sqrt{\frac{h}{0.75}}\right)^2$$

$$11.9561^2 = \frac{h}{0.75}$$

$$142.95(0.75) = h$$

$$107.21 = h\text{ (span)}$$

Calibrate the transmitter from 0% to 100% and 4 to 20 mA.

We only need 107.21 in H_2O, but you may want to calibrate the transmitter in some even span of measurement, such as the range of 0–110 in H_2O. A setting of 110 in H_2O will provide a full-scale flow of 101.29 gpm.

Steam Sample Problem: Dry saturated steam is carried in an 8-inch, schedule 80 pipe (ID = 7.625 in). A flow nozzle is used to measure the flow. If the beta ratio is 0.450 and the static pressure is 345.3 psig, what is the flow rate with a differential head pressure of 200 in H_2O across the meter?

$$W\,(\text{lb/h}) = 359SD^2\sqrt{h\gamma_f}$$

Find the density from Table A9, "Properties of Saturated Steam." A gauge pressure of 345.3 gives a specific volume of 1.2895.

$$\text{density in }\frac{\text{lb}}{\text{ft}^3} = \frac{1}{\text{specific volume in }\dfrac{\text{ft}^3}{\text{lb}}}$$

$$\gamma_f = \frac{1}{1.2895} = 0.7755$$

From Table 9-1, the Spink factor: beta = 0.4500, $S = 0.2026$

$$W\,(\text{lb/h}) = 359(0.2026)(7.625)^2\sqrt{(200)(0.7755)} = 52{,}664.68\text{ lb/h}$$

Gas Sample Problem: Natural gas is carried in a 6-inch, schedule 40 pipe (ID = 6.065 in). Flowing temperature is 60°F at 30 psig pressure. A concentric sharp-edged orifice plate, with flange taps, is used to measure the flow. If maximum flow rate is 4,000,000 standard cubic feet per day (scfd), SG G_f is 0.60, and the differential head of the flowmeter transmitter is 50 in H_2O, what is the orifice hole bore diameter?

$$Q\,(\text{scfh}) = 218.4SD^2 \frac{T_{abs}}{P_{abs}} \sqrt{\frac{hP_f}{T_f G_f}}$$

Change flow from per day to per hour and temperature and pressure to absolute:

$$\left(\frac{4{,}000{,}000 \text{ scf}}{\text{day}}\right)\left(\frac{1 \text{ day}}{24 \text{ h}}\right) = 166{,}666.7 \text{ scfh}$$

$$166{,}666.7 = 218.4S(6.065)^2 \frac{520}{14.7} \sqrt{\frac{(50)(30+14.7)}{(520)(0.60)}}$$

$$166{,}667 = 218.4S(6.065)^2 \frac{520}{14.7} \sqrt{\frac{(50)(44.7)}{(520)(0.60)}} = 759{,}216.398(S)$$

Find the S sizing factor for Table 9-1 in this section as follows:

$$\frac{166{,}666.7}{759{,}216.398} = S = 0.2195$$

From Table 9-1:

$beta = 0.575\ S = 0.2144$
$beta = 0.600\ S = 0.2369$

This will require interpolation:

$$beta = \left[\left(\frac{S\ desired - S\ lower\ value}{S\ upper\ value - S\ lower\ value}\right)(beta\ upper\ value - beta\ lower\ value)\right] + beta\ lower\ value$$

$$beta = \left[\left(\frac{0.2195 - 0.2144}{0.2369 - 0.2144}\right)(0.600 - 0.575)\right] + 0.575 = 0.5807$$

Find the orifice hole diameter:

$d = beta \bullet pipe\ ID = hole\ size$
$d = 0.5807 \bullet 6.065 = 3.522 \text{ in}$

For the calibrated range of the transmitter 0–50 in H_2O, and a flow rate of 166,666.7 scfh or 4,000,000 scfd, the orifice hole bore diameter = 3.522 in.

We can further simplify the equation for standard temperature of 60°F and standard pressure of 14.7 psia (STP).

We can reduce the equation to the constant multiplied by scfh and the molecular weight of the gas to get lbm per hour mass flow rate.

W (lb/h) *or* m = scfh • MW • 0.002634 (From the *ABB Flowmeter Handbook*)

Convert Actual Cubic Feet per Hour to Standard Cubic Feet per Hour

Note: acfh = scfh if both calculations are at 60°F and 14.7 psia. To convert acfh to scfh, multiply acfh by the temperature and pressure correction factors below.

$$\text{scfh} = \text{acfh}\left(\frac{P_f}{P_s}\right)\left(\frac{T_s}{T_f}\right)$$

where

T_f = temperature of flowing gas in °R (°F + 460)
T_s = standard temperature of gas in °R (60°F + 460) = 520°R
P_f = pressure of flowing gas in psia (psig + 14.7)
P_s = standard pressure of gas in psia (14.7)
acfh = actual cubic feet per hour

Note: Different standards for pressure and temperature are used in industry, but 14.7 psia (atmospheric pressure) and 60°F are the most common and are used in this review guide for sizing flow elements and control valves. Most of the ISA standard equations use this reference standard.

Note: Using the Rosemount 3051S with built-in advanced diagnostics capabilities, the following can be detected for alarms:

- Cavitation in the process stream
- A plugged impulse line
- Water in the transmitter housing
- A power supply that may be failing or faulty
- A signal line with a high impedance due to corrosion in the terminations or connections of the electrical signal loop

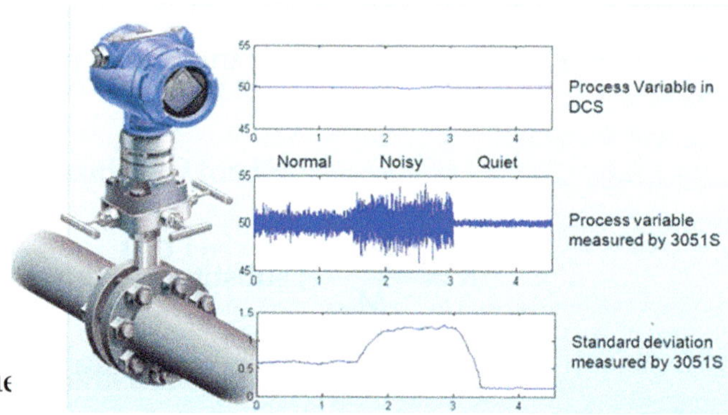

The 3051S has a microprocessor capable of statistical analysis of the process measurements.

These alarms can be read through the digital HART signal or by putting the transmitter in burst mode and splitting out three analog signals through a Rosemount Tri-Loop 333.

Note: The alarm signals cannot be used to meet the requirements of safety instrumented system applications; they are only for quality control and maintenance.

Applying Mass Flow Measurement with an Orifice – Worked Example

Note: The measurements are in English units for this application.

Mass Sample Problem: Natural gas is carried in a 6-inch, schedule 40 pipe (ID = 6.065 in). Flowing temperature is 85 °F at 325 psig pressure. A concentric sharp-edged orifice plate, with flange taps, is used to measure the flow. The specific gravity Gf = 0.657 and the $M_r = 19$. The differential head of the flowmeter transmitter is 0–50 in H_2O. The Spink factor is 0.2191 and beta ratio is 0.5802.

Find the scfh and the mass flow rate in lbm per hour and lbm per day.

$$\text{scfh} = 218.4(0.2191)(36.784)(35.374)\sqrt{\frac{(50)(325+14.7)}{(85+460)(0.657)}}$$

$$\text{scfh} = 62{,}264.161\sqrt{\frac{(50)(325+14.7)}{(85+460)(0.657)}}$$

scfh = (62,264.161)(6.8873) = 428,831.956 scfh

Find the mass for the scfh of gas knowing the relative molecular mass (M_r):

$$m = \frac{Q \bullet 4.07 \bullet M_w}{Z \bullet 1545.35} = (428{,}831.956)(0.002634)(19) = 21{,}461.324\ \frac{\text{lbm}}{\text{h}}$$

Find the mass using the simplified formula. This is the formula from the *ABB Flowmeter Handbook* for standard pressure and temperature (14.7 and 60°F).

$$m = scfh \bullet MW \bullet 0.002634 = (428{,}831.956)(19)(0.002634) = 21{,}461.324\ \frac{\text{lbm}}{\text{h}}$$

We now must convert lbm per hour to lbm per day:

$$\left(\frac{\text{lbm}}{\text{h}}\right)\left(\frac{24\ \text{h}}{\text{day}}\right) = (21{,}461.324)(24) = 515{,}071.776\ \frac{\text{lbm}}{\text{day}}$$

Real-World Application in a Computer (DCS or programmable logic controller (PLC))

The computer will read three signals from the field: pressure (psig), temperature (°F), and differential pressure (in H_2O).

Note: Do *not* extract the square root in the transmitter. This will be done in the computer calculation.

TT 100 = 0–120°F and 4–20 mA (the gas temperature)
PT 100 = 0–500 psig and 4–20 mA (the gas pressure)
PT 101 = 0–100 in H_2O and 4–20 mA (the gas flow rate as velocity)

The calculation in the computer will be some constant times the square root of the orifice equation. First calculate the scfh flow; this has already been defined in the previous example at a standard pressure and temperature of 14.7 and 60°F.

$$\text{scfh} = 62{,}264.161\sqrt{\frac{(\text{in H}_2\text{0})(\text{psig}+14.7)}{(F+460)(G_f)}}$$

Take the SG out of the square root; it is a constant:

$$\text{scfh} = 62{,}264.161\sqrt{\frac{1}{G_f}}\sqrt{\frac{(\text{in H}_2\text{0})(\text{psig}+14.7)}{(F+460)}}$$

For an SG of 0.657, the equation becomes:

$$\text{scfh} = 76{,}816.6663\sqrt{\frac{(\text{in H}_2\text{0})(\text{psig}+14.7)}{(F+460)}}$$

Change scfh to lbm per hour, and then lbm per hour to lbm per day:

$$\frac{\text{lbm}}{\text{h}} = (\text{scfh})(0.002634)(M_w);\quad \frac{\text{lbm}}{\text{day}} = (\text{scfh})(0.002634)(M_w)\left(\frac{24\text{h}}{\text{day}}\right)$$

Plugging this into the equation above using scfh and $M_r = 19$, we get:

$$\frac{\text{lbm}}{\text{day}} = 76{,}816.6663(0.002634)(19)(24)\sqrt{\frac{(\text{in H}_2\text{0})(\text{psig}+14.7)}{(F+460)}} = 92{,}264.8\sqrt{\frac{(\text{in H}_2\text{0})(\text{psig}+14.7)}{(F+460)}}$$

Testing the PLC equation, we get: 515,075.1161 lbm/day
That is an error of 3.34015 lbm difference between the previous calculation and the PLC calculation. This is well within significant figures.

Turbine Meter Applications

How Turbine Flowmeters Work

Turbine flowmeters use the energy of the fluid to mechanically rotate a "pinwheel" (rotor) in the flow stream. Blades on the rotor are angled to transform energy from the flow stream into rotational energy. The rotor shaft spins on bearings. When the fluid moves faster, the rotor spins proportionally faster.

Shaft rotation can be sensed mechanically or by detecting the movement of the blades. Blade movement is often detected magnetically, with each blade or embedded piece of metal generating a pulse. When the fluid moves faster, more pulses are generated. If a transmitter is used, it processes the pulse signal to determine the flow of the fluid. Transmitters and sensing systems are available to sense flow in both the forward and reverse flow directions.

Industries Where Used

In order of largest industry to smallest industry of use, turbine meters are used in oil and gas, water and wastewater, gas utility, chemical, power, food and beverage, aerospace, pharmaceutical, metals and mining, and pulp and paper.

The Pluses and Minuses of Turbine Meters

The cost of turbine meters is moderate. The turbine meter is very good at measuring clean, low-viscosity fluids of moderate velocity and a steady rate. Turndown is very good as turbine meters can read very low compared to the maximum flow. They are reliable if put in a clean fluid, especially if the fluid has some lubrication properties. They are American Gas Association (AGA) and American Petroleum Institute (API) approved for custody transfers. They do cause some pressure drop in the piping system and should be considered as a factor in gravity flowing systems. They are not reliable for steam, and their bearings eventually wear out.

How to Use Turbine Flowmeters

Turbine flowmeters measure the velocity of liquids, gases, and vapors in pipes, such as hydrocarbons, chemicals, water, cryogenic liquids, air, and industrial gases. High-accuracy turbine flowmeters are available for custody transfer of hydrocarbons and natural gas. These flowmeters often incorporate the functionality of a flow computer to correct for pressure, temperature, and fluid properties to achieve the desired accuracy for the application.

Be careful when using turbine flowmeters on fluids that are nonlubricating, because the flowmeter can become inaccurate and fail if its bearings prematurely wear. Some turbine flowmeters have grease fittings for use with nonlubricating fluids. In addition, turbine flowmeters that are designed for a specific purpose, such as for natural gas service, can often be limited to an operating upper range of temperature, such as 60°C. Operating above the upper temperature limit can cause damage to the flowmeter.

Smaller turbine flowmeters can be installed directly in the piping, but the size and weight of larger turbine flowmeters may require the installation of substantial concrete foundations and supports. The flow of corrosive liquids can be measured with proper attention to the materials of construction of all wetted parts, such as the body, rotor, bearings, and fittings.

Applications for turbine flowmeters are found in the water, petroleum, and chemical industries. Water applications include distribution systems within and between water districts. Petroleum applications include the custody transfer of hydrocarbons. Miscellaneous applications are found in the food and beverage, and chemical industries.

Application Cautions for Turbine Flowmeters

Turbine flowmeters are less accurate at low flow rates due to rotor/bearing drag that slows the rotor. Make sure to not operate these flowmeters below approximately 5% of maximum flow. Turbine flowmeters should not be operated at high velocity because premature bearing wear and/or damage can occur. Be careful when measuring fluids that are nonlubricating because bearing wear can cause the flowmeter to become inaccurate and fail. In some applications, bearing replacement may need to be performed routinely and increases maintenance costs. Application in dirty fluids should generally be avoided so as to reduce the possibility of flowmeter wear and bearing damage. In summary, turbine flowmeters have moving parts that are subject to degradation with time and use.

Abrupt transitions from gas flow to liquid flow should be avoided because they can mechanically stress the flowmeter, degrade accuracy, and/or damage the flowmeter. These conditions generally occur when filling the pipe and under slug flow conditions. Two-phase flow conditions can also cause turbine flowmeters to measure inaccurately.

Mass Flow Measurement with Turbine Meters

If the turbine flowmeter has one rotor only, it can only measure the velocity of the fluid. Temperature compensation will be needed for accuracy in mass flow measurements such as natural gas sales. Two rotor meters can be used to measure the angular momentum of the fluid for accurate mass flow measurements.

Actual versus Standard Units of Gas Measure

- Turbine flowmeters measure the actual volume of gas passing through the meter at the operating temperature and pressure. They are therefore sized in actual liters per minute (alpm) or actual cubic feet per minute (acfm).
- Standard liters per minute (slpm) or standard cubic feet per minute (scfm) are the equivalent volume of gas referenced back to standard temperature and pressure as described previously in this section.
- The ideal gas law used to convert between alpm and slpm is:

$$\text{alpm} = \text{slpm}\left(\frac{101.7\text{ bar}}{P1} \bullet \frac{T1}{288.15\text{ °K}}\right)$$

$P1$ = operating pressure in bara

$T1$ = operating temperature in degrees kelvin

Standard Air Calibration

The most common and economical calibration is in air at standard conditions. Data can be presented in actual units of measure or converted to standard units using the operating temperature and pressure.

Actual air volume flow is often termed *acfm*. Unfortunately, in real life *actual conditions* are seldom *standard conditions*. The scfm will change based on the flowing conditions:

- Pressure is applied to a volume of air—it gets smaller
- Vacuum is applied to a volume of air—it expands

The acfm depend on the following variables of the actual air.

- Pressure
- Temperature
- Humidity

The conversion from scfm to acfm can be expressed as:

$$\text{acfm} = \text{scfm}\left(\frac{P_{\text{standard}}}{P_{\text{actual}} - P_{\text{saturation}}}\right)\left(\frac{T_{\text{actual}}}{T_{\text{standard}}}\right)$$

where

$$\begin{aligned}
\text{acfm} &= \text{actual cubic feet per minute}\\
\text{scfm} &= \text{standard cubic feet per minute}\\
P_{\text{standard}} &= \text{standard absolute air pressure (psia)}\\
P_{\text{actual}} &= \text{absolute pressure at the actual level (psia)}\\
P_{\text{saturation}} &= \text{saturation pressure at the actual temperature (psi)}\\
T_{\text{actual}} &= \text{actual ambient air temperature (°R)}\\
T_{\text{standard}} &= \text{standard temperature (°R)}
\end{aligned}$$

The term $P_{\text{saturation}}$ is used with air due to the moisture content. It is not needed with process gases.

Turbine Flowmeter – Worked Example

The basic equation for flow through a turbine meter is:

$$V = KN$$

where

V = volume
K = volume per pulse
N = number of pulses

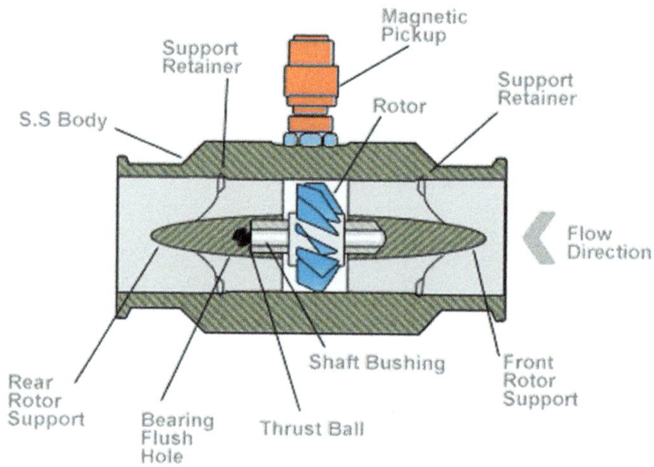

The average flow rate (Q_{avg}) is equal to the total volume divided by the time interval.

$$Q_{avg} = \frac{V}{\Delta t} = K\frac{N}{\Delta t}$$

$f = \frac{N}{\Delta t}$ Number of pulses per unit time...

$$Q_{avg} = Kf$$

Note: The turbine flowmeter can measure the flow rate in units of cubic inches or gallons per pulse.

where

Q_{avg} = average flow rate
Δt = change in time
f = frequency

Sample Problem: The turbine meter has a K value of 1.22 in³ per pulse.

a) Determine the liquid volume transferred for a pulse count of 6,400.

b) Determine the flow rate if the 6,400 pulses are counted in a duration of 40 s.

c) What would be the totalized flow after 15 min at the current pulse rate of flow?

d) What is the frequency of the signal?

a) Liquid volume:

$$V = KN$$
$$V = (1.22\,\text{in}^3)(6400) = 7808\,\text{in}^3$$
$$\text{Gallons} = 7808\ \text{in}^3 \bullet \frac{1\,\text{gal}}{231\,\text{in}^3} = 33.8\ \text{gal}$$

b) Flow rate:

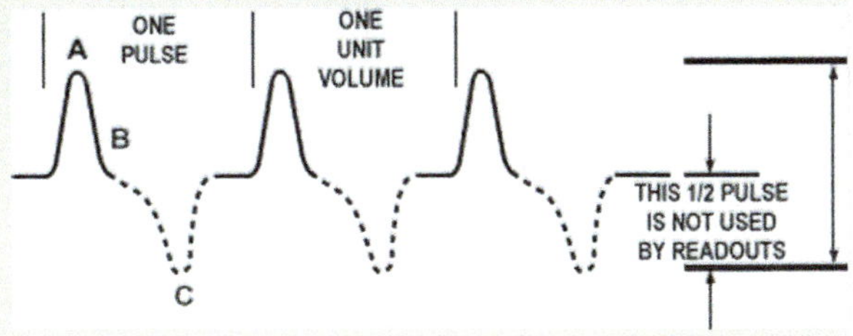

$$Q = \frac{V}{\Delta t}$$

$$Q = \frac{7808 \text{ in}^3}{40 \text{ s}} = \frac{195.2 \text{ in}^3}{\text{s}}$$

$$Q = \frac{195.2 \text{ in}^3}{\text{s}} \bullet \frac{60 \text{ s}}{1 \text{ min}} \bullet \frac{1 \text{ gal}}{231 \text{ in}^3} = 50.7 \frac{\text{gal}}{\text{min}}$$

c) Totalized flow after 15 minutes:

$$Q = 50.7 \frac{\text{gal}}{\text{min}} \bullet 15 \text{ min} = 760.5 \text{ gal}$$

d. Find the frequency.

Note: Frequency in Hz is frequency per second, so the Hz equation is: count (frequency)/s = Hz.

$$f = \frac{N}{\Delta t} = \frac{6400 \text{ count}}{40 \text{ s}} = \frac{6400 \text{ count}}{40 \text{ s}} = 160 \text{ Hz}$$

Sample Problem: A turbine flowmeter has a K value of 127 pulses per gallon.

a) Determine the liquid volume in gallons transferred for a pulse count of 7,300.

b) Determine the flow rate, if the flowmeter sends a pulse count of 86,500 pulses in 6.8 min.

c) What are the total gallons transferred in 8 hours for question (b)?

d) What is the frequency of the signal for question (b)?

a) Liquid volume:

$$V = KN$$

$$V = \left(\frac{1 \text{ gallons}}{127 \text{ pulses}}\right)(7300 \text{ pulses}) = 57.5 \text{ gallons}$$

b) Flow rate:

$$Q = \frac{V}{\Delta t}$$

$$V = \left(\frac{1\ \text{gallons}}{127\ \text{pulses}}\right)(86{,}500\ \text{pulses}) = 681.1\ \text{gallons}$$

$$Q = \frac{681.1\ \text{gallons}}{6.8\ \text{min}} = 100.16\ \text{gpm}$$

c) Totalized flow after 8 hours for the flow rate in question (b):

$$Q = 100.16\frac{\text{gal}}{\text{min}} \bullet \frac{60\ \text{min}}{1\,\text{h}} \bullet 8\,\text{h} = 48{,}076.8\ \text{gallons}$$

d) Find the frequency for the flow rate in question (b):

Note: Frequency in Hz is frequency per second; therefore, Hz = (count)/s.

$$f = \frac{\text{N}}{\Delta\text{t}} = \frac{86{,}500\ \text{count}}{6.8\ \text{min}} \bullet \frac{1\ \text{min}}{60\ \text{s}} = 212\ \text{Hz}$$

Vortex Flowmeter Applications

The basic equation for flow through a vortex meter is:

$$St = f(d/v)$$

where

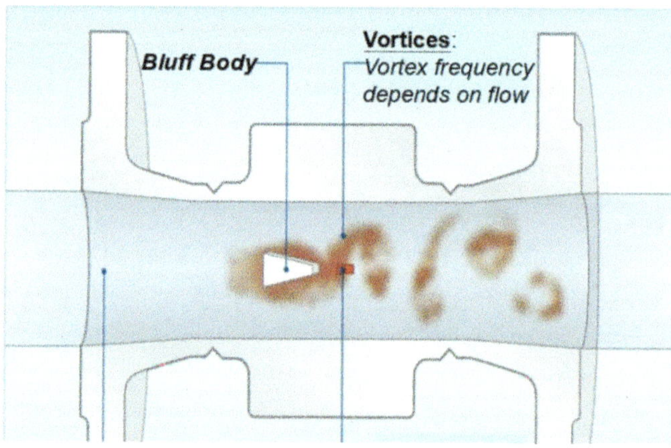

St = Strouhal number
f = vortex sheading frequency
d = width of the bluff body
D = inside pipe diameter
v = average fluid velocity
Q = volume

Vortex meters use piezoelectric or capacitance-type sensors to detect the pressure oscillation around the bluff body and respond to the pressure oscillation with a low voltage output signal that has the same frequency as the oscillation. Vortex shedding frequency is directly proportional to the velocity of the fluid in the pipe, and therefore to volumetric flow rate. The shedding frequency is independent of fluid properties such as density, viscosity, and conductivity, except that the flow must be turbulent for vortex shedding to occur.

Von Karman found that if the frequency of these vortices was measured, the frequency is proportional to the flow velocity that is generating the vortices. This frequency is called the Karman vortex frequency. The relationship of the frequency and the flow velocity can be mathematically expressed with the following formula:

$$Q = (\text{velocity} \bullet \text{area}) \text{ and } v = \frac{fd}{St}$$

St is the Strouhal number, determined experimentally.

Here is a table of some common Strouhal numbers.

d/D (bluff body)	*St*
0.1	0.18
0.3	0.26
0.5	0.44

10 Weight Measurement and Calibration

Weight Measurement Devices and Calibration

Weight measurements are typically made with load cells (strain gauges attached to metal bars). The bending moment of the bar causes the strain gauge to elongate, resulting in an increase of resistance in the strain gauge. This variable resistance is connected to a bridge circuit, and a voltage is measured across the bridge. The voltage is proportional to the weight applied to the measuring bar.

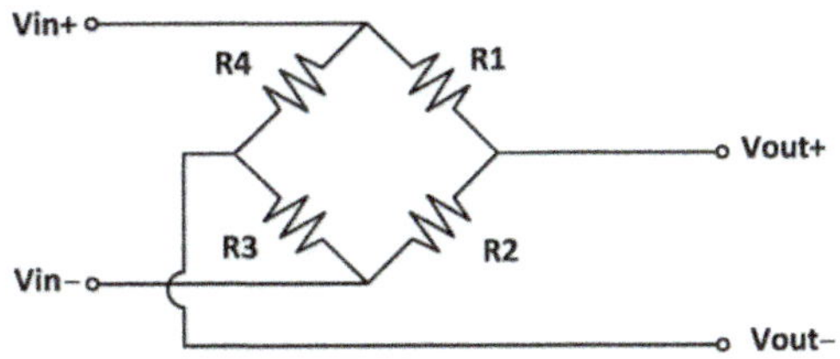

This strain gauge technology is used in measuring the weight in tanks, the weight in screw conveyors, and the weight on conveyor belts. The tare weight (tank weight) is nulled out and the voltage is set to zero or 0% in the bridge circuit. Then the maximum weight to be measured is applied. These weights are National Institute of Standards and Technology (NIST) certified. The span voltage is then calibrated to a maximum of 100%. This measurement is the net weight.

Note: Remember that all calibration processes should be repeated at least three times.

Load Cells

When outfitting a tank for batching, you must select a method for measuring the amounts of each ingredient added. A flowmeter or level meter may be the first device that comes to mind. However, the load cell might be the better solution.

A flowmeter may come in many forms, but it is usually an in-line device that measures the rate or flow of a fluid, either in volume or mass. On the other hand, a load cell is a device that allows for the contents of a tank or vessel to be weighed. This can be better than a level measurement, in cases such as pounds of sugar or pounds of flour.

Because a flowmeter is located in the tubing upstream from a tank, there is a difference between what the meter reads and what is truly in the tank. Programming changes can correct for the delays in measurement caused by this difference, but they can be imprecise and require frequent calibration.

However, because a load cell measures the weight of the tank itself, delays are minimized. Load cells offer a particular benefit when dealing with products that do not usually work well with flowmeters. Liquids with entrained air or bubbles may give meters trouble, but they are no problem for load cells. Even dry ingredients like powders can be measured just like liquid mass and volumes using load cells.

Load Cells for (Flow, Level, Force) Applications in Process

Mass Flowmeter

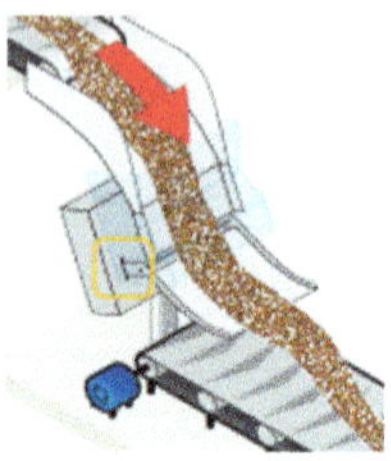

Batch Weighing

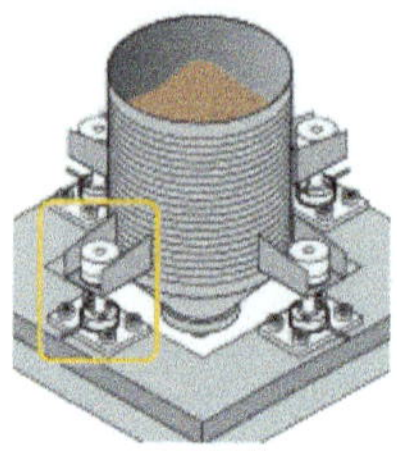

Weighbridge Measurements

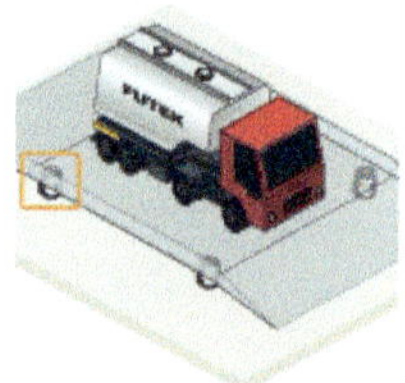

Tank Dispensing

Silo Measurement

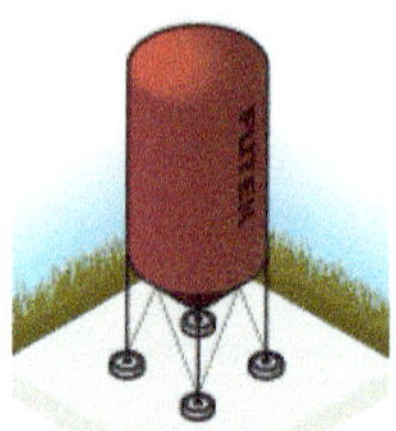

Automation Container Filling

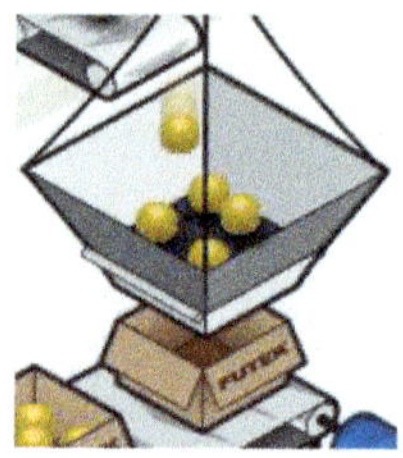

Hydraulic/Pneumatic Press Force

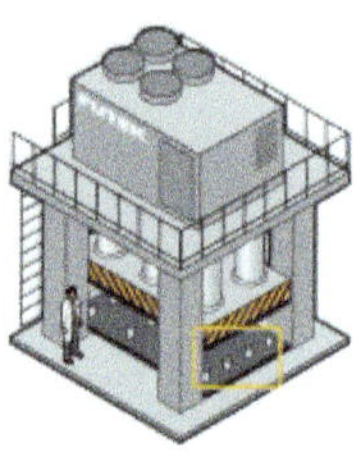

Wire Tension Measurement

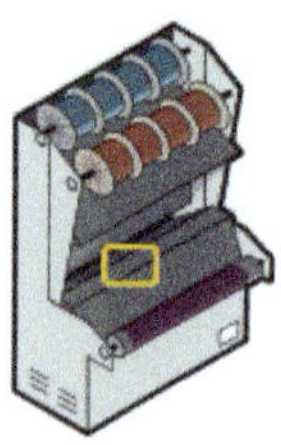

Robotic Tactile Sensing

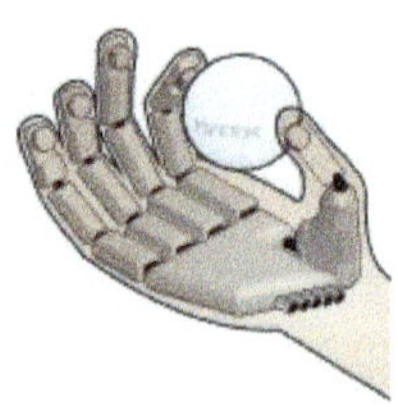

11 Process Analyzers

Electrical Conductivity and pH Correction

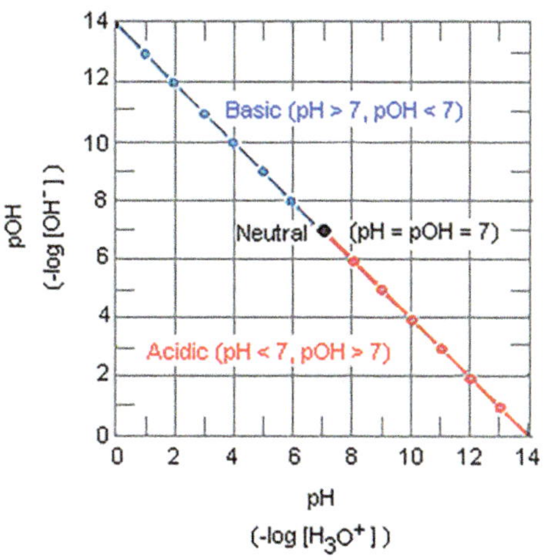

It is often useful to characterize an environment, such as a body of water, by measuring its pH and electrical conductivity (EC). pH is a measure of the acidity of the water or soil based on its hydrogen ion concentration and is mathematically defined as the negative logarithm of the hydrogen ion concentration:

$$pH = -\log[H^+]$$

The brackets around the H^+ symbolize *concentration. The pH of a material ranges on a logarithmic scale from 1 to 14, where pH 1–6 are acidic, pH 7 is neutral, and pH 8–14 are basic.* Lower pH corresponds with higher $[H^+]$, while higher pH is associated with lower $[H^+]$.

EC is a measurement of the dissolved material in an aqueous solution, which relates to the ability of the material to conduct electrical current through it. EC is measured in units called *Siemens per unit area* (e.g., *mS/cm*, or *milliSiemens per centimeter*), and the higher the dissolved material in a water or soil sample, the higher the EC will be in that material.

How Are pH and Electrical Conductivity Measured?

A meter and probe or litmus paper can be used to measure the pH of a sample. The more accurate and expensive of these methods is the meter and probe. pH meters are calibrated using special solutions, or buffers with a known pH value.

Electrical conductivity can be measured using a meter and probe as well. The probe consists of two metal electrodes spaced 1 cm apart (thus the unit of measurement is microSiemens or miliSiemens per centimeter, uS/cm or mS/cm). A constant voltage is applied across the electrodes, resulting in an electrical current flowing through the aqueous sample. Because the current flowing through the water is proportional to the concentration of dissolved ions in the water, the electrical conductivity can be measured. The higher the dissolved salt/ion concentration, the more conductive the sample and therefore the higher the conductivity reading.

Boiling Point Analyzer

A boiling point or distillation analyzer is typically used to control the 5% or 95% evaporated temperatures of the atmospheric tower side fraction products, the 95% evaporated temperature of the kerosene stream, and the 95% evaporated temperature of the diesel fraction. Depending on the individual refiner's circumstance, boiling point analyzers might be profitably employed to monitor or control appropriate percentage evaporated temperature for other streams in the crude distillation unit.

Vacuum Distillation Analyzer

This analyzer can be used to control the appropriate percentage of evaporated temperatures on gas oil streams from the vacuum tower. Refiners are more inclined to use this unit in refineries that have a hydrocracker or a lube oil manufacturing facility.

Flash Point Analyzer

This analyzer is most commonly used to monitor or control the flash point of the kerosene products. In some instances, it might be installed on the naphtha or diesel stream.

Cloud Point Analyzer

This instrument is normally used to monitor or control the diesel product from the crude distillation unit. Closer control of the cloud point causes the refiners to reduce the amount of kerosene that must be used to produce specification heating oil during the winter.

Freeze Point Analyzer

This instrument can monitor or control the freeze point of the naphtha or kerosene streams in refineries that produce jet fuel, thereby maximizing yields of these normally high-priced products.

Pour Point Analyzer

This analyzer is normally used to monitor or control the pour point of the diesel stream from the crude distillation unit. This will decrease the amount of kerosene that must be added to produce specification product during the winter, and permit the kerosene to be added to the normally higher-priced jet fuel feeds.

Color Analyzer

Some refiners employ crude oil to cool and condense the crude tower overhead vapor stream. If a leak develops in this heat exchanger, crude oil can leak into the naphtha or light straight run product. This could have a serious effect on the catalyst in the downstreaming processing units. A color analyzer installed on these streams can warn of leakage of crude oil into the light straight run or naphtha streams. A similar application could be made on the crude/pump-around reflux or crude/product exchangers. A color analyzer can also be installed on the atmospheric gas oil stream to warn unit operators if residual material is being entrained into the stream.

Combustion and Analyzers

Combustion furnaces, such as heating furnaces and boilers in plants, come in various sizes and types and serve as energy sources, that is, they are core components in all production activities. A large amount of fuel, such as gas or fuel oil, is consumed in plants, so the furnace's combustion efficiency directly affects the performance and running cost of the plants. Because combustion furnaces generate large amounts of exhaust gas, in recent years it has become important to reduce various greenhouse gases, including CO_2, in addition to coping with pollution caused by compounds such as nitrogen oxide and sulfur oxide. Consequently, not only are the proper measurement and control of O_2 and CO required, but also solutions to multi-faceted issues.

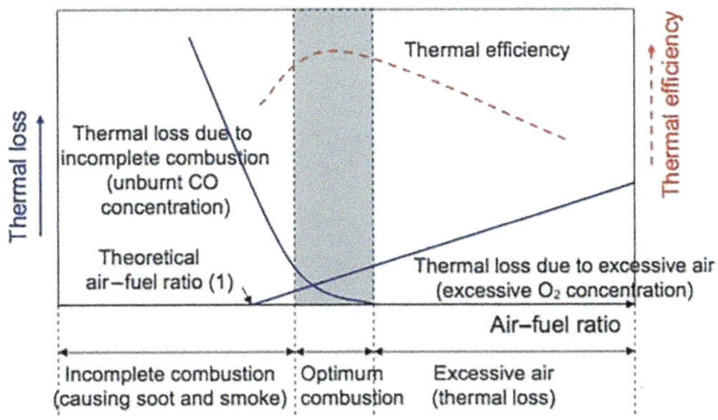

Relationship between air–fuel ratio and heat efficiency (combustion).

Combustion Furnace and Air–fuel Ratio Control

Combustion requires fuel and air (oxygen), and insufficient air causes fuel residue, resulting in incomplete combustion with soot and smoke. On the other hand, excessive air causes problems, such as a larger amount of exhaust gas and heating of excessive air, resulting in lower fuel efficiency. The figure above shows the principle of the air–fuel ratio and state of combustion. The air–fuel ratio plotted on the horizontal axis shows the ratio of actual supply air to the theoretical amount of air required for fuel combustion (theoretical air amount).

For combustion furnaces such as heating furnaces and boilers in plants and factories, small-scale controllers such as single-loop controllers or programmable logic controllers (PLCs) are employed to optimize the air–fuel control ratio for improving the combustion efficiency. In large combustion furnaces, distributed control systems (DCSs) and advanced control (e.g., multivariable predictive control) are used. These mainly control the air–fuel ratio and internal pressure of the furnace to prevent CO, CO_2, and NO_x (nitrogen oxide) from being emitted and to apply a cross limit circuit to prevent incomplete combustion while controlling combustion to maximize efficiency.

Air–Fuel Ratio Control Utilizing CO and O_2 Concentrations

According to Lyman F. Gilbert, the CO concentration in the optimum combustion zone (having the highest heat efficiency per unit amount of fuel) is around 200 ppm (parts per million) irrespective of fuel types and devices. However, CO increases rapidly once it has begun to increase. Thus, either a stable combustion must be kept with sufficient supply of air or a control system must monitor the CO concentration in real time and keep it constant at a relatively low level.

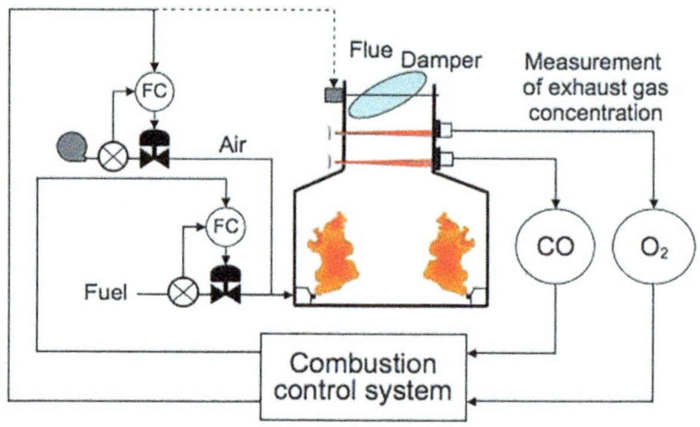

The amount of air supplied to a burner is controlled by two methods: using a forced draft fan (FDF) and damper as shown in the figure at the right, or using natural air intake by controlling the opening degree of the damper of an induced draft fan. The O_2 and CO concentrations are measured by a concentration meter at the entrance of the flue and then supplied to the control system. The measured CO concentration can be used for combustion control by two methods: (1) controlling O_2 when the O_2 concentration exceeds a prescribed value and overriding to CO control when the O_2 concentration falls below the value, or (2) giving a CO concentration bias (compensation) to the O_2 concentration.

Burner Management System Safety

A burner management system (BMS) safely controls the burner of the combustion furnace and includes an interlock mechanism and a safety shutoff mechanism to prevent explosion. The BMS must comply with safety standards based on risk assessment such as the international standard (ISO 12100) and the EU, US and Japanese standards (EU standard: EN746, US standard: NFPA 86, Japanese standard: JIS B9700).

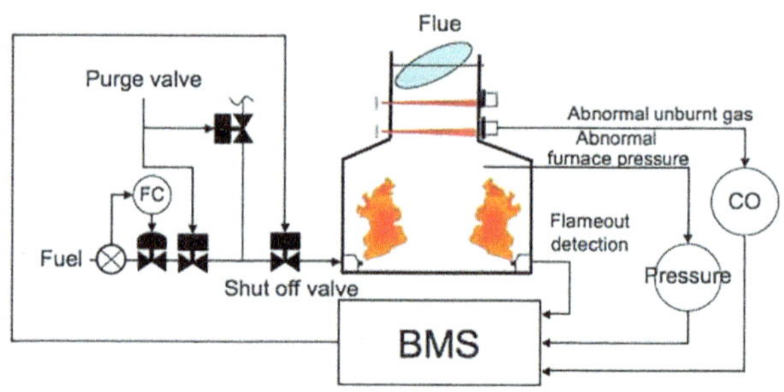

Because the analyzer monitors the CO concentration in near real time and detects the generation of toxic gas due to incomplete combustion, it increases the reliability of the safety system by inputting a signal of the detected CO concentration to the BMS. The CO concentration is expected for implementing the defined safety requirements. The figure to the right shows a typical combustion system in which CO (carbon monoxide) measurement capability is added to the BMS burner shut-off system.

OSHA Requirements

The Occupational Safety and Health Act of 1970 states: The employer must furnish a place of employment which is free from recognized hazards that could cause death or serious physical harm to employees if employees were exposed to fire and explosion hazards.

1. The burner management systems on boilers, furnaces, ovens and flare systems located within the facility must be in compliance with NFPA 85, *Boiler and Combustion Systems Hazards Code*, sections 5.3.7.1 and 5.7.2, as well as manufacturer's and design specifications.

2. The employer must perform a risk assessment to identify, evaluate and control the hazards involved with the process in accordance with ANSI/ISA S.84.001 and IEC 61511.

Carbon Dioxide Reading

The CO_2 level in the flue gas provides an indication of the efficiency of the combustion process. If the production of CO_2 is as high as possible with slight excess air (complete combustion), the flue gas heat losses are at their lowest. The CO_2 reading is calculated from the O_2 reading by the analyzer. This maximum theoretical level is never reached in practice. Each fuel has a maximum possible CO_2 level (CO_2 max), which is determined by the chemical composition of the fuel:

- Light fuel oil: 15.4% by volume CO_2
- Natural gas: 11.8% by volume CO_2

Examples of Process Analyzers

Analytical analyzers that are specifically designed for continuous emission monitoring (CEM) and process control.

Often process gas cannot be taken directly from the process to the analyzer. Most systems demand a level of application-specific sample handling.

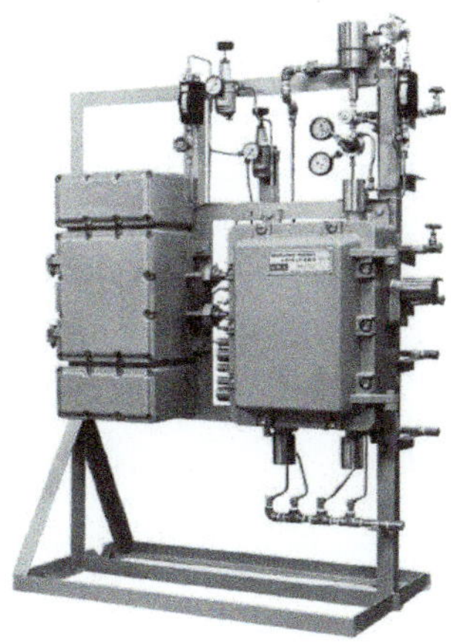

Some common streams measured:

- Sulfur dioxide (SO_2)
- Carbon monoxide (CO)
- Nitrogen monoxide (NO)
- Total hydrocarbon (THC)
- Nitrogen dioxide (NO_2)
- Hydrogen sulfide (H_2S)
- Carbon dioxide (CO_2)
- Opacity
- Oxygen (O_2)

Select the Appropriate Analyzer and Configuration

Off-Line Lab Analysis

- Manual sampling
- Sample transport to the lab
- Sample registration

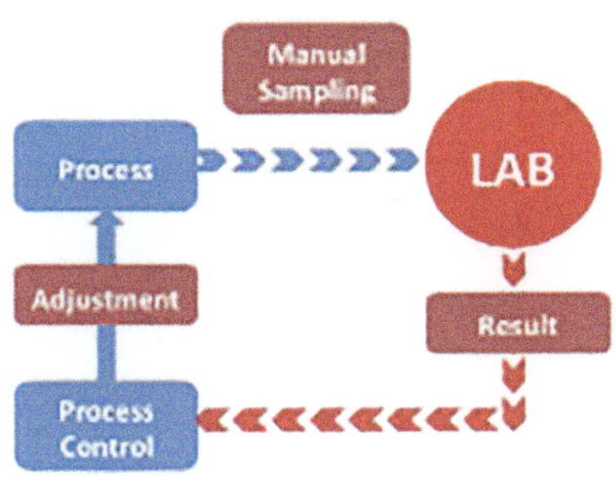

At-Line Analysis

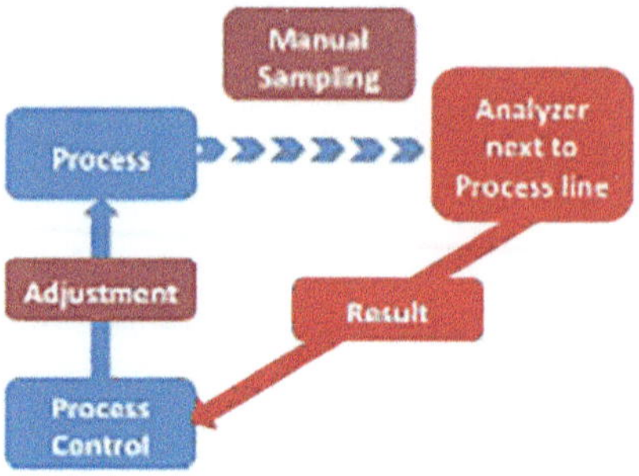

- Manual sampling
- Continuous manpower needed
- Automated lab analysis close to sampling point inside plant
- Ideal where multiple samples must be taken at several sampling points along the process

Online Analysis

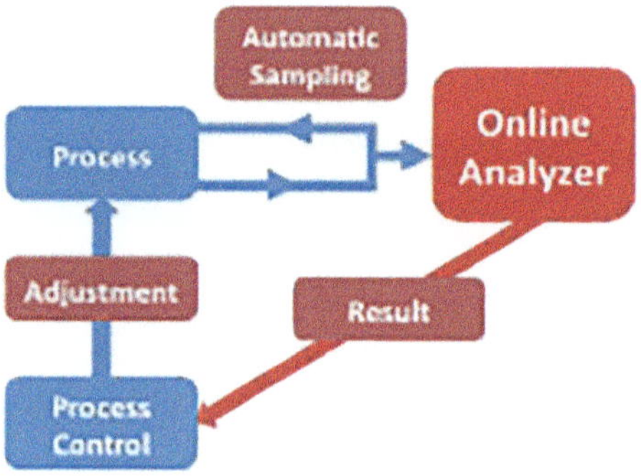

- In-process measurement and response
- Automated sampling and registration
- Automated sample preconditioning
- Automated lab analysis
- Fast feedback of results
- Very limited manpower needed
- Close loop control

12 Sensors for Fire and Rotating Equipment

Sensors for Fire Protection and Flame Monitoring

Optical flame detectors have multiple applications in a variety of hazardous environments, ranging from industrial heating and drying systems and industrial gas turbines to petrochemical oil and gas facilities. They can reduce the risk in a given environment by quickly detecting an unwanted flame. Therefore, they are more reliable, especially in outdoor settings, where they can respond faster to the presence of a flame than a heat or a smoke detector can. In industry, two basic types of flame or fire detection are used:

1. Detection of fire and flames for protection, such as in a compressor building (fire prevention)
2. Detection of fire and flames for process monitoring, such as a flare or oven (burner management)

Fire Prevention

In fire prevention, fire and flames are detected using an ultraviolet (UV) sensor or infrared (IR) sensor. A resistance temperature detector (RTD) may be used in heating, ventilation, and air conditioning (HVAC) ductwork to detect fire in the ductwork. A UV/IR flame detector consists of UV and IR sensors that are combined in a single apparatus. UV sensors work by detecting the UV radiation emitted by the flame and are sensitive to a wide range of flammable fuels including hydrocarbons, sulfur, hydrazine, and ammonia.

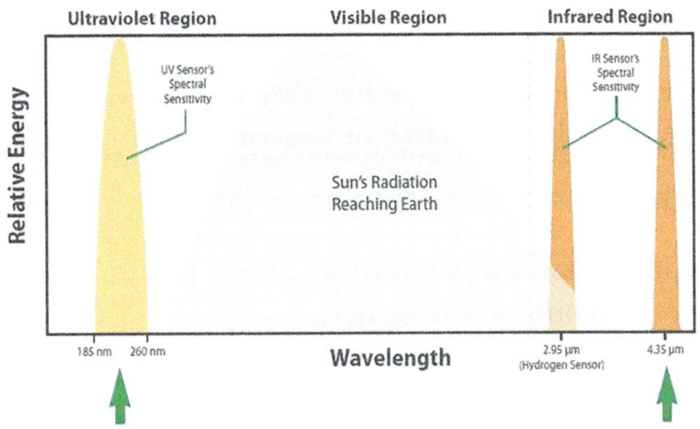

- Electrical generators
- Compressors, centrifugal and reciprocating
- Crushers and grinders
- Vibrating conveyors

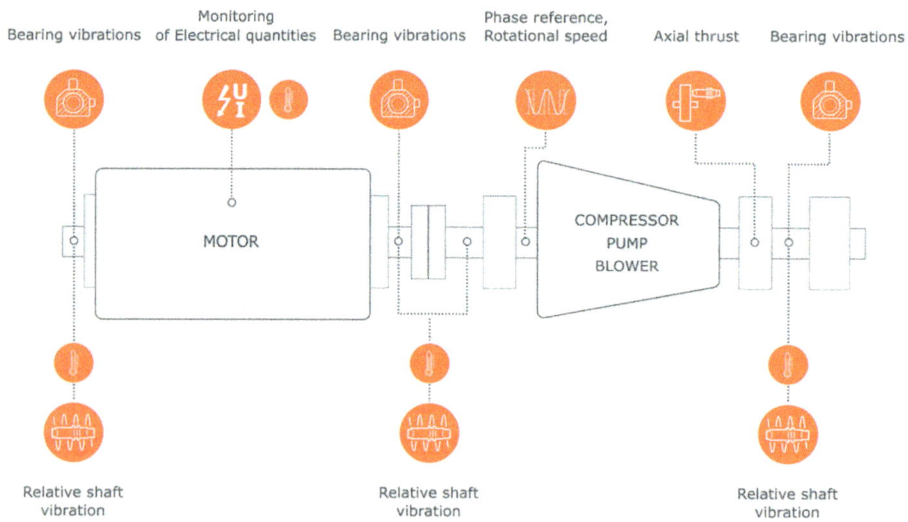

As stated, vibration analysts are looking for unwanted vibrations and high temperatures in the bearings. We are also interested in axial shifting at the compressor, pump, or blower (axial thrust). Some axial shafts use two vibration monitors, one on the x-axis and one on the y-axis at a 90° displacement. Some units combine the variables of acceleration, vibration, and temperature into one sensor. These signals are then broken out or split into separate individual signals in a device such as the monitoring equipment.

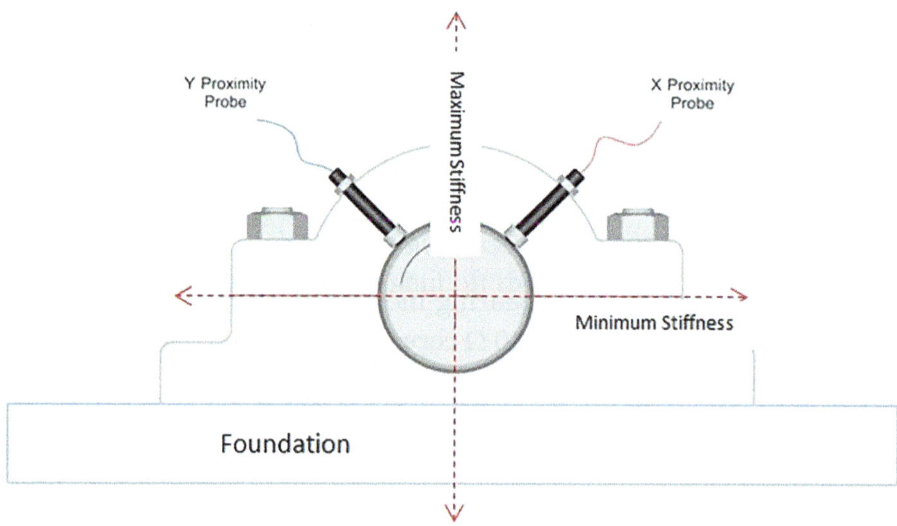

Some large compressors and electrical generators use jet engines mounted horizontally and connected by a shaft. These jet engines are referred to as *turbines*. Turbines can be fuel-fired or steam-driven. If the turbine is fuel-fired, it will have a BMS as well.

A compressor is used to compress gas from a lower pressure at the intake to a higher pressure at the exhaust by reducing the volume of the gas. Applications are wide ranging, with the greatest applications in the refrigeration, chemical and petrochemical, food processing, pulp and paper, power generation, natural gas, and government and military industries. For tool and instrument air systems, air is often compressed, whereas for chemical processing, hydrogen or oxygen are often compressed.

Failures

The most common cause of failure in a screw compressor are its bearings. A screw compressor has two types of bearings. The first type, the journal bearing, consists of two rings that when working correctly never touch each other. The bearing instead relies on a thin covering of oil in order to turn the compressor's rotor. This type of bearing can fail when the oil separating the two rings becomes contaminated, breaks down, or is overloaded.

The second type of bearing is a rolling element bearing. This type of bearing consists of two rings riding on ball bearings. Rolling element bearings can fail for multiple reasons, including:

- "Lubrication failure" (low oil pressure, low viscosity, or condensing oil)
- Spalling in bearings
- Debris buildup in bearings and rotors
- Mechanical mounting and coupling problems
- Piping strain
- Excessive oil injection
- Discharge line/vessel resonance
- Mechanical interference

What Should Be Measured

The most typical vibration monitoring locations are located at radial bearing positions for the motor and the compressor, and axial positions adjacent to the bearings on the motor and compressor. These sensor locations will indicate bearing problems most accurately on the motor and bearing, and compressor problems on the compressor. There are complete systems for monitoring vibration of compressors. The monitoring equipment can then send this information to the programmable logic controller (PLC) or the distributed control system (DCS) over a serial communications connection.

13 Process Control Valves and Actuators

Process Control Valves

A wide variety of valve types exist. The most widely used types for process control systems and other industrial fluid applications are those that have linear stem and rotary spindle movement:

- Linear stem movement type valves include globe valves and slide valves.
- Rotary spindle type valves include ball valves, butterfly valves, plug valves, and their variants.

The first choice to be made is between two-port and three-port valves:

- Two-port valves "throttle" (restrict) the fluid passing through them.
- Three-port valves can be used to "mix" or "divert" liquid passing through them.

Globe valves are frequently used for control applications because of their suitability for throttling flow and the ease with which they can be given a specific "characteristic," relating the valve opening to flow. For any given valve orifice size, the greater the differential pressure, the greater the flow rate. The maximum value of the valve flow coefficient C_v is defined as the number of US gallons of water per minute (at standard pressure and temperature) that will flow through a wide-open valve when there is a 1 psig pressure drop across the valve. The flow rate can be determined by the following equation:

$$1\ \text{gpm} = 1\,C_v \bullet \sqrt{1\ \text{DP}_{\text{psig}}}$$

Control valve sizing will be discussed in detail for liquid, steam, gas, vapor, and two-phase applications in this section of the manual. We will also look at the accessories used on common valves and valve sizing limitations that must be considered. The style of the valve trim (internal makeup), the characteristics (the gain of the flow per signal), and the DP across the valve all affect whether the valve will function in the process piping loop. There may be limits of the maximum and minimum DP across the valve to adhere to. The DP can limit the flow through the valve and can cause excessive noise and even destruction of the valve.

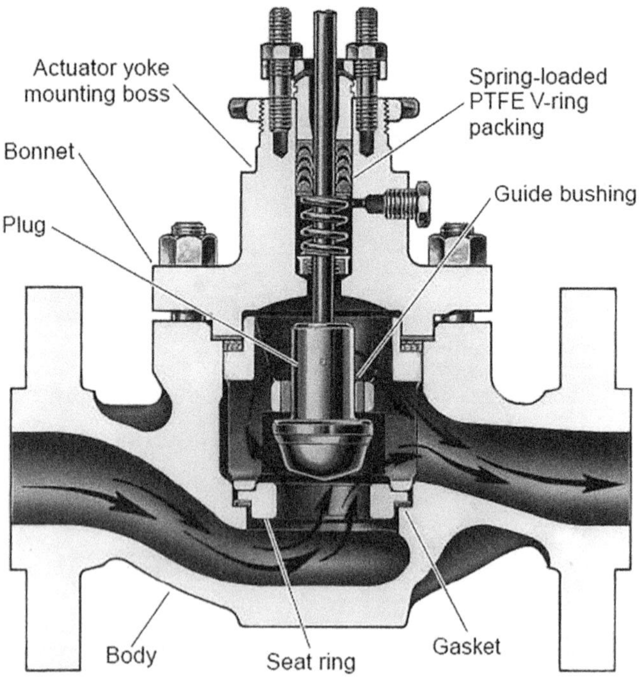

Cross sectional view of a globe valve showing the basic structure

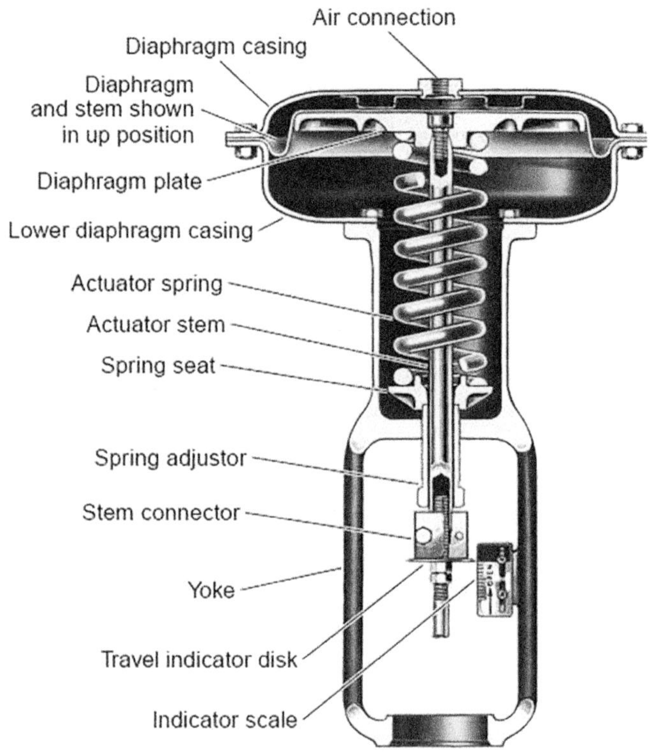

Typical arrangement of a spring and diaphragm type actuator

Considerations When Sizing a Control Valve

The following should be considered when sizing and selecting process control valves for a project.

$$1\text{ gpm} = 1\,C_v \bullet \sqrt{\frac{1\text{ DP}_{\text{psig}}}{1\text{ SG}}}$$

Flow Coefficient C_v

The use of the flow coefficient, C_v, first introduced by Masoneilan in 1944, quickly became accepted as the universal yardstick of valve capacity. So useful has C_v become that practically all discussions of valve design and characteristics or flow behavior now employ this coefficient. By definition, the valve flow coefficient is the number of US gallons per minute of water that will pass through a given flow restriction with a pressure drop of 1 psi.

For example, a control valve with a maximum flow coefficient of $C_v = 12$ has an effective port area in the full open position such that it passes 12 gpm of water with 1 psi pressure drop. Basically, it is a capacity index on which the engineer can rapidly and accurately estimate the required size of a valve restriction for desired flow in any fluid system. The flow is dependent on other variables as well, and there may be limitations of the maximum DP that can be applied across the valve.

Specific Gravity

In the flow formulas, the specific gravity (SG) is a square root function; therefore, small differences in gravity will have a minor effect on valve capacity. If the SG is not known accurately, a reasonable assumption will suffice. The use of SG equal to 0.9, for example, instead of SG equal to 0.8, will cause an error of less than 5% in valve capacity. Sqrt [1/(0.9–0.8)] = 3.16% error in flow.

Operating Conditions

The selection of a correct valve size, as determined by formula, is always premised on the assumption of full knowledge of the actual flowing conditions. Frequently, one or more of these conditions is arbitrarily assumed. The evaluation of these arbitrary data determines the final valve size.

No formulas, only good common sense combined with experience, can solve this problem. There is no substitute for good engineering judgment.

Most errors in sizing are due to incorrect assumptions as to actual flowing conditions. The tendency is to make the valve too large and to be on the "safe" side (commonly referred to as *oversizing*). A combination of several of these "safety factors" can result in a valve so greatly oversized it tends to be troublesome.

ISA Standard Valve Symbols

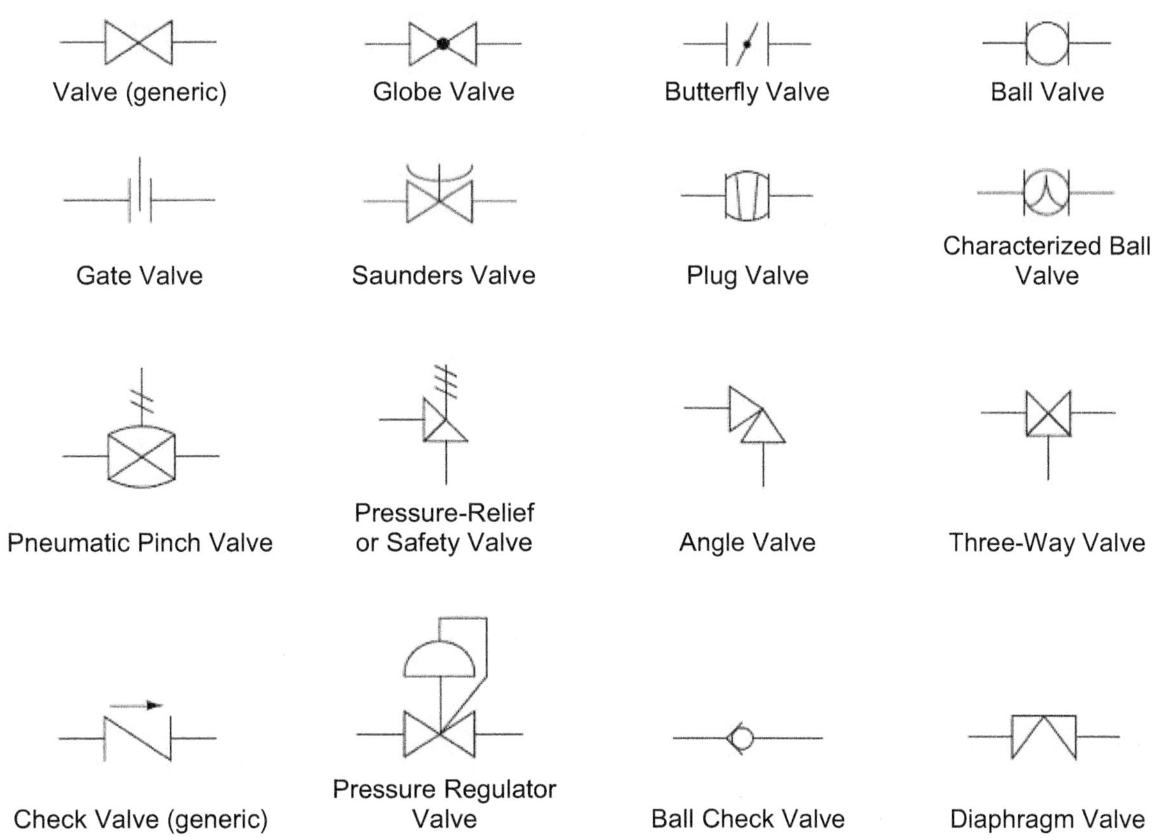

ISA Standard Pressure-Regulating Valve Symbols

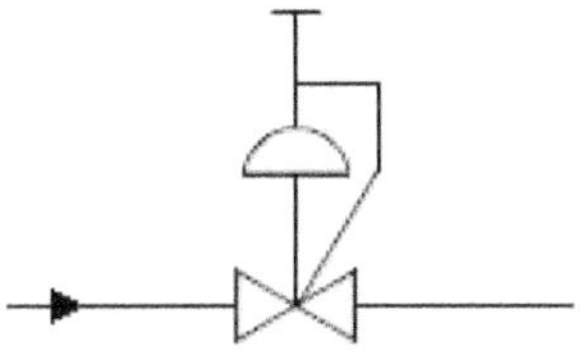

Pressure-Reducing Regulator
Self-Contained with Hand Wheel

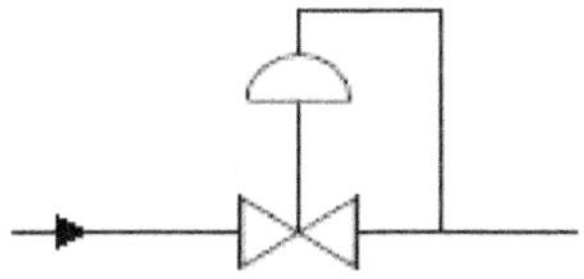

Pressure-Reducing Regulator
External Pressure Tap

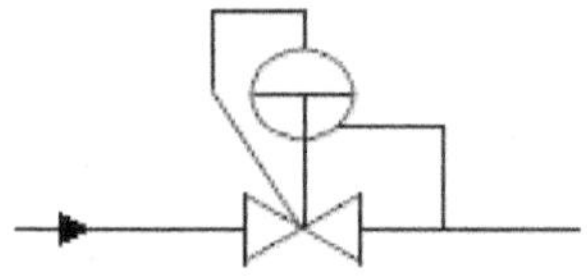

Pressure-reducing Differential Regulator
External and Internal Pressure Tap

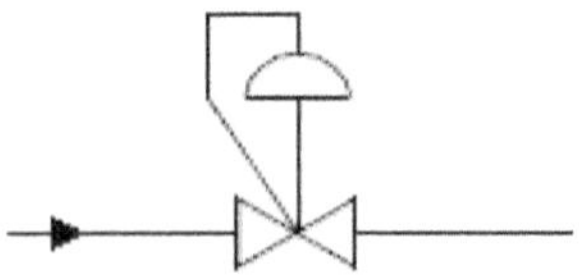

Back Pressure Regulator
Self-Contained

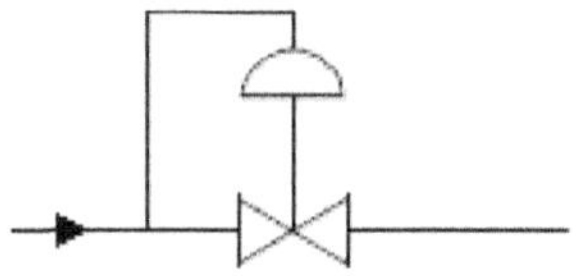

Back Pressure Regulator

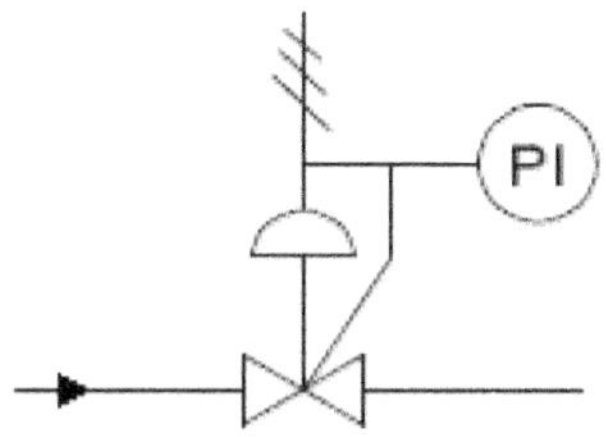

Pressure-Reducing Regulator with
Integral Outlet Pressure Relief

Valve Actuators

The operation of a control valve involves positioning its movable part (the plug, ball, or vane) relative to the stationary seat of the valve. The purpose of the valve actuator is to accurately locate the valve plug in a position dictated by the control signal.

The actuator accepts a signal from the control system and, in response, moves the valve to a fully open or fully closed position, or a more open or a more closed position (depending on whether on/off or continuous control action is used).

There are several ways of providing this actuation, but the two main ways are by:

1. Pneumatic actuator
2. Electric actuator

Other significant actuators include the hydraulic and the direct-acting types. Note that pneumatic actuators do not operate on a standard 3–15 psig pressure output from a current to a pneumatic convertor (I/P). This is a misconception. The standard pneumatic signal, 3–15 psig, is left over from the pneumatic controller days. The actuator operates on 0–15 psig, 0–30 psig, or 0–60 psig. The I/P may be calibrated from 1.5–12.8 psig or 8–20 psig to close the valve and overcome the pressure in the pipe pushing up against the valve seat trying to open the valve, as well as any friction of the valve stem packing trying to stop the valve stem from moving. This is called the *bench set calibration* of the valve's actuator.

ISA Standard Actuator Symbols

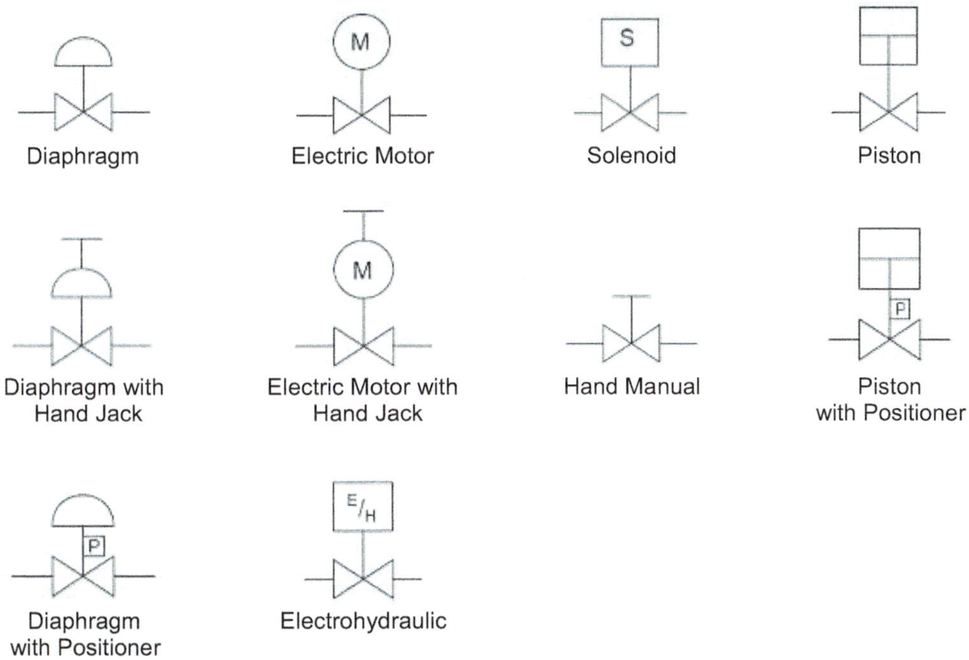

Limit Switches on a Valve – ISA Standard Symbol

A typical application of a valve for a gas service is shown below. Limit switches are attached to the actuator to verify the valve position status. The limit switches send a full open signal (ZSO) or full closed signal (ZSC). If neither signal is received by the PLC or DCS within a reasonable time, the limit switches provide a valve stuck or malfunctioning indication. The solenoid is a safety shutdown, lock-out type mechanism. The diamond symbol with the *R* indicates a manual reset of the solenoid valve in the field is necessary to provide instrument air to the gas valve for operation. This ensures that personnel inspect the furnace or heater before restoring the gas to prevent explosion or fire.

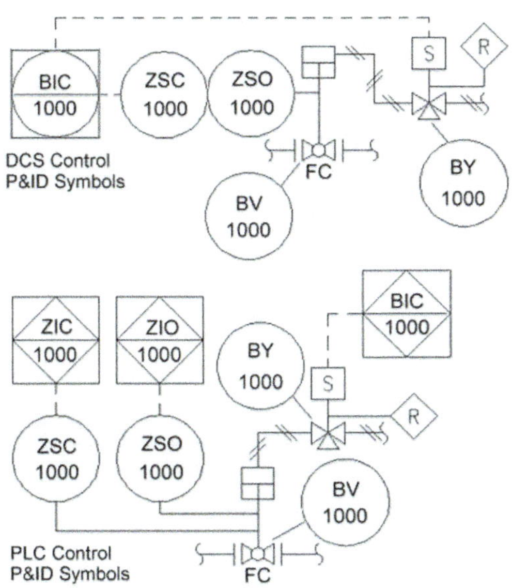

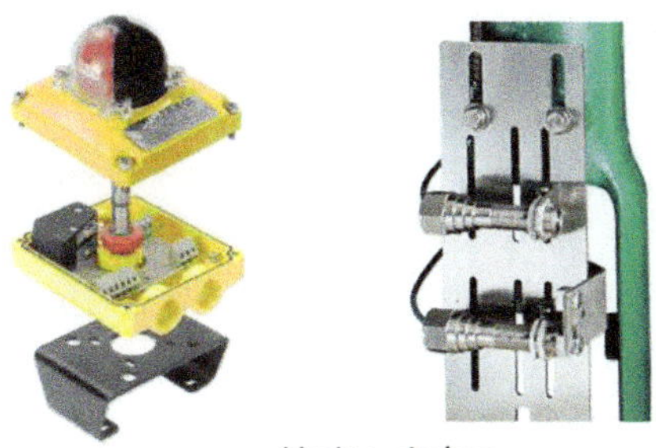

Limit switches

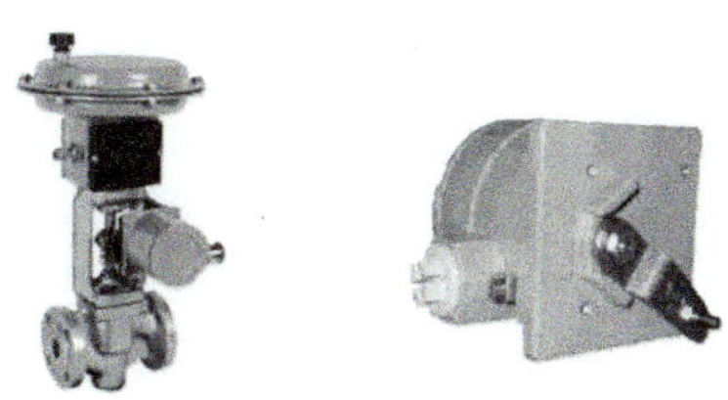

Position transmitter

Calculating the Size of the Actuator

Reference the figure of the air to open (ATO) valve below. The process fluid flows through the valve from left to right, exerting a force upward because of the process fluid's pressure multiplied by the seating area on the valve trim (the globe type "plug" against the valve seat). The actuator spring must be sized to hold the valve closed against the differential pressure exerted upward on the plug by the process fluid's pressure, and to add extra seating force to the valve to prevent leakage of the process fluid between the valve's plug and the seat. This leakage is measured in what is called *bubbles per minute*. Also, extra force on the spring may be required to overcome the friction of the packing. The spring is usually oversized for the application, because standard size springs are used for various applications and process fluid pressures.

The actuator must be sized for the total forces needed to move the valve stem into position.

- The force upward (F_p) – Process fluid pressure (psig) multiplied by the area of the plug (in^2) in pounds force (lbf)
- The force downward by the spring (F_k) – Force in (lbf) varies with spring size, Hooke's constant
- The force upward of the diaphragm (F_d) – The I/P supplied device pressure (psig) multiplied by the area of the valve diaphragm (in^2) in pounds force (lbf)
- The force to overcome stem friction (F_f) – To move the valve against the friction of the stem packing
- The force applied to the seat (F_s) – The force applied to the plug to prevent leakage through the valve seat, in (lbf) per lineal inch of seat around the circumference

For an ATO valve, the diaphragm force (F_d) must be in the opposite direction of the spring force (F_k) and equal to the sum of the force exerted by the spring (F_k) and the force exerted by the stem packing (F_f), minus the force of the process fluid (F_p) pushing up on the valve seat, before the spring will start compressing and the valve plug will start moving open.

For an air to close (ATC) valve, the diaphragm force (F_d) must be in the opposite direction of the spring force (F_k) and equal to the sum of the force exerted by the spring (F_k) and the force exerted by the stem packing (F_f), before the spring will start compressing and the valve stem will start moving closed.

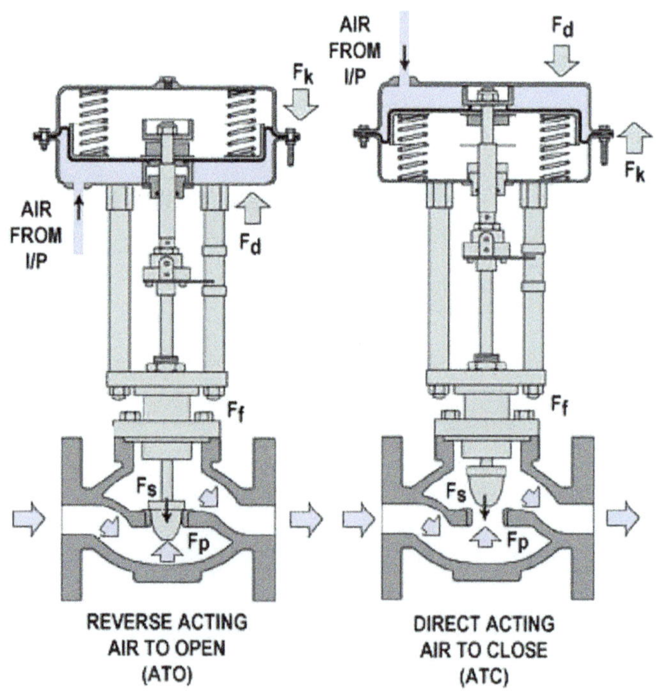

Once it starts closing, the diaphragm force (F_d) must be increased to equal of the compressed spring force (F_k) and equal to the process fluid force (F_p) pushing up on the seat and the force exerted by the stem packing (F_f) and the extra seating force (F_s) is needed to meet the bubble leakage specification (e.g., 150 lbf per linear inch of seat) before the diaphragm force will fully seat the valve plug. This force (F_d) may equal 8 psig • 100 in^2 for the diaphragm, to equal 800 lbf exerted downward by the actuator diaphragm.

The diaphragm pressure must be increased to force the stem downward more. The spring is already forcing up against the diaphragm plate and must be compressed more (see Hooke's constant); the process fluid (F_p) is forcing up against the seat trying to open the valve; plus we must add the force for seating (F_s) the valve and the still-required added resisting force of the packing friction (F_f), which will all add up to the required force that must be generated by the diaphragm and diaphragm plate connected to the plug stem.

The I/P supplied device pressure (psig) multiplied by the area of the valve diaphragm (in^2) in pounds force (lbf) will produce the minimum diaphragm force (F_d) needed to overcome the total resistive forces against the diaphragm. The spring will start compressing and the valve will start moving toward the fully closed position. If the I/P (current to pressure convertor) exerts 15 psig to the diaphragm, the diaphragm force downward (F_d) will be 15 psig • 100 in^2, which equals 1500 lbf. A force of 1500 lbf exerted by the diaphragm is the force needed to compress the spring all the way, causing the valve trim plug to move to the fully closed and seated position, and provide safe and effective operation.

All valve I/Ps are not calibrated for a range of 3–15 psig. When a large pressure exists in the process piping system, the valve's actuator will be calibrated to a range to produce sufficient force to overcome the force of the process fluid and seat the plug. In our example, the valve I/P was calibrated for a range of 8–15 psig.

Example Actuator Sizing

Sample Problem: We will now size a reverse acting valve actuator (ATO) for a process having the following data:

Single-seated globe valve with flow under the plug (ATO)
Delta pressure across the valve: 25 psig
Stem travel: 1.5 in
Actuator area: 78.5 in^2
Port diameter: 2-5/16 in
Plug seating class: II (20 lbf per lineal inch)
Stem friction: $F_f = 125$ lbf (graphite packing)
Spring force: $F_k = 500$ lbf/in

First, calculate the force exerted by the process fluid (F_p): 25 psig • area of plug.

$$F_p = 25 \bullet \frac{\pi\ 2.3125^2}{4} = 105.0 \text{ lbf}$$

Find the seating force for plug for a class II shutoff.

$$F_s = \pi \bullet D \bullet 20 \text{ lbf/in} = \pi \bullet 2.3125 \bullet 20 = 145.29 \text{ lbf}$$

To unseat the valve and start movement of the stem toward open, add the stem friction force to the spring force and subtract the process fluid force pushing upward:

$$F_{d(\min)} = F_k + F_f - F_p = 500 + 125 - 105.0 = 520.0 \text{ lbf}$$

Find the lower range value (LRV) of the I/P pressure.

$$P = \frac{F}{A} = \frac{520.0}{78.5} = 6.63 \text{ psi}$$

The force of the spring compressed when the valve is fully open:

$$F_x = F_k \bullet x(\text{inches of travel})$$

$$F_x = \frac{500 \text{ lbf}}{\text{in}} \bullet 1.5 \text{ in} = 750 \text{ lbf}$$

To open the valve fully, add the stem friction force to the spring force pushing down.

Note: The valve plug is already unseated, so there will no longer be a force *Fp* helping the spring to fully open, because there is practically no delta pressure being exerted on the plug.

$$F_{d(\max)} = F_x + F_f = 750 + 125 = 875 \text{ lbf}$$

Find the upper range value (URV) of the I/P pressure:

$$P = \frac{F}{A} = \frac{875}{78.5} = 11.15 \text{ psi}$$

The valve I/P transducer will be calibrated 6.63 psig to 11.15 psig.

Split Ranging Control Valves

In a split range control loop, output of the controller is split and sent to two or more control valves. The splitter defines how each valve is sequenced as the controller output changes from 0% to 100%. In most split range applications, the controller adjusts the opening of one of the valves when its output is in the range of 0%–50% (4–12 mA) and the other valve when its output is in the range of 50%–100% (12–20 mA).

When split ranging valves to handle flow rates varying from very large to very small, a 10% overlap in the valve signal is recommended. This causes the valves to have a smoother transition because of the difference in the C_v ranges of the two valves. The overlap provides a more linear process gain for the system.

In this example, when the gas pressure exceeds the pressure that the compressors can handle, the extra gas is sent to the flare to burn, relieving the pressure on the vessel.

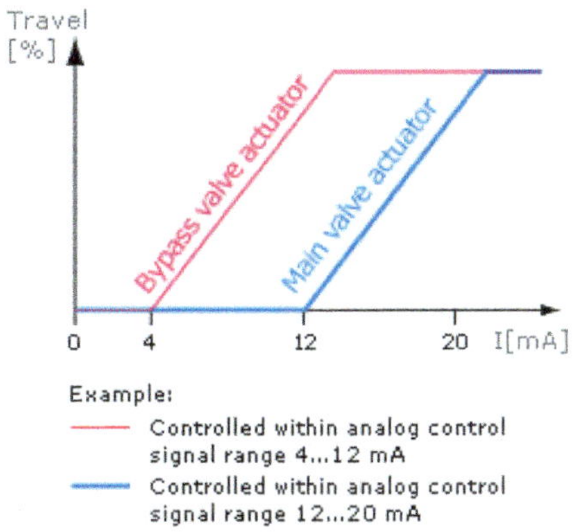

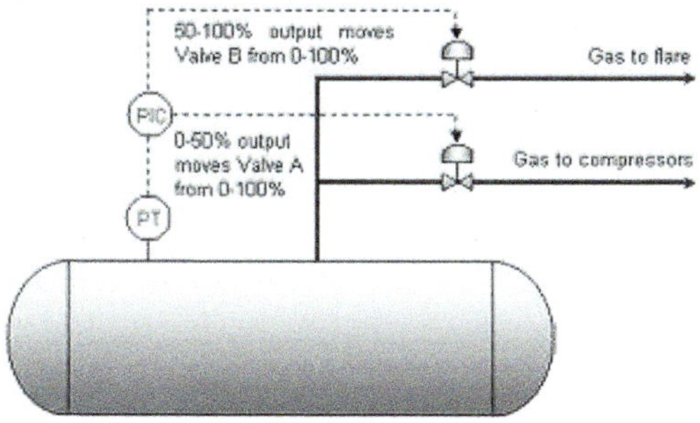

The reactor must maintain a specific temperature range in this example. This requires heating and cooling the jacket to regulate the temperature for the reaction.

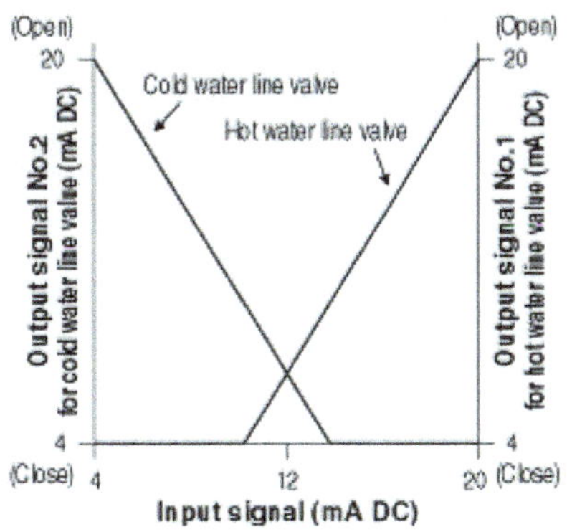

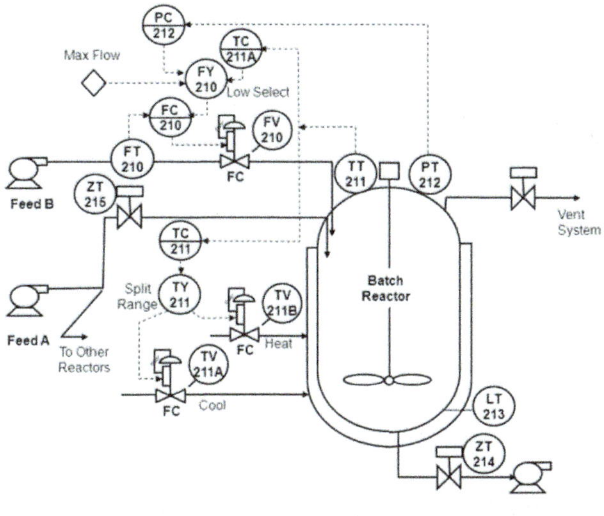

Valve Positioner Applications

A valve positioner takes an input signal from the DCS or PLC and positions the valve plug using a feedback signal from the position of the valve stem. The positioner will provide air pressure to the pneumatic actuator's diaphragm. The air pressure output signal to the actuator will be a percentage of the full-scale calibrated air output of the positioner. The percentage of full-scale air output will be proportional to the percentage of the full-scale input signal.

There may be a gain in the percentage of air output; this will result from the amplifier setting being greater or less than 1. The actuator moves the valve stem to a percentage of full stroke that is equal to the percentage of the input signal, for example, 50% open. The positioner receives a feedback signal from a lever arm connected to the valve plug stem. The positioner may also apply additional corrective pressure to the actuator diaphragm. This extra compensated pressure is proportional to the error of the position of the plug stem and will try to move the valve plug into the exact position being called for by the signal from the DCS or PLC. The positioner is being used as a cascade controller for the flow loop. It provides tighter and faster control of the valve stem position.

When a positioner is fitted to an ATO valve with an actuator, the spring range of the actuator may be increased to increase the closing (seating) force of the plug in the valve. A positioner can compensate for the extra pressure required to properly position the valve. The positioner will allow for an increase in the maximum differential pressure a particular valve can tolerate across the plug. Pulsations in this differential pressure across the plug will cause upward forces on the valve plug and can cause the valve to fluctuate in position. The positioner will compensate for these fluctuations with a feedback signal from the lever arm and send a proportional air signal to the diaphragm of the actuator to compensate for the error in the position of the plug and move the valve plug to the true desired position. The positioner also sends additional air pressure to the actuator when an error is measured in position, allowing the actuator to overcome the friction of the stem packing and reduce hysteresis effects.

Note that a positioner is a proportional device, and in the same way that a proportional controller will always give an offset, so does a positioner. On a typical positioner, the proportional band may be between 3% and 6%. The positioner sensitivity can usually be adjusted.

The following figures are examples of ISA standard valve positioner symbols.

Typical Positioner on a Valve

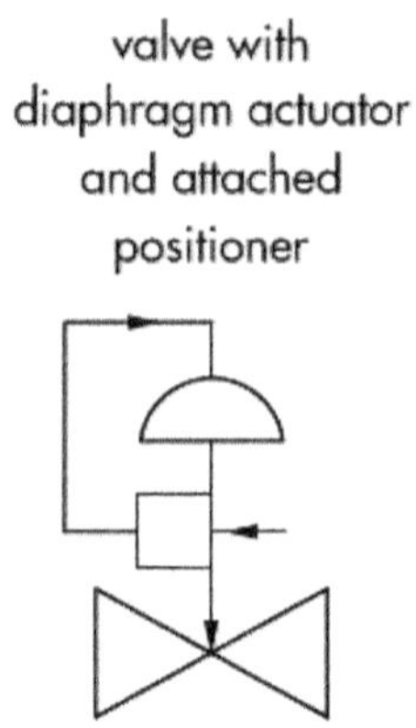

SIS System Application with Solenoid Interlock

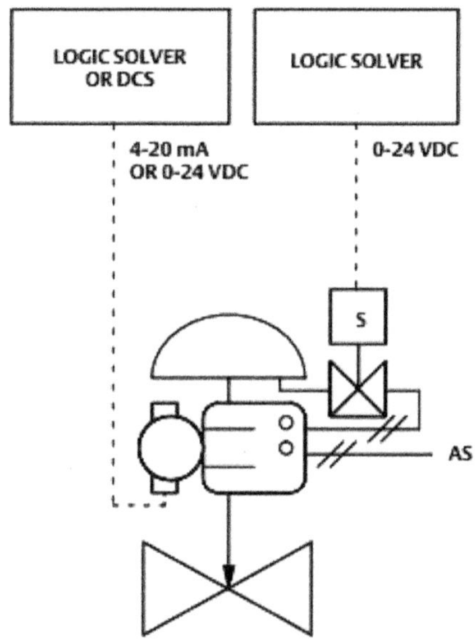

Summary of Positioners

- A positioner ensures that there is a linear relationship between the input signal from the control system and the position of the control valve. This means that for a given input signal, the valve will always attempt to maintain the same position regardless of changes in valve differential pressure, stem friction, diaphragm hysteresis, and so on.
- A positioner can sometimes modify the input signal to characterize the action of the valve trim. This is especially true of a digital valve positioner.
- A positioner may be used as a signal amplifier or booster. It accepts an input signal in the form of a low-pressure air control signal (3–15 psig). Using the positioner amplifier to add

gain to the input position signal, the positioner provides an amplified pressure output air signal to the actuator diaphragm to position the valve plug. This ensures that the valve reaches the desired position.

- Some positioners incorporate an electropneumatic converter so that an electrical input (typically 4–20 mA) can be used to control a pneumatic valve.
- Some positioners can also act as basic controllers, accepting input from sensors.

When Should a Positioner Be Used?

A positioner should be considered in the following circumstances:

- When accurate valve positioning is required.
- To speed up the valve response. The positioner uses higher pressure and greater airflow to adjust the valve position.
- To increase the pressure that a particular actuator and valve can close against (to act as an amplifier.)
- Where friction in the valve (especially the packing) would cause unacceptable hysteresis.
- To linearize a nonlinear actuator.
- Where varying differential pressures within the fluid would cause the plug position to vary.

Note: Positioners should not be used on fast-responding loops such as a flow loop with a lot of noise. The loop is much harder to stabilize, and cyclic oscillations may be continuous.

To ensure that the full differential pressure across the valve can be accepted, it is important to adjust the positioner zero setting so that no air pressure opposes the spring force when the valve is in its normal zero percent state or off-the-shelf position.

Electrical Positioners

Electric valve positioners send and receive electrical signals. These signals are typically a 4–20 mA positioning control signal and digital or discrete on/off position indication signals from limits switches inside the valve. There are three electric actuation types: single-phase and three-phase alternating current (AC), and direct current (DC) voltage. Electrical actuators are fairly responsive and can develop large torques for high-pressure applications.

Most actuators offer compatibility with a wide range of communication and process control systems. Network control of the actuator may be available using PROFIBUS, FOUNDATION Fieldbus, Modbus, DeviceNet, HART, and Rotork's Pakscan. More recent innovations of these actuators include the option of advanced displays, Bluetooth communications interfaces, and absolute encoders.

System Piping DP Pressure Drops

Look at the figure below. The section on fluid mechanics discussed the pump head, pressure head, kinetic head, and static head in a piping system. It showed how the pressure head changed to kinetic head to accelerate the fluid through the piping system and that work was done on the pipe, fittings, and elements.

On the right side of the figure below the pump head equals the pressure head at no flow. When the system is at minimum flow (the green Low Flow line), the avaliable pressure head produced by the pump is much less. This reduction in pressure head is the difference in the pump head minus the kinetic head, to accelerate the fluid and do work on the piping system. The reduction in head will be constant for this flow rate. As we increase the flow rate to maximum flow (the blue High Flow line), the pressure head drops even more to add more kinteic energy to the piping system to do even more work on the piping system. Moving from left to right on the graph, we see there is an additional drop in pressure for each section of the pipe and the elements in the piping system. At the end of the graph, there is still pressure head above zero. This is the head pressure of the tank level in feet of head.

Notice the DP across the valve in the middle of the graph. At low flow it is very large, just the opposite of the orifice element in our studies previously. The orifice or a head device consumes energy to do work passing the fluid through the orifice. The control valve absorbs the leftover energy in the system, the difference in the pump head minus the kinetic head and minus the work done on the system (the energy consumed in the form of heat and vibration).

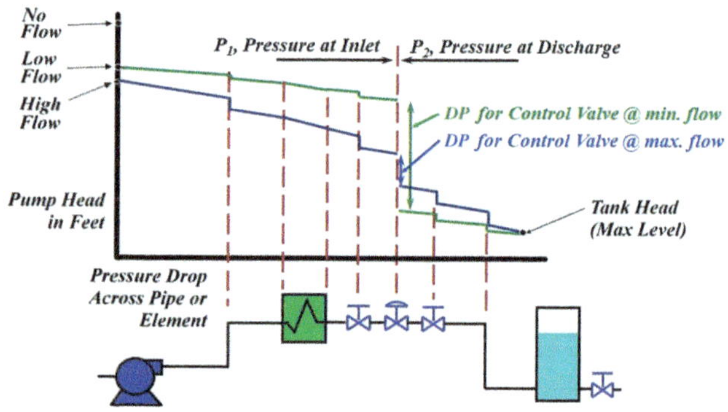

The control valve has to consume or burn up the extra head energy in the system; otherwise, the fluid will accelerate far too quickly. Recall that differential pressure multiplied by area is equal to force (DP$\bullet A$ (area) $= F$). As such, there is a change in force across the piping system.

Work $= \Delta F \bullet D$ (distance), also $\Delta F/m$ (mass) $= g$ (acceleration).

The valve has to burn up the extra DP in the system or the fluid will be accelerating far too quickly for the demand of flow. By burning up some of the pressure head across the valve, the fluid decelerates. This allows us to control the flow rate of the process fluid. Notice there is less DP across the valve at the higher flow rate. This is because we need a greater differential pressure across the piping system to accelerate the fluid more quickly and to do more work. Note that the hand valves typically do not change, so they act just like an orifice or head device and consume a fixed amount of energy proportional to the flow rate of the process fluid.

Control Valve DP Pressure Drop

The varying DP across the valve in the piping system discussed earlier is shown in the figure to the right. The graph shows the total energy put into the system and how it is divided up into its head energy components.

Let us do a breakdown of flow rates for a 2-inch linear valve with a maximum C_v equal to 39.2.

$$\text{gpm} = C_v \cdot \sqrt{\text{DP}_{\text{psig}}}$$

Graph of the Inherent Valve Characteristics (Off-the-Shelf)

The control valve characteristics refers to the relationship between the volumetric flow rate, C_v, or gpm (y-axis) through the valve *and* the valve travel or stem opening position, m, (x-axis), as the valve is opened from its closed position to various degrees of opening.

The control valve manufacturer can only state the inherent valve characteristics, which refer to the flow rate (C_v) versus the valve travel position (% open) relationship, at a constant pressure drop across the control valve (DP_V). Note that the pressure drop across the control valve (DP_V) can be almost constant only if the piping system pressure drop (DP_S) is concentrated at the control valve and not distributed along the pipeline. This means that the pipeline must be very short with minimum pressure-reducing devices or equipment in series with the control valve. Such a condition rarely exists in actual installations, with the exception of control valve manufacturer testing facilities, where the control valve characteristics are obtained using very short lengths of piping.

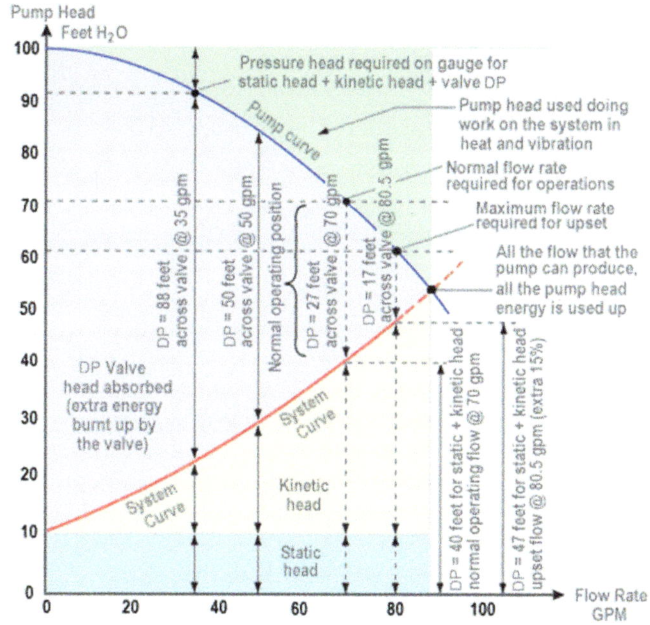

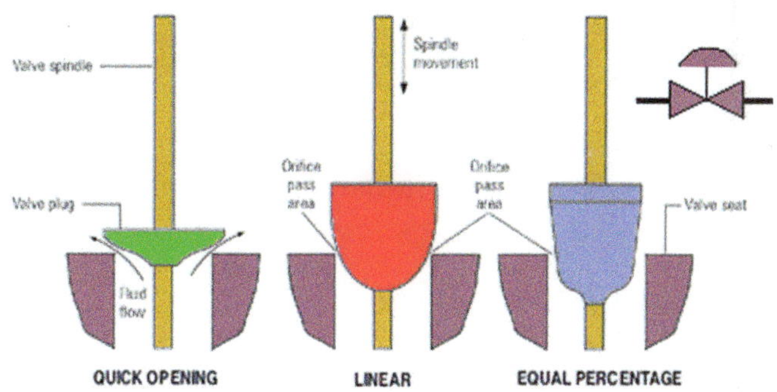

Flow Rate gpm	C_v Required for Flow	DP Valve ft/psi	% Stem Stroke (open)
35.0	5.66	88 ft/38.26 psi	14.54%
50.0	10.72	50 ft/21.74 psi	27.35%
70.0	20.43	27 ft/11.74 psi	52.12%
80.5	29.62	17 ft/7.39 psi	75.54%

QUICK OPENING LINEAR EQUAL PERCENTAGE

Cage valves are shown to the left.

The valve characteristics chart is shown below.

Choosing the Valve Characteristic

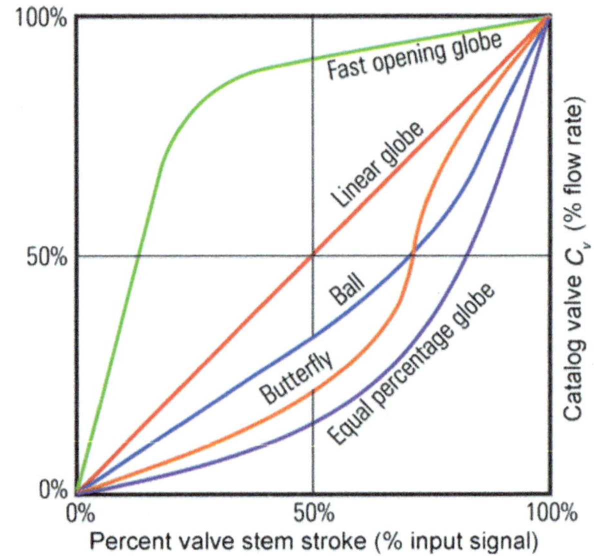

Because it is desirable to achieve and maintain process stability, the proper inherent valve characteristic must be selected to compensate for process changes. The first step is to determine the controlled process variable. There are four main classes: liquid level, pressure, flow, and temperature.

Controlled Variable: Liquid Level

If the valve pressure drop is constant, the inherent valve characteristic should be linear. This is because the process gain does not change. For varying pressure drops, the following guidelines are recommended: use an equal percentage characteristic if the drop at maximum flow is less than 20% of the drop at minimum flow, and a quick opening characteristic if the drop at minimum flow is greater than twice the drop at minimum flow. The equal percentage characteristic is selected because as the load changes, the variation in valve gain is minimized. The quick-opening characteristic is selected to negate the effects of the large pressure drop at high flows.

Controlled Variable: Pressure

Processes that control pressure are a function of the process time constant. These processes can be "slow" (e.g., large volumes, receivers, or long distribution lines) or "fast" (e.g., liquid flowing in a pipe). For the same reasons as liquid level, a linear characteristic is recommended for a slow process with a constant pressure drop. If the pressure drop at maximum flow is less than 20% of the drop at minimum flow in a slow process, then an equal percentage characteristic is used. For fast processes under any pressure drop, an equal percentage is again recommended.

Controlled Variable: Flow

Using flow as the controlled process variable offers additional challenges because the flow does not vary with the load. A load change in a flow loop does not change the flow rate but usually results in a change in pressure. The flow is controlled by the set point. An additional aspect is that transmitters are not always linear. Some are proportional to flow, and some are proportional to the square of the flow rate. The selection of the valve characteristic then becomes a function of the type of signal sent to the controller. Whether the valve is used in series or a bypass also makes a difference.

Controlled Variable: Temperature

Generally, in temperature control loops, the time constants are large and the characteristic frequency of the system changes as the load varies. Numerous temperature control loops exist in chemical plants and refineries, and experience has shown the best inherent characteristic is equal percentage.

Valve Selection for Liquid Level	
Valve DP	**Characteristic**
Approximately Constant	Linear
$DP_{Qmax} < 0.20\ DP_{Qmin}$	Equal Percentage
$DP_{Qmax} > 2.0\ DP_{Qmin}$	Quick-Opening

Valve Selection for Pressure		
Process	**Valve DP**	**Characteristic**
Fast	Any	Equal Percentage
Slow	Constant or $DP_{Qmax} > 2.0\ DP_{Qmin}$	Linear
	$DP_{Qmax} < 0.20\ DP_{Qmin}$	Quick-Opening

Valve Selection for Flow			
Signal	**Valve Location**	**Characteristic if Set Point Varies**	**Characteristic if Load Varies**
Proportional to Q (% flow)	Series	Linear	Equal Percentage
	Bypass	Linear	Equal Percentage
Proportional to Q^2 (% flow squared)	Series	Linear	Equal Percentage
	Bypass	Equal Percentage	Equal Percentage

Gain and Rangeability (Turndown Ratio in Valves)

Turndown is the ratio of the maximum to the minimum controllable flow. For a pinch valve, turndown of 10:1 is typical, so if the maximum flow is 5,000 scfm, you can expect to maintain stable control down to 500 scfm. Of course, the valve can close or drop tight to zero flow, but it is difficult to maintain stable control between zero and the minimum controllable flow. Turndown says nothing about the response, speed of the valve, undershoot, overshoot, or duty cycle.

The conventional definition of *rangeability* is the ratio between the maximum and minimum "controllable" flows through the valve. Minimum controllable flow (F_{MIN}) is not the leakage flow (which occurs when the valve is closed) but the minimum flow that is still controllable, and it can be changed up or down as the valve is throttled.

Using this definition, manufacturers usually claim a 50:1 rangeability for equal percentage valves, 33:1 for linear valves, and about 20:1 for quick-opening valves. These claims suggest that the flow through these valves can be controlled down to 2%, 3%, and 5% of the maximum. However, these percentages are often exaggerated. Also, the F_{MIN} rises as the distortion coefficient (Dc) drops. Therefore, at a Dc of 0.1, the 50:1 rangeability of an equal percentage valve drops to about 10:1.

Consequently, the rangeability should be defined as the flow range over which the actual installed valve gain stays within ±25% of the theoretical valve gain. To illustrate the importance of this limitation, the following figures show that the actual gain of an equal percentage valve starts to deviate from its theoretical gain by more than 25% when the flow reaches about 65%.

Therefore, in determining the rangeability of such a valve, the maximum allowable flow should be 65%. Using this definition, the rangeability of an equal percentage valve is seldom more than 10:1. In such cases, the rangeability of a linear valve can be greater than that of an equal percentage valve. Also, the rangeability of some rotary valves can be higher because their clearance flow tends to be lower, and their body losses near the wide-open position also tend to be lower than those of other valve designs.

To stay within ±25% of the theoretical valve's gain, the maximum flow should not exceed 60% of the maximum C_v in a linear valve or 70% of the maximum C_v in an equal percentage valve. In terms of valve lift, these flow limits correspond to 85% of the maximum lift for equal percentage valves and 70% for linear valves.

Proper Control Valve Sizing

The symptom of a misapplied control valve flow characteristic is a control loop that gives good control only at one end of its operating range and is either sluggish or becomes unstable at the other end of its operating range. An example of a misapplied valve inherent flow characteristic is an equal percentage valve in a system with little piping.

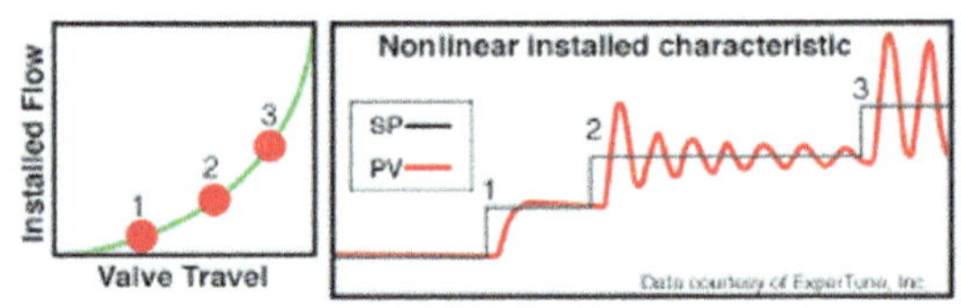

The nonlinear installed flow characteristic caused by a misapplied inherent control valve flow characteristic yields a system than cannot maintain proper controller tuning throughout the entire flow range.

The left-hand graph in the figure to the right shows the unstable oscillations as a result of a short run. The flow characteristic is caused by the fact that not enough nonlinear flow pressure drop is created by long runs of piping that would greatly reduce the DP across the valve at maximum flow. In the previous figure, the gain for an equal percentage valve is very high, near 100% of stroke, so when there is an upset and the valve opens, the flow rate accelerates quickly, overshooting the desired flow for correction. The control system (controller) must hunt, and as the valve starts to close down, the process loop gain becomes less, allowing the loop to stabilize.

On start-up of this system, if it was running at a low process load and the valve was around 25% open (Point 1), after going through the process of tuning a proportional-integral-derivative (PID) controller, the slope of the valve's installed characteristic curve is quite shallow. This means that the valve's gain (or the sensitivity of flow to changes in valve position) is quite low. To make up for this, the proportional gain of the controller would need to be set high. As expected, when you make a step change in the set point, there is a quick and stable response. Later, when the process load increases, the valve might be around 50% open (Point 2).

At this point, the slope of the valve's installed characteristic curve (its gain) is higher, meaning that the value of controller proportional gain previously selected is higher than it should be when the valve is 50% open. When the set point is stepped up from the current 50% position, there is an oscillatory response. When the process load increases to where the valve is around 75% open (Point 3), the slope of the valve's installed characteristic curve is quite steep, meaning that the valve's flow gain is quite high.

The value of controller gain originally selected when the valve was 25% open is now much too high, and when the set point again is stepped up, there is an unstable response. If the loop were instead tuned when the control valve was 75% open, and a lower value of proportional gain in the controller had been used, there would be a fast, stable response to a step change in the set point. However, if operated at lower loads, the response would be sluggish.

If you had misapplied a linear valve in a system with a lot of pipe, the situation would be the opposite of what is shown in the previous figure. Just as the system with a lot of pipe pushes the equal percentage inherent characteristic upward into a linear installed characteristic, it would push a linear inherent characteristic upward into a quick-opening characteristic. With a quick-opening characteristic, a small increment of valve position at small openings results in a large increase in flow capacity, while the same increment of valve position at large openings results in a small increase in flow capacity. This system would be very sensitive at low valve openings (high gain) and very insensitive at large openings (low gain). The system would still be difficult or impossible to tune to get a fast, stable response throughout the flow range.

On the other hand, a properly applied control valve inherent characteristic (an equal percentage valve in a system with a lot of pipe, or a linear valve in a system with little pipe) results in a linear, or at least nearly linear, installed characteristic, as shown in the left-hand graph of the figure to the right. A linear installed characteristic gives responsive and stable control throughout the entire flow range.

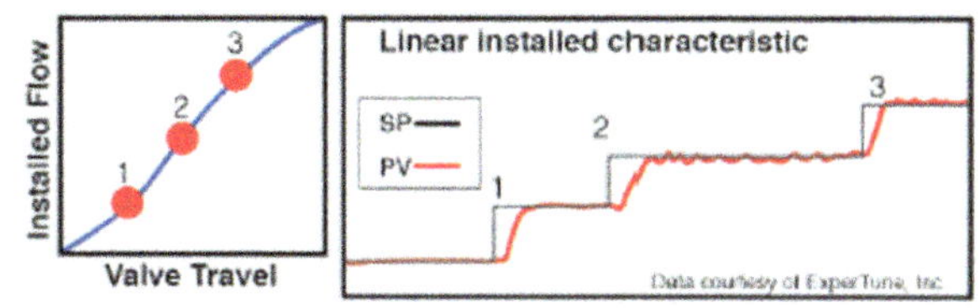

The linear or nearly linear installed flow characteristic that results from a properly applied inherent control valve flow characteristic gives good control with stability throughout the control range.

This properly applied valve's installed flow characteristic has a slope that is constant throughout the entire flow range, and the sensitivity of flow to changes in valve position is a nearly a linear flow gain. Regardless of what the operating point was during the controller tuning process, the PID tuning parameters selected would be appropriate at all points in the operating range.

The previous discussion can be summarized just by looking a heat exchanger application that has a significant amount of pressure drop, using an equal percentage control valve as shown below for flow control. When more heat or BTU per (lbm • °F) is called for, the flow must greatly increase so the gain of the heat exchanger drastically drops off near the top of the design flow rate. The pipe friction creates a greater nonlinear pressure drop, as shown below. The equal percentage valve has a gain that is the inverse of the heat exchanger, so when there is an increase in flow, the heat exchanger gain is dropping and the equal percentage valve gain is increasing, creating a fairly linear gain over the flow range and temperature control range of the process loop.

Equal Percentage Valves Provide Linear Gain in Flow Applications

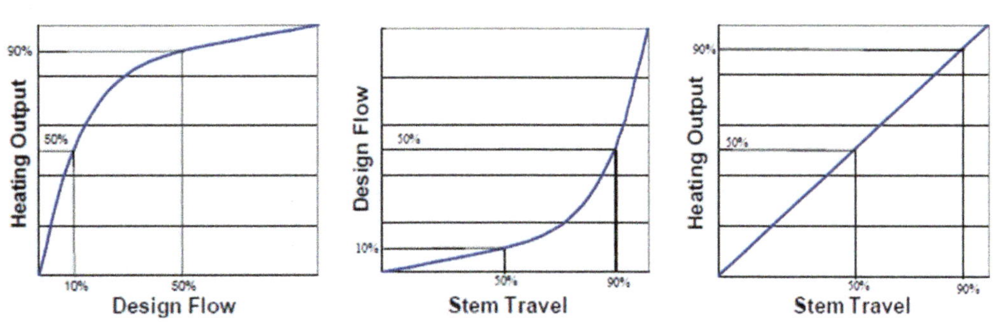

Control Valve Sizing

As demonstrated in the previous section, you cannot simply pick a differential pressure (DP) to be across the valve and size the valve for a flow control system. It is a system, and all the elements interact as a system. The pressure drops across the elements in the system are constantly changing as the flow rate changes.

The piping system can be thought of as similar to an electrical circuit. There is a battery (the pump) and fixed resistors in series to make up the system (the pipe, the fittings, the orifice plate or measuring restriction). The valve is a variable resistor in the system and is constantly changing resistance.

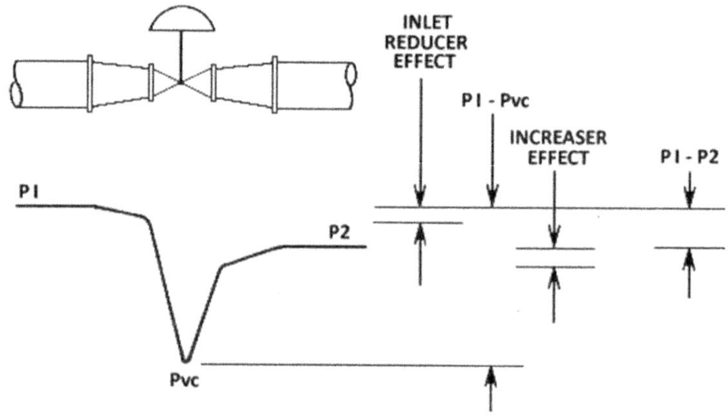

From electrical physics, you will remember Kirchhoff's voltage law. The sum of the voltage drops (pressure drops) equals the sum of the source (pressure produced by the pump). Unless we are pumping into a pressurized vessel or tank maintaining a level head, all the pressure is dropped as the fluid exits the pipe, and the fluid's pressure equals the atmosphere on exiting the pipe. All the pressure has been used up.

Remember that the pressure drop is "head pressure," and it is a component of the total head energy produced by the pump. The head pressure does not just drop off as it goes down the pipe; it is consumed to do work. This can be seen by looking at the pump motor. It absorbs a constant wattage to produce a constant horsepower (hp), which in turn produces the hydraulic head. The wattage or energy does not change with a constant pump rpm. Energy cannot be created or destroyed, so it just changes to heat and vibration to be dissipated into the environment.

Other advanced considerations when sizing the valve based on the DP across the valve include choking, cavitation, and flashing.

The Valve Sizing Equations

The basic equation for liquid flow is:

$$q = (N_1 F_p C_v)\sqrt{\frac{DP}{G_f}}$$

Note: N_1 is always equal to 1 for psia and gpm.

The basic equation for gas flow is:

$$q = (N_7 F_p C_v P_1 Y)\sqrt{\frac{x}{G_f T_1 Z}}$$

Note: N_7 is always equal to 1,360 for psia and scfh.

The basic equation for steam flow is:

$w = (N_6 F_p C_v Y)\sqrt{x P_1 \gamma_1}$ **Note:** N_6 is always equal to 63.3 for psia and lb/h.

where

G_f = SG at flowing condition, for gas $\dfrac{\text{(molecular weight of gas)}}{\text{28.967 (molecular weight of air)}}$

C_v = valve sizing coefficient

F_k = ratio of specific heat factors

F_p = piping geometric factor

K_1 = inlet velocity head loss coefficient

K_2 = outlet velocity head loss coefficient

K_i = inlet head loss coefficient; $K_1 + K_{B1}$

K_{B1} = inlet bernoulli coefficient

K_{B2} = outlet bernoulli coefficient

$\Sigma K = K_1 + K_2 + K_{B1} - K_{B2}$

N_1 = 1.00 (for psia, equation constant see Table A13 in the Appendix)

N_6 = 63.3 (for lb/h, equation constant see Table A13 in the Appendix)

N_7 = 1360 (for scfh, equation constant see Table A13 in the Appendix)

N_9 = 7320 (for scfh, equation constant see Table A13 in the Appendix)

DP = pressure in psid across the valve

P_1 = inlet pressure (psi absolute)

P_2 = outlet pressure (psi absolute)

q = volumetric flow (in gpm for liquid or scfh for gas)

T_1 = fluid operating temperature (psi absolute);(reference temp in °F + 460)

w = volumetric flow (in pounds per hour)

x = ratio of delta pressure to inlet pressure absolute

Y = expansion factor of the gas

Z = fluid compressibility

γ_f = specific weight of the steam or vapor in pounds per cubic foot(operating condition)

Note: For SI (metric) calculations $K_v = 0.856(C_v)$

Note: The *Fisher Control Valve Handbook*, the *Fisher Control Valve Catalog*, or Tables A11 and A12 in the Appendix of this manual can be used for C_v, F_L, and X_T references for problems in this guide and on the CSE exam.

All variables are discussed in detail. We will keep the equations simple and to the point for sizing. We will size for the correct size valve to be installed. *On the CSE exam, only obtaining the C_v of the valve may be of interest,* and not sizing for actual applications.

The other factors, such as piping geometry factor (F_p) for reducers in the piping and the Bernoulli factors (K_b), probably will not be used in the CSE exam. However, the exam may include questions about temperature limits and hardness of the materials that make up the valve for given applications.

Sizing Valves for Liquid – Worked Example

The basic equation for liquid flow through a control valve is:

$$q=(N_1F_pC_v)\sqrt{\frac{DP}{G_f}}$$ **Note:** N_1 is always equal to 1 for psia and gpm.

Solving for C_v we get:

$$C_v=\frac{q}{(N_1F_p)\sqrt{\frac{DP}{G_f}}}$$ **Note:** N_1 is always equal to 1 for psia and gpm.

$$F_p=\left[1+\frac{\Sigma K}{890}\left(\frac{C_v}{d^2}\right)^2\right]^{-1/2}$$ **Note:** F_p = piping geometric factor.

The piping geometric factor covers reducing fittings attached to each side of the valve body. See Table A11, "Properties and Sizing Coefficients of Globe Valves," and Table A12, "Properties and Sizing Coefficients of Rotary Valves," in the Appendix of this manual for use of C_v, X_T and F_L.

Now the equation becomes:

$$C_v=\frac{q}{N_1F_p\sqrt{\frac{DP}{G_f}}}$$

Liquid

Sample Problem: Assume an 8-inch pipe connected to a globe valve for a liquid propane service. Size the equal percentage valve using the following criteria.

$q = 800$ gpm	$T_1 = 70°F$	$G_f = 0.5$	DP = 25 psi
$P_1 = 300$ psig	$P_2 = 275$ psig		

Find the approximate C_v. The C_v is needed to find F_p (for now, set to $F_p = 1$).

$$C_v=\frac{q}{F_p\sqrt{\frac{DP}{G_f}}}=\frac{800}{\sqrt{\frac{25}{0.5}}}=113.13$$

Note: If the piping is the same size as the valve, the calculation is complete.
In Table A11 we find that a 3-inch globe valve (equal percentage) has a maximum C_v of 136 at full open. But we want to throttle at 50%, so pick a 4-inch globe valve with a C_v of 224. Now we will plug this C_v into the piping geometry equation to get the installed valve C_v.
$\Sigma K = K_1(\text{the entry factor}) + K_2(\text{the exit factor})$

$$\Sigma K = K_{1+2} = 1.5\left(1-\frac{d^2}{D^2}\right)^2$$ **Note:** $K_1 + K_{2,} = (0.5+1)\left(1-\frac{d^2}{D^2}\right)^2$ same size piping

$$\Sigma K = K_{1+2} = 1.5\left(1-\frac{4^2}{8^2}\right)^2 = 0.844$$ **Note:** 4 = valve size, 8 = pipe size

$$F_p = \left[1+\frac{\sum K}{890}\left(\frac{C_v}{d^2}\right)^2\right]^{-1/2}$$ **Note:** F_p = piping geometry factor

$$F_p = \left[1+\frac{0.844}{890}\left(\frac{224}{4^2}\right)^2\right]^{-1/2} = 1.1859^{-1/2} = 0.918$$

Find the corrected C_v for the installed valve.

$$C_v = \frac{q}{F_p\sqrt{\frac{DP}{G_f}}}$$

$$C_v = \frac{800}{(0.918)\sqrt{\frac{25}{0.5}}} = \frac{800}{6.238} = 123.24 \text{ or } 124$$

This shows that a 3-inch valve is too small; it will require the 4-inch valve with the maximum C_v equal to 224.

$$\% = \frac{124}{224} = 55.4\% \text{ of maximum } C_v \text{ and about 75\% open}$$

In Table A11, a Fisher-type ED (equal percentage) valve is used. A 3-inch valve would be correct with a C_v of 136, but it is too small.

The valve C_v would be 124/136 or 91% of the maximum C_v, and you might not get the required flow through the valve for throttling. *This is acceptable for the CSE exam.* Remember, valves should only be about 60%–80% of the maximum C_v to throttle, so size the normal operating C_v to fit at about 50% of the maximum C_v. The industrial standard is to size valves for 200% of the normal operating C_v.

Sizing Valves for Gas – Worked Example

The basic equation for gas or vapor flow through a control valve is:

$$q = (N_7 F_p C_v P_1 Y)\sqrt{\frac{x}{G_f T_1 Z}}$$ **Note:** N_7 is always equal to 1360 for psia and scfh.

$$C_v = \frac{q\text{ (in scfh)}}{1360 F_p P_1 Y\sqrt{\frac{x}{G_f T_1 Z}}}$$ **Note:** For volumetric flow units of gas or vapor.

where

$$Y = 1-\left(\frac{x}{3F_k x_{TP}}\right)$$ **Note:** The expansion factor of the gas or vapor.

Note: The expansion factor must be between 1.0 and 0.667. The downstream velocity will always be greater than the upstream velocity.

$F_k = \frac{k}{1.4}$ **Note:** Ratio of specific heat factors for the valve.

k = ratio of specific heats

$x = \frac{\text{DP}}{P_1}$ **Note:** Pressure drop ratio of DP to inlet P_1 upstream pressure.

x_T = pressure drop ratio required to produce the maximum flow through the valve

when $F_k = 1.0$ (x_T can be found in the valve coefficients table)

$x_{TP} = \frac{x_T}{F_p^2}\left[1 + \frac{x_T K_i}{N_5}\left(\frac{C_v}{d^2}\right)^2\right]^{-1}$ **Note:** X_{TP} is the pressure drop ratio factor with installed fittings attached

where

$F_p = \left[1 + \frac{\Sigma K}{890}\left(\frac{C_v}{d^2}\right)^2\right]^{-1/2}$ **Note:** Piping geometric factor, used when the valve is smaller than the pipe.

$K_i = K_1 + K_{b1}$ **Note:** Inlet head loss coefficient based on the Bernoulli equation.

$$K_1 = 0.5\left(1 - \frac{d^2}{D^2}\right)^2$$

$$K_2 = 1\left(1 - \frac{d^2}{D^2}\right)^2$$

K_{B1} or $K_{B2} = 1 - \left(\frac{d}{D}\right)^4$ **Note:** Bernoulli coefficients used with piping geometric factor.

Gas

Sample Problem: Assume a 6-inch pipe connected to a globe valve for natural gas service. Size the equal percentage valve using the following criteria.

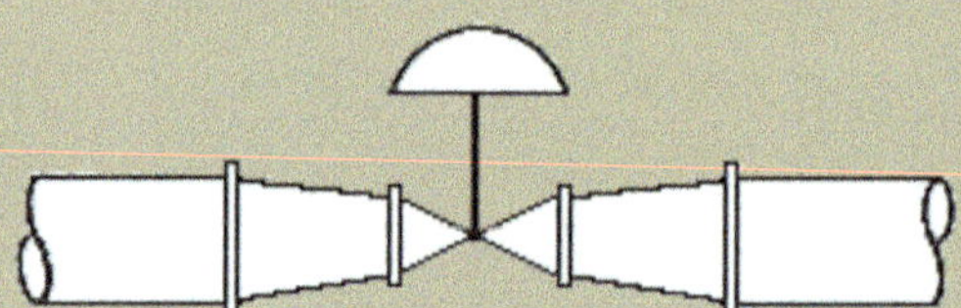

q = 800,000 scfh	T_1 = 60°F = 520°R	G_f = 0.60	DP = 150 psi
P_1 = 400 psig	P_2 = 250 psig	M_r = 17.38	k = 1.32

The molecular weight, M_r of gas/M_r of air (17.38/28.96), gives the specific gravity (SG) of G_f = 0.60. We will use SG and N_7 equal to 1360.

$C_v = \frac{q\,(\text{in scfh})}{1360 F_p P_1 Y \sqrt{\frac{x}{G_f T_1 Z}}}$ **Note:** For volumetric flow units of gas or vapors.

First find the approximate valve size and C_v to use in the sizing formulas.

$$F_p = 1, Y = 1, Z = 1$$

$$x = \frac{DP}{P_1} = \frac{150}{(400 + 14.7)} = 0.362; \ F_k x_T = \left(\frac{1.32}{1.4}\right) 0.68 = 0.641$$

Use the lesser value of the two equations above for x in the valve sizing formula.

$$C_v = \frac{q\,(\text{in scfh})}{1360 P_1 \sqrt{\dfrac{x}{G_f T_1}}} = \frac{800{,}000}{1360(400 + 14.7)\sqrt{\dfrac{0.362}{(0.60)(60 + 460)}}} = 41.64 \text{ or } 42$$

Note: If the piping is the same size as the valve in the CSE exam, the calculation is complete. When the pressure differential ratio x reaches a value of $F_k\, x_T$, the limiting value of x is defined as the critical differential pressure ratio. The value of x used in any of the sizing equations and in the relationship for Y must be held to this limit, even if the actual pressure differential ratio is greater. Thus, the numerical value of Y may range from 0.667, when $x = F_k\, x_T$, to a value of 1.0 for very low differential pressures. The x_T comes from the valve coefficient tables in the Appendix. (Calculate the value for 200% C_v for throttling applications.)

From Table A11, we determine that we want to throttle at about 50% of the maximum C_v, so double the C_v of the initial equation. In the globe valve coefficients table, we see a 3-inch valve with the $C_v = 136$.

Calculate for piping geometric factors.

inlet = 6-inch schedule 40 pipe

outlet = 6-inch schedule 40 pipe

$$\Sigma K = K_1 + K_2 + K_{B1} - K_{B2}$$

$$K_1 = 0.5\left(1 - \frac{d^2}{D^2}\right)^2 = 0.5\left(1 - \frac{3^2}{6^2}\right)^2 = 0.281$$ **Note:** 3 = valve size, 6 = pipe size

$$K_2 = 1\left(1 - \frac{d^2}{D^2}\right)^2 = 1\left(1 - \frac{3^2}{6^2}\right)^2 = 0.5625$$

$$K_{B1} = 1 - \left(\frac{d^2}{D^2}\right)^2 = 1 - \left(\frac{3}{6}\right)^4 = 0.9375$$

$$K_{B2} = 1 - \left(\frac{d^2}{D^2}\right)^2 = 1 - \left(\frac{3}{6}\right)^4 = 0.9375$$

Sum the resistance coefficients and Bernoulli coefficients to get the piping geometry factor.

$$\Sigma K = 0.281 + 0.5625 + 0.9375 - 0.9375 = 0.8435$$

$$F_p = \left[1 + \frac{\Sigma K}{890}\left(\frac{C_v}{d^2}\right)^2\right]^{-1/2} = \frac{1}{\sqrt{1 + \dfrac{0.8435}{890}\left(\dfrac{136}{3^2}\right)^2}} = 0.9067$$

Find the pressure drop ratio for the installed fitting attached to the 3-inch valve.

$$K_i = K_I + K_{B1} = 0.281 + 0.9375 = 1.2185$$

From Table A11 and Table A13 in the Appendix, we find: $N_5 = 1000$ and $x_T = 0.68$.

$$x_{TP} = \frac{x_T}{F_p^2}\left[1+\frac{x_T K_i}{N_5}\left(\frac{C_v}{d^2}\right)^2\right]^{-1} = \frac{x_T}{F_p^2\left[1+\frac{x_T K_i}{N_5}\left(\frac{C_v}{d^2}\right)^2\right]}$$

$$x_{TP} = \frac{0.68}{0.9067^2\left[1+\frac{(0.68)(1.2185)}{1000}\left(\frac{136}{3^2}\right)^2\right]} = 0.6955$$

Find the expansion factor (Y). It must be between 0.667 and 1.0.

$$F_k = \frac{1.32}{1.4} = 0.943 \quad \textbf{Note:} \text{ Ratio of specific heat factors for the valve.}$$

$$Y = 1-\left(\frac{x}{3F_k x_{TP}}\right) = 1-\left(\frac{0.362}{3\,(0.943)(0.6955)}\right) = 0.816$$

$$C_v = \frac{q\,(\text{in scfh})}{1360 F_p P_1 Y\sqrt{\frac{x}{G_f T_1}}} = \frac{800{,}000}{1360(0.9067)(414.7)(0.816)\sqrt{\frac{0.362}{(0.60)(520)}}} = 56.284 \text{ or } 57$$

We want to throttle at around 50%; a 3-inch valve has a C_v of 136. Using a 2-inch valve, the calculation would have required a C_v of 55.89, but the 2-inch valve has a C_v of only 50.7 at 100% open.

$$\% = \frac{57}{136} = 42\% \text{ of maximum } C_v \text{ and about } 64\% \text{ open}$$

$$C_g = 40\,C_v\sqrt{x_T}\,; \text{ if needed to convert } C_v \text{ to } C_g \text{ (asin the FCVH)}$$

Sizing Valves for Vapor and Steam – Worked Example

The basic equation for vapor or steam flow through a control valve is:

$$w = (N_6 F_p C_v Y)\sqrt{x P_1 \gamma_1}$$ **Note:** N_6 is always equal to 63.3 when psia and lb/h are used in calculation.

$$C_v = \frac{w(\text{lb/h})}{63.3 F_p Y\sqrt{x P_1 \gamma_1}}$$ **Note:** For mass flow units in lb/h.

Steam – Worked Example

Sample Problem: Assume that a 6-inch pipe inlet and 8-inch pipe outlet of schedule 40 are connected to a type ED globe (equal percentage) valve, for process steam service. Size the valve using the following criteria.

Note: $1/(\gamma_1)$ can be found in Table A9, "Properties of Saturated Steam," in the Appendix of this manual.

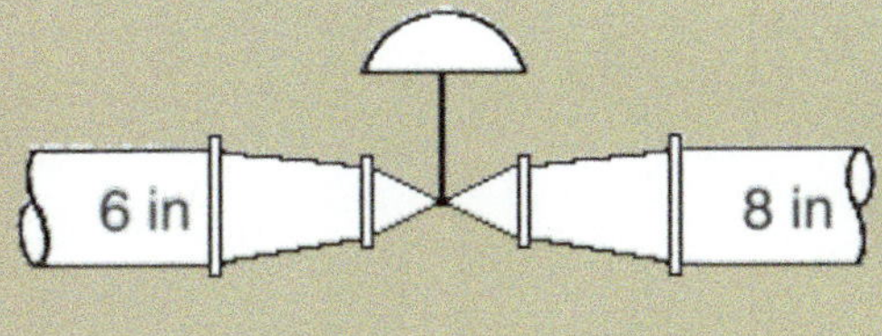

$q = 125{,}000$ lb/h $T_1 = 470°F$ $G_f = 0.60$ DP = 250 psi

$P_1 = 500$ psig $P_2 = 250$ psig $\gamma_f = 1.089$ $k = 1.31$

First find the approximate valve size and C_v for the formulas.

$F_p = 1$, $Y = 1$.

Find γ_f:

$$\gamma_f = \text{Specific weight is the reciprocal of specific volume of steam } \frac{1}{(\text{ft}^3/\text{lb})} = \frac{lb}{\text{ft}^3}$$

In Table A9 we can find that the specific volume of the steam at a pressure of 514.7 psia equals 0.9182 ft³/lb.

$$\frac{\text{ft}^3}{\text{lb}} = \left[\left(\frac{P\ desired - P\ lower\ value}{P\ upper\ value - P\ lower\ value}\right)\left(\frac{\text{ft}^3}{\text{lb}}\ upper\ value - \frac{\text{ft}^3}{\text{lb}}\ lower\ value\right)\right] + \frac{\text{ft}^3}{\text{lb}}\ low\ value$$

$$\frac{\text{ft}^3}{\text{lb}} = \left[\left(\frac{514.7 - 500}{520 - 500}\right)(0.9278 - 0.8915)\right] + 0.8915 = 0.9182\frac{\text{ft}^3}{\text{lb}}$$

$$\gamma_f = \text{Specific weight is the reciprocal of specific volume of steam } \frac{1}{(\text{ft}^3/\text{lb})} = \frac{1}{0.9182} = 1.089\frac{\text{lb}}{\text{ft}^3}$$

$$F_k = \frac{1.31}{1.4} = 0.936 \quad \textbf{Note: } \text{Ratio of specific heat factors for the valve.}$$

$$[x = \text{DP}/P_1 = 0.486] < [x = F_k x_T = (0.936)(0.69) = 0.646]$$

The pressure ratio is smaller than critical limits, so we will use $x = 0.486$.

Find the C_v:

$$C_v = \frac{w\ (\text{in lb/h})}{63.3 F_p Y \sqrt{x P_1 \gamma_f}} = \frac{125{,}000}{63.3(1)(1)\sqrt{(0.486)(514.7)(1.089)}} = 119.65 \text{ or } 120$$

Note: If the piping is the same size as the valve in the CSE exam, the calculation is complete. When the pressure differential ratio x reaches a value of $F_k x_T$, the limiting value of x is defined as the critical differential pressure ratio. The value of x used in any of the sizing equations, and in the relationship for Y, must be held to this limit, even if the actual pressure differential ratio is greater.

Thus, the numerical value of Y may range from 0.667, when $x = F_k x_T$, to a value of 1.0 for very low differential pressures. The x_T comes from Table A11.

The table shows a 3-inch valve with a C_v of 136, but we want to throttle around 50% (200% of 120 is 240), so a 4-inch valve with a C_v of 224 is too small. Doing the calculation with a 4-inch valve will prove that we are already at 71% C_v at normal flow and will probably be choking already. You should select a 6-inch valve with a C_v of 394.

Calculate for piping geometric factors.

inlet = 6 in, schedule 40 pipe

outlet = 8 in, schedule 40 pipe

$$K_1 = 0.5\left(1-\frac{d^2}{D^2}\right)^2 = 0.5\left(1-\frac{6^2}{6^2}\right)^2 = 0.0$$

$$K_2 = 1\left(1-\frac{d^2}{D^2}\right)^2 = 1\left(1-\frac{6^2}{8^2}\right)^2 = 0.1914$$

$$K_{B1} = 1-\left(\frac{d^2}{D^2}\right)^2 = 1-\left(\frac{6}{6}\right)^4 = 0.0$$

$$K_{B2} = 1-\left(\frac{d^2}{D^2}\right)^2 = 1-\left(\frac{6}{8}\right)^4 = 0.6836$$

Sum the resistance coefficients and Bernoulli coefficients to get the piping geometry factor.

$$\Sigma K = K_1 + K_2 + K_{B1} - K_{B2}$$

$$\Sigma K = 0.0 + 0.1914 + 0.0 - 0.6836 = -0.4922$$

$$F_p = \left[1+\frac{\Sigma K}{890}\left(\frac{C_v}{d^2}\right)^2\right]^{-1/2} = \frac{1}{\sqrt{1+\frac{-0.4922}{890}\left(\frac{394}{6^2}\right)^2}} = 0.9338$$

C: Find the pressure drop ratio for the installed fitting attached to the valve.

$$K_i = K_1 + K_{B1} = 0.0 + 0.0 + 0.0$$

Table A11 and Table A13, "Numerical Constants for Valve Sizing Formulas," in the Appendix show: $N_5 = 1000$ and $x_T = 0.78$.

$$x_{TP} = \frac{x_T}{F_p^2}\left[1+\frac{x_T K_i}{N_5}\left(\frac{C_v}{d^2}\right)^2\right]^{-1} = \frac{x_T}{F_p^2\left[1+\frac{x_T K_i}{N_5}\left(\frac{C_v}{d^2}\right)^2\right]}$$

$$x_{TP} = \frac{0.78}{0.9338^2\left[1+\frac{(0.78)(0.0)}{1000}\left(\frac{394}{6^2}\right)^2\right]} = 0.8945$$

D: Find the expansion factor (Y); it must be between 0.667 and 1.0.

$$Y = 1-\left(\frac{x}{3F_k x_{TP}}\right) = 1-\left(\frac{0.486}{3(0.936)(0.8945)}\right) = 0.8065$$

Note: Replace x_{TP} with x_T if the pipe size, inlet, and outlet, are the same size as the valve.

$$C_v = \frac{w\text{ (in lb/h)}}{63.3F_p Y\sqrt{xP_1\gamma_f}} = \frac{125{,}000}{63.3(0.9338)(0.8065)\sqrt{(0.486)(514.7)(1.089)}} = 158.87 \text{ or } 159$$

This shows that a 6-inch valve is the correct size.

$$\% = \frac{159}{394} = 41\% \text{ of maximum } C_v \text{ and about 63\% open}$$

Note: A 6-inch valve with a $C_v = 394$ should be used for this application.

$C_g = 40\,C_v\sqrt{x_T}$; if needed to convert C_v *to* C_g (as in the Fisher control valve handbook)

Sizing Valves for Two-Phase Flow – Worked Example

Two-phase flow is a flow comprised of liquid and vapor or liquid and gas in part ratios of mass. The quality of the gas or vapor and liquid must be known to size the valve. Recall that the quality of gas or vapor is quality (vapor) = vapor mass/total mass and quality (liquid) = liquid mass/total mass.

The basic equations for two-phase flow through a control valve are:

$$w = (N_6 F_p C_v Y)\sqrt{x P_1 \gamma_f} \quad \textbf{Note: } N_6 = 63.3$$

$$w = (N_6 F_p C_v Y)\sqrt{(P_1 - P_2)\gamma_f} \quad \textbf{Note: } N_6 = 63.3$$

Note: Y^2 applies only to the gas portion and not to the liquid portion of the mass flow.

$$F_p C_v = \frac{w(\text{lb/h})}{N_6\sqrt{\text{DP}\gamma_f Y^2}}$$

$\frac{1}{v_e} = \gamma_f Y^2$ **Note:** Specific volume is the reciprocal of density.

$$v_e = f_g v_g / Y^2 + f_f v_f$$

$F_p C_v = \frac{w(\text{lb/h})}{N_6}\sqrt{\frac{v_e}{\text{DP}}}$ **Note:** Y^2 is in the coefficient v_e.

$$Y = 1 - \frac{x}{3F_k x_t}$$

$$x = \frac{\text{DP}}{P_1}$$

$$F_k = \frac{k}{1.40}$$

Note: v_g is the specific volume of the gas, and M is the molecular weight of air.

$$v_g = \frac{RT}{M \bullet P_1(\textit{conversion factor } \text{in}^2 \textit{ to } \text{ft}^2)}$$

$$v_g = \frac{\left(1545\frac{\text{ft} \bullet \text{lbf}}{\text{lb} \bullet \text{mol} \bullet R}\right)(\text{deg } R)}{\left(28.97\frac{\text{lbm}}{\text{lb} \bullet \text{mol}}\right)\left(P_1\frac{\text{lbf}}{\text{in}^2}\right)\left(144\frac{\text{in}^2}{\text{ft}^2}\right)} = \frac{\text{ft}^3}{\text{lbm}}$$

Two-Phase Flow

Sample Problem: The goal in this example is to find the required valve capacity (F_pC_v) for the following conditions:

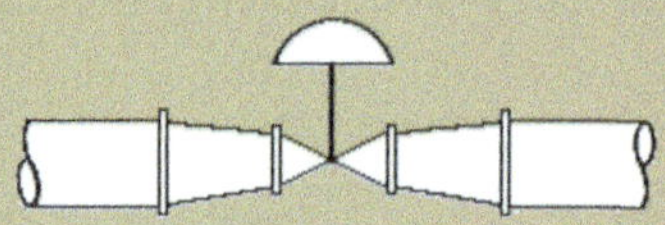

Air flow rate = 600 lb/h

Water flow rate = 26,000 lb/h

Upstream pressure, P_1 = 150

psia Pressure drop, DP = 50 psi

Temperature = 90°F (550°R)

Line size = 3-inch schedule 40

Step 1: Determine the relative mass fractions of gas and liquid, f_g and f_f. The total mass flow rate is $w = 600 + 26{,}000 = 26{,}600$ lb/h.

The fraction of gas:

$$f_g = \frac{600}{26{,}600} = 0.0226$$

The fraction of fluid:

$$f_f = \frac{26{,}000}{26{,}600} = 0.9774$$

Step 2: Make a preliminary selection of valve type and determine the critical pressure drop ratio factor (x_T) for expansion factor (Y) of the valve. Assume a single-seated globe valve with a contoured plug with flow under the plug (to open).

$$C_v = \frac{\text{gpm}}{\sqrt{\text{DP}}} = \frac{26{,}000\,\text{lb}}{1} \bullet \frac{1\,\text{gpm}}{8.33\,\text{lb}} \bullet \frac{1\,\text{min}}{60\,\text{s} = \dfrac{52\,\text{gpm}}{\sqrt{\text{DP}}} = \dfrac{52\,\text{gpm}}{\sqrt{50}} = 7.35}$$

Double the C_v for an approximation of the vapor through the valve.

Approximate $C_v = 14.7$

Using the manufacturer's catalog tables, for valves ranging from 0.75 in to 1.5 in, we obtain an average estimate of $x_T = 0.68$.

Find the expansion factor (Y).

$$Y = 1 - \frac{x}{3F_k x_t} = 1 - \frac{0.334}{3(1)(0.72)} = 0.8454$$

Step3: Calculate the pressure drop ratio (x).

$$x = \frac{50}{150} = 0.334$$

Because $x < x_T$, and the gas flow is not choked.

Find the ratio of specific heat factor (F_k):

$$F_k = \frac{1.40}{1.40} = 1.0$$

Step 4: Determine the effective specific volume of the mixture at upstream conditions. The specific volume of the air can be calculated from the gas law equation:

$$v_g = \frac{\left(1545 \frac{\text{ft} \bullet \text{lbf}}{\text{lb} \bullet \text{mol} \bullet \text{R}}\right)(550\ ^\circ R)}{\left(28.97 \frac{\text{lbm}}{\text{lb} \bullet \text{mol}}\right)\left(150 \frac{\text{lbf}}{\text{in}^2}\right)\left(144 \frac{\text{in}^2}{\text{ft}^2}\right)} = 1.358 \frac{\text{ft}^3}{\text{lbm}}$$

From Table A7, "Properties of Water Specific Volume and Density," at 90°F:

The liquid specific volume is:

$$v_f = 0.01610 \frac{\text{ft}^3}{\text{lbm}}$$

The mixture effective specific volume v_e is:

$$v_e = \frac{f_g v_g}{Y^2} + f_f v_f = \frac{(0.0226)(1.358)}{0.8454^2} + (0.9774)(0.01610) = 0.0587 \frac{\text{ft}^3}{\text{lbm}}$$

Step 5: Calculate the valve capacity from:

$$F_p C_v = \frac{\text{w(lb/h)}}{N_6} \sqrt{\frac{v_e}{\text{DP}}} = \frac{26,600}{63.3} \sqrt{\frac{0.0587}{50}} = 14.39$$

If the piping geometric factor (F_P) is equal to 1, then the C_v of the valve would be 14.39.

If the piping geometric factor (F_P) is equal to 0.98, then the C_v of the valve would be:

$$(0.98)(14.39) = 14.1$$

We want to throttle at 50%, so size the valve C_v for 200%.

$$14.1 \bullet 2.00 = 28.2\ C_v$$

Choose a 1-1/2-inch valve with a C_v of 35.8 and x_T equal to 0.68.
(The x_T matches our calculation, so no recalculation is necessary.)

$$\% = \frac{14.1}{35.8} = 39.4\% \text{ an } C_v \text{ and about } 63\% \text{ open}$$

Note: This is a 1.5 inch valve with a C_v of 35.8 and should be used for this application.

Flowing Quantity (Turndown Ratio of a Valve)

The selection of a control valve is based on the required flowing quantity of the process. The control valve must be selected to operate under several different conditions. The maximum quantity that a valve should be required to pass is 10%–15% above the specified maximum of normal flow. (This compensates for process upsets.) The normal flow and maximum flow used in sizing calculations should be based on actual operating conditions, whenever possible, without any factors having been applied.

On many systems, a reduction in flow means an increase in pressure drop, and the C_v ratio may be much greater than would be suspected. If, for example, the maximum operating conditions for a valve are 200 gpm at 25 psi drop, and the minimum conditions are 25 gpm at 100 psi drop, the C_v ratio is 16:1, not 8:1 as it would first seem. The required change in valve C_v is the product of the ratio of the maximum to the minimum flow and the square root of the ratio of the maximum to the minimum pressure drop. For example:

$$200\text{ gpm} = \sqrt{25\text{ psi}}\ \text{Same energy head} = \frac{200\text{ gpm}}{\sqrt{25\text{ psi}}} = \frac{25\text{ gpm}}{\sqrt{100\text{ psi}}}\ \text{Rearranging}\ \frac{200\text{ gpm}\sqrt{100\text{ psi}}}{25\text{ gpm}\sqrt{25\text{ psi}}} = \frac{16}{1}$$

Flashing

If the downstream pressure is equal to or less than the vapor pressure, then vapor bubbles created at the vena contracta do not collapse, resulting in a liquid–gas mixture downstream of the valve. This is commonly called *flashing*. When flashing of a liquid occurs, the inlet fluid is 100% liquid, which experiences pressures at the inlet and downstream of the control valve that are at or below vapor pressure. The result is a two-phase mixture (vapor and liquid) at the valve outlet and in the downstream piping. Velocity of this two-phase flow is usually very high and can result in erosion of the valve and piping components.

Joule-Thomson Effect – Auto-Refrigeration in Valves

The throttling process is commonly exploited in thermal expansion processes such as refrigerators, air conditioners, heat pumps, and liquefiers. To minimize or prevent the gas from liquefying and the valve from freezing, often a two-stage pressure reduction scheme is used to minimize the Joule-Thomson (J-T) effect. By taking the pressure drop in two stages, the total cooling effect is split between the two pressure-reducing valves, each of which may be able to absorb enough heat from the atmosphere to prevent the gas from liquefying. Also, heat can be applied to the piping before the first and second stage reducing valves, raising the gas temperature enough to prevent the gas from liquefying.

Choked Flow

Choked flow occurs in gases and vapors when the fluid velocity reaches sonic values at any point in the valve body, trim, or pipe. As the pressure in the valve or pipe is lowered, the specific volume increases to the point where sonic velocity is reached. In liquids, vapor is formed as the result of cavitation or flashing and increases the specific volume of the fluid at a faster rate than the increase in flow due to the pressure differential. Lowering the downstream pressure beyond this point in either case will not increase the flow rate for a constant upstream pressure. The velocity at any point in the valve or downstream piping is limited to sonic (Mach = 1). As a result, the flow rate will be limited to an amount which yields a sonic velocity in the valve trim or the pipe under the specified pressure conditions. Under choked conditions, DP_{max} (the allowable sizing pressure drop), is the choked pressure drop. **Note:** Opening the valve further will reduce the pressure drop across the valve and should stop the chocking.

14 Pressure-Relief Valves and Rupture Disks

Pressure-Relief Valves and Pressure Safety Valves

Gases and steams can be compressed. When gas reaches the disk in a valve, it compresses and builds up pressure before escaping through the valve. This compression can cause system pressure to build up rapidly. A liquid-type relief valve does not open fast enough to relieve gas or steam pressure. A gas system requires a valve that can open completely under excess pressure, or a pressure safety valve.

Pressure safety valve (PSV) and *pressure-relief valve* (PRV) are commonly used terms to identify pressure-relief devices on a vessel. These terms are often used interchangeably; depending on the project or company standards, pressure-relief devices may be identified as *safety valves* or as *relief valves*, or sometimes even as *safety relief valves*.

Although used freely and interchangeably, these terms differ:

- *Pressure-relief valve* is the term used to describe a relief device on a liquid-filled vessel. For such a valve, the opening is proportional to the increase in the vessel pressure. Therefore, the opening of the valve is not sudden but gradual, if the pressure is increased gradually.

  Note: Do not throttle pressure-relief valves.

- *Pressure safety valve* is the term used to describe a relief device on a compressible fluid or gas-filled vessel. For this type of valve, the opening is sudden. When the set pressure of the valve is reached, the valve opens almost fully.

A PRV should not be used to control the pressure out of a pump. The PRV will chatter, and the hammering action caused by the pulsations will quickly destroy the valve. A proportional control valve should be used for this application. Contact the manufacturer for more information about this application.

- Deposition or formation of solids between the disc and the seat
- Altered process fluid
- Corrosion
- Mechanical malfunction

Rupture discs (see Figure 14-1) are safety devices for one-time use. They consist of a membrane that bursts when the differential pressure between its two sides exceeds a set value. These devices are used alone or in combination with a PRV, providing a physical isolation layer between the process and the relief valve, especially on processes containing highly corrosive fluid. Some models are equipped with a sensor that indicates when the diaphragm is broken.

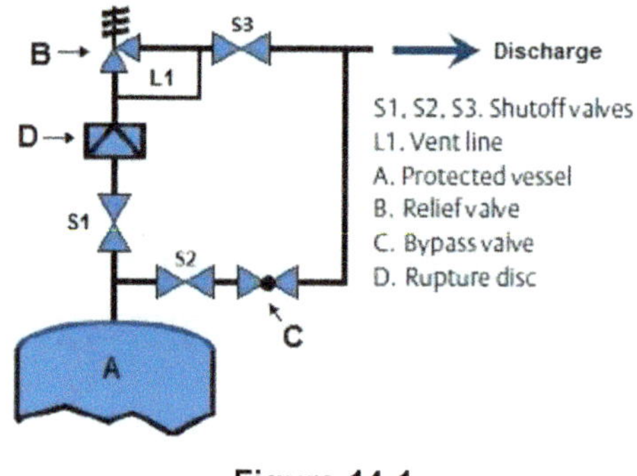

Figure 14-1

Rupture discs are simple devices, with no moving parts. Unlike PRVs or PSVs, the rupture disc will remain open until the ruptured diaphragm is replaced. Diaphragms are less likely to cause fugitive emissions, but there is always the possibility of pitting corrosion, which creates pinholes that lead to undetectable leakage.

Pilot-Operated Safety Valve

This type of safety valve uses the flowing medium itself, through a pilot valve, to apply the closing force on the safety valve disc. The pilot valve itself is a small safety valve.

There are two basic types of pilot-operated safety valves, diaphragm and piston. The diaphragm type is typically available only for low-pressure applications and produces a proportional type action, characteristic of relief valves used in liquid systems. This type of valve is therefore of little use in steam systems.

The piston-type valve consists of a main valve, which uses a piston-shaped closing device (or obturator), and an external pilot valve.

Bellow or Balanced Bellow and Diaphragm Valves

Some safety valves, most commonly those used for water applications, incorporate a flexible diaphragm or bellows to isolate the safety valve spring and upper chamber from dirty process fluid and contaminates.

Balanced bellows relief valves are typically used where the relief valves are piped to a closed flare system and the back pressure exceeds 10% of the set pressure which cannot be handled by conventional relief valves. Balanced bellows relief valves are also used in flow lines and multiphase lines, or for paraffinic or asphaltic crude which may plug the pilot-operated valves.

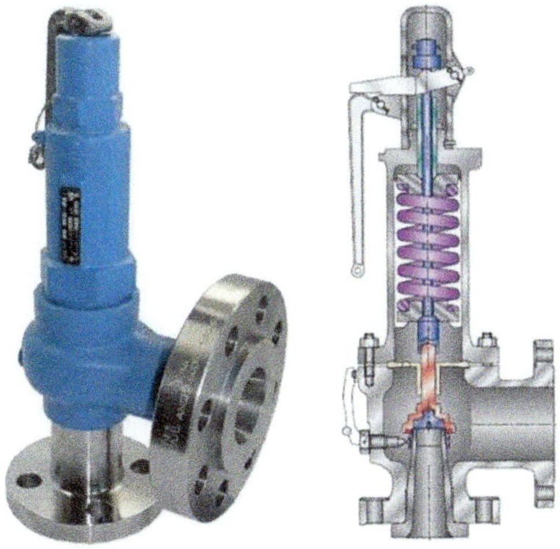

Standard relief valve guided stem.

When superimposed back pressure is variable, a balanced bellows or balanced piston design is recommended. The bellows or piston is designed with an effective pressure area equal to the seat area of the disc. The bonnet is vented to ensure that the pressure area of the bellows or piston will always be exposed to atmospheric pressure and to provide a telltale sign should the bellows or piston begin to leak. Variations in back pressure, therefore, will have no effect on set pressure. Back pressure may, however, affect flow.

The weight-loaded PRV is one of the simplest types of PRVs. It is a direct-acting valve. Because the dead weight pushes down on the valve's internal moving parts, the valve is held closed until the tank pressure equals the dead weight. These valves are often called *weighted pallet valves* because the set pressure can be varied by adding or removing weights on the top of a trim part called a *pallet*.

These weighted pallet valves are also known as *conservation vents* or *breather vents* because one of their primary uses is to protect low-pressure storage tanks that have fixed roofs. These storage tanks have very low design pressures in the inches of water column (mbar) range. Because the design pressures are so low, the simple pumping in of product or increased ambient temperatures can raise vapor pressures in the tank and cause these weight-loaded valves to "breathe" and discharge the pressure. The sizing and selection of these weight-loaded valves is often done per API Standard 2000 *Venting Atmospheric and Low-pressure Storage Tanks* or ISO 28300 *Petroleum, petrochemical and natural gas industries — Venting of atmospheric and low-pressure storage tanks.*

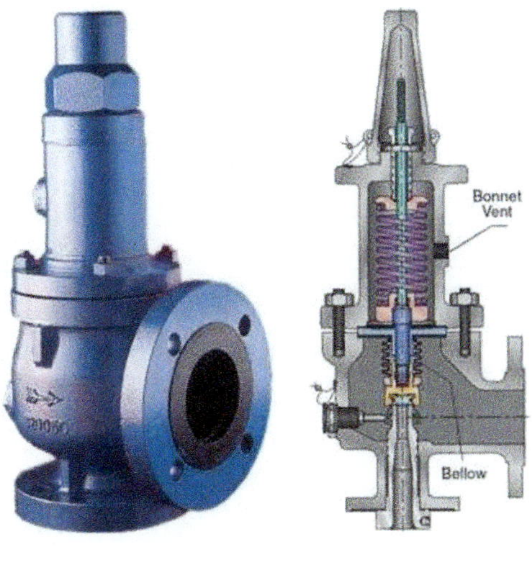

Balanced bellow relief.

Manway covers (PRVs), shown in the next figure, provide quick, easy access for tank inspection and maintenance. When the tank is exposed to abnormal internal pressure beyond the capability of the breather vent, the manway helps protect the tank against costly rupture. The most common cause of excessive internal pressure is fire exposure. Emergency venting capacity requirements depend on the wetted surface area of the tank. The venting requirements can be calculated using API Standard 2000.

Pressure-regulating devices for process and storage tanks.

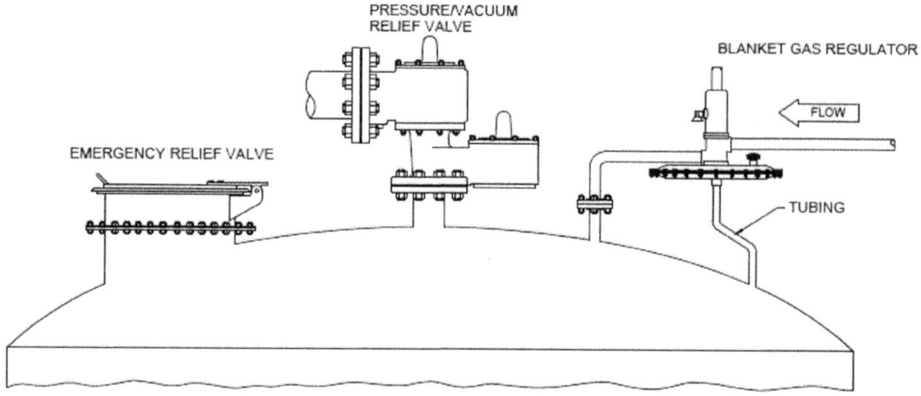

Side by side: pressure/vacuum vent.

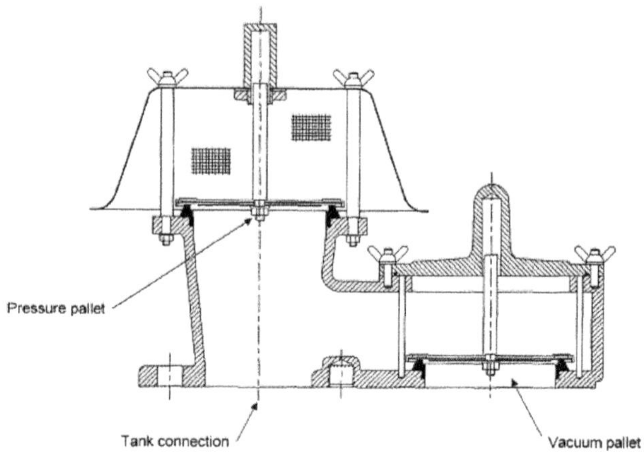

Large weight-loaded emergency vent.

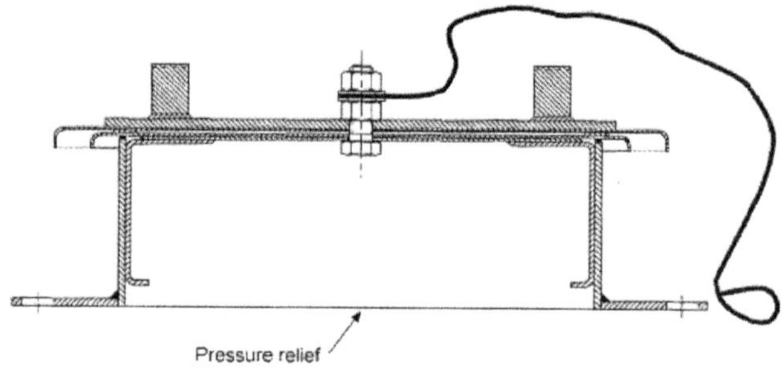

Blowdown is defined as the difference between opening and closing pressure. This pressure difference is expressed as pressure or as a percentage of the set pressure. Typical blowdowns are 0%–7%. A vent valve with 0% blowdown is known as a *throttling vent valve.* A throttling vent valve is similar to a direct-acting vent valve because it begins to open and close at almost the same pressure; however, unlike in a direct-acting vent valve, full lift of the seat in a throttling vent valve is obtained at or below 10% overpressure. Where tank operating pressures are very close to the maximum allowable tank pressure, this lift characteristic permits overpressure protection to be accomplished with smaller or fewer venting devices. One weighted PRV device may pass the same flow rate as two or more PRDs in parallel, such as a standard PRV and a rupture disk.

Venting Atmospheric and Low-Pressure Storage Tanks

Sizing and installing PRVs on low-pressure tanks is required by law per federal regulation 49 CFR 195.264 – API Standard 2000 *Venting Atmospheric and Low-Pressure Storage Tanks.* The following is from 49 CFR 195.264.

API Standard 2000 Table C-1—Operating Characteristics of Venting Devices		
Characteristic	Type of Venting Device	
	Direct-Acting	Pilot-Operated
Seat Tightness	Leakage rate increases with increasing pressure. Leakage may begin at 75% of set.	Leakage rate decreases with increasing pressure. Typically, no leakage above 30% of set. A small amount of leakage at pilot may begin at 90% of set.
Capacity/Overpressure (Refer to Figure C-5)	Rated capacity normally obtained at 200% of set, for pressure or vacuum.	Rated capacity obtained at 110% of set for pressure or vacuum.
Set Pressure Range (Typical)	Pressure—Weight Loaded 1/2 oz/in^2 to 16 oz/in^2 (0.865 in to 27.7 in wc) (2 mbarg to 69 mbarg)	Pressure—2 in wc to 15.0 psig (5 mbarg to 1.034 barg)
	Pressure—Spring Loaded 1.0 psig to 15.0 psig (69 mbarg to 1.034 barg)	Vacuum— –2 in wc to –14.7 psig (–5 mbarg to –1.013 barg)
	Vacuum—Weight Loaded 1/2 oz/in to –10 oz/in (–0.865 in to –17.3 in wc) (–2 mbarg to –43 mbarg)	
	Vacuum—Spring Loaded –10 oz/in^2 to –7 psig (–43 mbarg to 0.48 barg)	
Typical Blowdown	0%	0% to 7%

The weighted relief valve is typically available in either 0.5 psi incremental weights or 1.0 psi incremental weights. To adjust the valve, add or remove the weight plates. Below is a typical Republic blower PRV.

Weighted Relief Valve Installation and Operation

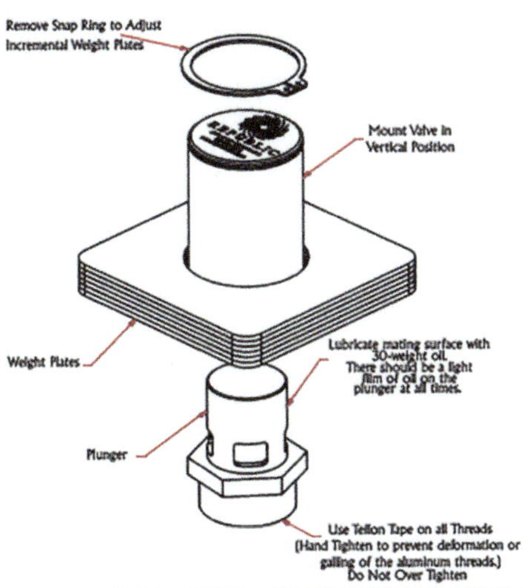

A	1in NPT	1.5 in NPT	2in NPT	3 in NPT	4in NPT
B	2.25	2.25	2.25	3.25	3.25
C	7.75	7.75	7.75	8.75	9.25
D*	8.07	6.00	8.50	12.75	16.50
E**	8.80	8.80	9.25	10.00	11.00
Part No.	210-3085	210-3086	210-3068	210-3069	210-3070

*0.5 psig Increment weights diagonal dimension

**Dimension when valve is fully open

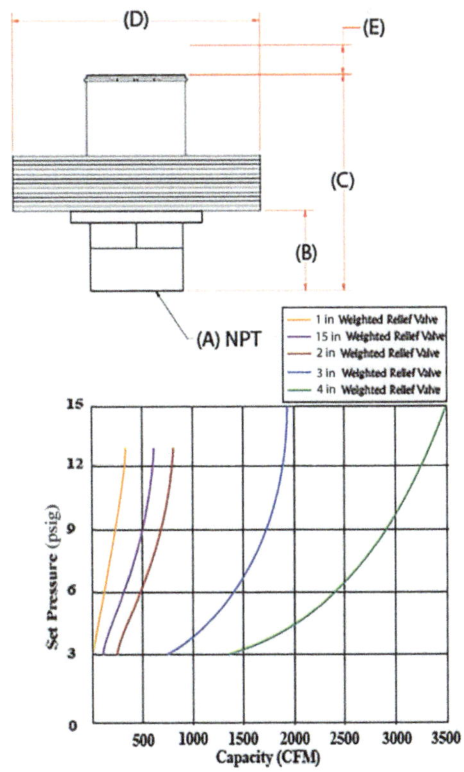

API Standards for Pressure-Relieving Systems

Here is a list of industry standards used in pressure relief sizing calculations and applicable codes.

Standard	Title	Description
API Standard 520 Part I	*Sizing, Selection, and Installation of Pressure-Relieving Devices in Refineries—Part I, Sizing and Selection*	Guide for sizing and selection of PRDs used in petroleum-related industries for equipment with maximum allowable working pressure of 15 psig or greater. The document is intended for protection of unfired pressure vessels and equipment against overpressure from operation or fire. PRVs or rupture disks may be used independently or in combination to provide the required protection against excessive pressure accumulation.
API Standard 520 Part II	*Sizing, Selection, and Installation of Pressure-Relieving Devices in Refineries—Part II, Installation*	Guide for installation of PRDs used in petroleum-related industries for equipment with a maximum allowable working pressure of 15 psig or greater.
API Standard 521	*Guide for Pressure-Relieving and Depressuring Systems Petroleum Petrochemical and Natural Gas Industries, Pressure-Relieving and Depressuring Systems*	Specifies requirements and gives guidelines for determining overpressure causes, relieving rates for pressure-relieving and vapor depressurizing systems in petroleum-related industries. The standard also discusses selection and design of disposal systems, including such component parts as piping, vessels, flares, and vent stacks.
API Standard 526	*Flanged Steel Pressure Relief Valves*	The standard covers specifications for flanged steel PRVs by presenting basic requirements such as orifice designation and area, valve size, pressure rating, and materials for direct spring-loaded PRVs and pilot-operated PRVs.
API Standard 527	*Seat Tightness of Pressure Relief Valves*	Describes methods of determining the seat tightness of metal and soft-seated PRVs, including those of conventional, bellows, and pilot-operated designs.
API Standard 576	*Inspection of Pressure-Relieving Devices*	Describes the inspection and repair practices for automatic PRVs commonly used in the oil and petrochemical industries. This API standard covers PRVs, pilot-operated pressure relief valves, rupture disks, and weight-loaded pressure-vacuum vents, with regard to inspection and repair.
API Standard 2000	*Venting Atmospheric and Low-Pressure Storage Tanks Petroleum, Petrochemical, and Natural Gas Industries—Venting of Atmospheric and Low-Pressure Storage Tanks*	Guide for normal and emergency vapor venting requirements for aboveground petroleum and petroleum product storage tanks, and aboveground and underground refrigerated storage tanks. The tanks discussed in the document are designed for low pressures ranging from full vacuum through 15 psig. This API standard discusses causes of overpressure and vacuum, venting requirements, means of venting, and breathing selection, as well as installation of venting devices and testing and marking relief devices for storage tanks.

CFR Standards for Pressure Relief Required by Federal Law

Standard	Title	Description
API Standard 2003	*Protection Against Ignitions Arising Out of Static, Lightning, and Stray Currents*	Describes the prevention of ignition of hydrocarbons by the release of energy charge generated by static electricity and stray currents, which can cause extreme pressures on tanks.
API Standard 2350	*Overfill Protection for Storage Tanks in Petroleum Facilities*	This standard is referenced by API 2000 and NFPA 30 and applies to storage tanks associated with marketing, refining, pipeline, and terminals operations.
API Standard 2510	*Design and Construction of LPG Installations*	This standard covers the design, construction, and location of liquefied petroleum gas (LPG) installations at marine and pipeline terminals, natural gas processing plants, refineries, petrochemical plants, and tank farms. This standard covers storage vessels, loading and unloading systems, piping, and related equipment.
NFPA 30	*Flammable and Combustible Liquids Code*	Enforceable under the Occupational Safety and Health Administration (OSHA) and many state and local regulations, NFPA 30 reduces the hazards associated with the storage, handling, and use of flammable and combustible liquids. Topics covered are explosion prevention and risk control, storage of liquids in containers, storage of liquids in tanks, piping systems, processing facilities, bulk loading-unloading, and wharves.
ASME VIII	*Boiler and Pressure Vessel Code Section VIII—Unfired Pressure Vessels*	Division 1: Provides requirements applicable to the design, fabrication, inspection, testing, and certification of pressure vessels operating at either internal or external pressures exceeding 15 psig. Division 2: Describes alternative rules and provides requirements for the design, fabrication, inspection, testing, and certification of pressure vessels operating at either internal or external pressures exceeding 15 psig. Division 3: Describes alternative rules for construction of high-pressure vessels and provides requirements for the design, fabrication, inspection, testing, and certification of pressure vessels operating at either internal or external pressures above 10,000 psi.

ASME VIII Code for Sizing Relief Valves and Rupture Disks (ASME boiler and pressure vessel code).

Note: Parts of the ASME VIII code may be on the CSE exam.

ASME VIII is required by federal law and the US Department of Homeland Security, not just by state and city code.

Introduction to ASME VIII

ASME code in installing PRVs on unfired pressure vessels is required by law per federal regulation 49 CFR 173.32(c)(4). The user, not the designer or manufacturer, is responsible for overpressure protection of the vessel. Therefore, the user must explicitly state the type of overpressure protection that will be provided once the vessel is installed. Finally, the user must state whether jurisdictional acceptance of the vessel is required prior to operation of the vessel per the authority having jurisdiction.

Overview Section VIII – Pressure Vessels

Division 1 provides requirements applicable to the design, fabrication, inspection, testing, and certification of pressure vessels operating at either internal or external pressures exceeding 15 psig. Such vessels may be fired or unfired. This pressure may be obtained from an external source or by the application of heat from a direct or indirect source, or any combination thereof. Specific requirements apply to several classes of material used in pressure vessel construction, and also to fabrication methods such as welding, forging and brazing.

Division 1 contains mandatory and nonmandatory appendices detailing supplementary design criteria, nondestructive examination, and inspection acceptance standards. Rules pertaining to the use of the single ASME certification mark with the U, UM, and UV designators are also included.

Division 2 requirements on materials, design, and nondestructive examination are more rigorous than in Division 1; however, higher design stress intensify values are permitted. These rules may also apply to human occupancy pressure vessels typically in the diving industry. Rules pertaining to the use of the single ASME certification mark with the U2 and UV designators are also included.

Division 3 requirements are applicable to pressure vessels operating at either internal or external pressures generally above 10,000 psi. Division 3 does not establish maximum pressure limits for Section VIII, Divisions 1 or 2, nor minimum pressure limits for this division. Rules pertaining to the use of the single ASME certification mark with the U3 and UV3 designator are also included.

Excerpts from the ASME VIII Code – Pressure-Relief Requirements

"UG-125(a) — All pressure vessels within the scope of this division, irrespective of size or pressure, shall be provided with pressure relief devices in accordance with the requirements of UG-125 through UG-137."

"(1) It is the responsibility of the user to ensure that the required pressure relief devices are properly installed prior to initial operation."

Excerpts from the ASME VIII Unfired Pressure Vessel Code

Some or all of this information may on a CSE exam question or used in a calculation for the exam.

"UG-125(c) – All pressure vessels other than unfired steam boilers shall be protected by a pressure relief device that shall prevent the pressure from rising more than 10% or 3 psi (20 kPa), whichever is greater, above the maximum allowable working pressure except as permitted in (1) and (2) below.

(See UG-134 for pressure settings.)"

"(1) When multiple pressure relief devices are provided and set in accordance with UG-134(a), they shall prevent the pressure from rising more than 16% or 4 psi (30 kPa), whichever is greater, above the maximum allowable working pressure."

"(2) When a pressure vessel can be exposed to fire or other unexpected sources of external heat, the pressure relief device(s) shall be capable of preventing the pressure from rising more than 21% above the maximum allowable working pressure. Supplemental pressure relief devices shall be installed to protect against this source of excessive pressure if the pressure relief devices used to satisfy the capacity requirements of UG-125(c) and UG- 125(c)(1) have insufficient capacity to provide the required protection."

"UG-125(d) – Where an additional hazard can be created by exposure of a pressure vessel to fire or other unexpected sources of external heat (for example, vessels used to store liquefied flammable gases), supplemental pressure-relieving devices shall be installed to protect against excessive pressure. Such supplemental pressure-relieving devices shall be capable of preventing the pressure from rising more than 20% above the maximum allowable working pressure of the vessel. A single pressure-relieving device may be used to satisfy the requirements of this paragraph and (c), provided it meets the requirements of both paragraphs."

"UG-133(f) — The set pressure tolerances, plus or minus, of safety or relief valves, shall not exceed 2 psi (13.8 kPA) for pressures up to and including 70 psig (483 kPa), and 3% for pressures above 70 psig (483 kPa). "

ASME VIII – Pressure Limits in Sizing

The ASME Code requires that when a rupture disk or PRV is used as the primary relief device, it must be sized to prevent the pressure from rising above 110% of the MAWP (UG-125(c)). MAWP stands for maximum allowable working pressure. If used as a secondary relief device or as multiple relief devices, the size must prevent the pressure from rising above 116% of the MAWP (UG-125(c)(1)). If used as a supplementary relief device for hazards external to the protected vessel or system, the size must prevent the pressure from rising above 121% of the MAWP (UG-125(c)(2)).

Table 14-1. ASME standard nozzle orifice data.

Relief Valve Nozzle Orifice Areas	
Size Designation	**Orifice Area, in^2**
D	0.110
E	0.196
F	0.307
G	0.503
H	0.785
J	1.280
K	1.840
L	2.850
M	3.600
N	4.340
P	6.380
Q	11.050
R	16.000
T	26.000

ISA Pressure-Relief Valve and Rupture Disc Symbols

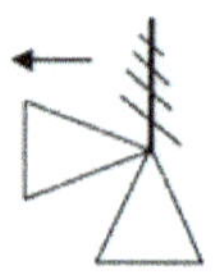

Pressure-Relief Valve

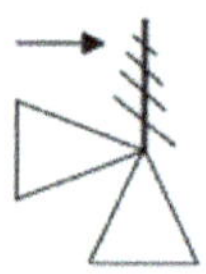

Vacuum-Relief Valve

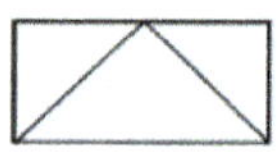

Pressure Rupture Disk

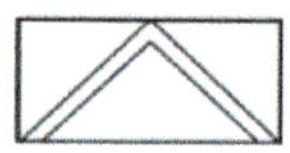

Vacuum Rupture Disc

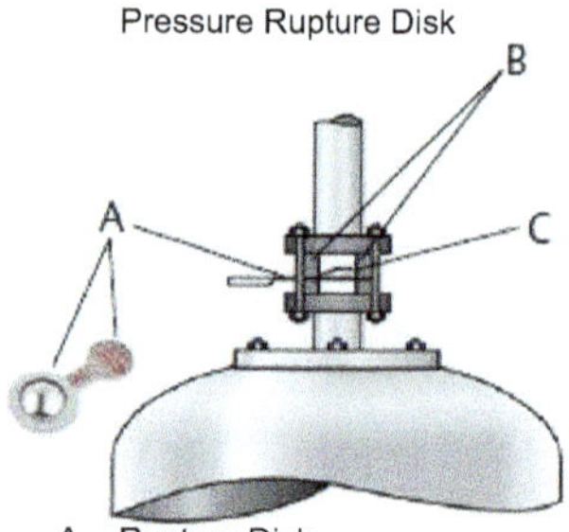

A. Rupture Disk
B. Disk Holder
C. Sensor

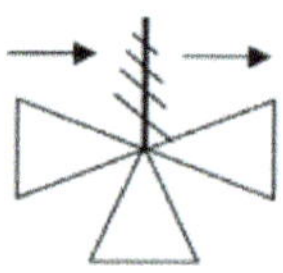

Breathing Valve or Pressure/Vacuum Relief Valve

Typical Rupture Disk and Holder

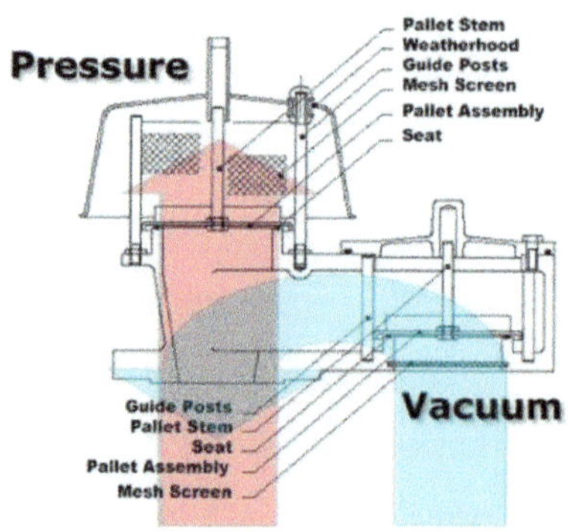

Operation of Breather/Vacuum Relief Valve

Sizing Equations for Relief Valves and Rupture Disks

ASME VIII Code Equations – USCS Units (United States Customary System Units)

The user should understand the symbols used in the sizing and capacity calculation formulas.

The basic equations for flow through a PRV or rupture disk are:

Vapor or Gases	**Vapor or Gases**
Mass Flow Rate Sizing (W = lb/h)	Volumetric Flow Rate Sizing (Q = scfm flow rate at 14.7 psia and 60°F)
$A = \dfrac{W\sqrt{T}\sqrt{Z}}{CKP_1K_b\sqrt{M_w}}$	$A = \dfrac{60Q\rho\sqrt{T}\sqrt{Z}}{CKP_1K_b\sqrt{M_w}}$
Steam	**Air**
Mass Flow Rate Sizing (W = lb/h)	Volumetric Flow Rate Sizing (Q = scfm flow rate at 14.7 psia and 60°F)
$A = \dfrac{W}{51.5KP_1K_b}$	$A = \dfrac{60Q(0.0763)\sqrt{T}\sqrt{Z}}{356KP_1K_b(5.3824)}$
Liquids	**Critical Pressure Ratio (r_c)**
Certified Volumetric Flow Rate Sizing (If Q = US gpm, K_u = 38) (If Q = cfh, K_u = 5.2143)	$r_c = \left[\dfrac{2}{k+1}\right]^{\frac{k}{k-1}}$
$A = \dfrac{Q\sqrt{G_f}}{K_uKK_v\sqrt{P_1 - P_2}}$	
Gas Constant (C) Sonic Flow (Typically 15 psig and above)	**Gas Constant (C) Subsonic Flow (Low-Pressure Flow)**
If (P_2/P_1) is **less** than r_c the flow will be sonic. Use this formula:	If (P_2/P_1) is **greater** than r_c, the flow will be subsonic. Use this formula:
$C = 520\sqrt{k\left(\dfrac{2}{k+1}\right)^{\frac{k+1}{k-1}}}$	$C = 735\sqrt{\dfrac{k}{k+1}\left[\left(\dfrac{P_2}{P_1}\right)^{\frac{2}{k}} - \left(\dfrac{P_2}{P_1}\right)^{\frac{k+1}{k}}\right]}$

A Note about Sonic or Choked Flow

The critical flow, also referred to as *sonic flow, choked flow,* or *Mach 1,* is the limitation point of a compressible fluid flow through an orifice. The critical flow can occur on a relief valve orifice or a choke. A choke is a condition where piping goes from a small branch into a larger header, where pipe size increases, or at the vent tip. The maximum flow occurs at sonic velocity, which exists as long as the pressure drop through the orifice is greater than the critical pressure drop.

The maximum choked flow can be less than the desired flow. Calculations are necessary to determine if the flow is choked below the desired or required flow rate that is needed for control of the process.

Variables for PRV and PSV Sizing Equations

A = actual nozzle area of valve in square inches.

C = gas constant (C = 315 if the ratio of specific heats is unknown).

G_f = specific gravity (SG) of flowing fluids (liquid/water) or (gas/air).

k = specific heats ratio.

K = coefficient of discharge (K_d • 0.9), (0.8775 for vapor, gas, or steam, and 0.67 for liquid).

K_b = back pressure correction factor, dimensionless (see Table 14-2, "Calculate K_b").

K_c = combination factor for installations with a rupture disc upstream of the valve. Use a 0.9 value for any rupture disc/PRV combination. Use a 1.0 value when a rupture disc is not installed.

K_d = dimensionless value relating the actual versus theoretical safety relief valve flow rate (0.975 for vapor, gas, or steam, and 0.744 for liquid).

K_p = overpressure correction for liquid (0.60 at 110%).

K_u = dimensionless factor used to adjust for the type of units used in the sizing equation (see liquid equation for value of K_u for gpm or cfh applications).

K_w = variable or constant back pressure factor for bellows-sealed valves only.

K_v = viscosity correction factor (use K_v = 1 except for very high viscous fluids).

M_r = relative mass (molecular weight).

P_1 = relieving pressure (psia). This is the set pressure (psig) + overpressure (psig) + atmospheric pressure (14.7 psia) – inlet pressure piping loss (psig).

P_2 = the pressure at the outlet of the valve in absolute pressure units (psia).

ρ = density of gas or vapor:

for vapors = (SG) • (density of air)

for liquids = (SG) • (density of water)

density of air = 0.0763 lb/ft^3 at 14.7 psia, and 60°F (USCS)

density of water = 62.305 lb/ft^3 at 70°F (USCS)

Q = capacity in volume per time units.

T = relieving temperature, absolute °R (°F + 460).

W = required relieving rate, mass flow.

Z = compressibility factor ($Z = 1$ for ideal gases).

Table 14-2. Calculate K_b.

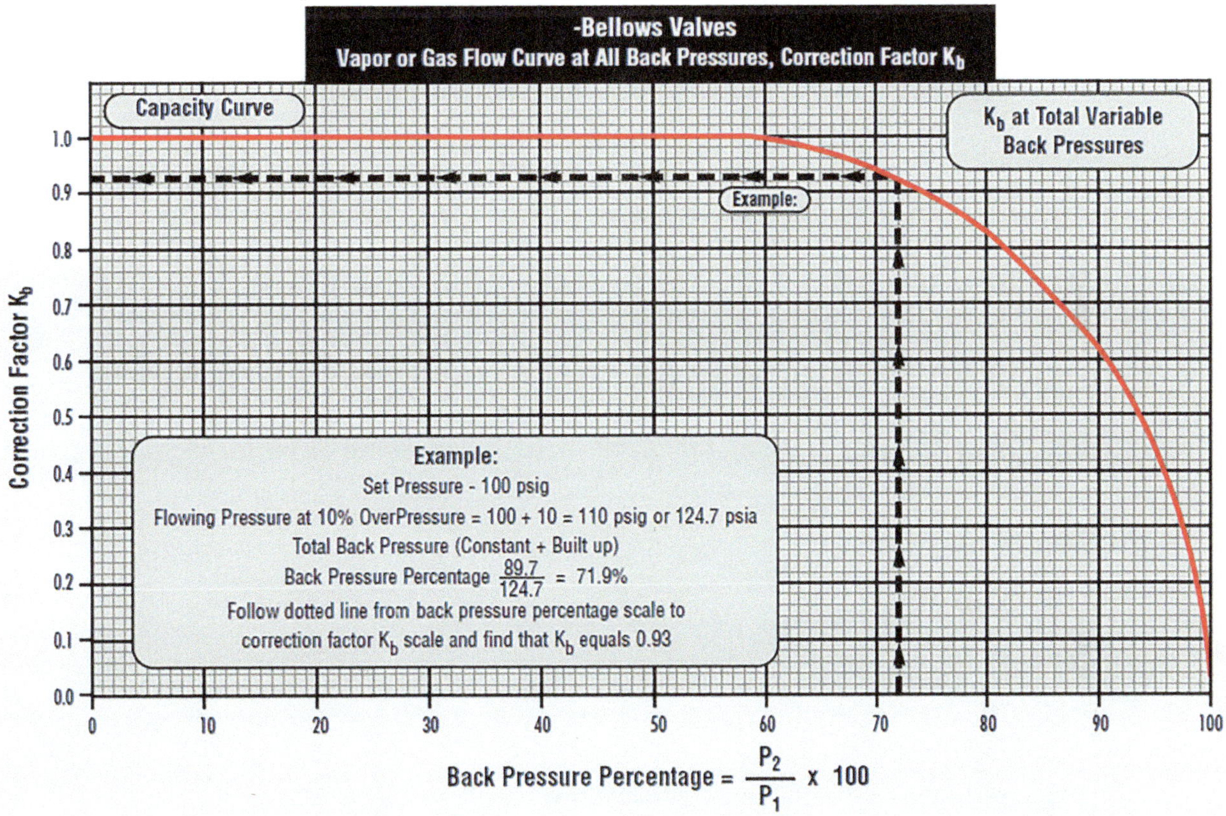

Sizing Rupture Disks – Worked Examples

The function of a rupture disk is to protect pressure vessels, piping systems, and other equipment from pressures exceeding their design pressure by more than a fixed, predetermined amount. The permissible amount of overpressure is covered by various codes and is a function of the type of equipment and the conditions causing the overpressure.

The aim of safety systems in processing plants is to prevent damage to equipment, avoid injury to personnel, and eliminate any risks of compromising the welfare of the community at large and the environment. Proper sizing, selection, manufacture, assembly, test, installation, and maintenance of a PRV are critical to obtaining maximum protection.

Note: Where rupture disks are installed upstream of a PRV, the rupture disc is usually the same pipe size as the PRV inlet nozzle.

Rupture Disk Sizing Example 1

Sample Problem: Size a rupture disk for liquid service using the following criteria. Application: primary relief.

$Q = 1500$ gpm (required) Vessel MAWP = 45 psig $Gf = 0.85$

P_2 (back pressure) = 5 psig

Use 10% overpressure as permitted by ASME code.

$$P_1 = (1.10)\text{MAWP} + 14.7$$

where

$K = 0.67$

$K_v = 1$ (except for very viscous applications)

$K_u = 38$ for gpm application

$P_1 = (1.1)(45) + 14.7 = 64.2$

$P_2 = 5 + 14.7 = 19.7$

$$A = \frac{Q\sqrt{G_f}}{K_u K K_v \sqrt{P_1 - P_2}} = \frac{1500\sqrt{0.85}}{(38)(0.67)(1)\sqrt{64.2 - 19.7}} = 8.14 \text{ in}^2$$

Use the manufacturer's catalog to find the actual disk size to order for your application.

Rupture Disk Sizing Example 2

Sample Problem: Size a rupture disk for the following service, GAS (type gas air) using the following criteria. Application: primary relief.

$Q = 5000$ scfm (required) Vessel MAWP = 150 psig $M_r = 28.9$ $k = 1.40$

P_2 (back pressure) = 20 psig Flow temperature = 250°F $Z = 1$

Use 10% overpressure as permitted by ASME code.

$$P_1 = (1.1)(\text{MAWP}) + 14.7 = 179.7$$

$$P_2 = 20 + 14.7 = 34.7$$

Flow pressure ratio:

$$\frac{P_2}{P_1} = \frac{20 + 14.7}{(1.1)(150) + 14.7} = 0.193$$

Critical pressure ratio:

$$r_c = \left[\frac{2}{k+1}\right]^{\frac{k}{k-1}} = \left[\frac{2}{1.40+1}\right]^{\frac{1.40}{1.40-1}} = 0.528$$

If (P_2/P_1) is less than r_c, use this formula:

$$C = 520\sqrt{k\left[\frac{2}{k+1}\right]^{\frac{k+1}{k-1}}} = 520\sqrt{1.40\left[\frac{2}{1.40+1}\right]^{\frac{1.40+1}{1.40-1}}} = 356$$

From Table 14-2, "Calculate K_b," we find that the value of K_b is 1.

$$K = 0.8775$$

Given the required flow in actual cubic feet per minute:

$$A = \frac{60Q(0.0763)\sqrt{T}\sqrt{Z}}{356KP_1(5.3824)K_b} = \frac{(60)(5000))(0.0763)\sqrt{(250+460)}\sqrt{1}}{(356)(0.8775)(179.7)(5.3824)(1)} = 2.02 \text{ in}^2$$

Rupture Disk Sizing Example 3

Sample Problem: Size a rupture disk for the following service, GAS (some process) using the following criteria. Application: primary relief.

$Q = 2000$ scfm (required)	Vessel MAWP = 15 psig	$G_r = 0.72$	$k = 1.26$
P_2 (back pressure) = 5 psig	Flow temperature = –40°F	$M_r = 20.808$	$Z = 0.95$

In this case, 10% of gauge pressure is less than 3 psi; therefore, 3 psi overpressure is permitted by ASME code.

$$P_1 = 3 + \text{MAWP} + 14.7 = 32.7$$

Flow pressure ratio:

$$\frac{P_2}{P_1} = \frac{5+14.7}{3+15+14.7} = 0.602$$

Critical pressure ratio:

$$r_c = \left[\frac{2}{k+1}\right]^{\frac{k}{k-1}} = \left[\frac{2}{1.26+1}\right]^{\frac{1.26}{1.26-1}} = 0.553$$

If P_2/P_1 is greater than r_c, use the low-pressure subsonic formula:

$$C = 735\sqrt{\frac{k}{k+1}\left[\left(\frac{P_2}{P_1}\right)^{\frac{2}{k}} - \left(\frac{P_2}{P_1}\right)^{\frac{k+1}{k}}\right]} = 735\sqrt{\frac{1.26}{1.26+1}\left[(0.602)^{\frac{2}{1.26}} - (0.602)^{\frac{1.26+1}{1.26}}\right]} = 115.68$$

Find the constants for the equation.

Use 10% overpressure as permitted by ASME code.

$$P_1 = \text{set pressure} + (0.10)\text{ set pressure} + 14.7$$

$$P_1 = (1.1)(325) + 14.7 = 372.2 \text{ (psia)}$$

$$P_2 = 15 + 14.7 = 29.7 \text{ (psia)}$$

Flow pressure ratio:

$$\frac{P_2}{P_1} = \frac{15+14.7}{(1.1)(325)+14.7} = 0.0798$$

Critical pressure ratio:
Note: The value of k can be found in Table A20.

$$r_c = \left[\frac{2}{k+1}\right]^{\frac{k}{k-1}} = \left[\frac{2}{1.30+1}\right]^{\frac{1.30}{1.30-1}} = 0.546$$

If P_2/P_1 is less than r_c, use this formula:

$$C = 520\sqrt{k\left[\frac{2}{k+1}\right]^{\frac{k+1}{k-1}}} = 520\sqrt{1.30\left[\frac{2}{1.30+1}\right]^{\frac{1.30+1}{1.30-1}}} = 346.98$$

From Table 14-2, for atmospheric pressure, K_b is 1.

Use formula:

Vapor or Gases
Mass Flow Rate Sizing

$$(W = \text{lb/h})$$

$$A = \frac{W\sqrt{T}\sqrt{Z}}{CKP_1K_b\sqrt{M_r}} = \frac{(15{,}000)\sqrt{(138+460)}\sqrt{1}}{(346.98)(0.8775)(372.2)(1)\sqrt{(17)}} = 0.785 \text{ in}^2$$

Use Table 14-1 to find the orifice size for the relief valve.

$$H = 0.785 \text{ in}^2$$

We will select an orifice size of H.

Air SCFH Application Example 3

Sample Problem: Size a PRV for air service using the following criteria. Application: primary relief.

$Q = 6{,}000$ scfh	Set pressure = 100 psig	$M_r = 28.97$	$k = 1.40$
P_2 (back pressure) = 15 psig	Relieving temperature = 138°F		$Z = 1$

Use 10% overpressure as permitted by ASME code. P_1 = set pressure + (0.10) set pressure + 14.7

$$P_1 = (1.1)(100) + 14.7 = 124.7 \text{ psia}$$
$$P_2 = 15 + 14.7 = 29.7 \text{ psia}$$

Flow pressure ratio:

$$\frac{P_2}{P_1} = \frac{15 + 14.7}{(1.1)(100) + 14.7} = 0.238$$

Critical pressure ratio:
Note: The value of k can be found in Table A20

$$r_c = \left[\frac{2}{k+1}\right]^{\frac{k}{k-1}} = \left[\frac{2}{1.40+1}\right]^{\frac{1.40}{1.40-1}} = 0.528$$

If P_2/P_1 is less than r_c, use this formula

$$C = 520\sqrt{k\left[\frac{2}{k+1}\right]^{\frac{k+1}{k-1}}} = 520\sqrt{1.40\left[\frac{2}{1.40+1}\right]^{\frac{1.40+1}{1.40-1}}} = 356$$

$C = 356$

We will use the air formula instead of the vapor/gas formula.
From Table 14-2, for atmospheric pressure, $K_b = 1$.

Use formula:

Air
Volumetric Flow Rate Sizing

(Q = scfm flow rate at 14.7 psia and 60°F)

$$A = \frac{60Q(0.0763)\sqrt{T}\sqrt{Z}}{356KP_1K_b(5.3824)} = \frac{(60)(6{,}000)(0.0763)\sqrt{(138+460)}\sqrt{1}}{(356)(0.8775)(124.7)(1)(5.3824)} = 3.2 \text{ in}^2$$

Use Table 14-1 to find the orifice size for the relief valve.

$$L = 2.850 \text{ in}^2$$
$$M = 3.600 \text{ in}^2$$

So we will select an orifice size of M.

Saturated Steam Application Example 4

Sample Problem: Size a PRV for saturated steam service using the following criteria. Application: primary relief.

$Q = 40{,}000$ lb/h Set pressure = 140 psig $M_r = 18$

P_2 (back pressure) = 14.7 psia

Use 10% overpressure as permitted by ASME code. P_1 = set pressure + (0.10) set pressure + 14.7

$$P_1 = (1.1)\,(140) + 14.7 = 168.7 \text{ (psia).}$$
$$P_2 = 14.7 \text{ (psia)}$$

Flow pressure ratio:

$$\frac{P_2}{P_1} = \frac{14.7}{(1.1)(140) + 14.7} = 0.087$$

From Table 14-2, for atmospheric pressure, K_b is 1.

Use formula:

STEAM

Mass Flow Rate Sizing

(W = lb/h)

$$A = \frac{W}{51.5KP_1K_b} = \frac{(40{,}000)}{(51.5)(.8775)(168.7)(1)} = 5.247 \text{ in}^2$$

Use Table 14-1 to find the orifice size for the relief valve.

$$N = 4.340 \text{ in}^2$$
$$P = 6.380 \text{ in}^2$$

So we will select an orifice size of P.

Notes

15 Review of Feedback Control Fundamentals

Compare Open-Loop Control to Closed-Loop Control

Open-Loop Example – A Mathematical Analysis

Most industries today use closed-loop control. It offers a faster and tighter response. That is, it can maintain the desired set point of a process almost exactly. Its output is almost perfect (exactly what is desired). Let us examine an everyday application, speed control of an automobile. Look at Figure 15-1. There are a desired speed (R); a controller, mechanical accelerator pedal mechanism or microprocessor controller and electronics, that provides a signal to the engine and transmission (u); a disturbance, the slope of the road (w); and a desired output, the actual speed of the automobile (Y).

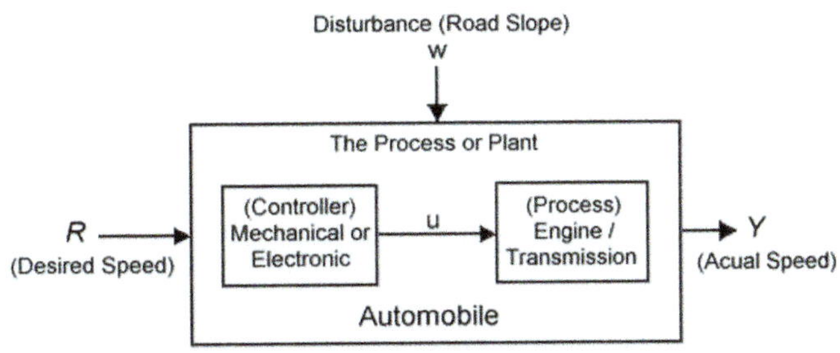

Automobile Speed Control

Figure 15-1

First let us examine open-loop control and its drawbacks. Open-loop control is cheap and can work in a circumstance where the output can vary, that is, the output can be in a range of speeds and does not have to be exact for the conditions of the process. This may not always be desirable. Look at Figure 15-2. Here we have variable R, desired speed, and variable Y_{ol}, output speed of the open-loop.

The automobile uses a mechanical linkage with an accelerator pedal to send a signal to the engine and transmission, which will control the speed of the automobile. The mechanical linkage combined with the accelerator pedal has a gain of 1/10. The accelerator pedal and mechanical linkage gain of 1/10 adds to the automobile's output speed. The road has a slope. This slope subtracts from the automobile's response of desired set-point speed (R), with a gain of 0.5. When the slope of the road is zero (for a level surface), the disturbance does not affect the output speed.

When the slope is greater than zero, for example, 1% or 10% grade, the automobile's actual speed is less than the desired speed. This can be seen when driving down a road and holding the accelerator pedal at a constant position. You will slow down going up a hill or slope (the rise versus the run or Y/X).

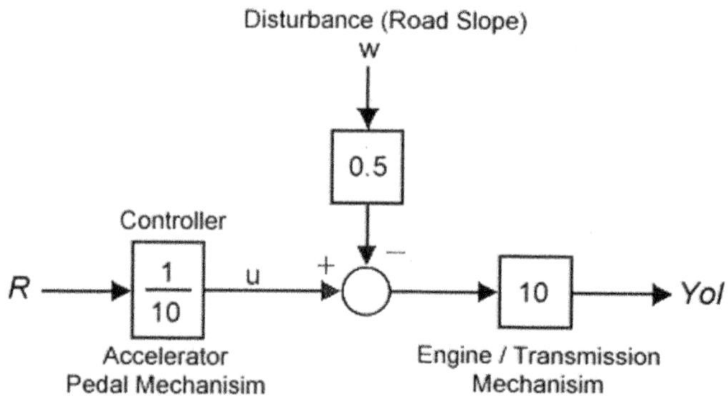

Open Loop Automotive Speed Control

Figure 15-2

where

R = desired or reference speed in miles per hour

u = throttle angle in degrees (sets engine speed)

Yol = actual open-loop speed of the automobile in miles per hour

w = road grade in percent

The set point (desired speed) is multiplied by the gain of the controller (1/10). The output of the controller is called the *manipulated variable* (u). The system disturbance, which models the effect of slope in the road, is multiplied by a gain of 0.5 and is subtracted from the manipulated variable. The manipulated variable, which is the throttle angle of the carburetor, sets the engine speed. The process final correction control device or element is the engine and transmission, which has a gain of 10. The manipulated variable, minus the system disturbance multiplied by a gain of 0.5, is then multiplied by the final control device or element gain of 10 to set the value of the final output, which is the actual speed of the process or plant (Y_{ol}). In this case, the process or plant is the automobile.

Let us look at the math to prove what is happening in the system.

The open-loop output speed is given by:

$$u = R\left(\frac{1}{10}\right)$$

$$Y_{ol} = (u - 0.5w)10$$

$$Y_{ol} = \left(\frac{R}{10} - 0.5w\right)10$$

$$Y_{ol} = R - 5w$$

So for a slope of 0%, if the set point is 55 mph, the output of the process is the actual automotive speed of 55 mph. This is only true if there is no disturbance.

$$55 \text{ mph} = 55 - 5(0); \quad \text{(slope of 0\%)}$$

If the slope is 1%, the output is 50 mph:

$$50 \text{ mph} = 55 - 5(1); \quad \text{(a slope of 1\%)}$$

If the slope is 10%, the output is 5 mph:

$$5 \text{ mph} = 55 - 5(10); \quad \text{(slope of 10\%)}$$

Closed-Loop Example – A Mathematical Analysis

For a large disturbance, open-loop control is not desirable. Let us look at the automobile with closed-loop control used, the speed control setting. Refer to Figure 15-3.

The controller uses a microprocessor combined with electronics to set the throttle angle setting of the engine's carburetor. This will set the speed of the engine to maintain the output of the process or plant, the actual speed of the automobile. The desired speed is reached and the speed control button is pushed. This is called the *set point* (R), the desired speed of the automobile. The closed-loop controller has a gain of 100. We will now illustrate the tight control of the final output of the process (Y_{cl}).

The set point or desired speed variable (R) is entered. Then the feedback signal, the process variable (Y_{cl}), is subtracted from the set-point variable. This is called the *error* or *set-point error signal* (e). The set-point error is multiplied by the controller gain of 100. This output is called the *manipulated variable* (u). The manipulated variable, which is the throttle angle of the carburetor, sets the engine speed. The process final correction control device or element is the engine and transmission, which has a gain of 10. The manipulated variable, minus the system disturbance multiplied by a gain of 0.5, is then multiplied by the final control device or element gain of 10, to determine the value of the final output, which is the actual speed of the process or plant (Y_{cl}). In this case, the process or plant is the automobile.

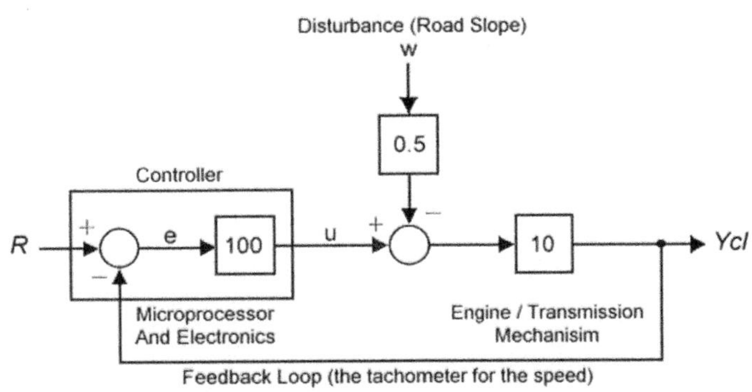

Closed Loop Automotive Speed Control

Figure 15-3

where

R = desired or reference speed in miles per hour

e = set-point error

u = throttle angle in degrees (sets engine speed)

Y_{cl} = actual closed-loop speed of the automobile in miles per hour

w = road grade in percent

Let us look at the math to prove what is happening to the system.

The closed-loop output speed is given by:

$$e = R - Y_{cl}$$

$$u = (e)100$$

$$u = (R - Y_{cl})100$$

$$Y_{cl} = (u - 0.5\,w)10$$

$$Y_{cl} = \left([100\,R - 100\,Y_{cl}] - 0.5w\right)10$$

$$Y_{cl} = 1000\,R - 1000\,Y_{cl} - 5w$$

$$1000\,Y_{cl} + Y_{cl} = 1000\,R - 5w$$

$$1001\,Y_{cl} = 1000\,R - 5w$$

$$Y_{cl} = \frac{1000\,R - 5w}{1001}$$

$$Y_{cl} = 0.999\,R - 0.005\,w$$

So, for a slope of 0%, if the set point is 55 mph, the output of the process is the actual automotive speed of 54.945 mph. This is only true if there is no disturbance or no slope in the road.

$$54.945 \text{ mph} = 0.999(55) - 0.005(0) \quad \text{(a slope of 0\%)}$$

If the slope is 1%, the output is 54.94 mph:

$$54.94 \text{ mph} = 0.999(55) - 0.005(1) \quad \text{(a slope of 1\%)}$$

If the slope is 5%, the output is 54.92 mph:

$$54.92 \text{ mph} = 0.999(55) - 0.005(5) \quad \text{(a slope of 5\%)}$$

If the slope is 10%, the output is 54.90 mph:

$$54.90 \text{ mph} = 0.999(55) - 0.005(10) \quad \text{(a slope of 10\%)}$$

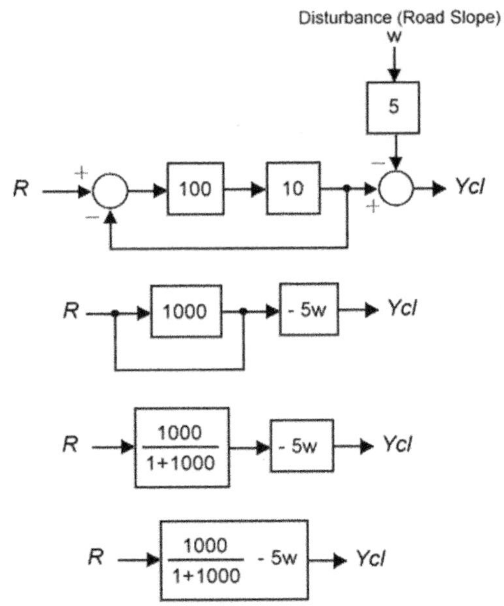

The Transfer Function for the Automobile

Figure 15-4

The Transfer Function for the Automobile

See the block diagram in Figure 15-4 for the process of deriving the transfer function for the automobile.

By using a more complex controller with additional modes of control, the process error can be removed completely and the process (plant) can respond quickly. We have just seen how proportional control has an offset from the set point. Proportional control will stop the upset or process error and try to return the process back to the set point. The proportional controller can have a significant error in the process output, if the disturbance is large.

By using the integral mode in a controller, the offset can be completely removed. This is sometimes called *reset action* because in the old days, the operator would make a manual change in the set point (reset the set point) to achieve the proper process output. With integral mode or reset action, the proportional output is increased (or repeated) every few seconds or minutes, depending on the controller design, until the process output equals the set point of the system.

By using derivative mode, the controller can respond quickly to a fast-changing process error or upset. The derivative mode, or "rate action," subtracts from the controller output to slow down a process that is increasing too quickly, such as a chemical reaction where the heat may increase so quickly it may explode.

Notes

16 Review of Frequency Response Fundamentals

Electrical Application – A First-Order System

Frequency response is a way to analyze what the output of the process or plant will be. We can calculate the output (e.g., volts or watts in power) for a given system gain and input (e.g., volts) at some frequency. Remember the capacitance reactance is varying with the change in frequency ($X_C = 1/2\pi f_C$). First, we will look at where the transfer function comes from. See Figure 16-1.

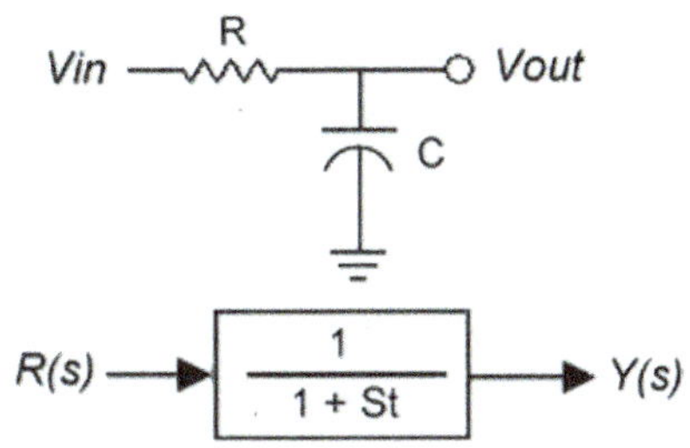

Figure 16-1

We will now derive the transfer function for this first-order system, where *R(S)* is the input signal at some frequency and *Y(S)* is the output voltage with some phase angle and amplitude.

Current equals the voltage drop across the resistor divided by the resistor value.

where

I = Current

V_R = Voltage across resistor

R = Resistor

C = Capacitor

$$I = \frac{V_R}{R}$$

$$I = \frac{Vin - Vout}{R}$$

Current also equals the drain out of the capacitor, from the voltage across the capacitor.

$$I = C\frac{dVout}{dt}$$

Substitute voltage drop divided by resistance for amps (*I*) and set the two equations equal to each other.

$$\frac{Vin - Vout}{R} = C\frac{dVout}{dt}$$

$$Vin - Vout = RC\frac{dVout}{dt}$$

$$S = \frac{d}{dt}$$

$$Vin - Vout = RCS(Vout)$$

$$Vin = Vout + RCS(Vout)$$

$$Vin = (1 + RCS)Vout$$

$$\frac{Vin}{1 + RCS} = Vout$$

$$\frac{1}{1 + RCS} = \frac{Vout}{Vin}$$

$$t = RC$$

The transfer function is equal to the gain of the system.

$$\frac{1}{1 + St} = \frac{Vout}{Vin}$$

Use the transfer function to calculate the voltage out of the system.

$$Vin\left(\frac{1}{1 + St}\right) = Vout$$

We have derived the transfer function for this first-order system. We can now plug in an input voltage and an angular frequency and calculate the attenuation of the output signal and the phase angle of the output signal.

Bode Plot of First-Order System

Make a Bode plot for a circuit with the following components.

resistor = 1000 Ω

capacitor = 2.65 μF

volts in = 10 V

$f_C = 60$ Hz (corner or cutoff frequency)

$$t_{(\text{in seconds})} = \frac{1}{2\pi f_C} = RC$$

t (time constant) = 1000 (Ω) · 0.00000265 (F) = 0.00265 s

$$Vin\left(\frac{1}{\sqrt{1^2 + [S(t)]^2}}\right) = Vout$$

$$S = \omega = 2\pi f$$

$$dB = 20\ \log\left(\frac{Vout}{Vin}\right)$$

Calculate the Data for the Bode Plot

Frequency (Hz)	Rads/s	Volts Out Phase Angle	Signal Attenuation
1	6.28	$10\text{ v}\left(\frac{1}{\sqrt{1^2 + [6.28\ (0.00265)]^2}}\right) = 9.9986\text{ v}$ $TAN^{-1}\left(\frac{6.28\ (0.00265)}{1}\right) = 0.95\theta$	$20\ \log\left(\frac{9.9986\text{ v}}{10\text{ v}}\right) = -0.0012\text{ dB}$
5	31.4	$10\text{ v}\left(\frac{1}{\sqrt{1^2 + [31.4\ (0.00265)]^2}}\right) = 9.9655\text{ v}$ $TAN^{-1}\left(\frac{31.4\ (0.00265)}{1}\right) = 4.76\theta$	$20\ \log\left(\frac{9.9655\text{ v}}{10\text{ v}}\right) = -0.030\text{ dB}$
10	62.8	$10\text{ v}\left(\frac{1}{\sqrt{1^2 + [62.8\ (0.00265)]^2}}\right) = 9.8643\text{ v}$ $TAN^{-1}\left(\frac{62.8\ (0.00265)}{1}\right) = 9.5\theta$	$20\ \log\left(\frac{9.8643\text{ v}}{10\text{ v}}\right) = -0.1187\text{ dB}$

Example 1: An Airplane

Variables	
Altitude	1
Latitude	1
Longitude	1
	3
Minus Constraints	0
Minus Equations	0
DOF =	3

DOF = 3 − (0 + 0) = 3

Three controllers are needed, one for each variable.

Example 2: A Train

Variables	
Altitude	1
Latitude	1
Longitude	1
	3
Minus Constraints	
Altitude	1
Latitude	1
Minus Equations	0
DOF =	1

DOF = 3 − (2 + 0) = 1

One controller is needed; one for longitude only.

Example 3: A Hot Water Heat Exchanger

Variables		
W_S	(flow rate of steam)	1
W_{CW}	(flow rate of cold water)	1
W_{HW}	(flow rate of hot water)	1
Q	(quantity of steam in cubic feet)	1
P_S	(supply pressure of steam)	1
T_{CW}	(temperature of cold water)	1
T_{HW}	(temperature of hot water)	1
		7
Minus Constraints or Constants		
Q	(quantity of steam)	1
P_S	(supply pressure of steam)	1
T_{CW}	(temperature of cold water)	1
		3

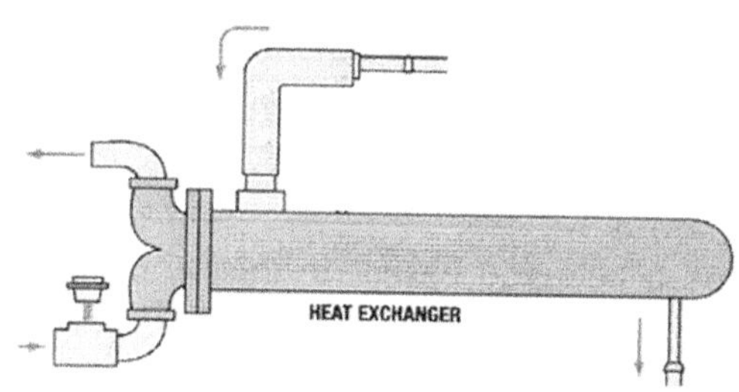

Minus Equations	
Material Balance (conservation of mass)	1
Energy Balance (conservation of energy)	1
	2

DOF = 7 − (3 + 2) = 2

Two controllers are needed.

1. One controller for steam flow.
2. One controller for the energy equation (mass ι C_p ι ΔT). The controller type will be a temperature controller, and it will control the outlet water temperature. It will provide a remote set point to the steam flow controller.

Controllers and Control Strategies (Models/Modes)

In general terms, a *control loop* is a group of components working together as a system to achieve and maintain the desired value of a system variable by manipulating the value of another variable in the control loop. Each control loop has at least one input and one output. There are two types of control loops: open loop and closed loop. Refer to Chapter 15, "Review of Feedback Control Fundamentals."

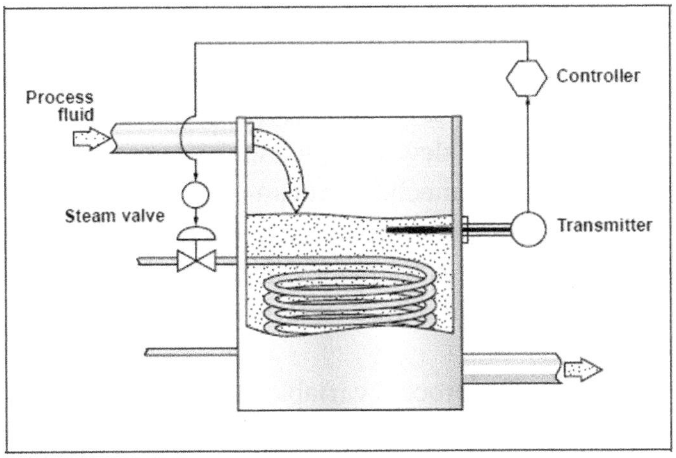

Feedback Loop

In this section, we use simple math to derive a compensated output signal for a control loop with a disturbance acting on the system, for a given set point using a process feedback signal to maintain the desired set point with proportional control having a minimal offset from the set point.

In an *open-loop* system, the controller *does not* have a feedback signal from the system. The controller has a set point and an output signal. The system process variable varies because of system disturbances regardless of the set point signal and fixed controller output signal.

An example of an open-loop system is a car when only the accelerator pedal is used. The accelerator pedal is held in fixed position. When the car goes up a hill, the car will tend to slow down. The decrease in speed is inversely proportional to the increase in slope.

(sample period), a proportional response is achieved. This type of control mode is used frequently in industrial heating applications with furnaces using silicon-controlled rectifiers and solid-state relays, as well as tank-filling applications and pH correction systems using pneumatic pumps. See the "Control of pH Values in Processes" section in Chapter 11 for more information on pH control.

Process Loop Gain

One goal of most controllers is control loop stability. This is achieved by making the loop response to load changes as linear as possible and by keeping the loop gain more or less constant. The gain of any device is the ratio of its output divided by its input. The loop gain (L_G) is the product of the gains of the loop components, the process gain (G_P), sensor gain (G_S), controller gain (G_C) and valve gain (G_V). Tuning the loop means that if our goal is quarter-amplitude damping, we adjust the controller so that the loop gain will be about 0.5. All gains put together are called the *total process gain* (TPG):

$$L_G = G_P \bullet G_S \bullet G_V \bullet G_C = \text{TPG} \approx 0.5$$

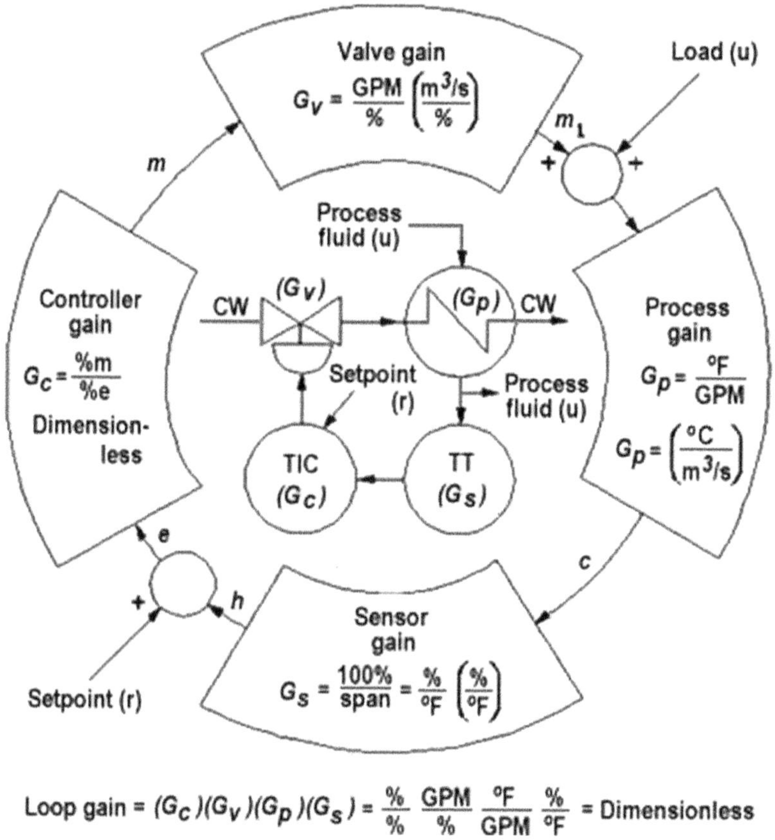

Linear Valve Application

If TPG is more or less constant and does not change much with the process load, linear valves should be used. In a linear, constant gain, valve, a 1% change in lift results in a 1% change in flow ($G_V = 1.0$). Linear valves are acceptable if TPG = (G_P)(G_S)(G_V) ≈ 0.5–2.0, as the load varies between its minimum and maximum limits. Linear valves are used in most process applications except temperature control and heat transfer.

See the figure to the right. If the process is nonlinear (G_P varies with the load), the product of the loop gain (L_G) should be held more or less constant by compensating for the variation in G_P by using a nonlinear valve with inverse G_V nonlinearity.

Nonlinear Valve Applications

If the valve gain rises as the valve opens, its characteristic is called *equal percentage*, and if its gain drops as it opens, it is called *quick-opening*. Special, custom-made valves can provide other nonlinearities, for example, characterized v-ball valves that have gains that are the inverse of the pump curves. Therefore, compensation is provided by using an inverse valve characteristic ($G_V = 1/G_P$)

so that the installed characteristics of the total process will be more or less linear (TPG ≈ 0.5–2.0). Naturally, G_V can never be the exact inverse of G_P, but as long as the selected valve characteristic keeps the TPG within 0.5 and 2.0, instability and limited cycling can usually be avoided by good controller tuning (assuming TPG = 1.0 when adjusting G_C).

Control Valve Linearization

The installed flow characteristic of a control valve directly affects the process gain. It is essential that the installed characteristic is linear. In the figure to the right, look at the install characteristic. The process gain is constant, regardless of the controller output. If the installed gradient of the curve varies by more than a factor of two, control loop performance will be noticeably affected. This means the controller must have different tuning parameters for different ranges in the controller output. If nothing is done to linearize the valve, the controller will have to be detuned to accommodate the maximum process gain. This leads to sluggish control loop response over much of the valve's operating range.

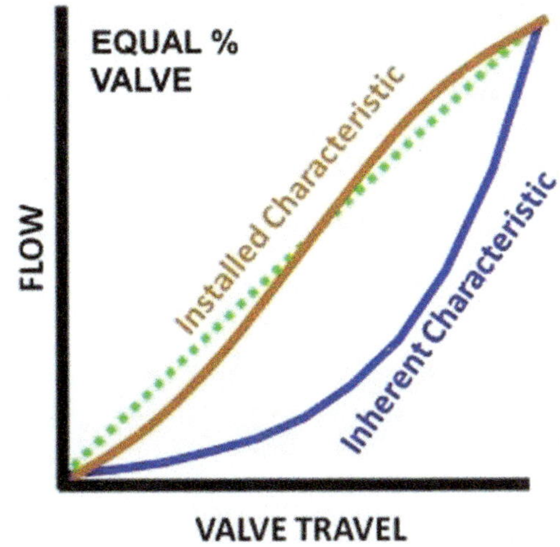

A nonlinear flow characteristic should be linearized to obtain good control performance throughout the valve's operating range. This is done with a linearizer (also called a *characterizer*). The linearizer is a control block, function generator, *f(x)* curve, or a lookup table placed between the controller and the valve. See the figure below. Although the linearization can be done in a digital positioner, the DCS or PLC is the best location for it. This enables replacement of the positioner without having to reprogram the linearization curve in the new positioner.

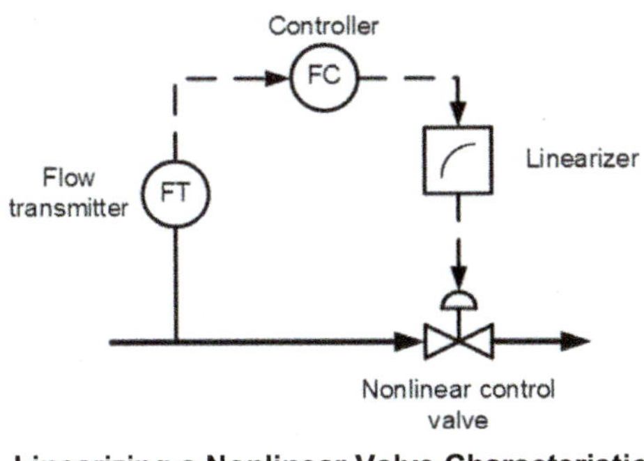

Linearizing a Nonlinear Valve Characteristic

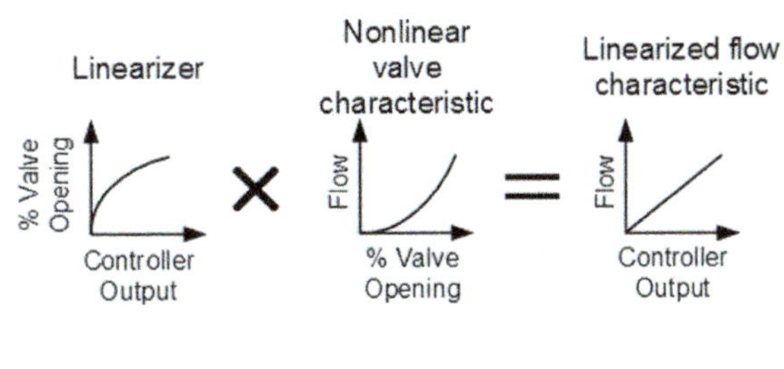

How a Linearizer Works

Linearization is done with an *X-Y* curve or function generator that is configured to represent the reciprocal (inverse) of the control element's flow curve. See the figure above to the right.

To design the linearizer, you have to first determine the flow characteristic curve of the valve operating in the actual process. For this, you should take readings of the flow or process variable (PV) and controller output (*CO* or *C*) under steady-state conditions at various controller output levels.

You need a minimum of three pairs of data from a closed-loop process response curve. More pairs of data would provide a much better response curve for characterizing a nonlinear relationship.

Make sure you span the entire operating range of the controller output, and try to obtain readings spaced equally across the controller output span. You can do process tests to obtain these values, or examine the data from your process historian. Then convert the process variable data from engineering units to a percentage of full scale of the measurement.

Sort the data pairs in ascending order, and enter them into a function generator. The PV readings in percentage become the *X* values (input side), and the *CO* readings in percentage become the *Y* values (output side). Include a zero point (0, 0) if you do not already have one in your data set and be sure to estimate a full span point (100, *Y*) also, if you do not have one. Also, if your valve opens as the *CO* decreases, your *Y* column will obviously have to reflect this.

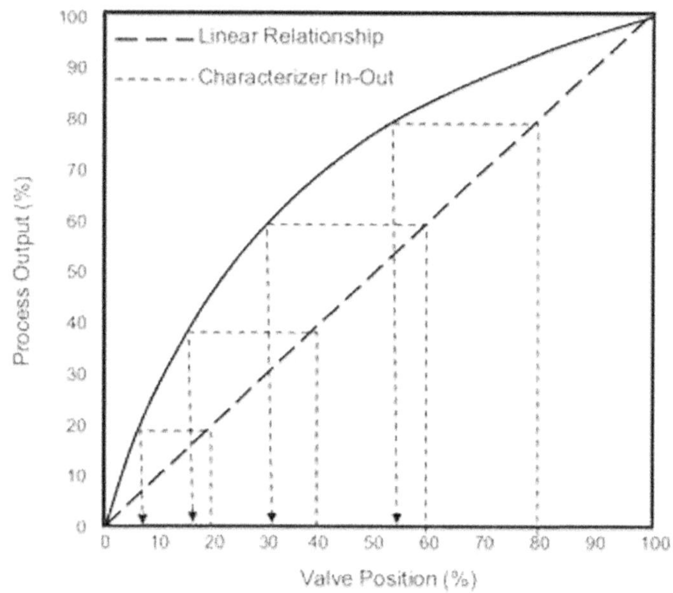

Characterizer

IN	OUT
0	0
5	1.5
10	3
15	5.5
20	8.5
25	10
30	11.5
35	13
40	17
45	20
50	22
55	26
60	31
65	36
70	42
75	48
80	54
85	63
90	74
95	86
100	100

Example of Characterizer Setup

Linearizing of the signal is typically seen in controllers, but some valve positioners and transmitters can have an 8- or 16-segment curve to correct for a nonlinear measurement and then send the input signal to the DCS or PLC. The points are fitted by a curve, and the output signal is interpolated between preset entered points. This is popular in digital valve controllers and transmitters of nonlinear measurements, such as radiometric or gamma radiation transmitters for levels or thickness, as well as power monitoring such as in a GE Multilin power monitor for compressors and very large motors.

Signal Filtering in Process Control

Signal noise is generated due to the pulsating characteristics of process control applications, such as fluctuations in the process streams composed of liquids, gases, powders, slurries, and melts.

These noises can be generated by pressure pulsations from the design of pumps or sloshing in agitated tanks or cavitation of fluids.

The derivative mode of a PID controller (rate action) can amplify the noise in the measured PV and make the controller output (CO) become erratic. Noise in the PV will be amplified by the CO signal and will produce "chatter" in the final control element. This extreme control action will increase the wear on a mechanical final control element, such as a valve, leading to increased maintenance and making it harder to stabilize the process. This higher frequency noise must be filtered out. First, look into the transmitter and the process equipment for a solution. If the noise cannot be reduced, a filter must be applied to the PV and/or controller signals.

Applying Signal Filters

External Filters in Control

There are three popular places to put external filters in the feedback loop. By *external,* we mean that the filters are designed, installed, and maintained separately from the controller.

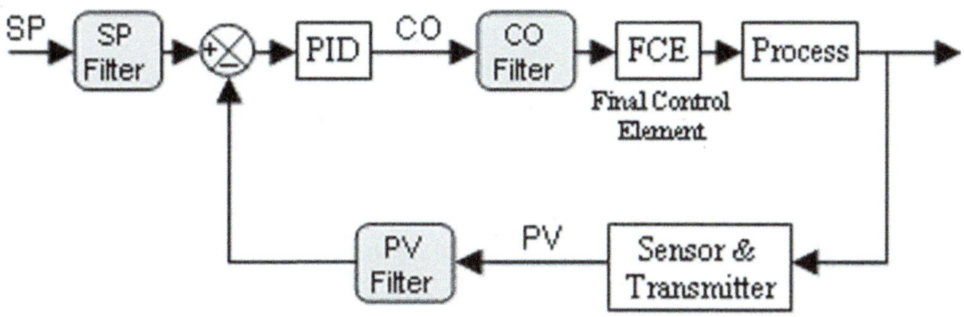

Internal Filters in Control

For feedback control, filtering need only be applied to the signal feeding the derivative term. As stated before, noise does not present a problem for proportional and integral action. These elements will perform best without the delay introduced from a signal filter.

First Order Filter

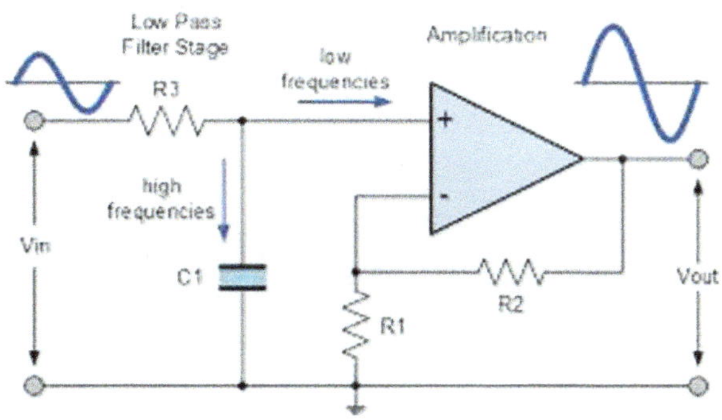

The Derivative Term of the PV Filtered

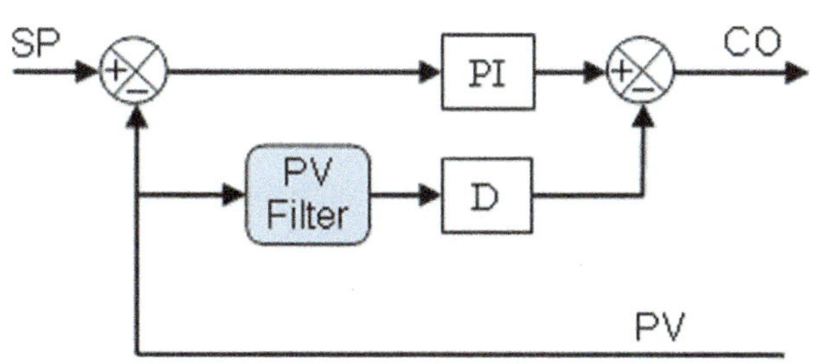

Plot of the Process Variable Signal Filtered

The plot below shows the random behavior of a raw (unfiltered) PV signal and the smoother trace of a filtered PV signal.

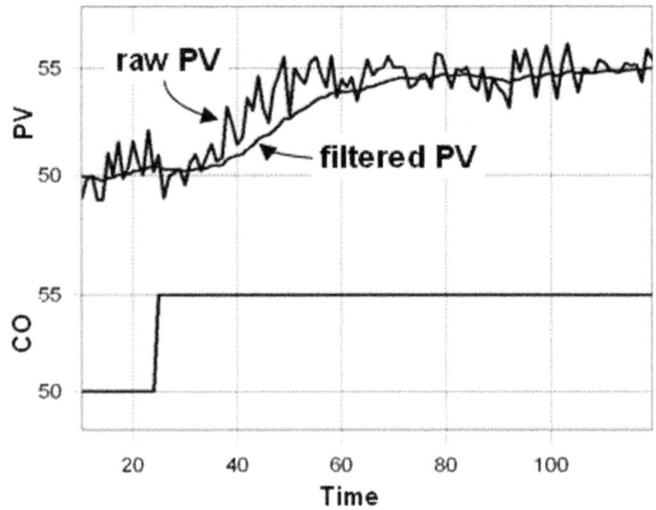

Filter Time Constant and Sample Time

The plot above shows that the derivative mode would add to the output tremendously without filtering. The derivative mode (rate action) would see a gain of about 10/1 compared to 0.5/1 in the signal filtered.

To select a filter time for attenuation of noise or to eliminate the noise in the process variable signal (PV), we would take the reciprocal of the angular frequency, 2π Hz or 2π(cps), of the noise signal and select a filter time constant that is equal to or greater than the time constant of the corner frequency.

For the first order filter, we must pick a corner frequency (f_C) that is smaller or less than that of the frequency of noise we wish to attenuate. This will cause the lower frequencies of the process signal (PV) to pass through the filter to the controller amplifier section, allowing the system to respond to the lower frequency upsets in the system. Remember that the time constant of the corner frequency is:

$$T_c = \frac{1}{2\pi f_c}$$

The time constant and frequency as used in the first order filter is:

$$T_c \bullet 2\pi f_c = 1 \text{ where } \tau = T_c \text{ and } S = \omega_c = 2\pi f_c \text{ used in the equation for } \frac{1}{1+S\tau}$$

$$-3\text{dB} = 0.707 = \frac{1}{\sqrt{1^2 + (2\pi f_c T_C)^2}} \text{ where } T_C \bullet 2\pi f_c = 1$$

In the transfer function for the first order filter, at corner frequency the noise signal will be attenuated to –3 dB or 70.7%. All frequencies above or greater than corner frequency will be drastically attenuated or fall off in amplitude ratio. The trick here is to pick a frequency as low as can be tolerated and still keep the process control system responsive.

If you do not understand how the first order filter works, review Chapter 16, "Review of Frequency Response Fundamentals."

$$\textit{signal in} \bullet \left(\frac{1}{1+St}\right) = \textit{signal out}$$

The Nyquist–Shannon sampling theorem states that the sampling time should be at least twice the highest frequency of the process signal. If the process signal varies every 10 s, the process frequency is 0.1 cps or Hz, which is equal to 1/10 s. The sampling frequency should be two times the process signal frequency to make the system responsive.

So, 2 ι 0.1 = 0.2 cps; therefore, the maximum sample time should be 1/0.2 cps = 5 s.

Example of Filter Time Selection

The process signal has a noise frequency of 5 cps (cycles per second or Hz). The process signal has periods of 10 s or greater. Make the acceptable choice between the time constant for the filter and the sample time for the DCS. Remember, the smaller the DCS sample time, the better the system response. Choose from the following selections:

	Corner Frequency, f_C	Filter Time Constant, T_f (s)	Sampling Time, T_S (s)	Attenuation of 5 Hz noise (dB)
A	10 Hz (62.8 rads/s)	0.016	0.5	–0.98
B	5 Hz (31.4 rads/s)	0.032	1	–3.01
C	2 Hz (12.56 rads/s)	0.080	5	–8.64
D	1 Hz (6.28 rads/s)	0.159	8	–14.14
E	0.5 Hz (3.14 rads/s)	3.140	10	–39.88

The best answer is C:

- Choice A will never attenuate the noise signal. The filter is low pass, so the noise is passed.
- Choice B, the corner frequency, is the noise frequency, so 70.7% of the noise will still pass. The DCS scan time is acceptable because it is smaller than the required 5 s period for samples.
- Choice C is the best answer; the noise will be attenuated by 63.02%. Only 36.98% of the 5 Hz noise in the PV signal will pass to the controller, and the DCS scan time is still fast enough to respond to the 5 s recommended sample time period of the process.

In closed loop, the controller uses the feedback signal in automatic mode. Gain is increased until a sustained oscillation is achieved. The plot of the oscillation is then used to retrieve the tuning parameters.

We will now look at two methods for tuning a controller, the ultimate gain (continuous cycling) and process reaction curve (step response) methods. We will use the Ziegler-Nichols method, but the Cohen-Coon or Integral Absolute Error (IAE) method could be used instead.

Closed-Loop Tuning of the Controller

Tuning Based on the Ultimate Gain Method (Ziegler-Nichols)

Essentially, this tuning method works by oscillating the process. Turn off the integral mode or set time to zero (0) and turn off the derivative mode. Increase the gain of the controller and make a slight set-point change. Repeat the process and gradually increase the gain of the controller each time, until a sustained oscillation is achieved as shown in the following figure.

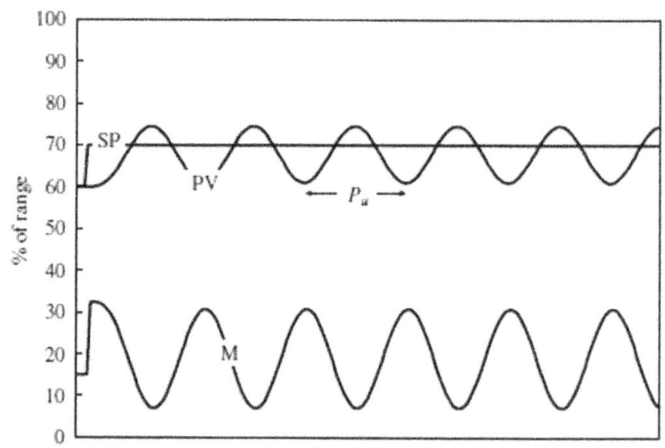

Sustained oscillation

This is called the *ultimate gain* (K_u). It is the gain of the controller necessary to make the process sustain oscillation. The *proportional band gain* (P_u) is the reciprocal of the K_u.

Tune the controller by entering the new Ziegler-Nichols values from the calculations in Table 17-1.

The table values are to be entered as *gain*. If you must convert gain to proportional band, then $Pu = 1/K_u$ and $K_u = 1/P_u$. If Pu is used for the controller, then convert back to proportional band *after* applying the table calculations.

Proportional band = 1/gain = 1/proportional band

The period or time constant equals T_u in minutes. The time calculation will be entered as *minutes per repeat* for integral time and as *minutes* for derivative time.

Remember that when entering the integral time:

Minutes per repeat = 1/repeats per minute
Repeats per minute = 1/minutes per repeat
Proportional band (PB) is typically displayed as a percentage. For example, 0.50 K_c is equal to 200% *PB*, and 2.00 K_c is equal to 50% *PB*.

Table 17-1. Tuning parameters for the closed-loop Ziegler-Nichols method.

Controller Type	Gain, K_c	Integral Time, T_I	Derivative Time, T_D
P	$0.5\ K_u$		
PI	$0.45\ K_u$	$\frac{T_u}{1.2}$	
PID	$0.6\ K_u$	$\frac{T_u}{2}$	$\frac{T_u}{8}$

Example: Tune PID Controller Using Ultimate Gain (Continuous Cycling)

Period Time, T_U: 12 min	Gain, Ku: 2.2
Note: T_I = minutes per repeat	$K_c = 0.6 \cdot K_u = 0.6 \cdot 2.2 = 1.32$
K_c = standard gain of controller (output/input)	$T_I = \frac{T_U}{2} = \frac{12}{2} = 6$ min
P_u = proportional gain of controller (input/output)	
K_u = gain necessary to make the process cycle	$T_D = \frac{T_u}{8} = \frac{12}{8} = 1.5$ min

Open-Loop Tuning of the Controller

Tuning Based on the Process Reaction Curve

In process control, the term *reaction curve* is sometimes used as a synonym for a step response curve. Many chemical processes are stable and well damped. For such systems, the step response curve can be approximated by a first-order plus dead-time (FOPDT) model. It is relatively straightforward to fit the model parameters to the observed step response. Look at the following reaction curve.

Essentially, the tuning method works by manually causing a step change in the process. This is accomplished by putting the controller in manual and forcing an output change of the controller. Record the step change process reaction curve on the chart recorder and follow the setup instructions below.

1. Locate the point where the curve stops curving upward from the bottom left and starts to complete the curve up to the right and settle at a new process measurement level. This will be about halfway up the reaction curve. It is the inflection point.
2. Draw an asymptote line tangential to the point of the inflection. Where the asymptote line crosses the bottom of the process reaction curve, the previous output is assumed to be zero. (It is the measurement before the set-point change was made, which is now zero to the measurement of the process change.) It may be equal to 50 psi or 500°, but set it to a live zero. The time between the start of the output step change and the start of the asymptote line at the live zero of the process measurement is the apparent time delay or dead time (T_D) of the system.

When the asymptote line reaches the steady-state value of 63.2% of delta measurement, the time difference between the end of the dead time measurement (T_D) and the end of the 63.2% of delta measurement is called the *time constant* for the process (τ). Draw a line straight down from the 63.2% point to the live zero line. These are the values of the time constant of the process (τ) and the dead time of the process (T_D).

3. The gain of the system (K_P) (the slope of the asymptote line) is given by:

$$K_P = \frac{\Delta measurement}{\Delta valve\ change} = \frac{\Delta measurement}{\Delta controller\ output}$$

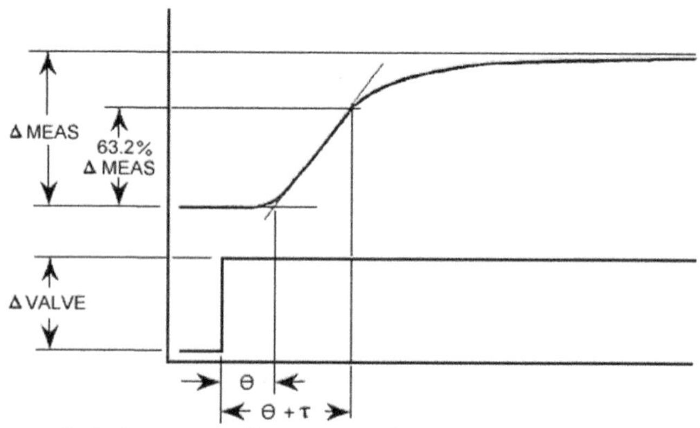

Typical process reactioncurve for tuning a controller in openloop.

Typical process reaction curve for tuning a controller in open loop.

Example: Tuning Using a Process Reaction Curve (Step Response)

We will use the following graph of the process reaction curve (the step response) to tune the controller for this worked example.

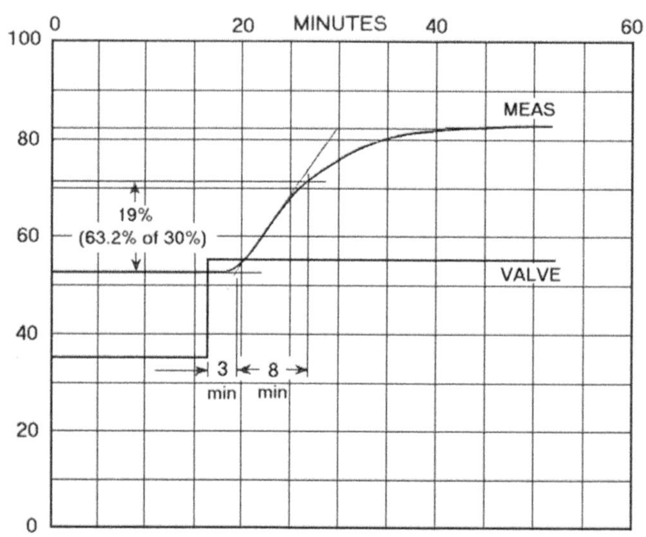

(See Table 17-2 for tuning formulas.)

Table 17-2. Tuning parameters for the open-loop method (Ziegler-Nichols).

Controller Type	Gain, K_c	Integral time, T_I	Derivative Time, T_D
P	$\frac{\tau}{K_P\theta}$		
PI	$\frac{0.9\,\tau}{K_P\theta}$	$\frac{\theta}{0.3}$	
PID	$\frac{1.2\,\tau}{K_P\theta}$	$\frac{\theta}{0.5}$	0.5θ

Note: This table of tuning parameters has τ equal to lag time in the equation for a lag time of 62.3% of delta process measurement (θ is equal to the dead time) of the process.

Table 17-3. Tuning parameters for the open-loop Cohen-Coon method.

Controller Type	Gain, K_P	Integral Time, T_I	Derivative Time, T_D
P	$\frac{1}{K_P}\frac{\tau}{\theta}\left(1+\frac{\theta}{3\tau}\right)$		
PI	$\frac{1}{K_P}\frac{\tau}{\theta}\left(\frac{9}{10}+\frac{\theta}{12\tau}\right)$	$\frac{\theta\left[30+3(\theta/\tau)\right]}{9+20(\theta/\tau)}$	
PID	$\frac{1}{K_P}\frac{\tau}{\theta}\left(\frac{4}{3}+\frac{\theta}{4\tau}\right)$	$\frac{\theta\left[32+6(\theta/\tau)\right]}{13+8(\theta/\tau)}$	$\frac{4\theta}{11+2(\theta/\tau)}$

Note: This table of tuning parameters has τ equal to lag time in the equation for a lag time of 62.3% of the delta process measurement (θ is equal to the dead time) of the process.

Data given:

Time constant τ: 8 min

Dead time θ: 3 min

Delta P_V: 82% − 52% = 30%

Delta output: 55% − 35% = 20%

Note: K_C = controller gain settings

T_I = minutes per repeat

T_I^{-1} = repeats per minute

T_D = minutes

Solve for tuning parameters:

$$K_P = \frac{\Delta P_v\%}{\Delta output\%} = \frac{30\%}{20\%} = 1.5$$

$$K_c = \frac{1.2\tau}{K_P\theta} = \frac{(1.2)(8)}{(1.5)(3)} = 2.134$$

$$T_I = \frac{\theta}{0.5} = \frac{3}{0.5} = 6 \text{ min}$$

$$T_D = 0.5\theta = (0.5)(3) = 1.5 \text{ min}$$

Advanced Tuning Methods for Controllers

The Integral Criteria Method

General definition: An integral criterion is a performance measure that is based on the integral of some function of the control error and on possibly other variables (such as time).

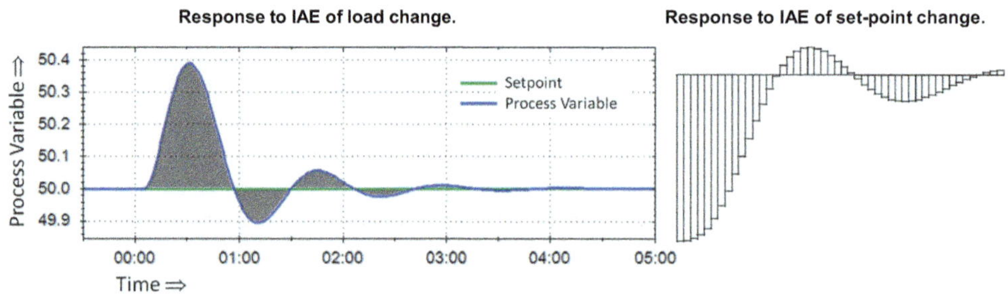

The integral criteria method is an open-loop method that calculates the controller tuning parameters from the coefficients in a time-constant-plus-dead-time model. The relationships provide the values of the tuning coefficients that minimize an integral criterion. Although other criteria are available, this section will present relationships for only the following two integral criteria:

$$\text{IAE} = \int_0^\infty |\text{E}|\, \text{dt}$$

$$\text{integral of time and absolute error (ITAE)} = \int_0^\infty |\text{E}|\, t\, \text{dt}$$

(See Tables 17-4 and 17-5 for the IAE tuning formulas.)

Trial and Error Tuning Method

In most plants, the engineers or technicians tune a control loop by "feel" or trial and error. They tune the loop using the Ziegler-Nichols equations and then they fine-tune the controller by nudging the parameters for the proportional, integral, and derivative.

Table 17-4. IAE1: Tuning parameters for the IAE (load change) method.

Controller Type	Gain, K_p		Integral time, T_I		Derivative time, T_D	
	$K_C = \dfrac{A\left[\dfrac{\theta}{\tau}\right]^B}{K_P}$		$T_I = \dfrac{\tau}{A\left[\dfrac{\theta}{\tau}\right]^B}$		$T_D = A\left[\dfrac{\theta}{\tau}\right]^B \tau$	
	A	B	A	B	A	B
P	0.902	–0.985				
PI	0.984	–0.986	0.608	–0.707		
PID	1.435	–0.921	0.878	–0.749	0.482	1.137

Note: This table of tuning parameters has τ equal to lag time in the equation for a lag time of 62.3% of the delta process measurement (θ is equal to the dead time) of the process.

Table 17-5. IAE2: Tuning parameters for the IAE (set-point change) method.

Controller Type	Gain, K_p		Integral Time, T_I		Derivative Time, T_D	
	$K_C = \dfrac{A\left[\dfrac{\theta}{\tau}\right]^B}{K_P}$		$T_I = \dfrac{\tau}{A + B\dfrac{\theta}{\tau}}$		$T_D = A\left[\dfrac{\theta}{\tau}\right]^B \tau$	
	A	B	A	B	A	B
P						
PI	0.758	−0.861	1.020	−0.323		
PID	1.086	−0.869	0.740	−0.130	0.348	0.914

Note: This table of tuning parameters has τ equal to lag time in the equation for a lag time of 62.3% of the delta process measurement (θ is equal to the dead time) of the process.

Dead Time and PID Control

The best way to get better control of a dead time process is to reduce the dead time. For example, you may be able to relocate the transmitter closer to the process. A PID controller with proper tuning gives a fast, stable response. There are other tricks that can help the response. For example, applying a small filter to the PV can smooth the response. Also, if the process has a small lag, you can use a little derivative very carefully. For a process with a larger lag, using a derivative can usually help response. The optimal tuning of a PID controller for a dead-time-only process can be performed as follows:

controller gain = 0.3/process gain
integral time = 0.42 ι process dead time

Process Characteristics from the Transfer Function

We will now look at the controller and control loop characteristics. Mathematically, we will describe the response of a control loop and calculate the overshoot and damping of a typical control loop.

If you do not understand what a transfer function is or where it comes from, refer to Chapter 15, "Review of Feedback Control Fundamentals." It explains how a feedback control loop works, derives the mathematics and calculates output will be based on the closed-loop and open-loop system gain. We will derive a block diagram of the transfer function.

If you do not understand frequency response and what it means or where the transfer function comes from, refer to Chapter 16, "Review of Frequency Response Fundamentals." It covers how the transfer function is derived, how the signal is attenuated and phase is shifted, and how the system response is plotted, so you may understand what is happing in the system.

In Chapter 16, an electrical RC circuit is introduced and the characteristics are discussed, including how a varying frequency changes the reactance of the circuit. Then a hydraulic circuit is discussed, including how a constant capacitance with a varying valve position changes the frequency of tank head in time. A varying time constant of RC also exists. It is a change of the valve position multiplied by the capacitance of the system.

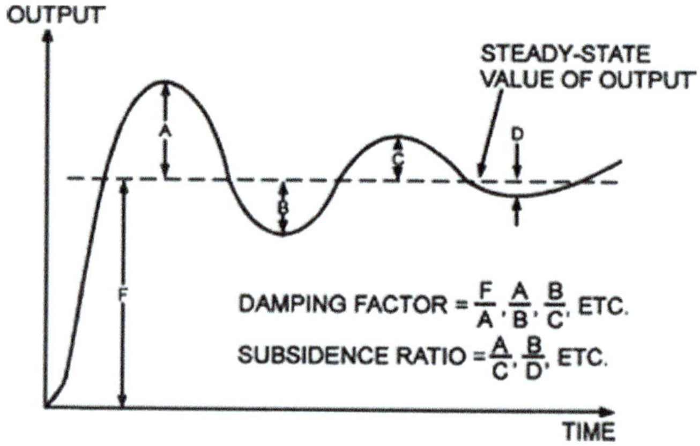

To the right is a graph showing a typical controller response to a set-point change. Most engineers use 0.25 amplitude damping to control loops in the process industry.

Let us find out how to solve for the above-mentioned criteria.

Poles, Zeros, and Dampening from the Transfer Function

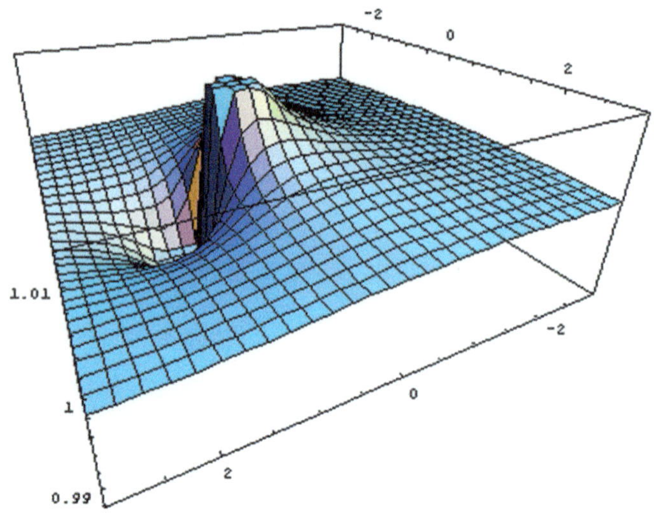

When the pole is rising upward, the output is trying to go to infinity and the zero is trying to pull the output back down to zero output.

For example, if you define *impedance of the circuit* as the transfer function for the plant, then, $Z(s) = N(s)/D(s)$.

This is just a way of saying that the function has a numerator and a denominator that both depend on s, some frequency.

Clearly, if $N(s)$ goes to zero, the transfer function $Z(s)$ will go to zero. If $D(s)$ goes to zero, the transfer function $Z(s)$ will go to infinity.

This means the impedance of the circuit varies at different frequencies, has phase shifts and oscillations, and tends to resonate at the pole frequency. When s (the frequency) goes to zero, we have only a pure DC gain and a steady-state output for control.

Find the Poles from the Given Function

where

s^2 = acceleration

ω_n = natural frequency

$G(s)$ = system gain

p_1 = first pole

p_2 = second pole

Equation given:	Find poles:
$G(s) = \frac{\omega_n^2}{s^2 + 2\xi\omega_n s + \omega_n^2}$	$p_1; p_2 = \frac{-b \pm \sqrt{b^2 - 4ac}}{2}$
$G(s) = \frac{25}{s^2 + 5s + 25}$	$p_1; p_2 = \frac{-5 \pm \sqrt{25 - 4(25)}}{2}$
Pole 1: $-2.5 + j4.33$ Pole 2: $-2.5 - j4.33$	$p_1; p_2 = \frac{-5 \pm \sqrt{25 - 100}}{2} = -2.5 \pm j4.33$

Find the Damping from the Given Function

where

s^2 = acceleration

ω_n = natural frequency

$G(s)$ = system gain

ξ = system damping ratio

Equation given:	Solve for the equation:
$G(s) = \frac{\omega_n^2}{s^2 + 2\xi\omega_n s + \omega_n^2}$	$s^2 + 5s + 25$ $s^2 + 2\xi\omega_n s + \omega_n^2;\ \omega_n = \sqrt{25}$
$G(s) = \frac{25}{s^2 + 5s + 25}$	$2\xi\omega_n s = 5s$
ξ = damping ratio = 0.5	$\xi = \frac{5}{2\omega_n}; = \frac{5}{2\sqrt{25}} = \frac{5}{10} = 0.5$

We will now calculate the rise time, period, natural frequency, and settling time. We will refer to the graph to the right and the previous graph for the peak amplitude designations.

Notice the rise time in the graph on the right. It rises in a vertical line from 10% to 90% of steady-state value. This is the definition of rise time.

Notice the step response in the graph on the right. It rises in a vertical line from 0% to 63.2% of peak value. This is the definition of the time constant.

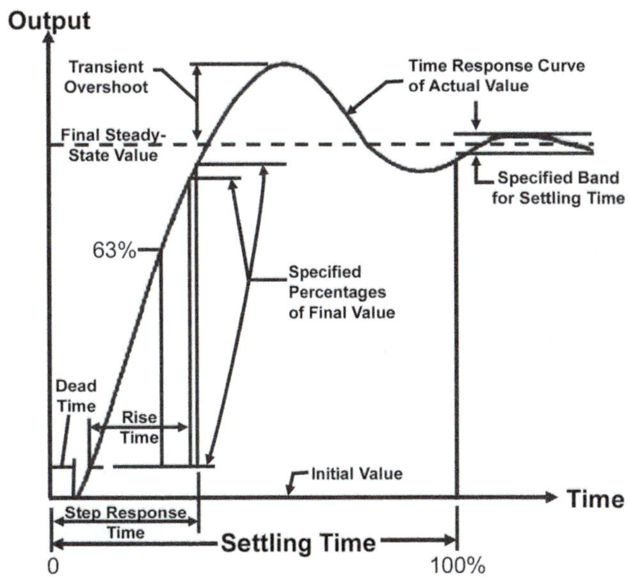

The time constant will be the step response time minus the dead time.

Find the Time Constant

Data given:	Solve for time constant for reaction curve, τ:
Step response time at 63%: 6 s Dead time: 1 s	$\tau = T_{sr} - T_d$ $\tau = 6 - 1$ $\tau = 5$ s

Find the Period

Data given:	Solve for period, P:
Step response time at 63%: 6 s Dead time: 1 s Time constant: 5 s Damping: 0.5	$P = \frac{2\pi\tau}{\sqrt{1-\xi^2}}$ $P = \frac{6.28(5)}{\sqrt{1-0.5^2}}$ $P = 32.26$ s

Find the Time Constant from the Period

Data given:	Solve for time constant from period:
Period: 36.26 s Damping: 0.5	$\tau = \frac{\sqrt{1-\xi^2}}{2\pi} P$ $\tau = \frac{\sqrt{1-0.5^2}}{6.28} 36.26$ $\tau = 5$ s

Find the Overshoot and Peak Value

where

$A\%$ = percent of pressure in first overshoot above desired target pressure when step change is made.

$C\%$ = percent of pressure in second overshoot above desired target pressure when step change is made.

e = the (natural) exponential function.

Process variables given: **$A\%$ = 50 psi; ζ (dampening) = 0.5**	The percentage overshoot and peak is:
The first overshoot is: $A\% = 100e^{-\pi\xi/\sqrt{1-\xi^2}}$ The second overshoot is: $C\% = 100e^{-3\pi\xi/\sqrt{1-\xi^2}}$	$A\% = 100e^{-\pi 0.5/\sqrt{1-0.5^2}}$ $A\% = 100e^{-1.57/\sqrt{0.75}}$ $A\% = 100e^{-1.812}$ $A\% = 100(0.163)$ $A\% = 16.3\%$ 50 psi • 0.163 = 8.15 psi overshoot 50 psi + 8.15 psi = 58.15 psi peak

Block Diagram Algebra

Simplification Method

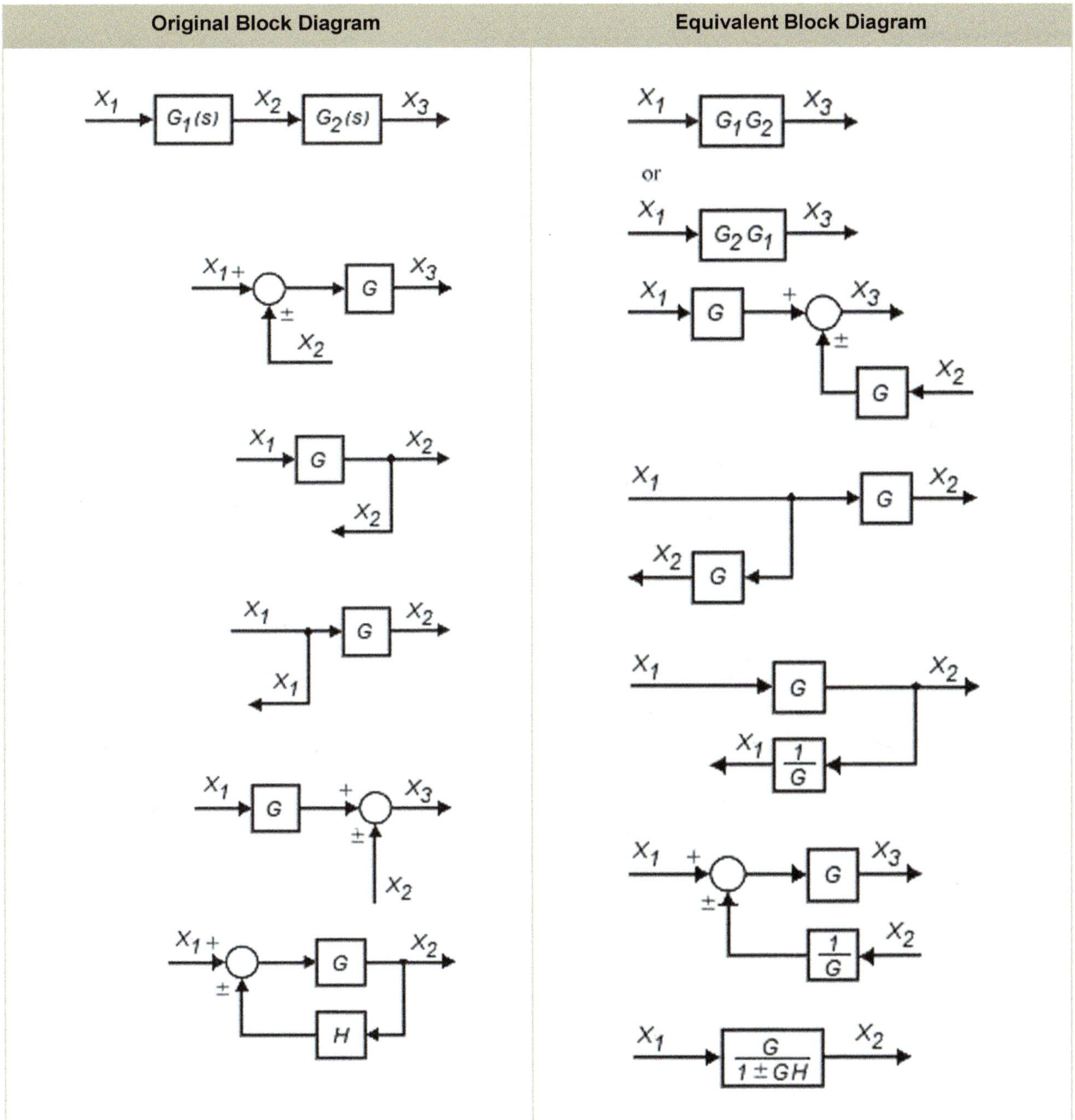

where

X_1 = set point signal
X_2 = process response
G = open loop gain of system
H = feedback gain

Example of Block Diagram Algebra Reduction

This may be on the CSE exam. Start at figure (a), the original multivariable diagram, and simplify.

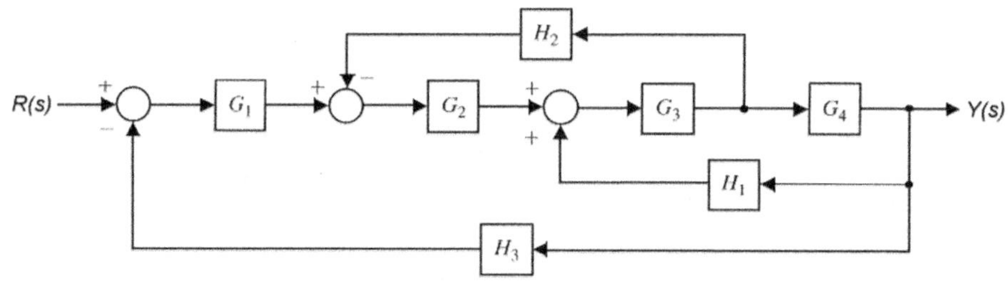

(a)

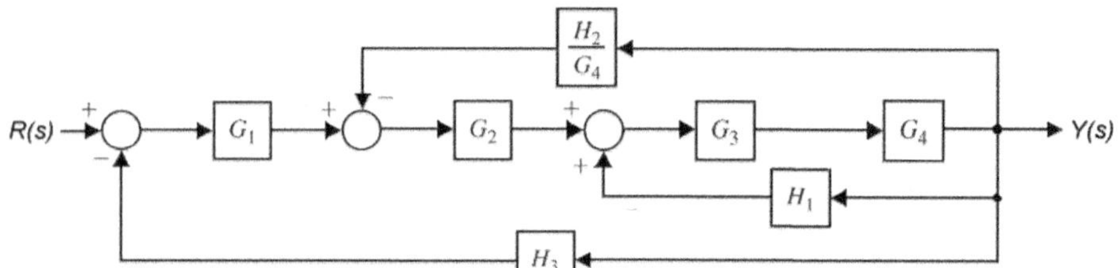

(b)

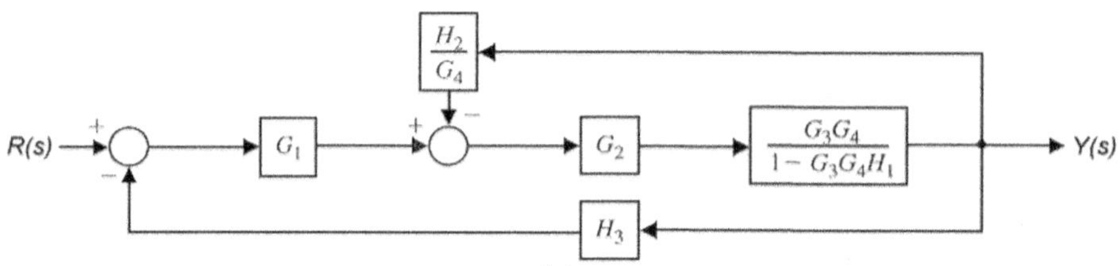

(c)

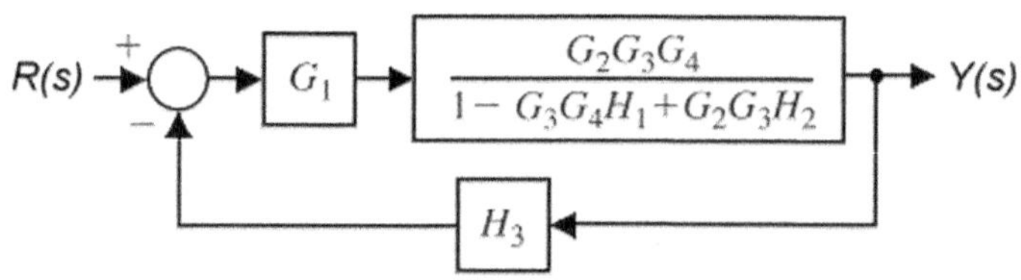

(d)

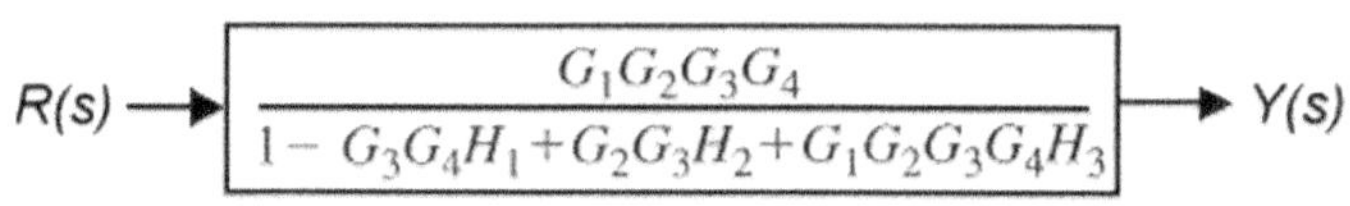

(e)

Nyquist Stability Criterion

This may be on the CSE exam. Most closed-loop systems are open-loop stable and do not have any poles (open-loop pole) in the right half of the *s* plane. Closed-loop systems that are stable will not have any roots in the right half plane. The Nyquist–Shannon diagram of an open-loop stable system does not encircle the stability point (–1, j0).

Note: The curve cannot encompass the stability point (–1, j0) in the polar plot or the system will become unstable. This can be seen in the last polar plot below. Encompassing the phase margin point (1 < –140°) or the gain margin point (–0.5, j0) makes the system marginally unstable.

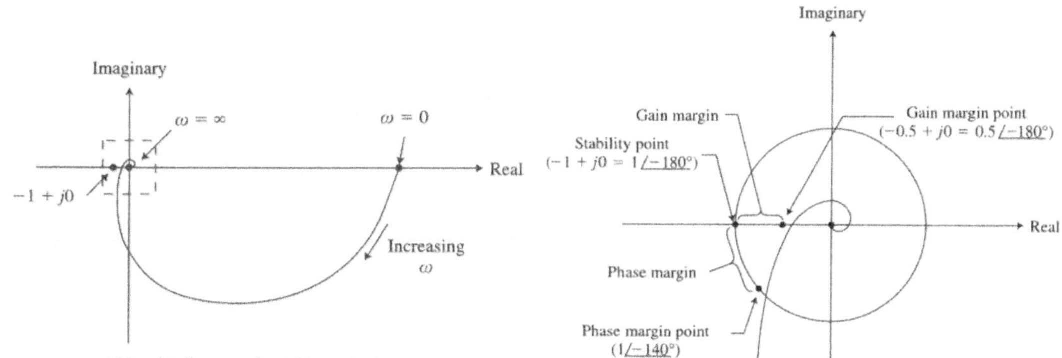

a) Nyquist diagram of a stable control system

b) Close-up view of the boxed region in part a

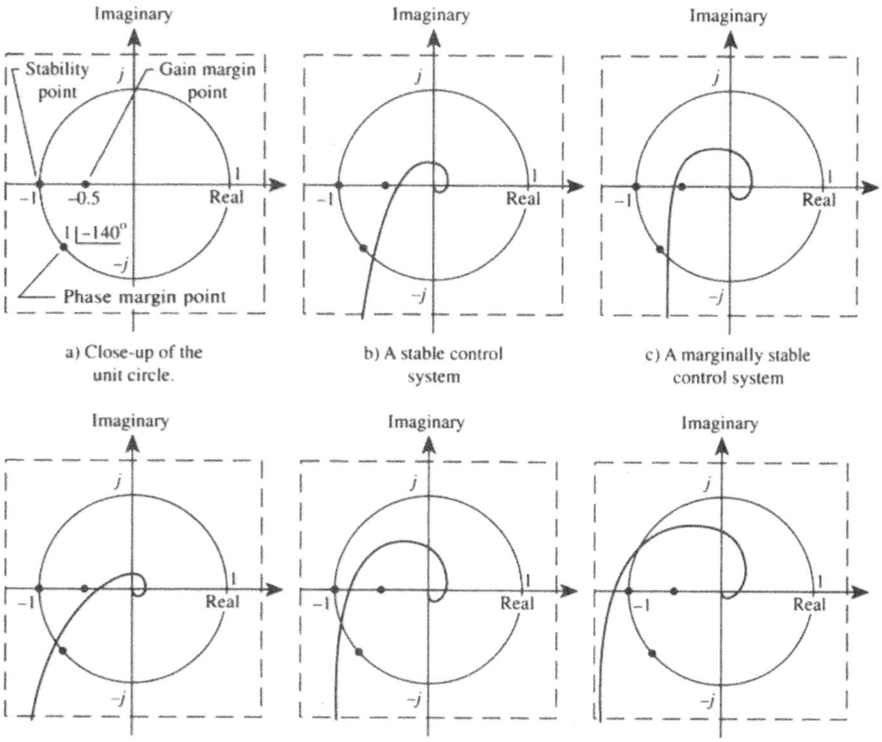

a) Close-up of the unit circle.

b) A stable control system

c) A marginally stable control system

d) A marginally stable control system

e) A marginally stable control system

f) An unstable control system

Routh Stability Criterion

This may be on the CSE exam. It will show a block diagram and give a transfer function for each block. We are interested only in the poles of the closed-loop transfer function of the system. Poles are at the bottom of the equation. The poles of the closed-loop transfer function equation will be used to evaluate the stability of the system using the Routh criterion.

Controller equation: $G = \dfrac{K(s+3)}{s(s+1)}$ Plant equation: $P = \dfrac{(s+4)}{(s^2+6s+13)}$

We want to find the closed-loop transfer function: $\dfrac{Ks^2 + 7Ks + 12K}{s^4 + 7s^3 + (19 + K)s^2 + (13 + 7K)s + 12K}$

where

$R(s)$ = set point
G = controller gain
P = plant gain
$Y(s)$ = process response
K = a variable for gain entered
s = Laplace transfer function
A = zeros in controller closed loop gain function block (a place holder to simplify the algebra)
B = poles in controller closed loop gain function block (a place holder to simplify the algebra)
C = zeros in plant closed loop gain function block (a place holder to simplify the algebra)
D = poles in plant closed loop gain function block (a place holder to simplify the algebra)

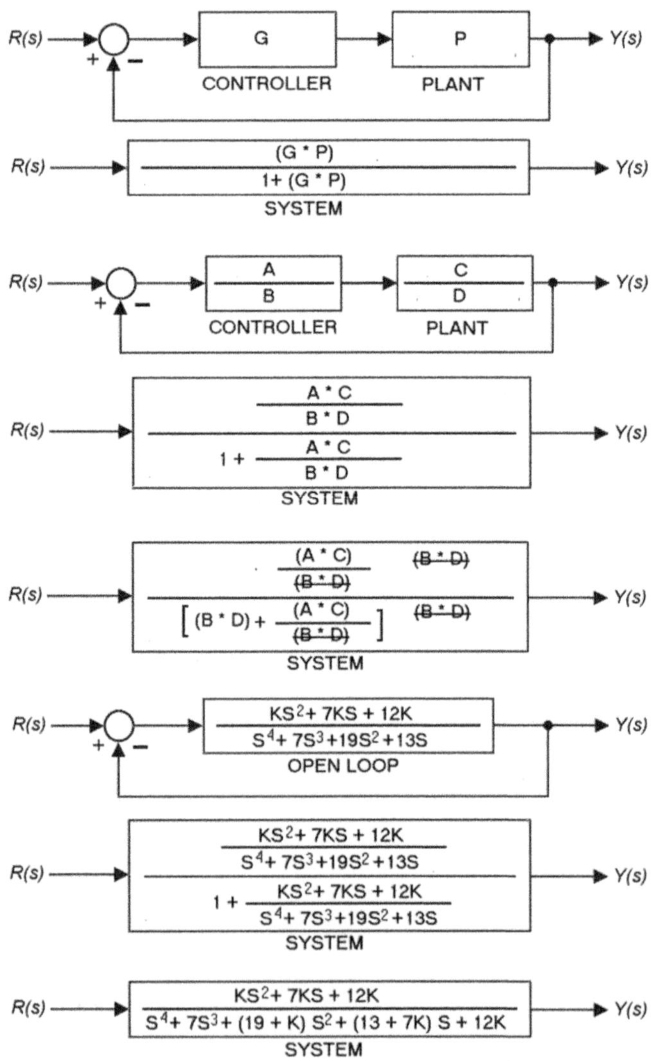

$$open\ loop = \frac{A}{B} \bullet \frac{C}{D} = \left(\frac{K(s+3)}{s(s+1)} \frac{(s+4)}{(s^2+6s+13)} \right) = \frac{(Ks+3K)(s+4)}{(s^2+s)(s^2+6s+13)}$$

$$open\ loop = \frac{A \bullet C}{B \bullet D} = \frac{(Ks+3K)(s+4)}{(s^2+s)(s^2+6s+13)} = \frac{Ks^2+4Ks+3Ks+12K}{s^4+6s^3+13s^2+s^3+6s^2+13s}$$

$$open\ loop = \frac{A \bullet C}{B \bullet D} = \frac{Ks^2+4Ks+3Ks+12K}{s^4+6s^3+13s^2+s^3+6s^2+13s} = \frac{Ks^2+7Ks+12K}{s^4+7s^3+19s^2+13s}$$

$$closed\ loop = \left(\frac{\left(\frac{A \bullet C}{B \bullet D}\right)}{1+\left(\frac{A \bullet C}{B \bullet D}\right)} \right) = \frac{\left(\frac{Ks^2+7Ks+12K}{s^4+7s^3+19s^2+13s}\right)}{1+\left(\frac{Ks^2+7Ks+12K}{s^4+7s^3+19s^2+13s}\right)}$$

$$closed\ loop = \frac{\left(\frac{Ks^2+7Ks+12K}{s^4+7s^3+19s^2+13s}\right)}{1+\left(\frac{Ks^2+7Ks+12K}{s^4+7s^3+19s^2+13s}\right)} = \frac{\left(\frac{Ks^2+7Ks+12K}{s^4+7s^3+19s^2+13s}\right)(s^4+7s^3+19s^2+13s)}{\left[1+\left(\frac{Ks^2+7Ks+12K}{s^4+7s^3+19s^2+13s}\right)\right](s^4+7s^3+19s^2+13s)}$$

$$closed\ loop = \frac{Ks^2+7Ks+12K\left(\frac{s^4+7s^3+19s^2+13s}{s^4+7s^3+19s^2+13s}\right)}{(s^4+7s^3+19s^2+13s)+Ks^2+7Ks+12K\left(\frac{s^4+7s^3+19s^2+13s}{s^4+7s^3+19s^2+13s}\right)}$$

$$closed\ loop = \frac{Ks^2+7Ks+12K}{s^4+7s^3+19s^2+13s+Ks^2+7Ks+12K} = \frac{Ks^2+7Ks+12K}{s^4+7s^3+(19+K)s^2+(13+7K)s+12K}$$

$$P_{(s)} = s^4+7s^3+(19+K)s^2+(13+7K)s+12K$$

The previous block diagrams and equations show the steps to calculate the closed-loop transfer function, needed for the CSE exam. We only need the poles in the bottom of the closed-loop system transfer function block diagram and equation. These poles will be evaluated for stability of the system in the Routh criterion as follows.

For given coefficients a_i of the *characteristic equation*, the method of Routh, which is an alternative to the method of Hurwitz, can be applied. Here the coefficients $a_i (i = 0,1,...,n)$ will be arranged in the first two rows of the Routh schema, which contains $n + 1$ rows:

Row n	s^n	a_o	a_2	a_4	a_6	...	...	0
Row n–1	s^{n-1}	a_1	a_3	a_5	a_7	...	...	0
Row n–2	s^{n-2}	b_1	b_2	b_3	b_4	...	0	
Row n–3	s^{n-3}	c_1	c_2	c_3	c_4	...	0	
Row 3	s^3	d_1	d_2	0				
Row 2	s^2	e_1	e_2	0				
Row 1	s^1	f_1						
Row 0	s^0	g_1						

Now the Routh criterion includes the following:

A polynomial $P(s)$ is Hurwitzian, if and only if the following three conditions are valid:

1. All coefficients $a_i (i = 0,1,...,n)$ are positive.
2. All coefficients $b_1, c_1, ...$ in the first column of the Routh schema are positive.
3. All coefficients $b_1, c_1, ...$ in the first column of the Routh schema are not zero.

As in the first row of the Routh schema (Row n), if a coefficient is negative, the system is unstable. For proving instability, it is sufficient to build the Routh schema only until a negative or zero value occurs in the first column. In the example, the given schema could have been stopped at the fifth row.

Another interesting property of the Routh schema says that the number of roots with positive real parts is equal to the number of changes of sign of the values in the first column.

Check for Stability Using Routh (Example)

$$P(s) = s^4 + 7s^3 + (19 + K)s^2 + (13 + 7K)s + 12K$$

Note: $\{P(s) = a_0 + a_1 + a_2 + a_3 + a_4\}$

The Routh schema is:

s^4	a_0	a_2	a_4	0
s^3	a_1	a_3	a_5	0
s^2	b_1	b_2	b_3	
s^1	c_1	c_2	0	
s^0	d_1			

Building the cross-products, start with the elements of the firstrow. The calculation of these b values will be continued until all remaining elements become zero. The coefficients $b_1, b_2, ...$ in the third row are the results from cross-multiplication of the first two rows according to:

$$b_1 = \frac{a_1 a_2 - a_0 a_3}{a_1} = \frac{(7)(19 + K) - (1)(13 + 7K)}{7}$$

$$b_1 = \frac{(133 + 7K) - (13 + 7K)}{7} = \frac{120}{7}$$

$$b_2 = \frac{a_1 a_4 - a_0 a_5}{a_1} = \frac{(7)(12K) - (1)(0)}{7}$$

$$b_2 = 12K$$

$$b_3 = \frac{a_1 a_6 - a_0 a_7}{a_1} = \frac{(7)(0) - (1)(0)}{7} = 0$$

Note: We do not have s^5 so a_5 will equal 0.

Note: We do not have a_6 or a_7 so they will equal 0.

The calculation of the c values are performed accordingly from the two previous rows as follows:

$$c_1 = \frac{b_1 a_3 - a_1 b_2}{b_1} = \frac{\left(\frac{120}{7}\right)(13+7K)-(7)(12K)}{\left(\frac{120}{7}\right)}$$

$$c_1 = \frac{\left(\frac{120}{7}\right)(13+7K)-(7)(12K)}{\left(\frac{120}{7}\right)} = \frac{13+7K}{} - \left[\frac{84K}{\left(\frac{120}{7}\right)}\right]$$

$$c_1 = 13 + 7K - 4.9K = 13 + 2.1K$$

$$c_2 = \frac{b_1 a_5 - a_1 b_3}{b_1} = \frac{\left(\frac{120}{7}\right)(0)-(7)(0)}{\left(\frac{120}{7}\right)} = 0$$

Note: We do not have a_5 so it will equal 0.

For our example, the last two rows are:

$$d_1 = \frac{c_1 b_2 - b_1 c_2}{c_1} = \frac{(13+2.1K)(12K) - \left(\frac{120}{7}\right)(0)}{(13+2.1K)}$$

$$d_1 = \frac{(13+2.1K)(12K)}{(13+2.1K)} = 12K$$

$$d_1 = b_2 = 12K$$

The Routh schema is:

s^4	1	(19 + K)	12K	0
s^3	7	(13 + 7K)	0	0
s^2	$\left(\frac{120}{7}\right)$	12K	0	
s^1	(13 + 2.1K)	0	0	
s^0	12K			

Substituting a value for the controller equal to K will let us evaluate the scheme for stability. It can be seen that any number greater than 0 will give a positive value.

Feedforward and the Lead/Lag Compensator with Heat Exchangers (Example)

Feedforward control is a powerful technique that can dramatically improve control variance in cases where load changes cause big deviations from the set point and the actual process dynamics are too slow to enable the feedback controller to operate fast enough to catch these disturbances. There are many instances in plants where feedforward could have been used successfully but has not been employed because most control people do not know how to implement it properly. An example would be when driving a car and approaching a hill, when do you step on the gas? That is feedforward.

The big problem with feedback control is that you must tune the feedback controller to match the dynamics of the actual process. In this case, the process will likely have a large first-order lag with a time constant that may be many minutes long. The integral in the controller will then have to be set close or equal to this time constant. This means that the controller can only respond slowly to changes, particularly to slow PV changes. Also remember that a feedback control system can only start making corrections after an error is picked up in the controller. The overall result, as shown in the figure below, is that a large deviation from the set point will occur before the feedback controller can get the process under control, stop the upset, and return the process variable back to the set point.

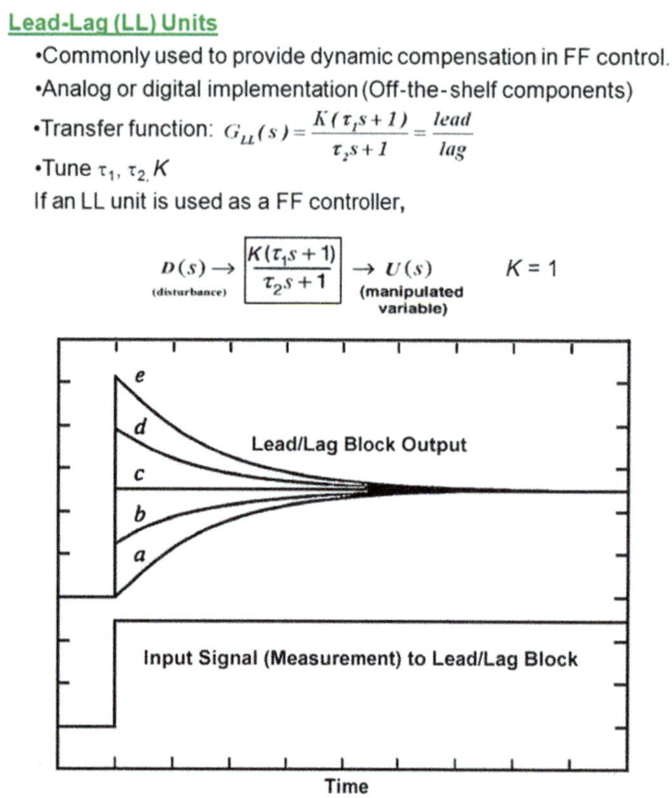

Examples of dynamic responses for a Lead/Lag block algorithm. Ratios of lead to lag times are:

(a) 0; (b) 0.5; (c) 1.0; (d) 1.5; (e) 2.0

Imagine that when the step load change occurs, we move the control valve in such a manner that it will increase the temperature in an exactly equal and opposite way, proportional to the drop in the temperature (PV) being experienced by the increased process flow through the heat exchanger. If this output could be achieved, then we would actually cancel out the effect of the load change, so that there would be no deviation at all from the set point. This is exactly what feedforward control attempts to do with a measurement of the disturbance and a model of the process. Theoretically, zero control variance is possible with a feedforward controller!

The figure below to the right shows how a basic feedforward controller scheme is being used to accomplish this task. (**Note:** The feedback controller system has been removed for simplicity.) The signal from the flowmeter measurement (the changing process load or fluid flow) is sent to a module known as a feedforward module or feedforward controller block. The output from the feedforward controller is then summed with the temperature control blocks and is sent to the feedback controller block, to manipulate the cooling or steam control valve.

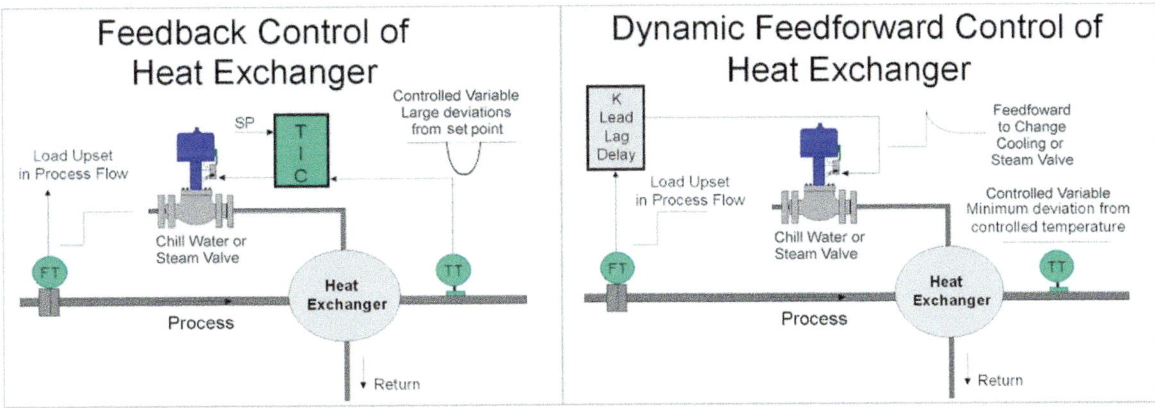

Notes

18 Communications and Control Networks

Overview of Corporate and Plant Networks

On the CSE exam, there may be a few questions on fieldbus, intelligent devices, and networks. We will briefly review the highlights of these subjects. For more information on fieldbus, contact your local distributor.

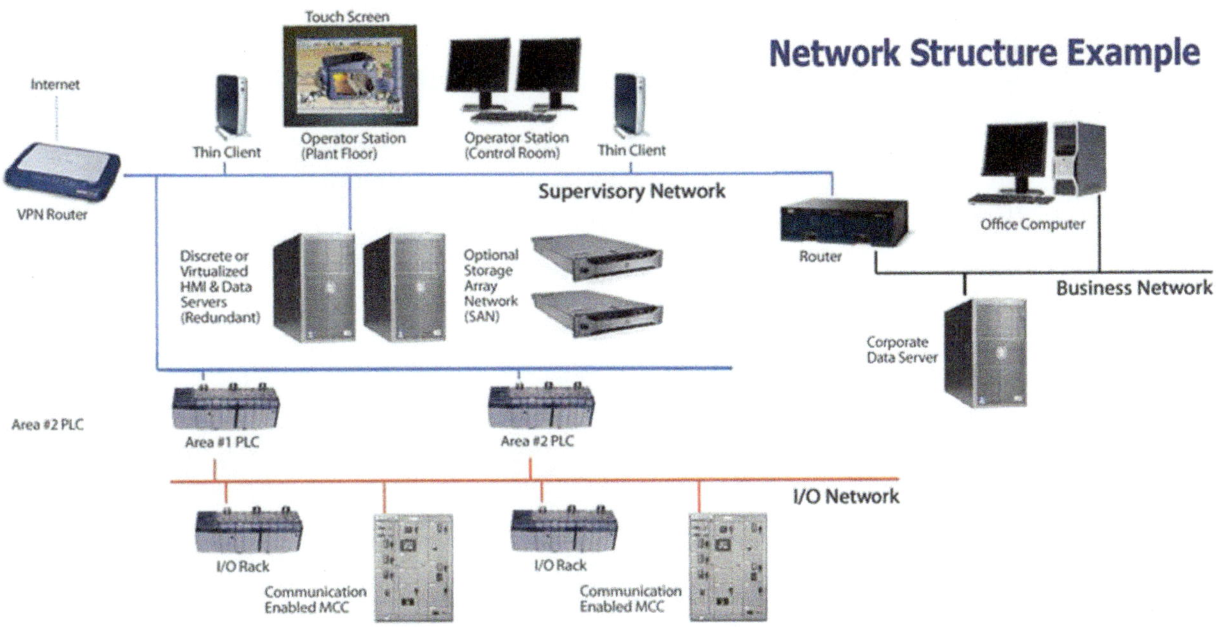

Fieldbus is a digital, two-way, multidrop communication link among intelligent control devices that replaces the 4–20 mA analog standard devices. *The key to fieldbus is that the device is digital, not analog.* There are numerous protocols on the international market. FOUNDATION Fieldbus, PROFIBUS, AS/i, ControlNet, DeviceNet, Modbus, and HART are the most popular in the process industry.

The most popular types of fieldbus typically use EIA-485 protocol with token passing and 31.25 kbps on a single twisted pair wire that can be run up to 1900 m. They can have 32 segments and 1024 intelligent devices per network.

The connected intelligent devices are not calibrated; the data is scaled in software. Intelligent devices may deliver from 1 up to 12 or more data variables of information from one instrument. The data is delivered in data packets to the intelligent control device or master. With FOUNDATION

Fieldbus, any intelligent device can be the controller. More than likely, the valve may be selected as the intelligent controller that will be responsible for the PID calculation for its control loop.

Intelligent devices must be configured when first installed. This is done through Electronic Device Description Language or Field Device Tool. Most of the intelligent devices are plug-and-play. PROFIBUS devices can even be changed out without reconfiguring the device once it is initially configured. The configuration data is stored by the master controller and is then automatically downloaded to the new device on connection to the network.

Networks can be connected by wire or fiber-optic cable, or can be wireless. There are three major categories of networks: LAN (local area network), WAN (wide area network), and MAN (metropolitan area network). The LAN is typically limited to 100 m (or 330 ft per segment) and 1024 nodes.

Industrial instruments typically communicate through a version of one of the three communication network protocols:

1. In a *fieldbus network,* they use one of the previously mentioned networks.
2. In a *serial network,* they use EIA/RS-232, EIA/RS-485, or EIA/RS-488.
3. In an *Ethernet network,* they use Ethernet/IEEE 802.3, Token Ring/IEEE 802.5, and the fiber distributed data interface (FDDI). A fiber backbone for the control system usually uses IEEE 802.1Q. This is a 1-gigabit Ethernet network.

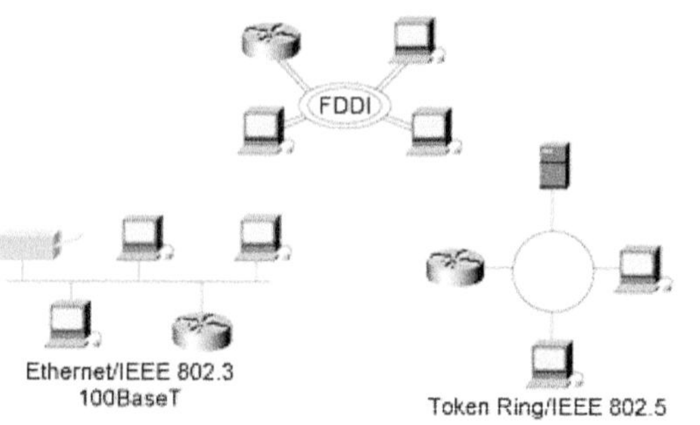

Three typical Ethernet networks.

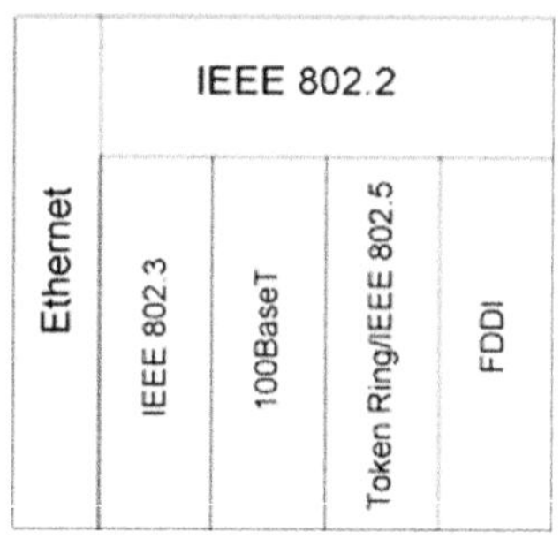

Ethernet protocols.

If the device communicates through Ethernet protocol, it typically has a *media access control* (MAC) address. Like a social security number, this number is unique to each device. For a device on one network to talk to a device on another network using a different protocol, a *protocol converter* or *gateway* is needed.

Open System Interconnect and TCP/IP Network Layer Model

Seven Layers of Networking in the Open System Interconnect Model

The network layers in the Open System Interconnect (OSI) model enable us to break down functional steps in communication by program protocols. You can think of protocols as the different steps in an assembly line. Computers communicate with each other through encapsulating or assembling data in packets. The receiving computer then disassembles these encapsulated packets. Remember the network layers using the following:

Layers from top to bottom (7–1): **A**ll **P**eople **S**eem to **N**eed **D**ata **P**rocessing
Layers from bottom to top (1–7): **P**lease **D**o **N**ot **T**hrow **S**ausage **P**izza **A**way

Physical (Layer 1)

This layer conveys the bitstream—electrical impulse, light or radio signal—through the network at the electrical and mechanical level. It provides the hardware means of sending and receiving data on a carrier, including defining cables, cards, and physical aspects. Fast Ethernet, RS232, and ATM are protocols with Physical layer components.

Physical layer examples include Ethernet, FDDI, B8ZS, V.35, V.24, and RJ45.

Data Link (Layer 2)

At this layer, data packets are encoded and decoded into bits. This layer furnishes transmission protocol knowledge and management, and handles errors in the Physical layer, flow control, and frame synchronization. The Data Link layer is divided into two sublayers: the MAC layer and the logical link control (LLC) layer. The MAC sublayer controls how a computer on the network gains access to the data and permission to transmit it. The LLC layer controls frame synchronization, flow control, and error checking.

Data Link layer examples include PPP, FDDI, ATM, IEEE 802.5/ 802.2, IEEE 802.3/802.2, HDLC, and Frame Relay.

Network (Layer 3)

This layer provides switching and routing technologies, creating logical paths, known as *virtual circuits*, for transmitting data from node to node. Routing and forwarding are functions of this layer, as well as addressing, internetworking, error handling, congestion control, and packet sequencing.

Network layer examples include AppleTalk DDP, IP, and IPX.

Transport (Layer 4)

This layer provides transparent transfer of data between end systems, or hosts, and is responsible for end-to-end error recovery and flow control. It ensures complete data transfer.

Transport layer examples include SPX, TCP, and UDP.

Session (Layer 5)

This layer establishes, manages, and terminates connections between applications. The session layer sets up, coordinates, and terminates conversations, exchanges, and dialogues between the applications at each end. It deals with session and connection coordination.

Session layer examples include NFS, NetBIOS names, RPC, and SQL.

Presentation (Layer 6)

This layer provides independence from differences in data representation (e.g., encryption) by translating from application to network format, and vice versa. The presentation layer works to transform data into the form that the application layer can accept. This layer formats and encrypts data to be sent across a network, providing freedom from compatibility problems. It is sometimes called the *syntax layer.*

Presentation layer examples include encryption, ASCII, EBCDIC, TIFF, GIF, PICT, JPEG, MPEG, and MIDI.

Application (Layer 7)

This layer supports application and end-user processes. Communication partners are identified, quality of service is identified, user authentication and privacy are considered, and any constraints on data syntax are identified. Everything at this layer is application specific. This layer provides application services for file transfers, email, and other network software services. Telnet and FTP are applications that exist entirely in the application level. Tiered application architectures are part of this layer.

Application layer examples include (www) browsers, NFS, SNMP, Telnet, HTTP, and FTP.

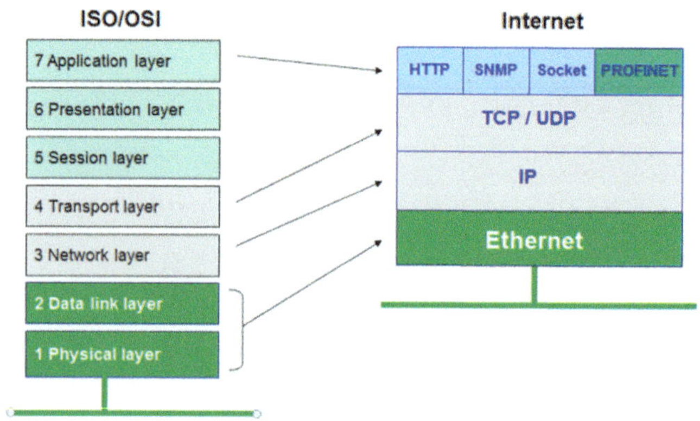

The Typical Network Model

When data flows through a network, a new Internet Protocol (IP) address is assigned to the device that passes the data to each new network. This is done in Layer 3 devices, as mentioned previously. Layer 3 devices are routers, smart switches, and web switches. In the flow of data through segments of a network, the source and destination MAC address is changed on its way to the destination device. This is done in bridging devices, Layer 2 switches.

If we want to travel to another network or through the Internet, we must change the IP address. This is done on the Layer 3 device, typically a router. When a change in the IP address happens in the Layer 3 smart or web switch or router, this is called a *hop.* The number of times the IP address changes through its route to the destination device is called the *hop count.*

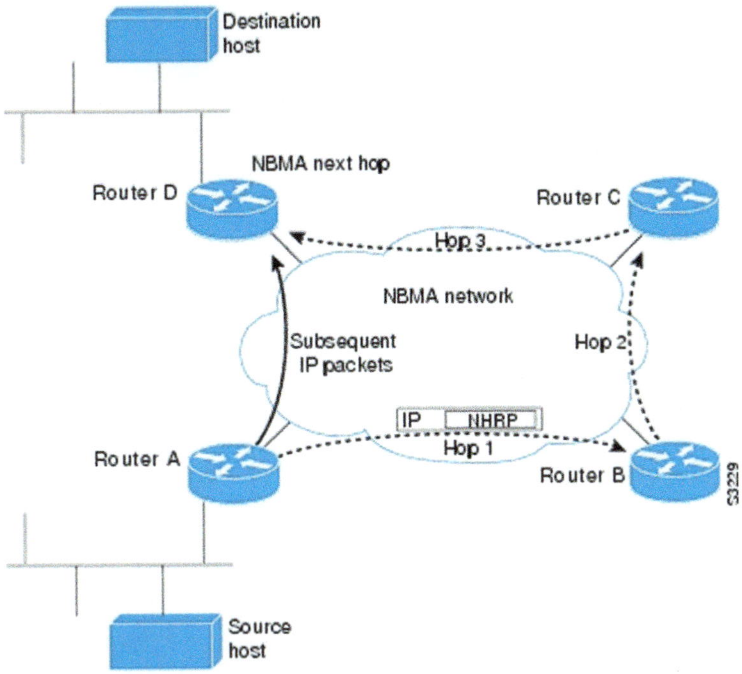

The figure below shows how the IP address is changed through the network from computer A to computer B.

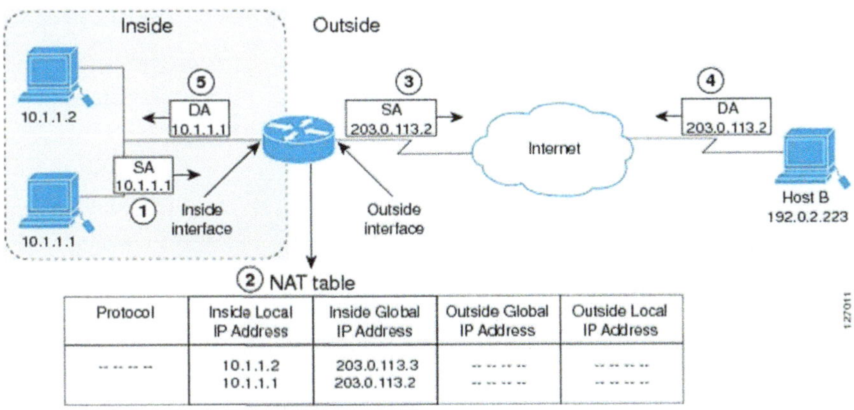

The following process describes how the IP address is changed as shown above:

NAT (network address translation)
DA (destination IP address)
SA (source IP address)

1. The user at host 10.1.1.1 opens a connection to host B.
2. The first packet that the router receives from host 10.1.1.1 causes the router to check its NAT table:
   – If a static translation entry was configured, the router goes to Step 3.
   – If no translation entry exists, the router determines that SA 10.1.1.1 must be translated dynamically; selects a legal, global address from the dynamic address pool; and creates a translation entry. This type of entry is called a *simple entry*.

3. The router replaces the inside local SA of host 10.1.1.1 with the global address of the translation (203.0.13.2) entry and forwards the packet.

4. Host B receives the packet and responds to host 10.1.1.1 using the inside global IP DA 203.0.13.2.

5. When the router receives the packet with the inside global IP address (203.0.13.2), it performs a NAT table lookup using the inside global address as a key. It then translates the address to the inside local address of host 10.1.1.1 and forwards the packet to host 10.1.1.1.

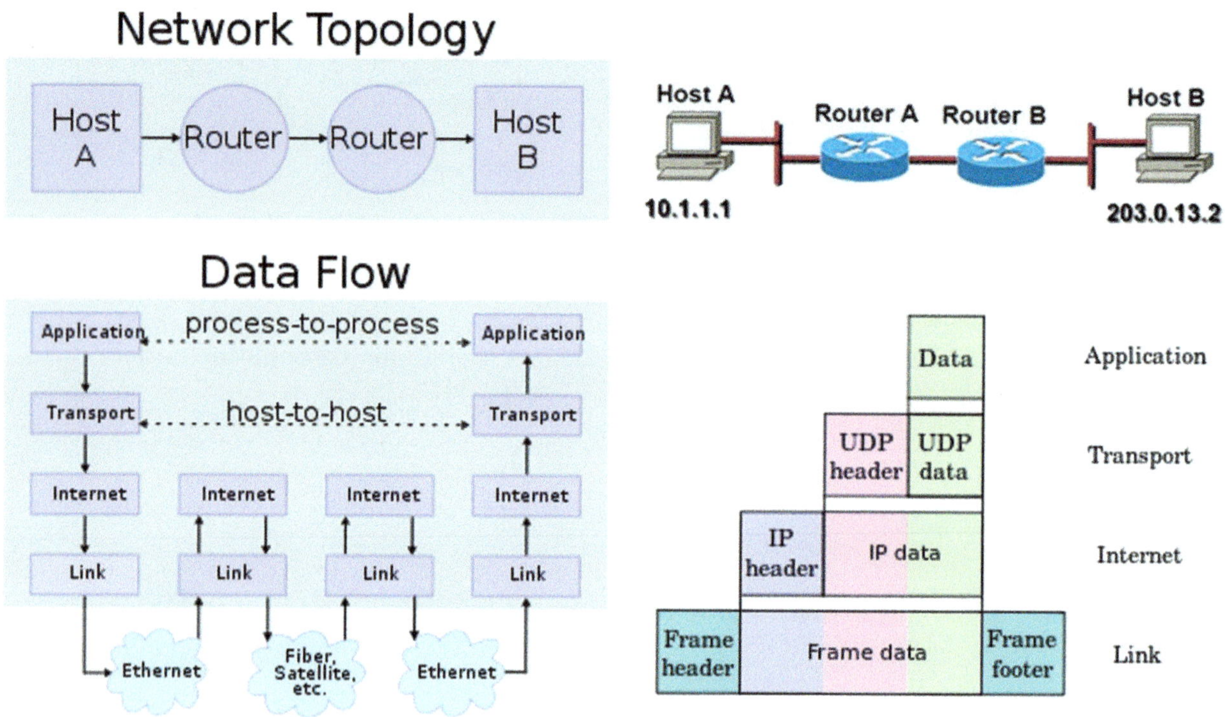

The figure above shows how data is transferred between applications, through direct connections and Ethernet/Internet communications. The closer we come to the Physical layer (Layer 1), the more data that must be added to the information packet for the devices to understand how to communicate the data to the destination device or application.

Overview of Industrial Networks

We will now briefly discuss some of the more popular industrial networks used in plants today.

HART Networks

Traditional HART Network

The HART Communication Protocol (Highway Addressable Remote Transducer Protocol) is an early implementation of fieldbus, a digital industrial automation protocol, although it is typically not used as a fieldbus system. Its most notable advantage is that it can communicate over legacy 4–20 mA analog instrumentation wiring, sharing the pair of wires used by the older system.

The HART Communication Protocol is based on the Bell 202 telephone communication standard and operates using the frequency shift keying (FSK) principle. The digital signal is made up of two frequencies: 1200 Hz and 2200 Hz representing bits 1 and 0, respectively. Sine waves of these two frequencies are superimposed on the direct current (DC) analog signal cables to provide simultaneous analog and digital communications. See the figure below.

Because the average value of the FSK signal is always zero, the 4–20 mA analog signal is not affected. The digital communication signal has a response time of approximately 2–3 data updates per second without interrupting the analog signal. A recommended minimum loop impedance of 250 Ω is required for communication.

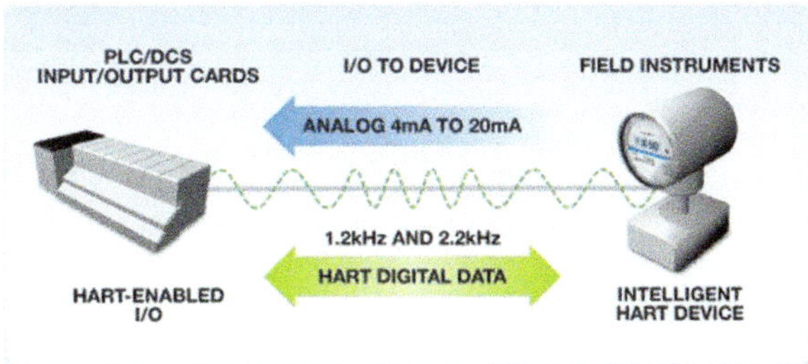

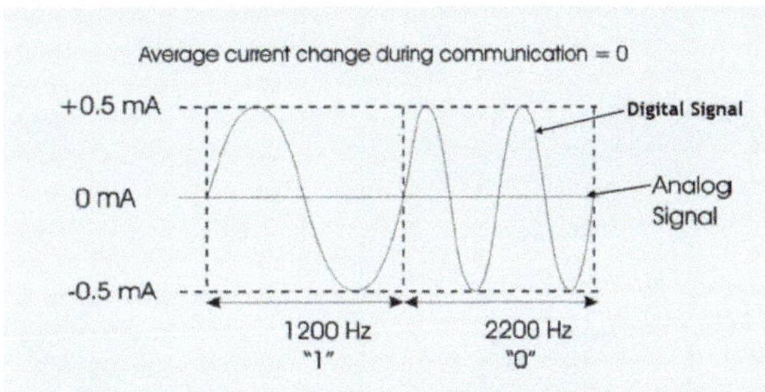

HART digital signal for information and calibration superimposed on the 4–20 mA current process signal.

In multidrop mode (digital only) as shown below, the digital signal is used. The analog loop current is fixed at 4 mA. In multidrop mode, it is possible to have more than one instrument on one signal cable. HART revisions 3 through 5 allowed polling addresses of the instruments to be in the range 1–15. HART 6 and later allowed up to 63 addresses. Each instrument must have a unique address. Note that the device address will be 0 for non-multidrop and 1 or greater for multidrop.

Wired HART Network

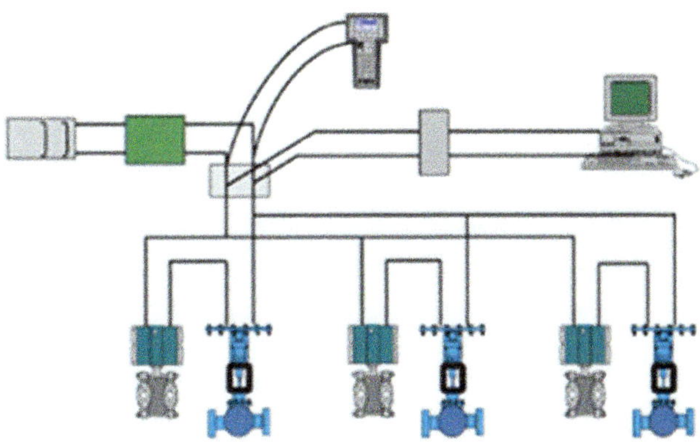

Wireless HART Network

In the WirelessHART network protocol, devices are addressable through multiple network gateways to ensure the data has an unobstructed path back to the controller. The transmitter may be powered by a power supply, battery, or solar panel. The battery typically has to be replaced every couple of years for an update every 1 s and every several years for an update every 60 s. Check with the manufacturer.

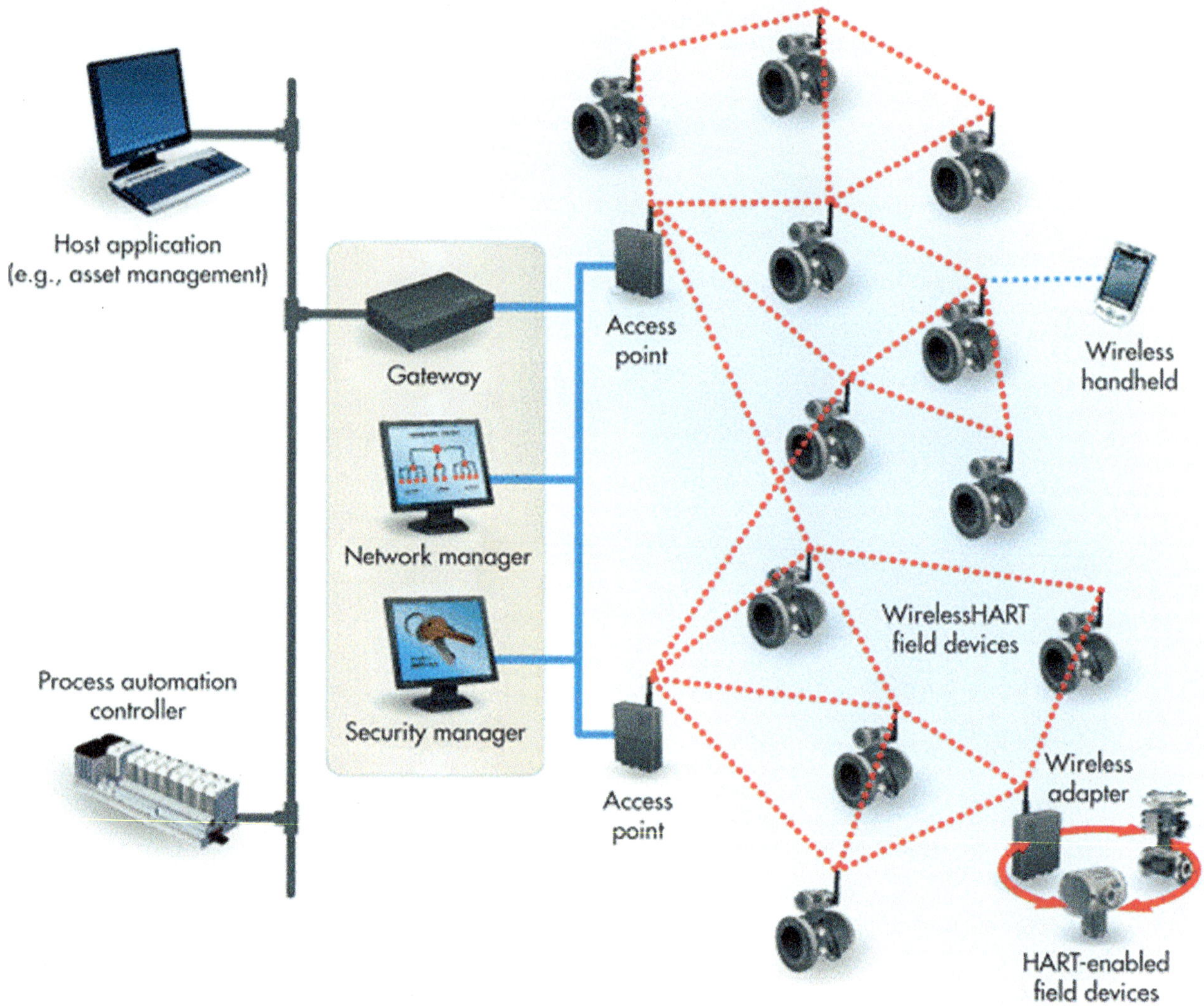

The ANSI/ISA-5.1-2009 symbols for electromagnetic waves are shown below. The sine wave is used for radar and sonic signals as used in instrumentation. The lightning bolt is used to show communications of wireless network devices and gateways.

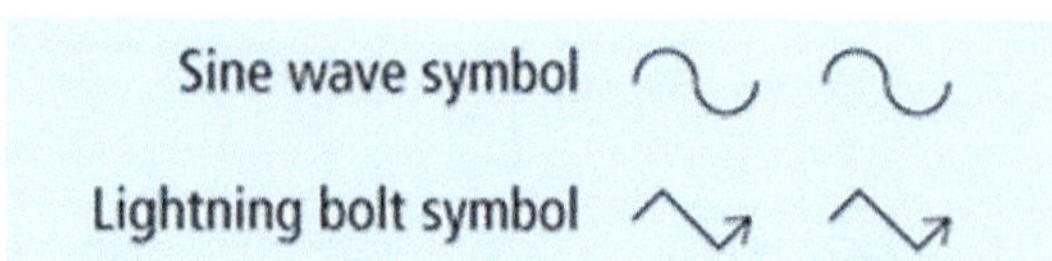

PROFIBUS and AS/i Networks

PROFIBUS (process field bus) is a standard for fieldbus communication in automation technology, not to be confused with the PROFINET standard for industrial Ethernet. PROFIBUS is openly published as part of IEC 61158. In excess of 30 million PROFIBUS nodes were installed as of 2009, and 5 million of these are in the process industries. PROFIBUS uses Manchester encoding, allowing for its use in safety applications.

Reasons for Choosing PROFIBUS

- Preferred fieldbus for most end users and used in the largest number of applications worldwide
- Openness and interoperability, allowing changes/updates at low cost
- Protocol optimized for factory and process control using standardized interfaces.
- Less hardware needed, fewer costs, and less space needed, leading to lowered installation and life-cycle costs
- Easy integration of functional safety and motion control for factory and process automation
- Flexible media redundancy to ensure maximized uptime
- Huge vendor and product choice

PROFIBUS DP

PROFIBUS DP (decentralized peripherals) is used to operate sensors and actuators via a centralized controller in production (factory) automation applications. The many standard diagnostic options, in particular, are discussed here. PROFIBUS DP uses two core screened cables with a violet sheath and runs at speeds between 9.6 kbit/s and 12 Mbit/s. A particular speed can be chosen for a network to give enough time for communication with all the devices present in the network. The RS485 balanced transmission used in PROFIBUS DP allows only 126 devices to be connected at once; however, more devices can be connected or the network can be expanded with the use of hubs or repeaters.

Application Layer

To utilize these functions, various service levels of the DP protocol were defined:

- DP-V0 for cyclic exchange of data and diagnosis
- DP-V1 for acyclic data exchange and alarm handling
- DP-V2 for isochronous mode and data exchange broadcast (slave-to-slave communication)

Security Layer

The security layer fieldbus data link (FDL) works with a hybrid access method that combines token passing with a master/slave method.

Bit-Transmission Layer

Three different methods are specified for the bit-transmission layer:

- With electrical transmission pursuant to EIA-485, twisted pair cables with impedances of 150 Ω are used in a bus topology. Bit rates from 9.6 Kbit/s to 12 Mbit/s can be used. The cable length between two repeaters is limited from 100 to 1200 m, depending on the bit rate used. This transmission method is primarily used with PROFIBUS DP.
- With optical transmission via fiber optics, star-, bus-, and ring-topologies are used. The distance between the repeaters can be up to 15 km. The ring topology can also be executed redundantly.
- With Manchester bus powered (MBP) transmission technology, data and fieldbus power are fed through the same cable. The power can be reduced in such a way that use in explosion-hazardous environments is possible. The bus topology can be up to 1900 m long and permits branching to field devices (maximum 60 m branches). The bit rate here is a fixed 31.25 Kbit/s. This technology was specially established for use in automation and control for PROFIBUS PA.

PROFIBUS PA

PROFIBUS PA (process automation) is used to monitor measuring equipment via a process control system in automation and control applications. This variant is designed for use in explosion/hazardous areas (Ex-zone 0 and 1). The Physical layer (i.e., the cable) conforms to IEC 61158-2, which causes power to be delivered over the bus to field instruments, while limiting current flows so that explosive conditions are not created, even if a malfunction occurs.

The number of devices attached to a PA segment is limited by this feature. PA has a data transmission rate of 31.25 Kbit/s. However, PA uses the same protocol as DP and can be linked to a DP network using a coupler device. The much faster DP acts as a backbone network for transmitting process signals to the controller. This means that DP and PA can work tightly together, especially in hybrid applications where process and factory automation networks operate side-by-side.

PROFINET

PROFINET is a standard for industrial Ethernet based on PROFIBUS. This model was largely inspired by IEC 61499, *Function Blocks for Embedded and Distributed Control Systems Design*.

Technology

Three protocol levels are defined:

1. TCP/IP for PROFINET CBA and the commissioning of a plan with reaction times in the range of 100 ms

2. RT (real-time) protocol for PROFINET CBA and PROFINET IO applications up to 10 ms cycle times
3. iRT (Isochronous Real-Time) for PROFINET IO applications in drive systems with cycles times of less than 1 ms

AS/i

AS-Interface (AS/i), the actuator sensor interface, is a lower level fieldbus system. Its primary application is for on/off sensor and actuators. It is used extensively in manufacturing and machine safety applications, by using asynchronous data transfer on two streams at the same time. All sensors and actuators have a network address and are powered by the two-wire network cable with a typical 30.5-volt DC power supply. It can control up to 62 slaves per master at data rates of 167 Kbit/s and up to lengths of 300 m. AS/i uses 4-bit digital input and 3-bit digital output data, up to 248-bit digital input and 186-bit digital output data, and 124-bit digital data for analog values. AS/i has a maximum bus cycle time of 10 ms.

PROFIBUS Fieldbus Message Specification

Fieldbus Message Specification (FMS) is the most universal Siemens Step 7 programmable logic controller (PLC) communications protocol. FMS uses the OSI Application layer (Layer 7) to establish the *lower layer interface* (LLI). LLI establishes various communication relationships and provides FMS with device-independent access to the OSI Data Link layer (Layer 2).

PROFIBUS

PROFIBUS can be tailored to specific needs using application profiles. Many profiles combine standards for transmission media, communication protocol (FMS, DP-V0, etc.), and unique protocols. Each application profile is tailored to a specific use, and new profiles are developed regularly. Some application profiles are widespread; three examples are PROFISafe, PROFISync, and PROFIDrive.

PROFISafe

PROFISafe uses additional software to create a high-integrity network. This network is useful in situations where high safety is a requirement. For suppliers and manufacturers to be certified in PROFISafe, they must maintain high standards in quality.

PROFIDrive

PROFIDrive was created for motion control applications. Software added to the PROFIBUS DP specification causes the network to achieve precise control of servomotors and other equipment. Therefore, PROFIDrive can achieve synchronization across the network.

Use of the OSI Networking Layers in PROFIBUS

The Physical layer corresponds to OSI Layer 1, which receives encoded messages from the upper layers and converts the messages to physical signals on the fieldbus transmission medium and vice versa.

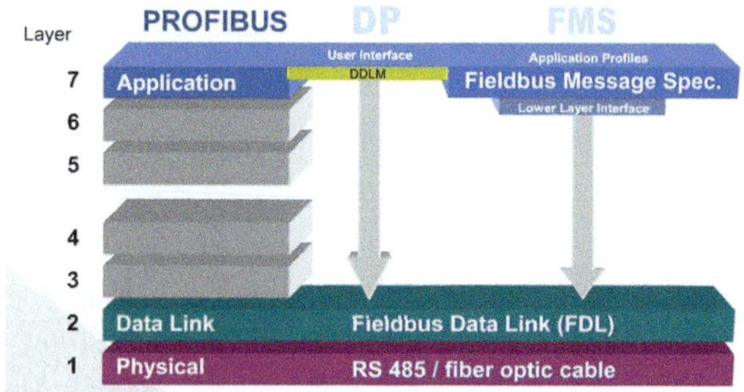

Layers 3, 4, and 5 are not developed for PROFIBUS. Layer 7 is not used for the PROFIBUS DP protocol. This lean architecture ensures high-speed data transmission. The *Direct Data Link Mapper* (DDLM) provides direct access to the functions of Layer 2 for the PROFIBUS DP user interface.

FOUNDATION Fieldbus

FOUNDATION Fieldbus was originally intended as a replacement for the 4–20 mA standard, and today it coexists alongside other technologies such as Modbus, PROFIBUS, and industrial Ethernet. FOUNDATION Fieldbus is used in many heavy process applications, such as refining, petrochemicals, and power generation, and even in food and beverage, pharmaceutical, and nuclear applications.

FOUNDATION Fieldbus was developed over a period of many years by ISA as SP50. In 1996, the first H1 (31.25 Kbit/s) specifications were released. In 1999, the first high speed Ethernet (HSE) specifications were released. The International Electrotechnical Commission (IEC) standard on fieldbus, including FOUNDATION Fieldbus, is IEC 61158-1:2019 *Industrial communication networks - Fieldbus specifications - Part 1: Overview and guidance for the IEC 61158 and IEC 61784 series.*

Reasons for Choosing FOUNDATION Fieldbus

- Reduced number of wires and marshalling cabinets
- Reduced number of intrinsic safety barriers
- Reduced number of input/output convertors
- Remote configuration of devices
- Increased sophistication and flexibility of instruments
- Increased uptime due to less equipment and better self-diagnostic and remote diagnostics
- H1 technology included in the international IEC standard IEC 61158

H2 or HSE

HSE, also known as the H2 protocol, is ideally suited for use as a control backbone. Running at 100 Mbit/s, the technology is designed for device, subsystem, and enterprise integration. It supports the entire range of fieldbus capabilities, including standard function blocks and device descriptions, as well as application-specific flexible function blocks for advanced process and discrete/hybrid/batch applications.

HSE supports complex logic functions, such as those performed by PLCs, or data-intensive process devices, such as analyzers and gateways to other networks. HSE enhances access to H1 fieldbus technology via linking devices, while providing expanded capabilities for high-speed automation devices and hybrid/batch applications

FOUNDATION Fieldbus H1

FOUNDATION Fieldbus H1 is intended primarily for process control, field-level interface, and device integration. Running at 31.25 Kbit/s, the technology interconnects devices such as transmitters and actuators on a field network. H1 is designed to operate on existing twisted pair instrument cabling with power and signal on the same wire. Fiber-optic media is optional. It also supports intrinsic safety (IS) applications. A typical fieldbus segment consists of the following components.

FOUNDATION Fieldbus H1 devices comprise a function block application, act as a publisher and subscriber of process variables, transmit alarms and trends, and provide server functionality for host access and management functions.

Devices can act as schedulers and timing masters for regulating communication on a fieldbus segment. They are also used for bus interfaces in process control systems or in linking devices. Capable of controlling bus communications and many connections to multiple devices, they support both client and server applications. Communications are scheduled via the Link Active Scheduler (LAS). The LAS bridges between the HSE and H1 networks.

Typical FOUNDATION Fieldbus Segments

- **H1 card** – Fieldbus interface card. It is common practice to have redundant H1 cards, but ultimately this is application specific.
- **PS** – Bulk power (volts, direct current—VDC) to the fieldbus power supply.
- **FPS** – Fieldbus power supply and signal conditioner. Integrated power supplies and conditioners have become the standard.
- **T** – Terminators. Exactly two terminators are used per fieldbus segment, one at the FPS and one at the furthest point of a segment at the device coupler.
- **LD** – Linking device, alternatively used with HSE networks to terminate four to eight H1 segments acting as a gateway to an HSE backbone network.
- **Fieldbus devices** – For example, transmitters, transducers, and valves.

Like PROFIBUS, FOUNDATION Fieldbus uses a modular wiring connection, offering quick installation and maintenance of devices. There is no need to unwire anything. Each device should have at least 9 VDC, and it is assumed that most devices have at least 12 VDC. The fieldbus power supply is normally a redundant, 24 V bulk unit that provides power for multiple fieldbus segments.

The standard color code for wiring includes the following:

- Orange PVC jacket – General-purpose or Class 1, Division 2 applications
- Blue PVC jacket – Intrinsically safe applications
- 18 AWG wires – Orange for positive and blue for negative

Use of the OSI Networking Layers

The fieldbus physical layer corresponds to the OSI Physical Layer (Layer 1), which is the Foundation Fieldbus Physical layer, which receives encoded messages from the upper layers and converts the messages to physical signals on the fieldbus transmission medium and vice versa.

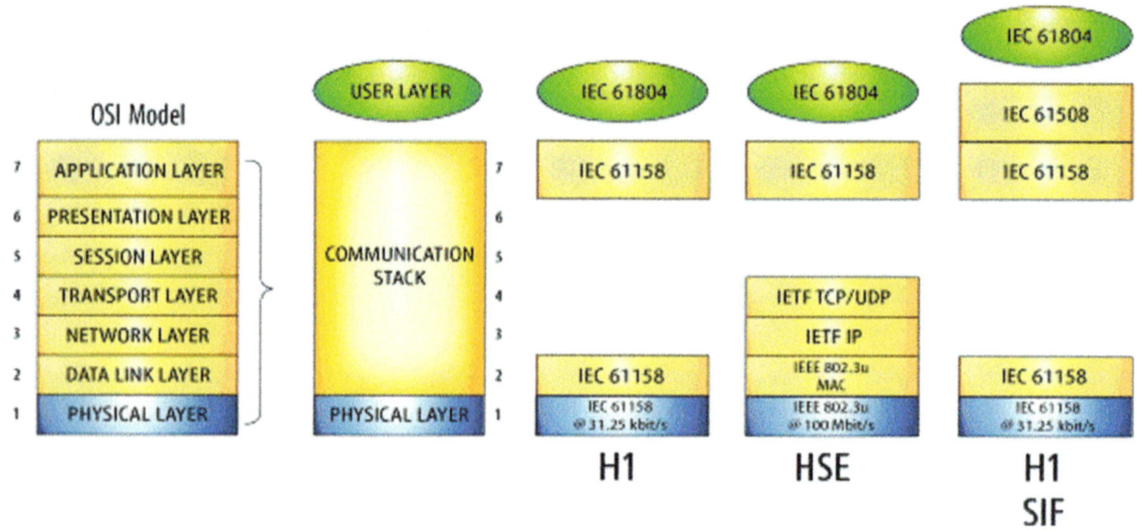

The communication "stack" corresponds to Layers 2 through 7 in the OSI model. The Application layer (Layer 7), encodes and decodes user layer commands. The Data Link layer (Layer 2), controls transmission of messages onto the fieldbus through the Physical layer (Layer 1). The Data Link layer also manages access to the fieldbus through a deterministic, centralized bus scheduler called the LAS. The LAS is used for scheduling transmissions of deterministic messages and authorizing the exchange of data between devices. The fieldbus does not use OSI Layers 3, 4, 5, and 6.

ODVA (CIP) Networks

ODVA

ODVA supports and prefers the use of the Common Industrial Protocol (CIP) networks. CIP is an industrial protocol for industrial automation applications. ODVA is the organization that supports network technologies built on the CIP. These technologies also currently include application extensions to CIP: CIP Safety, CIP Motion, and CIP Sync.

Previously known as the *Control and Information Protocol*, CIP encompasses a suite of messages and services for manufacturing automation applications, including control, safety, synchronization, motion, configuration, and information. It enables users to integrate these manufacturing applications with enterprise-level Ethernet networks and the Internet. It is supported by hundreds of vendors around the world and is media-independent. CIP provides a unified communication architecture throughout the manufacturing enterprise. It is used in EtherNet/IP, DeviceNet, CompoNet, and ControlNet.

ControlNet

ControlNet is an open industrial network protocol for industrial automation applications, also known as a fieldbus. ControlNet was earlier supported by ControlNet International, but in 2008 support and management of ControlNet was transferred to ODVA, which now manages all protocols in the CIP family.

Features that set ControlNet apart from other fieldbuses include the built-in support for fully redundant cables and the fact that communication on ControlNet can be strictly scheduled and highly deterministic.

Reasons for Choosing ControlNet

- Highly deterministic and repeatable network
- Remains unaffected as devices are connected or disconnected from it
- Dependable, synchronized, and coordinated real-time performance
- Can be installed as a redundant system for mission critical applications
- Can be used as a default network for the ControlLogix and Honeywell PLC/DCS platforms
- Can be used as backbone to multiple distributed DeviceNet networks
- Can be used in explosion-proof/hazardous locations with intrinsically safe Flex input/output (I/O) modules

ControlNet Physical Layer

ControlNet cables consist of RG-6 coaxial cable with BNC connectors, though optical fiber is sometimes used for long distances. The network topology is a bus structure with short taps. ControlNet also supports a star topology if used with the appropriate hardware. ControlNet can operate with a single RG-6 coaxial cable bus, or a dual RG-6 coaxial cable bus for cable redundancy.

In all cases, the RG-6 should be of quad-shield variety. The maximum cable length without repeaters is 1000 m, and the maximum number of nodes on the bus is 99. However, there is a trade-off between the number of devices on the bus and the total cable length. Repeaters can be used to further extend the cable length. The network can support up to 5 repeaters (10 when used for redundant networks). The repeaters do not utilize network node numbers and are available in copper or fiber-optic choices. The OSI Physical Layer signaling uses Manchester code at 5 Mbit/s.

CompoNet

CompoNet provides users with a bit-level network to control small, high-speed machines and the CIP network services to connect to the plant and the enterprise. CompoNet fulfills all the requirements of applications using large numbers of simple sensors and actuators by providing high-speed communications with configuration tools and combining this with efficient construction, simple setup, and high availability, all on a single network.

CompoNet uses time division multiple access (TDMA) in its network layer. This approach combines with a physical layer designed to minimize transmission delays and automatically compensates for potential collisions within the CompoNet hierarchy to provide a deterministic network with the ability to update large numbers of nodes at state-of-the-art update rates. In addition, CompoNet includes an option for power (24 VDC, 5 A) and signal in the same cable with the ability to remove and replace nodes under power.

Rockwell Automation (Allen-Bradley) DH485, DH+, RIO

These are legacy (very old) networks and are of little concern today unless you run into one of them in an old plant. DH485 (Data Highway) and DH+ (Data Highway Plus) are industrial bus protocols developed by Rockwell Automation/Allen-Bradley. DH485 and DH+ allow 64 nodes, and the communications protocol used to interconnect the network link and a personal computer (PC) is called DF1. DH uses a trunk cable that runs up to 10,000 ft and drop cables to each node that extend to 100 ft. This system uses peer-to-peer communication in which each node bids on being the floating master.

DH+ is appropriate for smaller networks. It uses peer-to-peer communication implementing token passing. Nominal voltages on the bus are 8–12 V peer-to-peer, and the bus is +/–200 mV sensitive over the two differential lines. Each node on the bus is transformer coupled onto the bus. The bus should be terminated to 150 Ω at each end of the bus. All messages on the bus are either a command or a reply. Also developed by Rockwell Automation/Allen-Bradley is the DH485 protocol, a LAN design for factory-floor applications. DH485 allows for the connection of up to 32 devices, including controllers, color graphics systems, and PCs. RIO (remote I/O) is rarely seen today; for the most part, it has been replaced.

Modbus Networks

Traditional Modbus Networks

Modbus is a serial communications protocol originally published by Modicon (now Schneider Electric) in 1979 for use with its PLCs. Simple and robust, it has since become a de facto standard communication protocol, and it is now a commonly available means of connecting industrial electronic devices. Modbus is used extensively in variable frequency drives and motion control applications for interactive coordination of drives or master/slave communications.

Reasons for Choosing Modbus in the industrial environment are:

- Developed with industrial applications in mind
- Openly published and royalty-free
- Easy to deploy and maintain
- Moves raw bits or words without placing many restrictions on vendors

Modbus enables communication among many (approximately 240) devices connected to the same network, for example, a system that measures temperature and humidity and communicates the results to a computer. Modbus is often used to connect a supervisory computer with a remote terminal unit (RTU) in supervisory control and data acquisition (SCADA) systems. Many of the data types are named from its use in driving relays: a single-bit physical output is called a *coil*, and a single-bit physical input is called a *discrete input* or a *contact*.

Communication and Devices

Each device intended to communicate using Modbus is given a unique address. In serial and MB+ networks, only the node assigned as the master may initiate a command. On Ethernet, any device can send out a Modbus command, although usually only one master device does so. A Modbus command contains the Modbus address of the device it is intended for. Only the intended device will act on the command, even though other devices might receive it. The basic Modbus commands can instruct an RTU to change the value in one of its registers, control or read an I/O port, and command the device to send back one or more values contained in its registers.

Protocols

Modbus Plus

Despite the name, Modbus Plus is not a variant of Modbus. It is a different protocol, involving token passing.

Modbus TCP/IP

Modbus TCP/IP has become ubiquitous because of its openness, simplicity, low-cost development, and minimum hardware required to support it. There are several hundred Modbus TCP/IP devices available in the market, with more being developed each year. It is used to exchange information between devices, monitor them, and program them. It is also used to manage distributed I/Os and is the preferred protocol of manufacturers of this type of devices.

Enron Modbus

Enron Modbus is a modification to the standard Modicon Modbus communication protocol developed by Enron Corporation. The numbering of the register addresses is different, it supports 32-bit registers as well as 16-bit, and it has the ability to transmit event logs and historical data.

EtherCAT

EtherCAT (Ethernet for Control Automation Technology) is an Ethernet-based fieldbus system invented by Beckhoff Automation. The protocol is standardized in IEC 61158-1:2019 *Industrial communication networks - Fieldbus specifications - Part 1: Overview and guidance for the IEC 61158 and IEC 61784 series* and is suitable for both hard and soft real-time requirements in automation technology. The goal during development of EtherCAT was to apply Ethernet for automation applications requiring short data update times (also called *cycle times*; ≤100 μs) with low communication jitter (for precise synchronization purposes; ≤1 μs) and reduced hardware costs.

With EtherCAT, the standard Ethernet packet or frame (according to IEEE 802.3) is no longer received, interpreted, and copied as process data at every node. The EtherCAT slave devices read the data addressed to them while the telegram passes through the device, processing data "on the fly." Similarly, input data are inserted while the telegram passes through. A frame is not completely received before being processed; instead, processing starts as soon as possible. Sending also is conducted with a minimum delay of small bit times. Typically, the entire network can be addressed with just one frame.

SERCOS

The SERCOS (serial real-time communication system) interface is a globally standardized open digital interface for the communication between industrial controls, motion devices (drives), and input output devices (I/Os). It is classified as the standard IEC 61491, *Electrical Equipment of Industrial Machines - Serial Data Link for Real-Time Communications between Controls and Drives.* The SERCOS interface is designed to provide hard real-time, high-performance communications between industrial motion controls and digital servo drives.

Tight coordination is desirable especially in motion controls, where directing the movement of an individual axis of motion must be precisely coordinated so that the motion of the entire system follows a desired path. Types of equipment requiring such coordination are, for example, metal cutting machine tools, metal forming equipment, assembly machinery, packaging machinery, robotics, printing machinery, and material handling equipment. Typically, SERCOS is used for these applications.

Summary – Automation and Process Control Networks

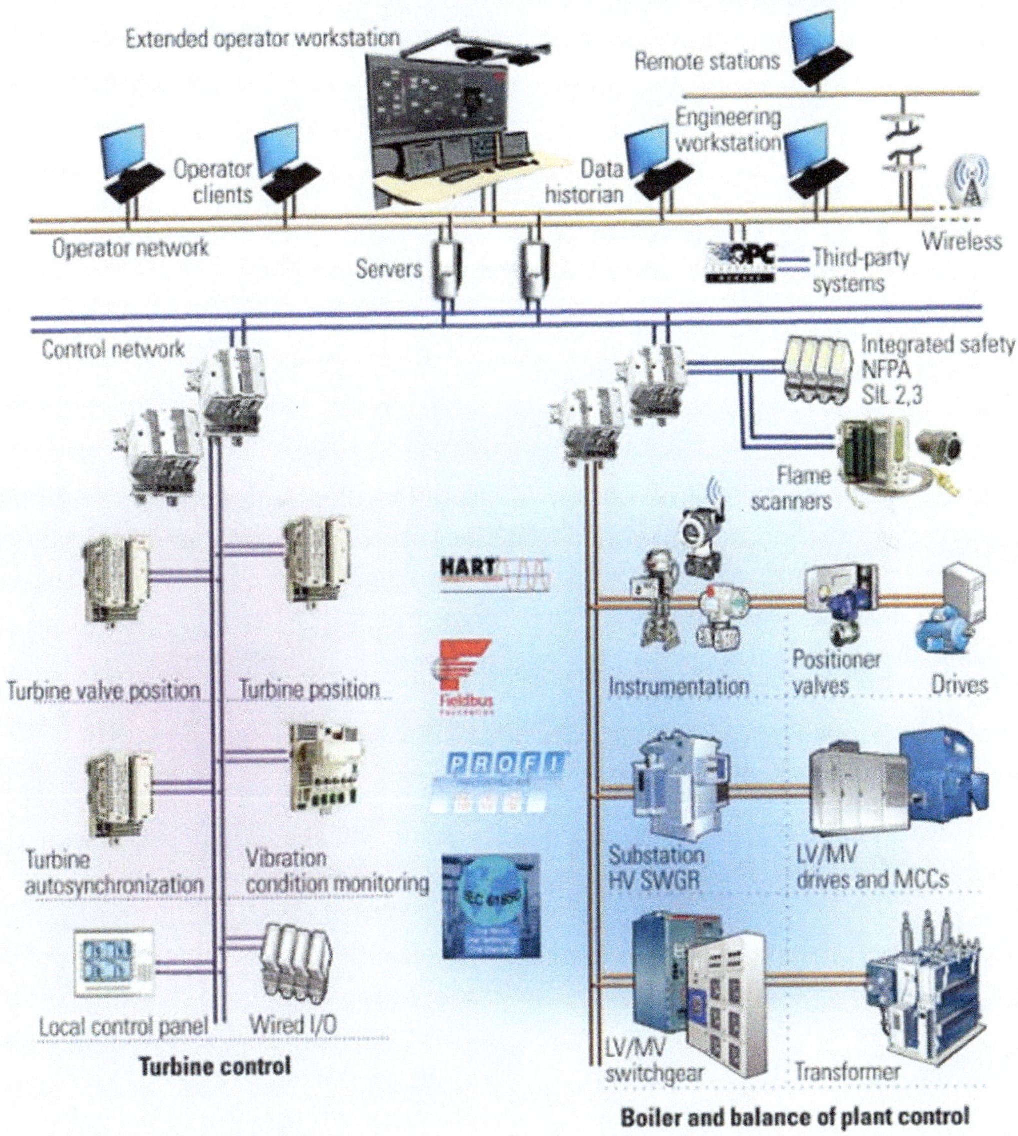

Plant Facility Monitoring and Control Systems (FMCS)

BACnet

BACnet (Building Automation and Control Network) is a communications protocol for building automation and control networks. It is an ASHRAE, ANSI, and ISO standard protocol. BACnet was designed to enable communication of building automation and control systems for applications such as heating, ventilating, and air conditioning control; lighting control; access control; and fire detection systems and their associated equipment. The BACnet protocol provides mechanisms for computerized building automation devices to exchange information, regardless of the particular building service they perform. Proper communication between building automation devices is critical for maximizing building energy efficiency, indoor air quality, and other aspects of "green" buildings.

LonWorks

LonWorks (Local Operating Network) is a networking platform specifically created to address the needs of control applications. The platform is built on a protocol created by Echelon Corporation for networking devices over media such as twisted pair, power lines, fiber optics, and RF.

By 2010, approximately 90 million devices were installed with LonWorks technology, including such diverse functions as embedded machine control, municipal and highway/tunnel/street lighting, heating and air-conditioning systems, intelligent electricity metering, subway train control, building lighting, stadium lighting and speaker control, security systems, fire detection and suppression, and alarming.

Networked Intelligent and Smart Devices

Intelligent and smart devices (instruments) are typically attached to a network and/or are addressable by digital communications, such as a computer, PLC, DCS, or a communicator.

Smart devices are calibrated by way of digital communications with words and bytes of digital data, used by the device's microprocessor. Bits are set to configure the variables in the device. The microprocessor is connected to the outside world by an analog-to-digital convertor and a digital-to-analog convertor. The microprocessor can scale the input and outputs by manipulating bits in the words sent on the data lines. The smart device can perform calculations and nonlinear algorithms.

Although some may disagree, an intelligent device is not a smart device. Smart devices, such as level transmitters, are capable of being programmed or calibrated with a communicator or software over the network. A device that is neither smart nor intelligent must be calibrated or be commissioned by hand (set the zero and span).

An intelligent device is not calibrated in the field or shop. It is calibrated at the factory and left alone. The user chooses what part of the device range to use for the output scale. Standard devices and smart devices typically deliver only one variable, for example, temperature, pressure, or mass flow rate. But an intelligent device can deliver, for example, temperature, pressure, delta pressure, mass flow rate, and viscosity all in one data stream (digital signal). Intelligent devices can perform diagnostics and compensate for errors, and report and warn of failure or predicted failure.

The information is sent in framed data packets to the PLC, DCS, or host computer, which then extracts the multiple data variables for use from the data packets. The information is typically delivered in 1 byte per data variable. The data packet itself may be 8–40 bytes long or longer. A frame can be from 64 to 1,518 bytes long in total.

sPID Control in Intelligent Networked Devices

PROFIBUS and FOUNDATION Fieldbus devices are intelligent devices. FOUNDATION Fieldbus is an example of how the intelligent device has moved the PID controller algorithm into the instrument to control the process loops in the field in place of the DCS or PLC. Most of the new HART devices are intelligent, offering diagnostics and multiple variables from the same measurement device.

For instance, if you would like a transmitter to be at a span of 10–100 in H_2O, you can just enter that amount and then the transmitter can send that information as a scaled psi, percentage of maximum range, or some other measurement scale all at the same time from the same sensor. For a mass flowmeter, just set the scale of maximum flow in the processor of the meter to be 50 gpm max or 10 gpm max. All the other variables will follow: volume, density, viscosity, and temperature.

19 Digital Logic in Control Systems

Overview of Digital Logic

Discrete control plays a vital role in the process control industry. Discrete control is used for material handling, lockouts, and safety controls of systems, indicators, alarms, and switching applications. Discrete control usually takes the form of Relay Ladder Logic (RLL) or digital logic combined with some type of mechanical apparatus. The programmable logic controller (PLC) is the workhorse of the industry today and is covered on the CSE exam with International Society of Automation (ISA) binary logic and Relay Ladder Logic.

Digital Logic Gate Symbols

Familiarize yourself with the following binary logic table and its functions. ISA binary logic is the same in function as digital logic, although the symbols are slightly different. Familiarize yourself with ISA-5.2-1976 (R1992), *Binary Logic Diagrams for Process Operations.* ISA logic is used in the CSE exam. Look at some examples of its use, such as in ISA's *Control Systems Engineer Study Guide.*

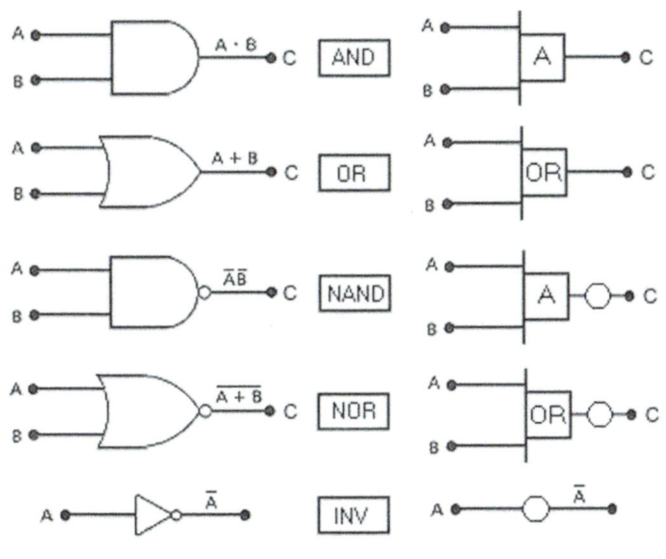

BINARY LOGIC SYMBOLS ISA LOGIC SYMBOLS

Digital Logic Gate Truth Tables

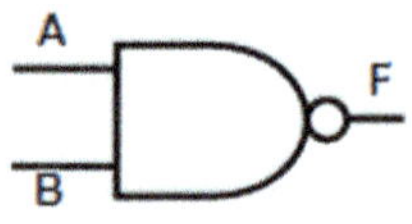

NAND

Inputs		Output
A	B	F
0	0	1
1	0	1
0	1	1
1	1	0

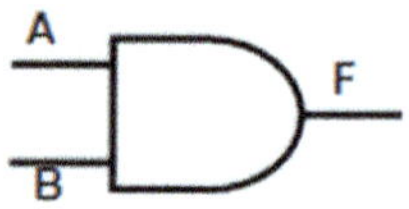

AND

Inputs		Output
A	B	F
0	0	0
1	0	0
0	1	0
1	1	1

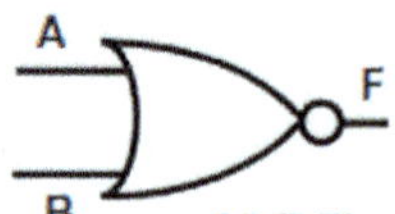

NOR

Inputs		Output
A	B	F
0	0	1
1	0	0
0	1	0
1	1	0

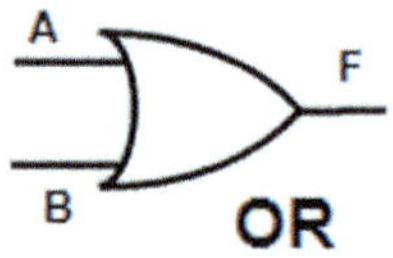

OR

Inputs		Output
A	B	F
0	0	0
1	0	1
0	1	1
1	1	1

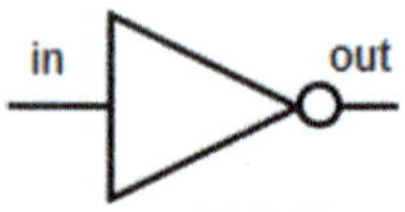

NOT

Input	Output
I	F
0	1
1	0

ISA Binary Logic

This is an example of the use of ISA logic symbols and SAMA (Scientific Apparatus Makers Association) symbols. Familiarize yourself with ISA-5.2-1976 (R1992), *Binary Logic Diagrams for Process Operations.*

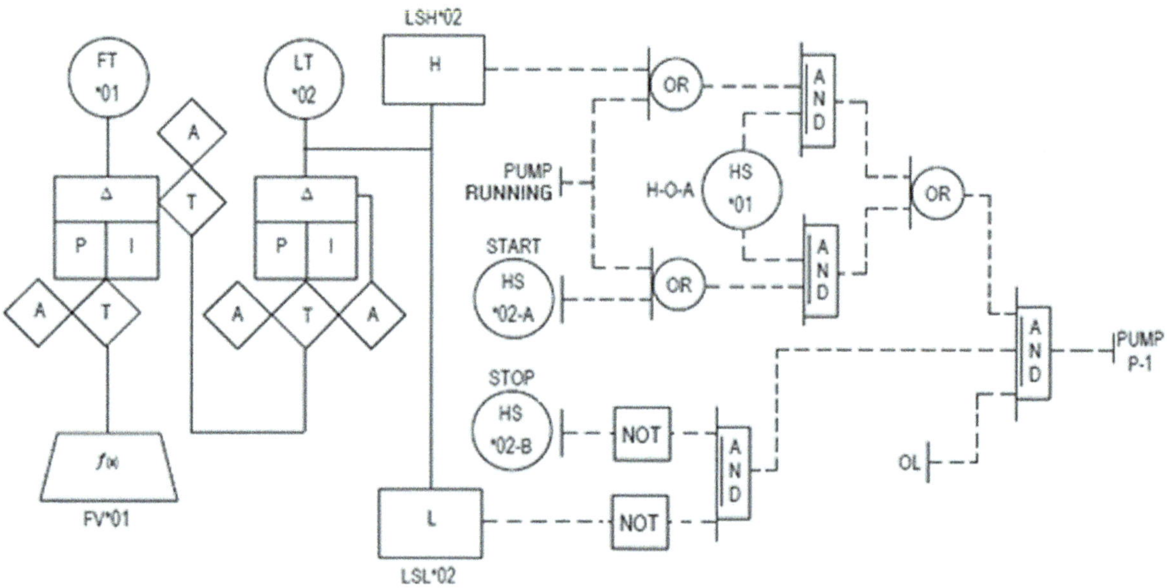

The CSE exam may have a diagram similar to the one below. Questions will be asked as to the state or outcome of the logic, and whether certain states occur in the process. Familiarize yourself with this type of logic and control diagram.

Tank-filling interlock logic diagram.

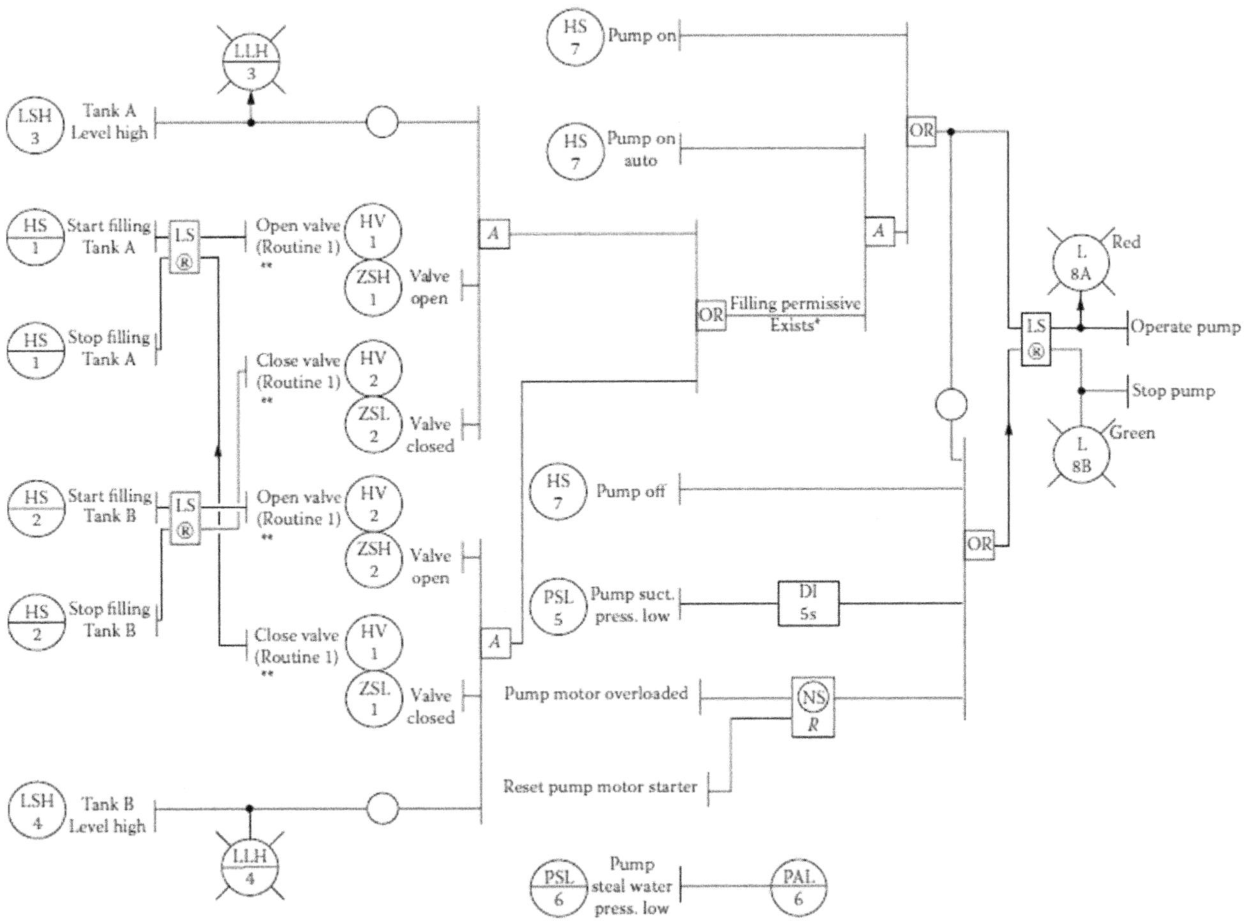

**This information describing solenoid, actuator and port design is required for detailed design work

		Solenoid valve	Control valve	
		HY-1	HV-1	
		HY-2	HV-2	
			Actuator	Port
Operation	Open valve	De-Energized	Vented	Open
	Close valve	Energized	Pressurized	Closed

Relay Ladder Logic

The CSE exam may have a diagram similar to the one below. Questions will be asked as to the state or outcome of the logic, and whether certain states occur in the process. Familiarize yourself with this type of logic and control diagram. You may be asked on the CSE exam which combination of buttons and switches is needed to start the motor (contactor *M*) or the pump.

DCS Plant-Wide Control System Architecture – Networked

A DCS is a computerized control system for a process or plant, usually with many control loops, in which autonomous controllers are distributed throughout the system but there is central operator supervisory control. This is in contrast to systems that use centralized controllers, either discrete controllers located in a central control room or within a central computer. The DCS concept increases reliability and reduces installation costs by localizing control functions near the process plant with remote monitoring and supervision.

DCSs first emerged in large, high-value, safety-critical process industries and were attractive because the DCS manufacturer supplied both the local control level and central supervisory equipment as an integrated package, thus reducing design integration risk. Today the functionality of SCADA and DCS systems is very similar, but DCS tends to be used in large, continuous process plants where high reliability and security is important and the control room is not geographically remote.

Note: All instrument grounding is typically done in the marshalling cabinet. Take note of this later on in Chapter 27, "ISA Standards for Documentation"; Chapter 29, "Overview of NEC/NFPA and Other Codes"; and when studying loop diagrams. You will implement a *single point ground* system, not an isolated ground!

Undergoing a Factory Acceptance Test for the Customer

The factory acceptance test (FAT) is performed to verify the control and logic functions of the DCS/PLC programming by simulating inputs and outputs that would occur in the plant when the equipment is installed.

If the programming and hardware are working as expected, the client will "authorize acceptance" of the equipment and programming.

PLC Control System Architecture

The basic architecture of a PLC consists of several main components—the processor module, the power supply, and the input/output (I/O) modules. The processor module consists of the central processing unit (CPU) and memory. In addition to a microprocessor, the CPU also contains at least one interface to a programming device and may contain interfaces to remote I/O and other communication networks. The power supply is usually a separate module, and the I/O modules are separate from the processor. The types of I/O modules include discrete (on/off), analog (continuous variable), and special modules such as motion control or high-speed counters. The field devices are connected to the I/O modules.

Programmable Logic Controller versus Process Automation Controller

If the application includes monitoring and control of a large number of analog and digital I/O points (each wire termination is called a *point* in the industry), then a PAC is generally the better solution. This is also the case when the application encompasses an entire plant or factory floor. This type of application typically calls for several distributed I/O modules, along with extensive process control analog loop functions. For this application, a PAC or large DCS is better suited than a PLC.

Be aware that a small factory line, such as for making cookies or cat food, may have around 500 points. A large plant will have more than 10,000 points.

The confusion arises when an application lies somewhere between simple and complex operations. In these circumstances a high-end PLC and a low-end PAC platform will work equally well. Ultimately, a choice between the two will be defined strictly by factors outside of specific application requirements. These factors include, but are not limited to, past experience with each platform, price, the level of local support, and anticipated future growth and changes within the plant. Reliability of the operations should also be considered when specifying and purchasing the equipment. Millions can be lost if the plant shuts down due to equipment failure. Other considerations include access to process and production data from the controllers for corporate marketing, as well as cost and quality control.

Once a decision is made between a PLC or a PAC for the control architecture scheme, users typically have a wide range of products to choose from for various applications, even if only a single vendor is being considered. This is because PLCs and PACs are typically designed with systems of scale in mind, meaning that there is a family of controllers to choose from that range from a lower I/O count to larger system capacity, with correspondingly more features and functions as I/O counts and prices increase.

Controller application function comparison chart.

PLC	DCS
Manufacturing or assembly of specific items	Involves the combination and/or transformation of raw material
Product is visible as it moves through the process	Impossible to visually see the product as it moves through the process
High-speed logic control	Discrete logic with redundancy
Small number (less than 500) of regulatory/analog loop control functions	Large number (up to 1000s) of regulatory/analog loop control functions
Simple batch control	Complex batch control

PLC Programming Languages

IEC 1131-3 defines the basic programming languages and is platform independent. The basic languages are as follows:

- **Function Block Diagram (FBD)** – A graphical dataflow programming method
- **Ladder Diagram (LD) or RLL** – Relay logic diagram-based programming
- **Structured Text (ST)** – A BASIC-like programming language
- **Sequential Function Chart (SFC)** – A graphical method for structuring program execution
- **Instruction List (IL)** –Mnemonic programming

The program images below will be shown in detail in the following examples.

FBD, SFC, and ST languages are typically an advanced feature in programming packages and can be purchased for an additional charge. FBD is the common programming language in DCS architectures.

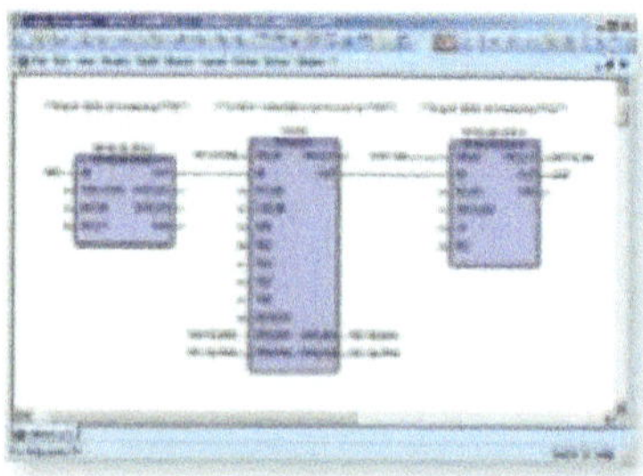

FBD

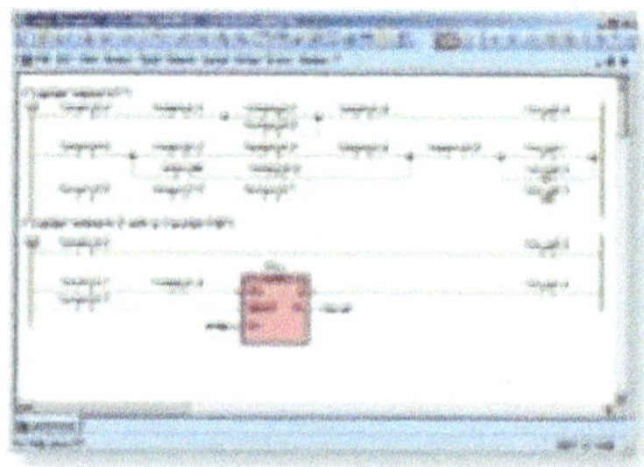

LD, RLL

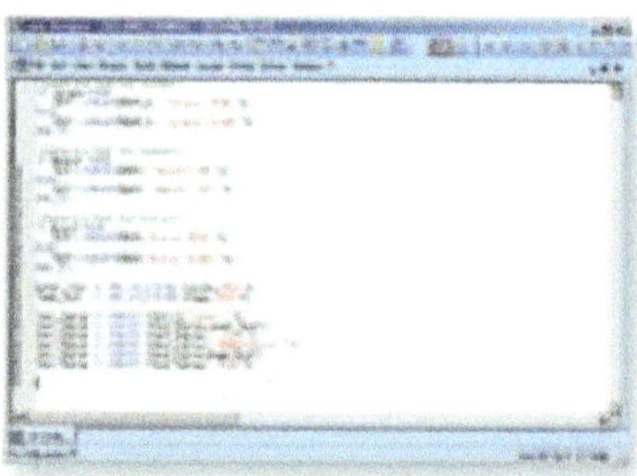

ST

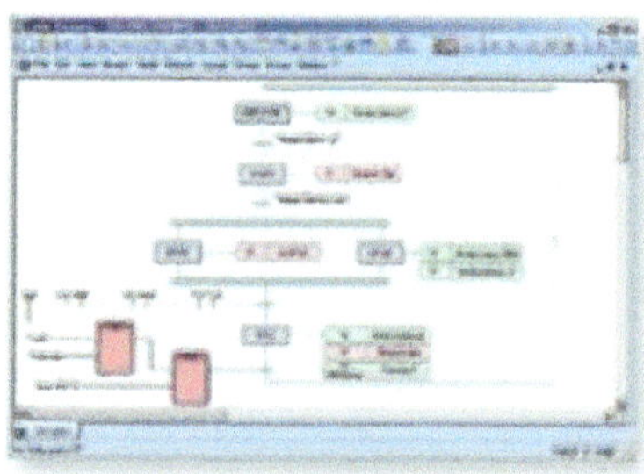

SFC

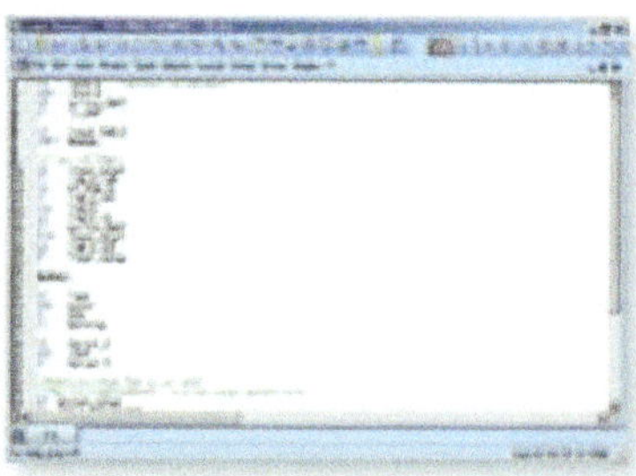

IL

LD or RLL

The following is a typical PLC program that might be on the CSE exam:

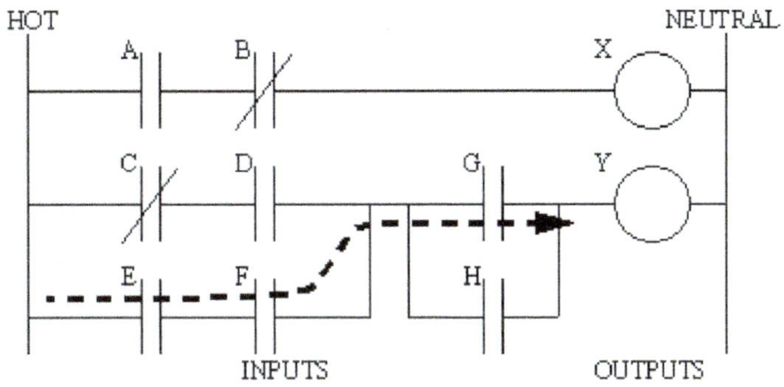

Note: Power needs to flow through some combination of the inputs (A,B,C,D,E,F,G,H) to turn on outputs (X,Y).

Contacts *A* and *Not B* are needed to turn on or (energize) output *X*. Contacts *Not C* and *D* and *G* or *H* are needed to turn on output *Y*. We can also use contacts *E* and *F* and *G* or *H* to turn on output *Y*. Shown path is stated as contacts *E* and *F*, and *G* will energize output *Y*.

Structured Text

The following is a typical PLC program (ST) that might be on the CSE exam. The ST program is called as a subroutine by the main ladder logic program.

```
PROGRAM TrafficLights
VAR
        i : INT;
END_VAR
i := 0;
REPEAT
        i := i + 1;
        UNTIL i >= 10;
END_REPEAT;
END_PROGRAM
```

Note: Allen Bradley does not implement the standard so that the programs can be written with text only. When programming in RSLogix, only the section indicated to the left would be entered. The variable 'i' would be defined as a tag, and the program would be defined as a task.

Functional Block Diagram

The following is a typical PLC program (FBD) that might be on the CSE exam. The FBD program is called as a subroutine by the main ladder logic program.

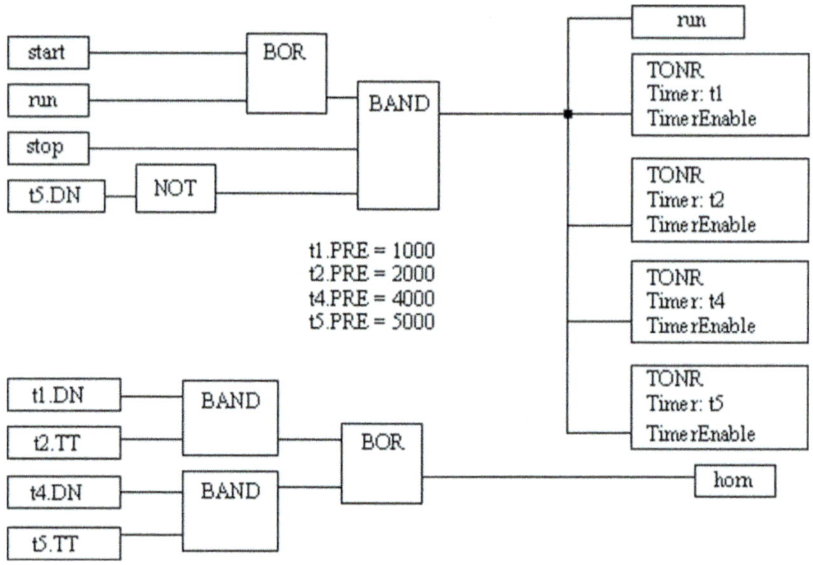

Sequential Function Chart

The following is a typical PLC program (SFC) that might be on the CSE exam. This is typically used in batch processes.

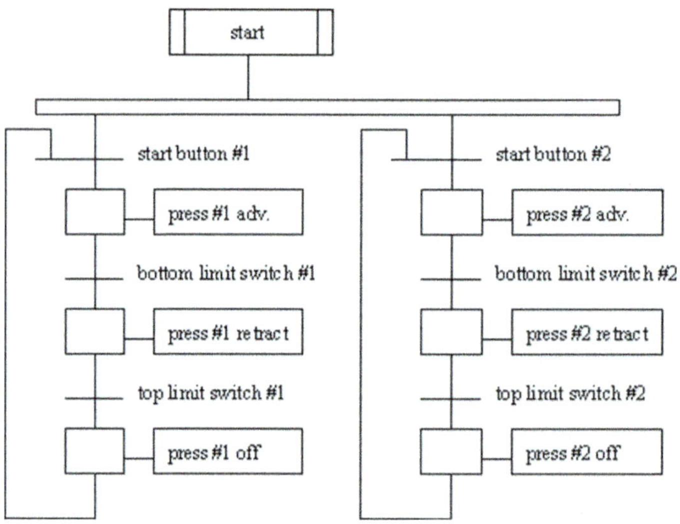

20 Motor Control and Logic Functions

The figures below show how a motor control center (MCC) is implemented in a plant electrical system.

Plant Electrical System

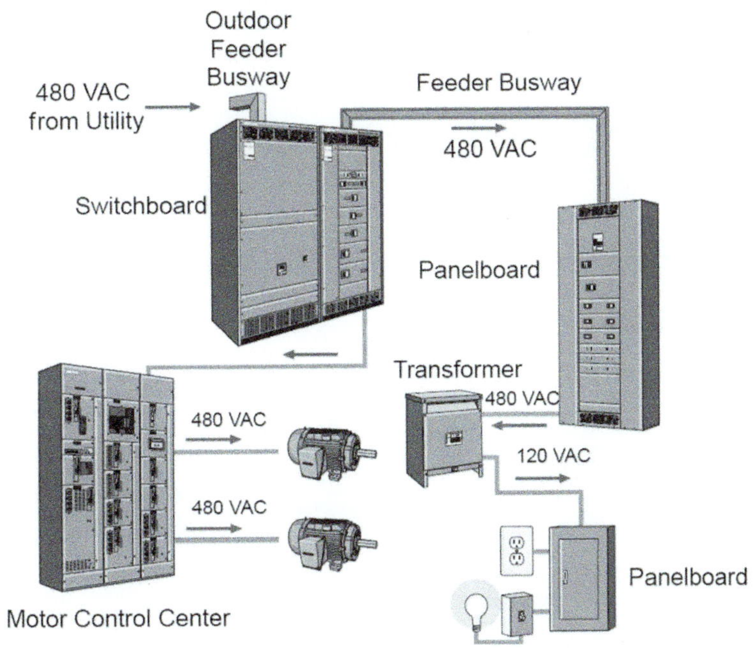

Motor Control Center (MCC)

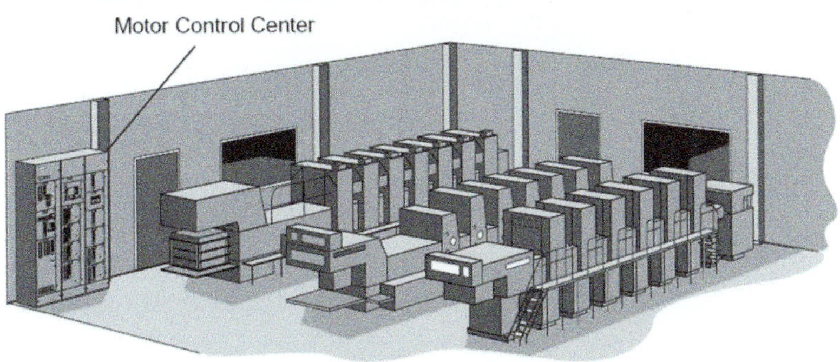

Typical MCC Design

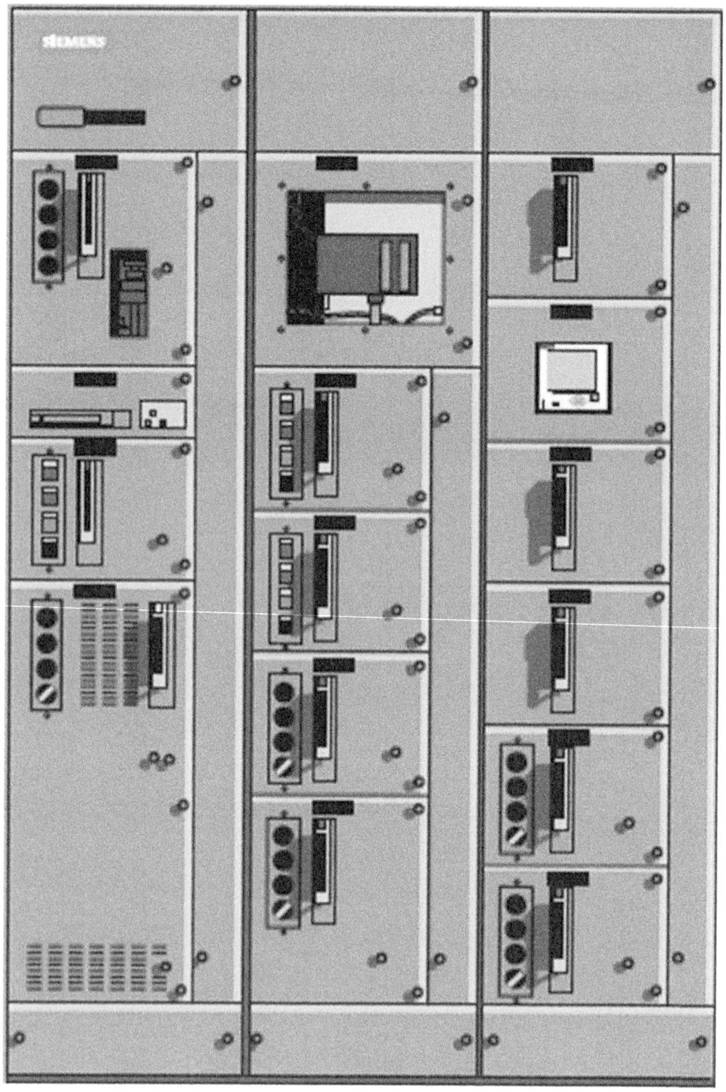

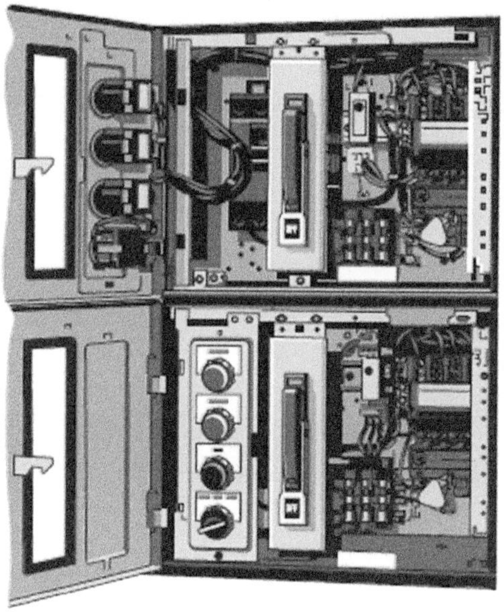

This a typical MCC found in most plants. It is a modular and expandable unit. The smaller squares are the motor starter compartments (see right unit).

The compartments are called *buckets*, and the wiring can be quickly disconnected and the bucket removed and replaced for fast repair and turnaround during shutdowns and design expansions.

The buckets typically slide in or plug into a vertical busbar with spring-clamped contacts, this is called *stab-in design*. But some buckets are wired to the busbar. They take up different vertical dimensions based on the motor starter size. These dimensions are called *space factors*.

The MCC can come with 42 circuit panelboards, small power transformers, variable frequency drives (VFDs), heater and lighting contactors, breakers, and switches. Spare units and blank covers (called *spaces*) are available for future expansion and can also be installed.

Most MCCs today are equipped with a programmable logic controller (PLC) or human-machine interface (HMI) and a communications network for controlling the MCC remotely as well as power usage monitoring. The networks are typically DeviceNet or Ethernet or an older technology such as Modbus RS-485 serial networks.

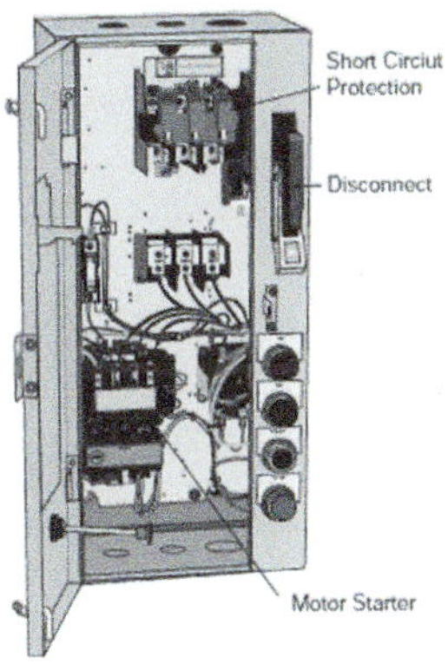

Stand-alone starter in NEMA enclosure.

Typical motor controller.

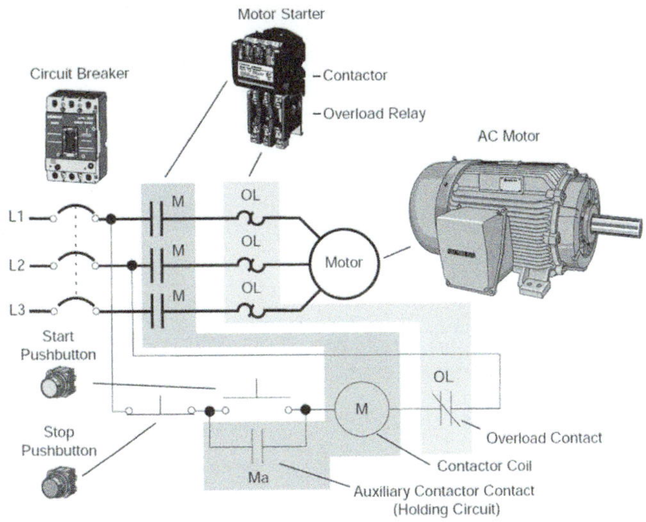

Minimal components to make a motor stop-start station.

How to Control a Motor

Motor Stop-Start Station

We will start with the stop-start station. This is the workhorse of the control industry. It used in MCCs to control motors, lights, and heaters and is also used every day in PLCs to control the program logic.

Types of Motor Control Schematics
Ladder Diagram

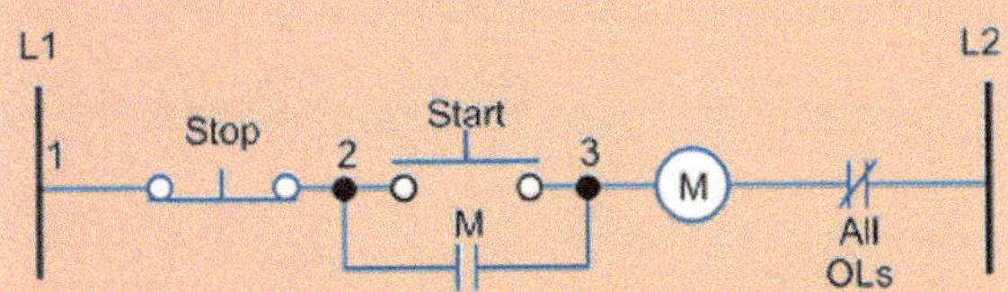

Ladder Diagram of a Standard Stop-Start Station

Ladder diagrams are also called "line" or "elementary" diagrams. They're used to represent the function of the control circuit and associated devices.

Be very familiar with the stop-start station. It is the fundamental application for sealing in and removing or disconnecting power to automated devices and circuits. It used with PLC or distributed control system (DCS) input and output interactions, whether it is a physical button, a software graphic symbol setting a bit in memory of a computer, or a level or pressure switch to start a pump or shut off a valve.

You will need expertise in understanding logic functions in Relay Ladder Logic (RLL) and Boolean logic functions. These subjects were briefly discussed in Chapter 19, "Digital Logic in Control Systems."

Overload and Fault

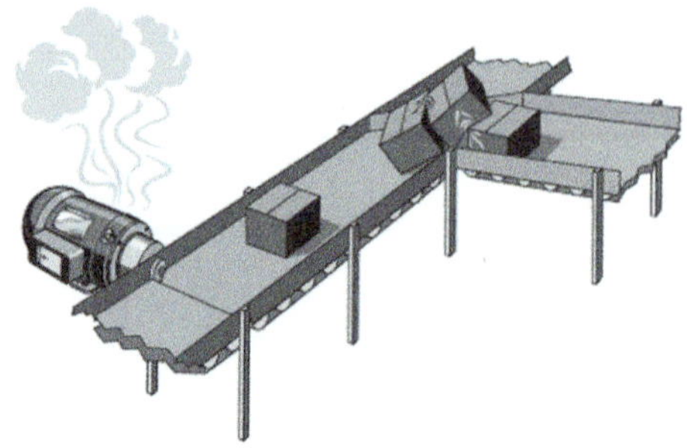

Overload protection must be used to protect the motor per the National Electrical Code (NEC), which is usually adopted into state and city codes. Some people use electronic or smart networked overload protection units.

The Basic National Electrical Manufacturers Association Stop-Start Station

The terminal number designations used below are standard on all National Electrical Manufacturers Association (NEMA) motor starters. Most cities require that all electrical control equipment have a UL (Underwriters Laboratory) listing for safety requirements and also to meet insurance requirements. Note that registration by the CSA Group (formerly the Canadian Standards Association) usually will *not* be accepted.

Types of Motor Control Schematics
Ladder Diagram

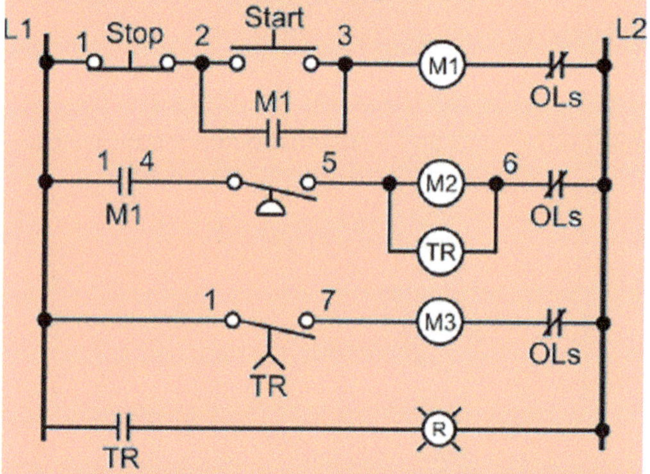

The horizontal lines in this ladder diagram are similar to the rungs of a ladder.

Typical Motor Control Schematic

The schematic above is a typical control schematic for pump or fan applications. We could use it to start a blower for a pneumatic system that we studied earlier.

The start button is pressed. M1 (motor-1 starter coil) energizes and NO auxiliary contact M1 (1,4) closes. M1 could be the blower motor. Only if pressure has developed in the piping system, the pressure switch (4,5) closes. This energizes M2 (motor-2 starter coil) and TR (timing relay). M2 could be the exhaust fan motor. The TR timer starts timing for a predetermined period to make sure air is flowing and the blower is running correctly. (Possible faults might be a loose fan on the shaft, a valve that is closed, or a pipe that is open or loose.)

Note the arrow up in the TR symbol. This is a "time on" timer, and the contact will close only after the timer is done timing. When the TR timer is done, the contact (1,7) closes and the M3 (motor-3 starter coil) energizes. M3 could be the rotary vane airlock valve motor. The TR NO contact on the last rung closes, and the system run light comes on. (The color R, red, usually stands for "stopped" or "danger.") So the airlock valve starts turning, dropping product into the airstream to be conveyed through the pipe. Only if the stop button is pressed, the pressure switch drops out because of an open pipe, or the fan comes loose from the blower shaft (stops spinning) does the system stop and reset itself to the off-the-shelf state we started with.

NEMA and IEC Terminal Designations

Relays and Contacts

DIN stands for *Deutsches Institut für Normung*, which means "German Institute for Standardization."

Per DIN standards, the terminals of auxiliary contacts on contactors and control devices are marked with a two-digit number. Terminals that belong together are marked with the same location digit (first digit).

The second digit (called the *function digit*) identifies the function of each contact per the following designations.

Type of Contact	Function Digits
Normally Open	3 and 4
Normally Closed	1 and 2
Normally Open (Special Function) Normally Closed (Special Function)	5 and 6 i.e. Time-Delay or Overload 7 and 8 Contacts

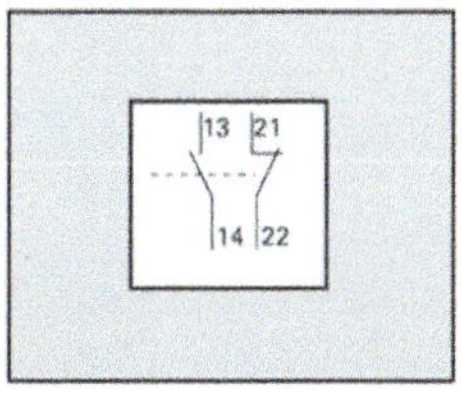

The numbers 13 and 14 represent an auxiliary contact.

- The number 1 identifies that this is the first contact in the sequence.
- The numbers 3 and 4 identify this as a normally open contact.

The numbers 21 and 22 represent another auxiliary contact.

- The number 2 identifies that this is the second contact in the sequence.
- The numbers 1 and 2 identify this as a normally closed contact.

Coil Lettering and Relay Socket Numbers (NEMA and IEC Numbers)

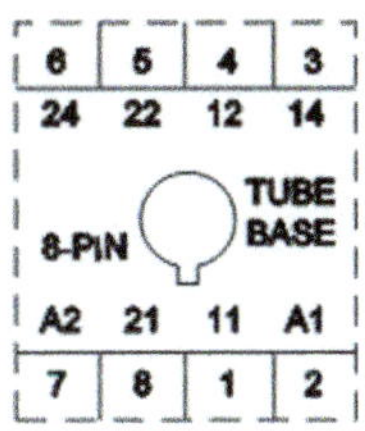

COIL AND CONTACT DESIGNATIONS

Function	Designation
Closing coil	CC
Holding coil	HC
Latch coil	LC
Time-delay closing contacts	TC or TDC
Time-delay opening contacts	TO or TDO
Trip coil	TC
Unlatch coil	ULC

NEMA ICS 19-2002 (R2007) Definitions

Block diagram – A diagram made up of a group of interconnected blocks, each of which represents a device or subsystem.

Connection diagram – See *Wiring diagram.*

Construction diagram – A diagram that shows the physical arrangement of parts, such as wiring, buses, and resistor units (e.g., a diagram showing the arrangement of grids and terminals in a grid-type resistor).

Control sequence diagram – A portrayal of the contact positions or connections that are made for each successive step of the control action.

Control system diagram – A conceptual diagram of the functional interrelationship of subsystems, usually in block form that does not include the process equipment or details of circuits and device elements.

Controller diagram – A diagram that shows the electrical connections between the parts comprising the controller and that shows electrical connections.

Dimension drawing; outline drawing – A drawing (base plan, floor plan, etc.) that shows the physical space and mounting requirements of a piece of equipment. It shall be permitted to also indicate ventilation requirements and space provided for connections or the location to which connections are to be made.

Graphic symbol – Symbols used on single-line (one-line) diagrams, on schematic or elementary diagrams, or, as applicable, on connection or wiring diagrams. Graphic symbols are correlated with parts lists, descriptions, or instructions by means of device designations.

Interconnection diagram – A diagram that shows only the external connections between controllers and associated machinery and equipment.

One-line diagram; single-line diagram – A diagram that shows, by means of single lines and graphic symbols, the course of an electrical circuit or circuits and the component devices or parts used therein. Physical relationships are usually disregarded.

Process diagram; flow diagram – A conceptual diagram of the functional interrelationship of subsystems in block or pictorial form that shows process equipment such as machinery for proper understanding.

Schematic diagram; elementary diagram – A diagram that shows all circuits and device elements of an equipment and its associated apparatus or any clearly defined functional portion thereof.

Wiring diagram; connection diagram – A diagram that locates and identifies electrical devices, terminals, and interconnecting wiring in an assembly. A wiring diagram may be: In a form showing interconnecting wiring by lines or indicating interconnecting wiring only by terminal designations (wireless diagram), or a panel layout diagram showing the physical location of devices plus

a) The elementary wiring diagram, or

b) A wiring table, or

c) A computer wiring chart, or

d) A machine command tape or cards.

The term does not include mechanical drawings, commonly referred to as wiring templates, wiring assemblies, cable assemblies, etc.

Standard Symbols

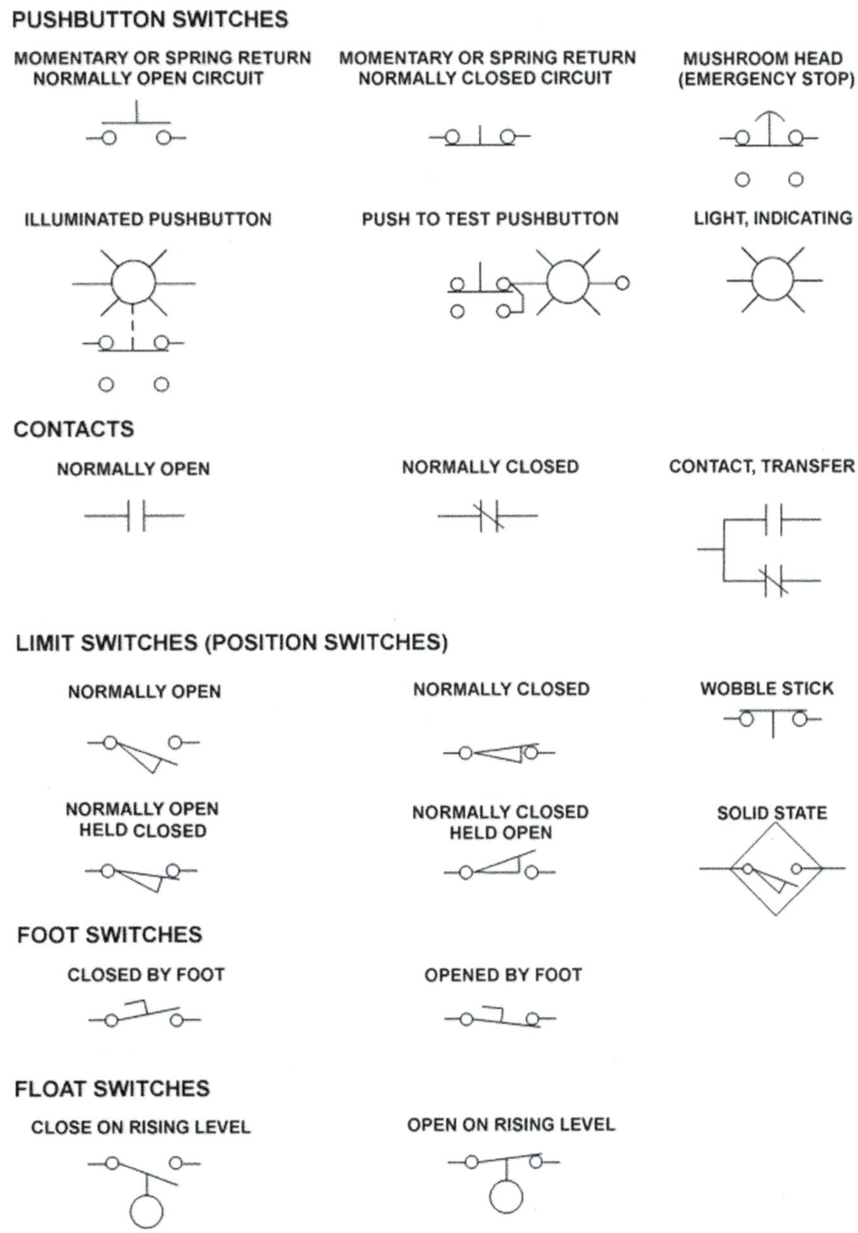

PRESSURE SWITCHES

CLOSED ON RISING PRESSURE

OPENED ON RISING PRESSURE

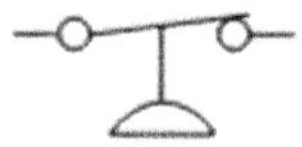

TEMPERATURE SWITCHES

CLOSE ON RISING TEMPERATURE

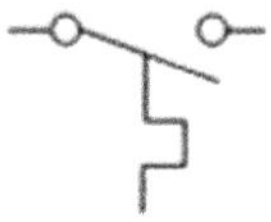

OPEN ON RISING TEMPERATURE

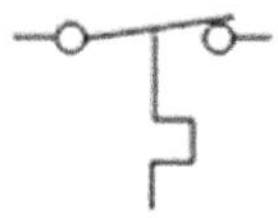

CONTACTS, TIME DELAY

The point of the arrow indicates the direction of the switch operation in which action is delayed. (Time-delay closing contacts and Time-delay opening contacts)

NORMALLY OPEN WITH TIME DELAY CLOSING

NORMALLY CLOSED WITH TIME DELAY OPEN

NORMALLY OPEN WITH TIME DELAY OPEN

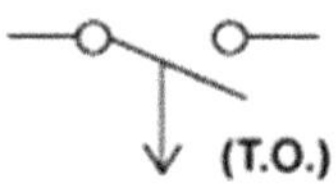

NORMALLY CLOSED WITH TIME DELAY CLOSED

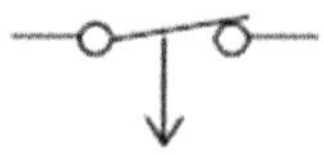

SELECTOR PUSHBUTTON (SELECTOR SWITCH) TWO OR MORE POSITIONS

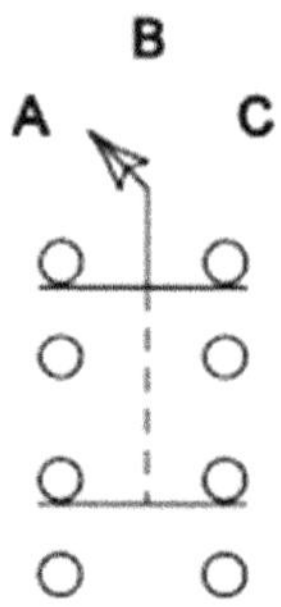

CONTACTS	SELECTOR POSITION								
	A			B			C		
	BUTTON			BUTTON			BUTTON		
	IN	NORMAL	OUT	IN	NORMAL	OUT	IN	NORMAL	OUT
1 - 2		X		X	X		X		
3 - 4	X					X		X	X
5 - 6	X	X	X		X				
7 - 8							X	X	X
X - INDICATES CONTACT CLOSED									

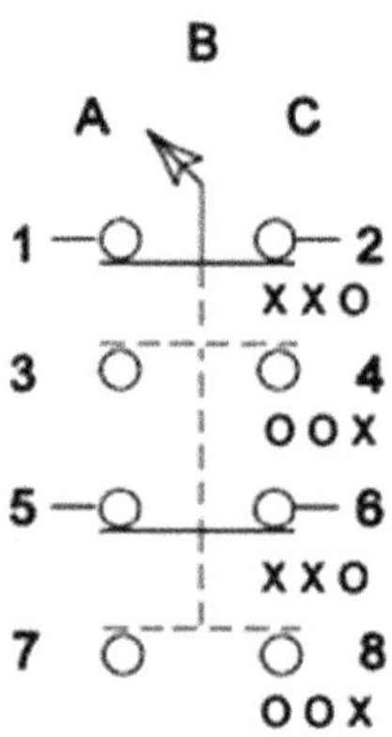

OPTIONAL METHOD TO SHOW FUNCTIONS OF A CAMMED SELECTOR SWITCH

NOTE: The convention of using "O O X" at the terminals is not NEMA standards, but is used in the industry and is easier to understand, plus no logic state table is needed. (O= Open contact and X = Closed contact)

NOTE: "A - B - C" could be the function of a Stop-Start Station in the field at a pump or the starter bucket of a MCC. Example:

"HAND - OFF - AUTO"
or
"LOCAL - OFF - REMOTE"

NEMA and IEC Comparisons

NEMA and IEC Comparisons Contactor/Starter Markings

	NEMA		IEC
Coils	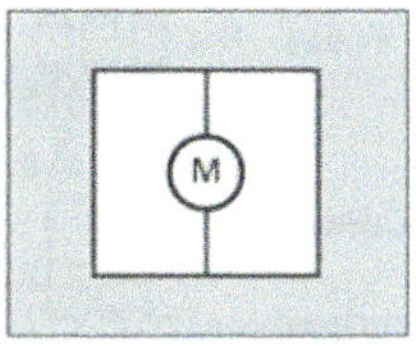	M — Defined by type of coil. No Standard terminal designation.	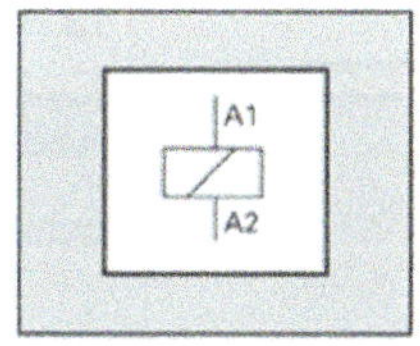
Power Contacts 3-Pole Device	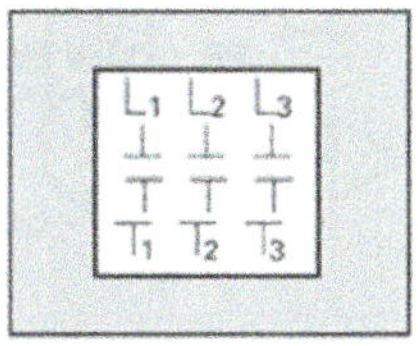	Line Side Connections Load Side Connections	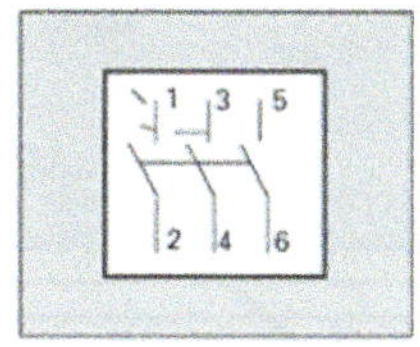
Overload Relay	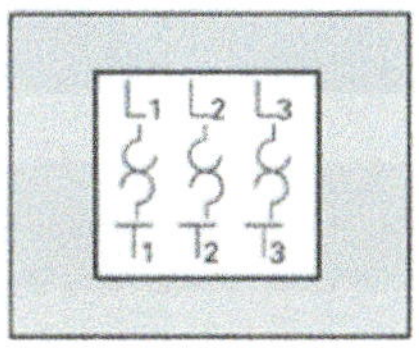	Line Side Connections Load Side Connections	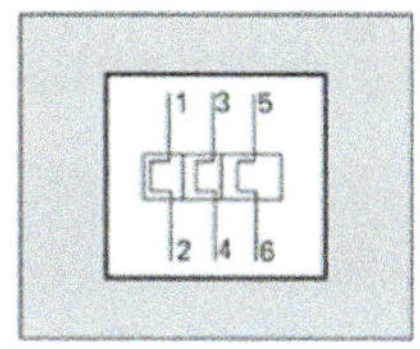

Symbols and Terminal Markings

Control Circuits	NEMA	IEC
Normally Open (NO)		
Normally Closed (NC)		
Time Delay Circuits		
On Delay Normally Open (Timed Closed)		
Normally Closed (Timed Open)		
Off Delay Normally Open (Timed Open)		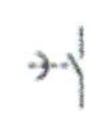
Normally Closed (Timed Closed)		

Stop-Start Station Control Circuit Schematic

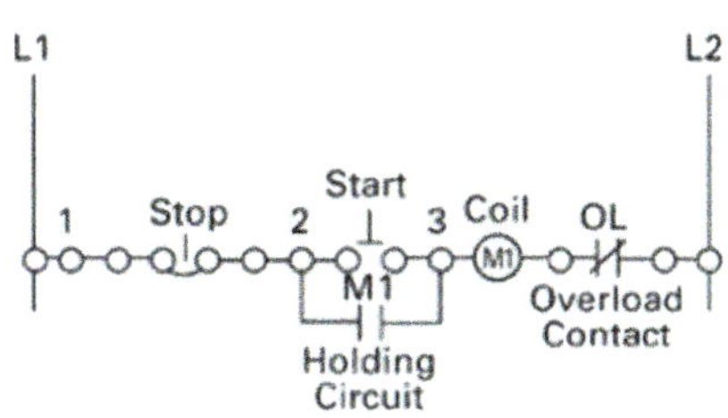

NEMA

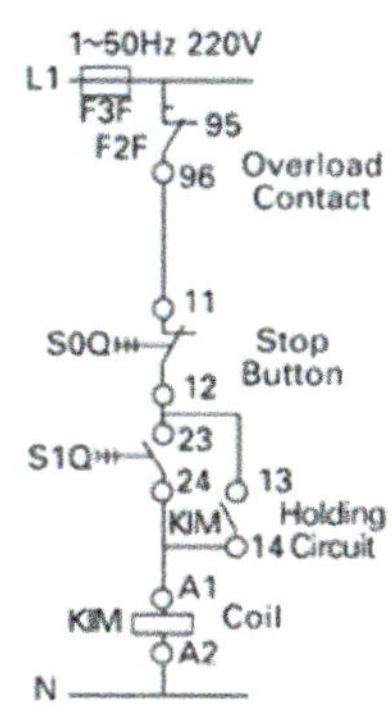

IEC

The 120 V sensors are "ANDed" together to energize the waste valve for safety purposes.

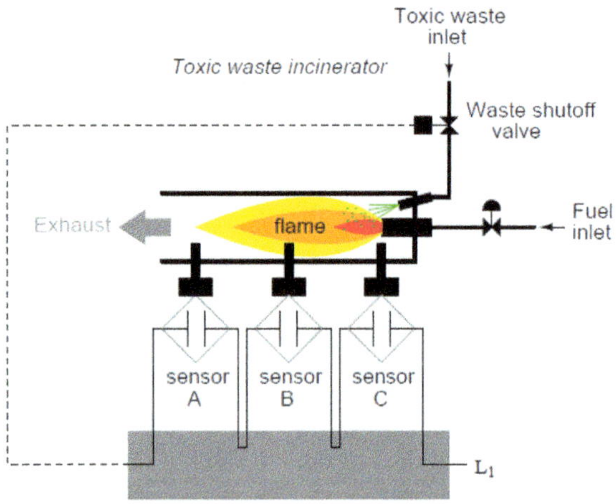

Here we can use logic to drive an output device (e.g., a triac) to energize the waste valve.

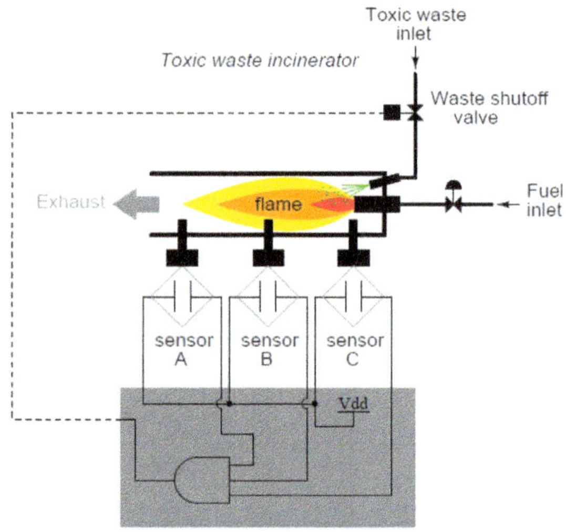

We can use voting to make the system more reliable for operation of this hazardous process. We will use 2 out of 3 voting (2oo3). This means as long as minimum of two of the three sensors are working, the system is operating in a safe manner.

sensor inputs

A	B	C	Output
0	0	0	0
0	0	1	0
0	1	0	0
0	1	1	1
1	0	0	0
1	0	1	1
1	1	0	1
1	1	1	1

Output = 0 (close valve)

Output = 1 (open valve)

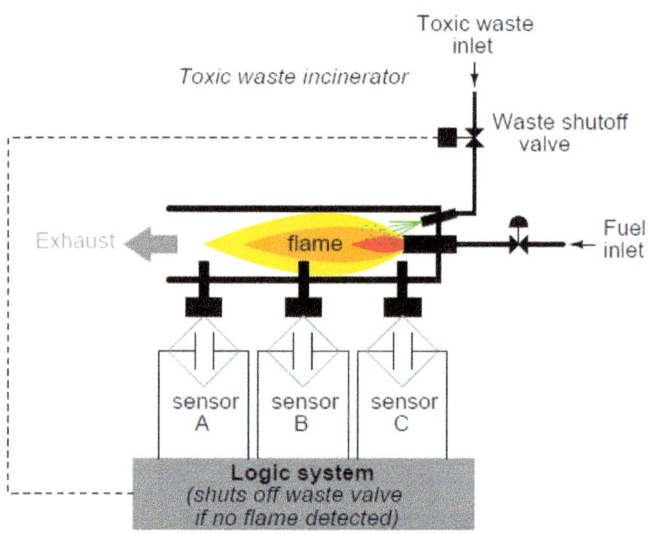

Safety Logic in the PLC

We can also use voting in the PLC ladder rung logic to do 2 out of 3 voting (2oo3). This means as long as two of the three sensors are working, the system is operating in a safe manner and the waste valve stays energized.

Note: The bar over the input variable represents a NOT function or a de-energized state.

sensor inputs

A	B	C	Output	
0	0	0	0	
0	0	1	0	
0	1	0	0	
0	1	1	1	$\overline{A}BC = 1$
1	0	0	0	
1	0	1	1	$A\overline{B}C = 1$
1	1	0	1	$AB\overline{C} = 1$
1	1	1	1	$ABC = 1$

$\text{Output} = \overline{A}BC + A\overline{B}C + AB\overline{C} + ABC$

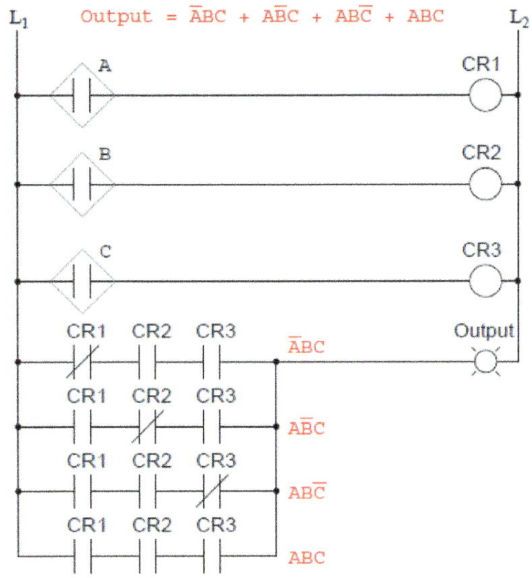

Alarming on Sensor Input Failure

If all three sensors are operating properly, they should detect flame with equal accuracy. Thus, they should either all register "low" (000: no flame) or all register "high" (111: good flame). Any other output combination (001, 010, 011, 100, 101, or 110) constitutes a disagreement between sensors, and may therefore serve as an indicator of a potential sensor failure.

Output = 0 (close valve) Output = 0 (sensors agree)

Output = 1 (open valve) Output = 1 (sensors disagree)

sensor inputs			**Good flame**	**Sensor disagreement**	
A	B	C	Output	Output	
0	0	0	0	0	
0	0	1	0	1	$\overline{A}\overline{B}C$
0	1	0	0	1	$\overline{A}B\overline{C}$
0	1	1	1	1	$\overline{A}BC$
1	0	0	0	1	$A\overline{B}\overline{C}$
1	0	1	1	1	$A\overline{B}C$
1	1	0	1	1	$AB\overline{C}$
1	1	1	1	0	

$\text{Output} = \overline{A}\overline{B}C + \overline{A}B\overline{C} + \overline{A}BC + A\overline{B}\overline{C} + A\overline{B}C + AB\overline{C}$

If we added circuitry to detect any one of the six "sensor disagreement" conditions, we could use the output of that circuitry to activate an alarm. Whoever is monitoring the incinerator would then exercise judgment in either continuing to operate with a possible failed sensor (inputs 011, 101, or 110), or shutting down the incinerator to be absolutely safe. Also, if the incinerator is shut down (no flame) and one or more of the sensors still indicates flame (001, 010, 011, 100, 101, or 110) while the other(s) indicate(s) no flame, that will signal a sensor problem.

The first step to take in designing the logic for the "sensor disagreement" detection circuit is to write a truth table describing its behavior. Because we already have a truth table describing the output of the "good flame" logic circuit, we can simply add another output column to the table to represent the second circuit and make a table representing the entire logic system.

The PLC Logic for Valve and Alarm Monitoring

The top rung in the PLC determines if 2 out of 3 (2oo3) sensors are good and lets the waste valve open. The bottom rung determines if any one of the sensors failed and turns on the "disagreement" alarm lamp. The waste solenoid continues to operate in the alarmed state. The PLC will only turn off the waste valve if two sensors fail.

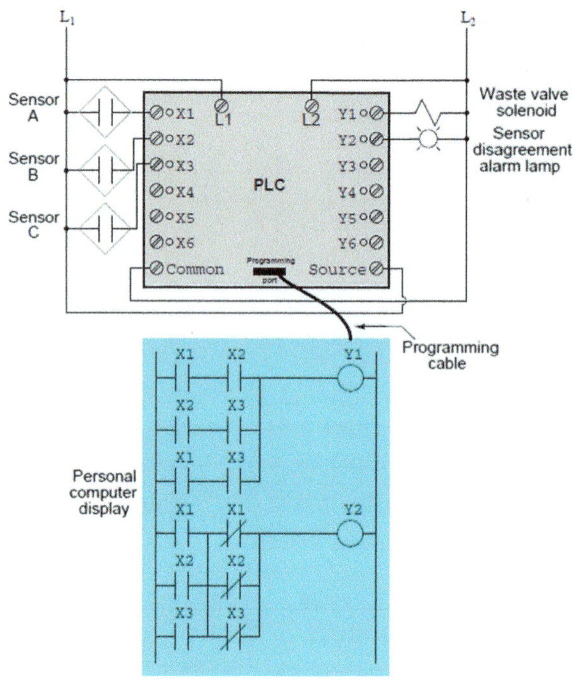

Output = 0 (close valve) | Output = 0 (sensors agree)
Output = 1 (open valve) | Output = 1 (sensors disagree)

sensor inputs | **Good flame** | **Sensor disagreement**

A	B	C	Output	Output
0	0	0	0	0
0	0	1	0	1
0	1	0	0	1
0	1	1	1	1
1	0	0	0	1
1	0	1	1	1
1	1	0	1	1
1	1	1	1	0

Schematic to Programming Languages

Comparison of RLL, Function Block Diagram (FBD), and Structured Text Language (STL).

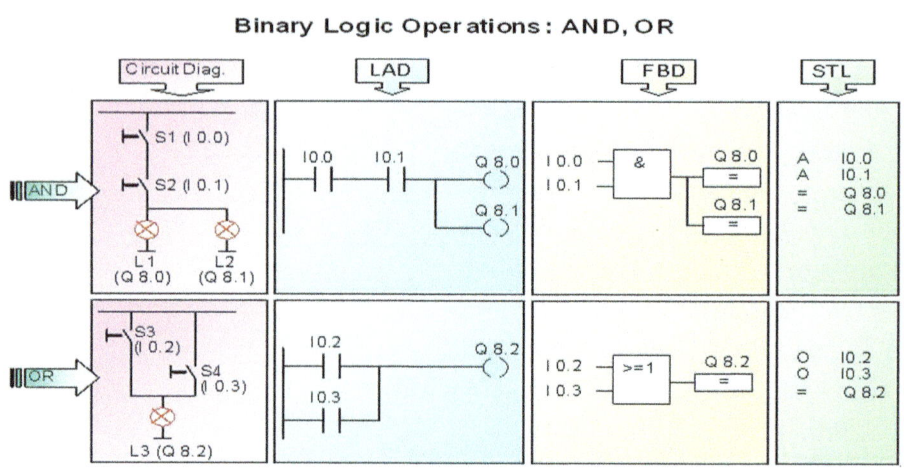

City of Galveston, Texas: Quadraplex lift station pump system by 5 Star Electric.

City of Houston, Texas: Wastewater treatment plant return activated sludge (RAS) pumps with stainless-steel air conditioners.

21 Analog Circuits in Control Systems

Overview of Analog Signals

On the CSE exam, there may be a few questions on International Society of Automation (ISA) symbols for electrical and pneumatic systems. Study the following ISA standards publications:

- ANSI/ISA-5.1-2009, *Instrumentation Symbols and Identification*
- ISA-5.2-1976 (R1992), *Binary Logic Diagrams for Process Operations*
- ISA-5.3-1983, *Graphic Symbols for Distributed Control/Shared Display Instrumentation, Logic, and Computer Systems*
- ISA-5.4-1991, *Standard Instrument Loop Diagrams*

The author of this book considers this information a required subject. Many problems can arise in dealing with the above standards. You will be tested on details, so do not feel comfortable with your company's standards. Only the ISA standard is correct. There may be exam questions from the documentation text, not just the symbols.

Typical Analog Loop Wiring Diagram

Most instrumentation operates on what is called a *4–20 mA loop* or a *loop-powered system*. The signal is sent as a 4–20 mA current loop (4 mA to 20 mA) because the signal does not degrade

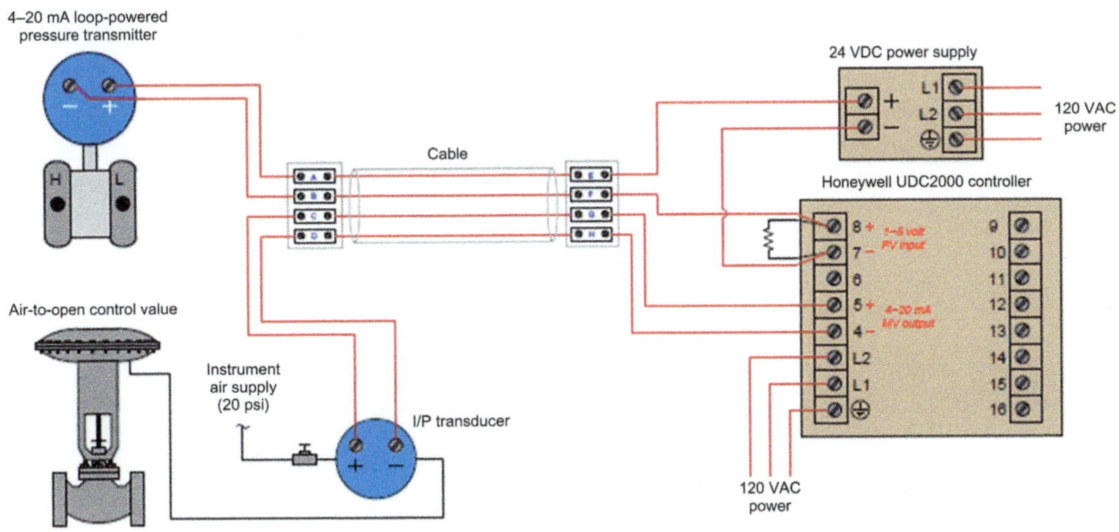

with changes in voltage or resistance in the signal wires. The 4–20 mA analog signal is also immune to electrical noise. Any noise that is injected into the signal loop, such as electromagnetic interference and radio frequency interference, is typically cancelled out by the data acquisition card or input signal card electronic circuitry by design using common mode rejection amplifiers.

It is important to understand that this is a current-based signal and *not* voltage-based! Ohm's law applies to the total voltage drops in the system and will have to be considered when designing your current loops in the plant control system.

See the equivalent schematic of a transmitter below. The transmitter acts as a current pump or a current sourcing device, not a voltage regulator. The measurement or control signal (4–20 mA) is directly proportional to the measurement process variable or controller percentage output.

A pneumatic analogy of the electronic transmitter and process variable transmission signal that is not affected by pressure drops in the signal line by varying hand valves DP1 or DP3 or length of pipe.

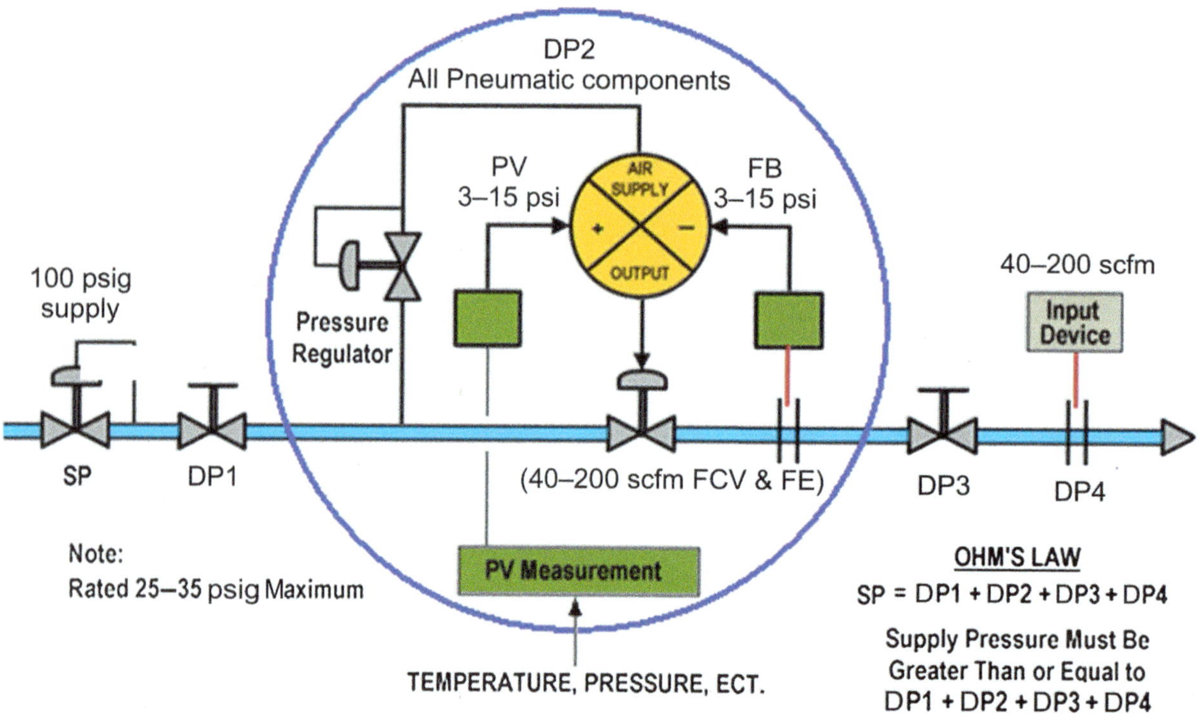

A standard 4-20 mA electronic current transmitter with a process variable transmission signal that is not affected by voltage (pressure) drops in the signal line.

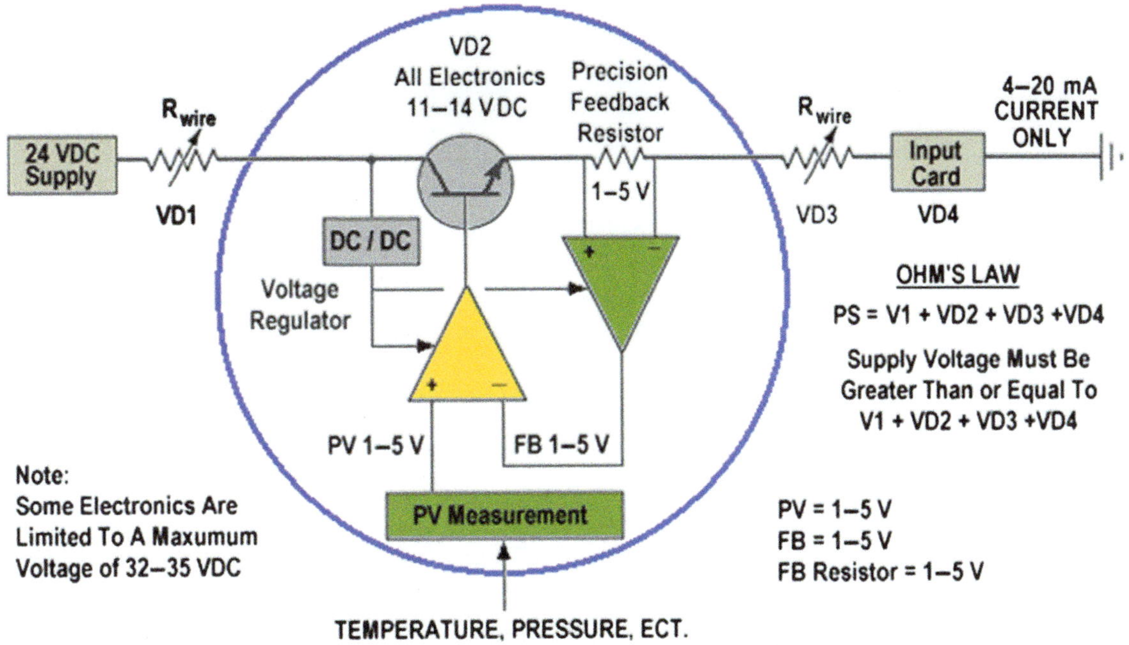

Above is a schematic example of simplified analog electronics circuit, such as in an analog transmitter.

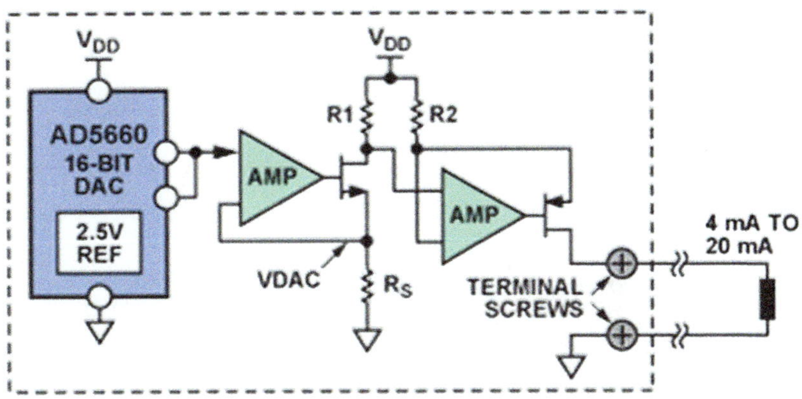

To the right is a transmitter circuit controller that works in a manner that is similar to the way most digital transmitters, such as a transmitter retrofitted with a smart electronics module, work today.

Transmitters will use ADC (analog-to-digital convertor) input greater than µP (microprocessor) to process signals greater than DAC (digital-to-analog convertor) output in the electronics. This allows for scaling and easy bit-level trimming of the input and output signals of the transmitter.

Constant Current Loops and Ohm's Law

Current Loop Fundamentals

Since the 1950s, current loops have been used to transmit transducer data for process monitoring and control. With their low implementation cost, inherent resistance to noise, and ability to carry signals long distances, current loops have proven particularly well-suited for industrial environments. The following figure shows the electronics in a calibrator, but controllers and transmitters work the same way. The flow direction is that of electron flow, not power or hole flow.

The 4–20 mA current loop.

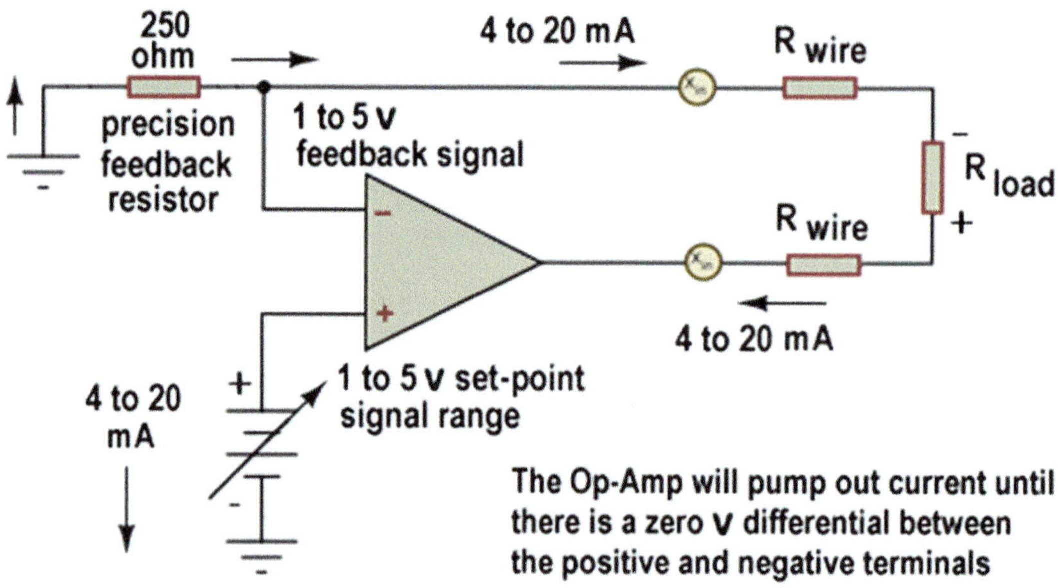

The input voltage to this circuit, the set point, is assumed to be coming from some type of physical transducer/amplifier arrangement, calibrated to produce 1 V at 0% of physical measurement and 5 V at 100% of physical measurement. The standard analog current signal range is 4–20 mA, signifying 0%–100% of the measurement range, respectively. For a 5 V input (set point), the 250 Ω (precision) resistor will have 5 V applied across it, from the result of 20 mA of current flowing through the loop circuit, with R (load) in series with the circuit's current loop.

It does not matter what resistance value R_{LOAD} is, or how much wire resistance is present in the current loop, so long as the op-amp has a high enough power supply voltage to output the voltage necessary to maintain the 20 mA flow through R_{LOAD} and R_{WIRE}. The 250 Ω precision resistor establishes the relationship between input voltage and output current. The input and output have a direct relationship of 1–5 V in/4–20 mA out. If we were converting the 1–5 V input (set point) signal to a 10–50 mA output signal (an older, obsolete instrumentation standard for industry), we would use a 100 Ω precision resistor instead.

Using Current to Transmit Transducer Data

Industrial-grade transducers commonly use current signals to transmit data. This is in contrast to most other transducers, such as thermocouples and resistive strain gauges, that use a voltage signal. While voltage-based transducers are sufficient in many environments, current-based transducers are at times preferred. For instance, the inherent disadvantage of using voltage to transmit signals in an industrial environment is voltage drops over long cable lengths due to wire resistance. You can use high input impedance devices to reduce the loop current and signal loss due to voltage drop. However, these devices are sensitive to the noise induced by nearby motors, conveyor belts, and radio transmissions.

According to Kirchhoff's current law, all current present at the beginning of a loop must reach the end. This is the basic principle on which a current loop operates. Measuring current at any position along a single loop produces the same result. By using current signals and low-impedance data acquisition devices, industrial applications benefit from better noise immunity and longer transmission cable lengths.

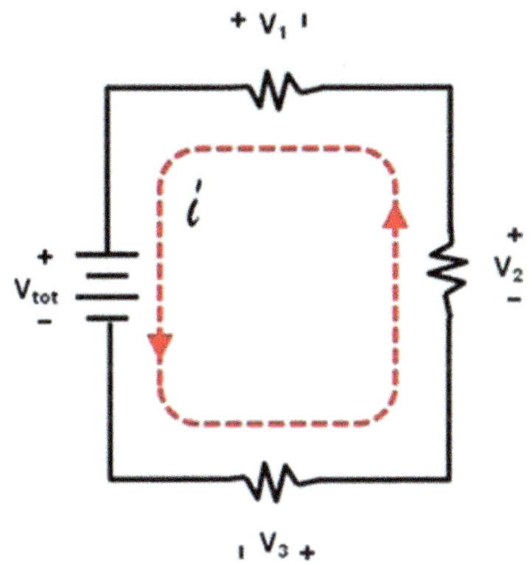

Current Loop Components

The main components of a current loop include a DC power supply, a transducer, a data acquisition device or input card, and wires connecting them together in a series, as shown in the figure below. The DC power supply provides the power to the system. The transducer regulates the flow of current through the wires at 4–20 mA, where 4 mA represents a live zero and 20 mA represents the maximum signal.

Note: A 0 mA, or –25% signal, signifies a break or fault in the loop.

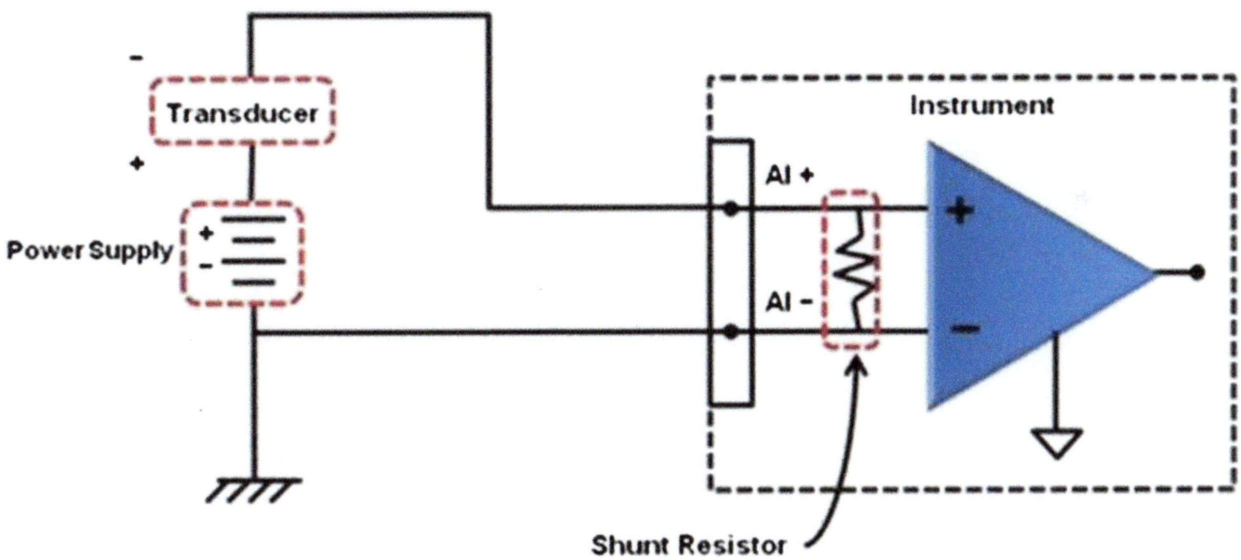

Current Loop System

The data acquisition device measures the regulated current flow (as in an input to a programmable logic controller—PLC or distributed control system—DCS). An effective and accurate method of measuring current is to place a precision shunt resistor between the input leads of the data acquisition device amplifier as in the figure above. The resistor will convert the current signal into a voltage measurement. This input voltage measurement is directly proportional to the transducer's output of the original measurement on a 0%–100% process measurement scale.

To help better understand current loops, consider the design of an example system using a temperature transducer throughout this example. System specifications include the following:

- The transducer must measure the temperature in a vessel.
- The transducer is located 2000 ft away from the PLC input.
- There is variable resistance in the loop and modifications, such as inserting meters, are possible.

Designing a Current Loop System

For the example 4–20 mA current loop system, the chosen transducer measures temperature and has an operating voltage of 12–32 V. The selected transducer is calibrated to the span of the temperature measurement to produce a 4–20 mA (0%–100%) signal, proportional to the measurement. The current signal is required to provide an accurate representation of the temperature measured by the transducer for the process. The input signal transducer (TY) is a precision resistor with a tolerance of 0.1% accuracy (250 Ω @ 0.001% = a resistance between 249.75 Ω and 250.25 Ω).

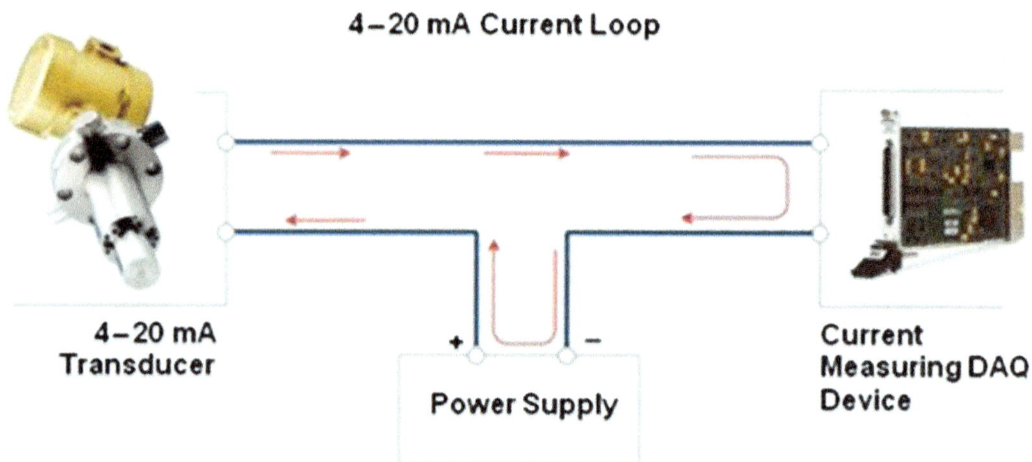

Choosing a Power Supply

Discovering which power supply best meets your requirements is simple. In a current loop, the power supply must supply voltage equal to or greater than all the voltage drops in the system combined. The data acquisition device for this example uses a precision shunt resistor to measure current. You must calculate the voltage drop across the resistor. A typical shunt resistor could be 249–251 Ω. Implementing basic calculations with the 4–20 mA current loop shows the following or better:

$$i \cdot R = \mathrm{V}$$

$$0.004\ \mathrm{A} \cdot 249\ \Omega = 0.996\ \mathrm{V}$$

$$0.020\ \mathrm{A} \cdot 249\ \Omega = 4.980\ \mathrm{V}$$

With this 249 Ω shunt resistor, you can directly relate current to a range of 1–5 V, allowing the data acquisition amplifier's voltage reading to be related to the temperature transducer's signal. The aforementioned temperature transducer requires a minimum operating voltage of 12 V with a maximum of 32 V. Adding the voltage drop of the precision shunt resistor to the transducer yields the following:

$$12\ \text{V} + 5\ \text{V} = 17\ \text{V}$$

This suggests that a 17 V power supply suffices. However, the current loop wire's internal resistance creates an additional load on the power supply. In applications where a transducer is far away from the measuring instruments, you must factor wire resistance into the loop power calculations. Copper wires have a DC resistance that is directly proportional to their lengths and diameters (gauge). With the temperature transducer from the example system, you must calculate the 2000 ft distance in the power requirements. With 24-gauge solid copper wire, the internal resistance is 2.62 Ω/100 ft.

Note: The wire resistance of the loop varies with the environment temperature.

This resistance calculation becomes:

$$2000\ \text{ft} \bullet \frac{2.62\ \Omega}{100\ \text{ft}} = 52.4\ \Omega$$

$$54.2\ \Omega \bullet 0.020\ \text{A} = 1.048\ \text{V}$$

At 2000 ft away, there is a 1.05 V drop along the wire. To complete the loop, you need two wires, effectively doubling the length to 4000 ft:

$$2 \cdot 54.2\ \Omega \bullet 0.020\ \text{A} = 2.096\ \text{V}$$

In total, approximately 2.1 V is lost due to wiring when the pressure transducer is 2,000 ft away from the data acquisition device. Combining all the voltage drops results in the following:

$$2.096\ \text{V} + 5\ \text{V} + 12\ \text{V} = 19.096\ \text{V}$$

If you used a 17 V power supply to drive the previous setup, the temperature transducer voltage would be below the minimum operating level for a transducer due to natural wire resistance and shunt resistor requirements. Choosing a common 24 V power supply usually meets the power requirements of the temperature control measurement loop and is the industrial standard for process installations.

Note: Always ground the negative of your process DC power supply loops. This is for short-circuit protection of the wiring in case there is a device failure. The loop power fuse will isolate the faulty instrument loop and cause the plant process systems to continue to function.

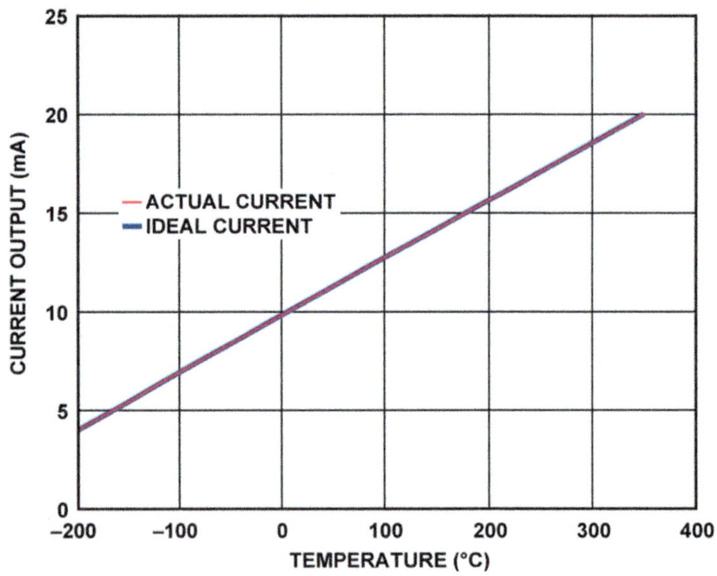

Adding More Transducers and Instruments

For larger applications, you can add more transducers and instruments in parallel to the power supply, as shown in the figure below. Please note these are *not* parallel loads as on AC 120 V circuits. Each process loop is an *individual series* loop. Only the power supply is in parallel, (common) to all process loops. They all have the same +24 V supply and –0 V (grounded negative) return to the power supply. Each loop is a separate series path or loop, made of the transducer and PLC/DCS input card.

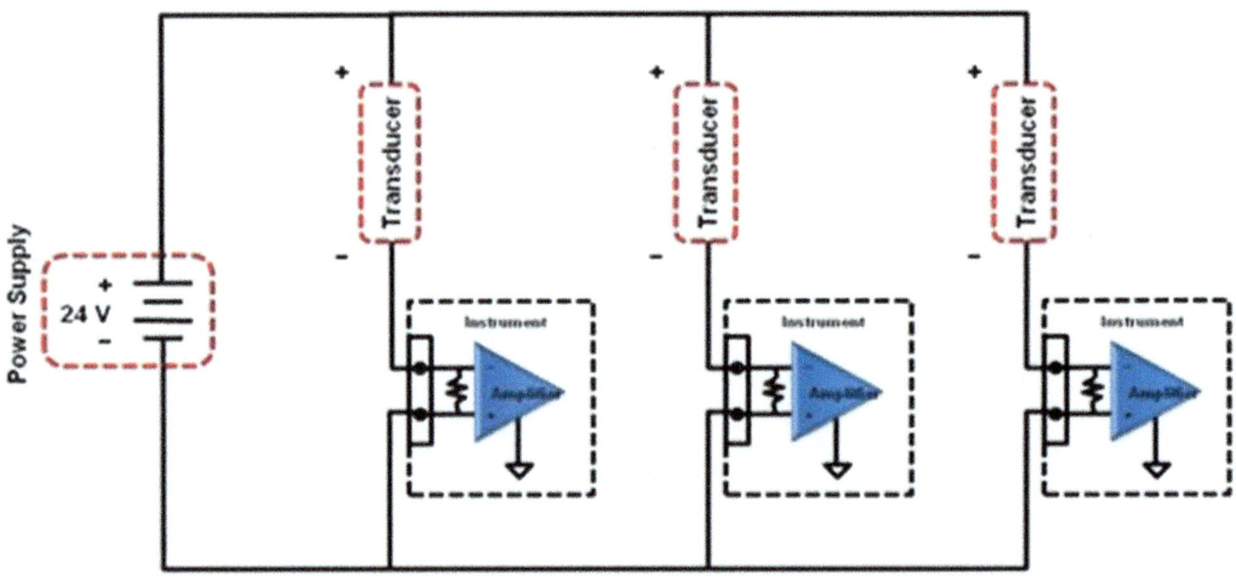

Because you add the extra current loops in parallel, the power supply's voltage does not need to change. You only must make sure it can source the current required by all combined process loops. You can mix and match different types of transducers as long as the total impedance is less than the supply voltage divided by the required current. This applies for each single loop on the 24 V supply.

Devices in Series (Only with 4–20mA Current Loop)

At times, you may need multiple current-measuring data acquisition devices to measure the same signal. To add another data acquisition device to the same transducer, you must place the new instrument in series with the process measurement current loop. For this example, you add a simple digital current loop meter with the existing data acquisition device, as shown in the figure below.

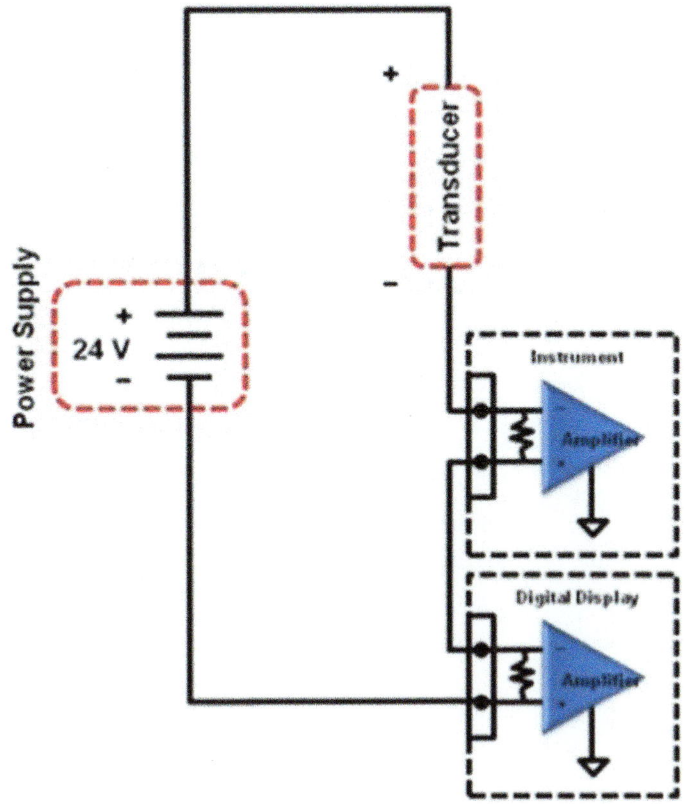

Adding an extra instrument to a process transmitter (transducer) Loop.

Each time you add a current-measuring data acquisition device in series with the transducer, you must recalculate the total voltage requirement to ensure that the power requirements of the loop are still within specification. In this example, the new current meter uses a 105 Ω precision resistor:

$$i \cdot R = \mathrm{V}$$

$$20\ \mathrm{mA} \cdot 105\ \Omega = 2.1\ \mathrm{V}$$

With a 105 Ω shunt resistor, the current meter requires a maximum of 2.1 V, so the total voltage requirement of the system becomes:

$$2.1\text{ V} + 2.096\text{ V} + 5\text{ V} + 12\text{ V} = 21.196\text{ V}$$

Because adding the meter brings the total voltage requirement to only about 21.2 V, the 24 V power supply previously recommended is still sufficient for the application. We may want to add a process meter display unit or chart recorder in the field or control room. Each of these could use a 250 Ω resistor for an extra 10 volts (5 V + 5 V) of supply voltage needed by the two additional devices.

In this case, a current repeater is needed. The temperature loop signal will then series through the repeater and the field display unit, and the chart recorder will then be powered in its own independent series loop with its own voltage drops. Now a 24 V power supply will work fine.

A Typical Current Loop Repeater

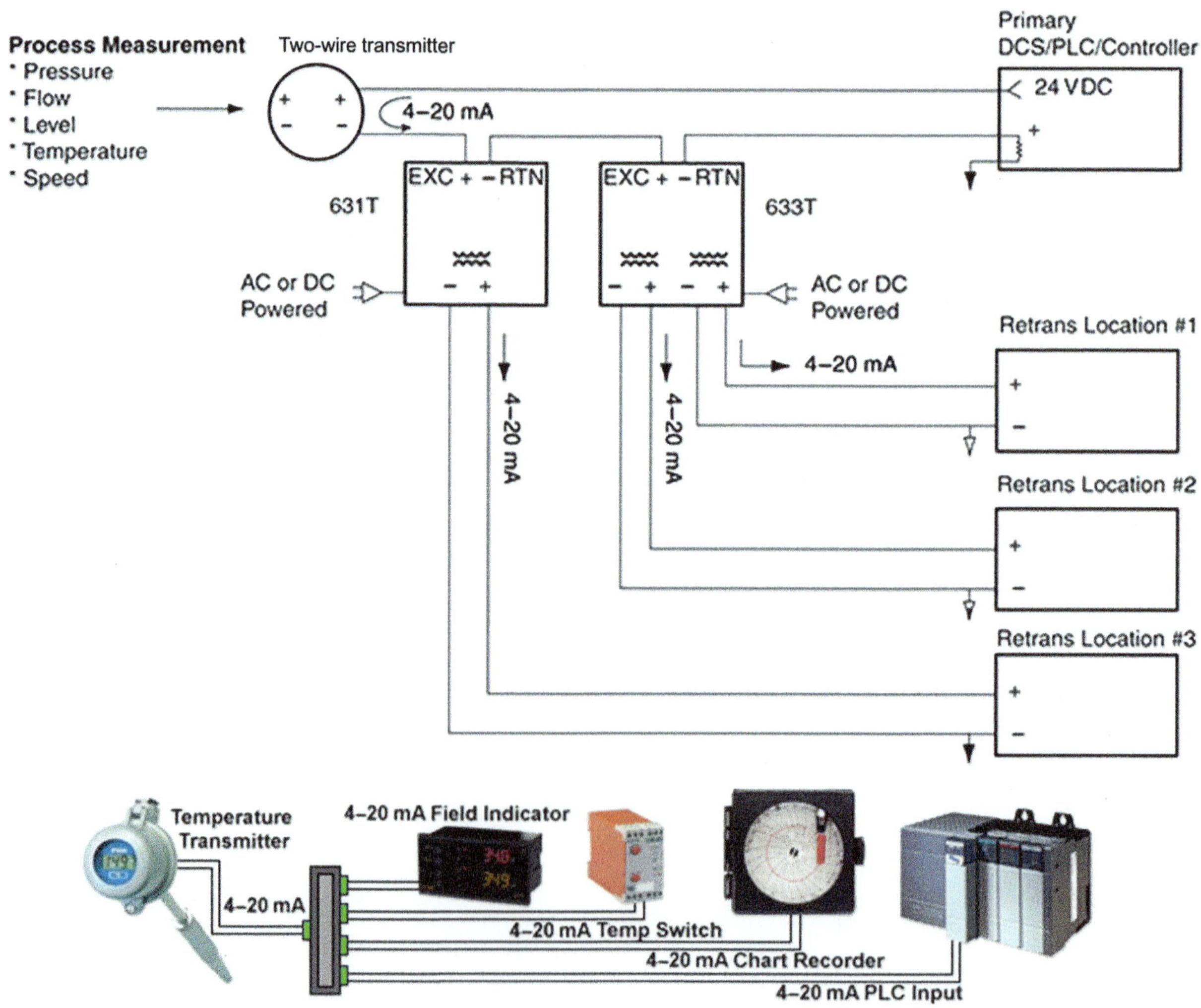

Active and passive current loops

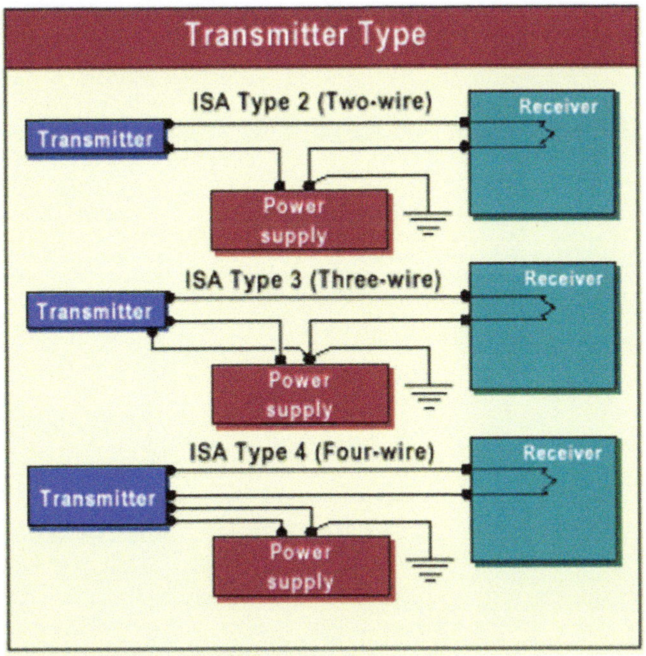

ISA S50.01 categorizes transmitters into two-wire, three-wire, or four-wire types.

Passive transmitters require a power supply to power up the loop and electronics. The Type 2 transmitter loop is called *passive.*

Type 3 and Type 4 transmitters are considered as *active* transmitters, but this can be misleading. The transmitter is powering up the current loop, but the transmitter's electronics are being powered up by an external power source. Some transmitters will have an internal power supply built in. Caution must be taken with these transmitters. Type 4 (four-wire) transmitters can be powered with 120 VAC power for the display, computer, and electronics. These transmitters *can be* or *may not be* active. You must consult the manufacturer's data first.

If the transmitter, instrument, or analyzer is *active* and you supply an external 24 VDC power source, you may damage the device. Typically, you will blow out a transistor in the current driving circuit.

Summary

If the device is Type 2 passive (two-wire), you will need a power supply to power the current loop.

If the device is Type 3 or Type 4 (three-wire or four-wire), you must verify if an external power supply is needed. An external power supply in series with the current loop may damage the internal electronics, if the device has an internal power supply. With an internal power supply, the voltage may equal 48 V (24 V+ 24 V). Most electronic components are rated for 32 to 42 V maximum.

Sinking and Sourcing Devices

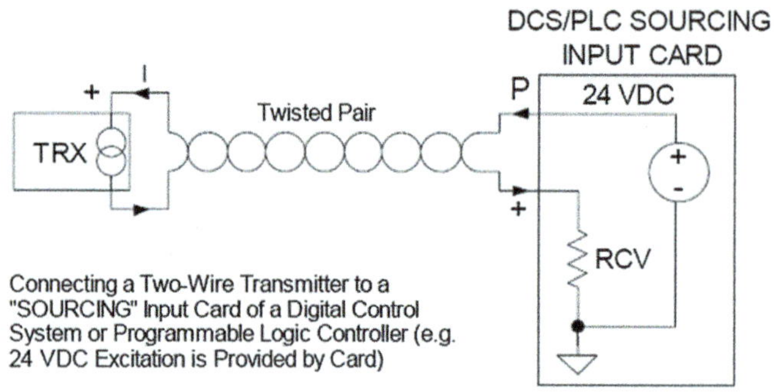

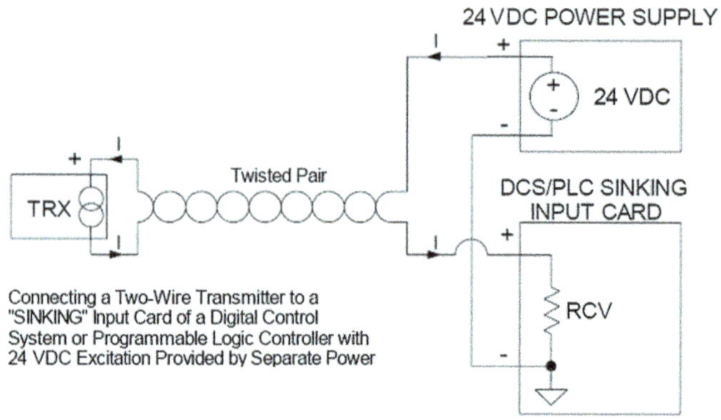

What Is the Difference Between PNP and NPN?

Most industrial proximity sensors (inductive, capacitive, ultrasonic, and photoelectric) are solid-state. The term *solid-state* refers to the type of components used within the sensor. *Solid-state* electronic components such as transistors are used to switch the output of the sensor upon detection of an object. With PNP and NPN, the P stands for positive and the N stands for negative.

Two specific types of three-wire sensors are available: PNP and NPN. The difference is a result of the internal circuit design and type of transistors used. A key point is that PNP and NPN have nothing to do with whether the sensor is normally open (NO) or normally closed (NC).

Example: Both a PNP sensor and an NPN sensor may be either NO or NC.

Note: The application of sinking and sourcing also applies to analog input and output wiring. You must understand this concept before ordering your analog cards and devices!

PNP Sensor versus NPN Sensor

The selection of a PNP sensor or an NPN sensor is determined by the nature of the circuit in which the device is to be used. In a traditional relay-type control circuit, usually either the PNP or the NPN type of sensor can be used, as shown below. However, PNP sensors tend to be more commonly used.

When selecting a sensor to be used with a PLC, it is important that the sensor matches the type of PLC input card or switching device to be used. Two types of input cards exist, those that "sink" current (the positive is applied to the load and the control device switches current flow to ground) and those that "source" current (the load is connected to ground and the positive is applied to the control device, which switches current flow to the load). Although the terms *sinking/sourcing* and *positive/negative logic* are well known in some industries, they are not commonly used in others. It is therefore important to identify the type of sensor to be used with the PLC card based on the PLC manufacturer's documentation and/or wiring diagrams.

Most common in Europe is the *sinking* type of input; these input cards will be used with the PNP sensor as shown below. Less common today are input cards that *source*; these were popular in Asia and require the NPN type of sensor to operate correctly. Many modern PLC input cards can be configured and wired to be either sinking or sourcing.

Note: The dashed lines in the second figure may be needed for the card electronics to work. The PNP sensor sources current to the card, the card completes the circuit to ground, and the NPN sensor receives current from the card and completes the circuit by sinking the current to ground.

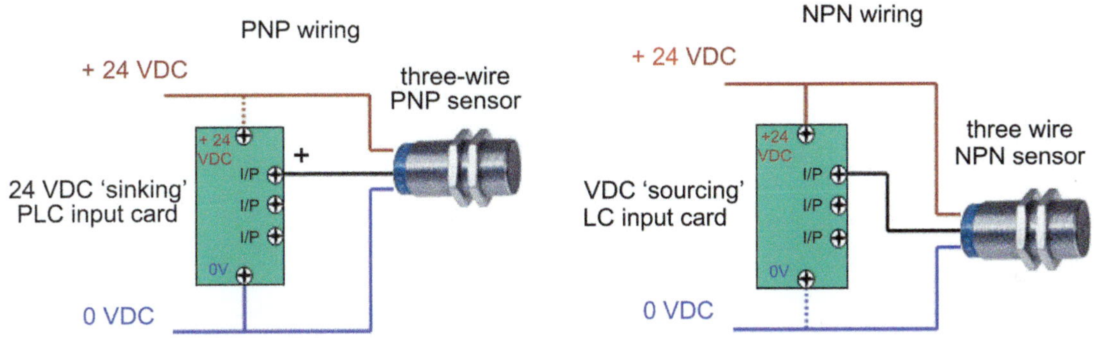

Summary of sourcing and sinking applications

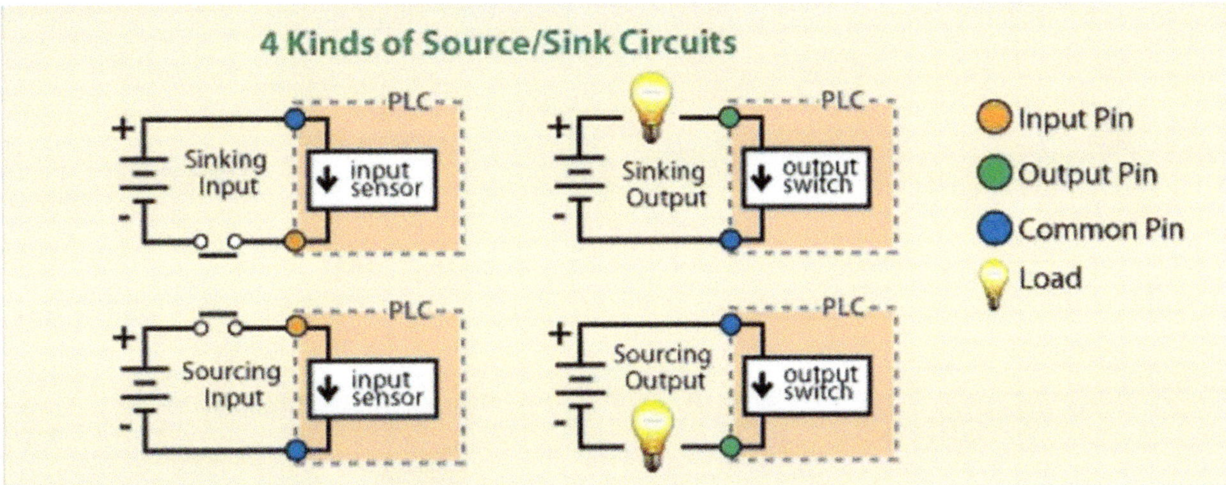

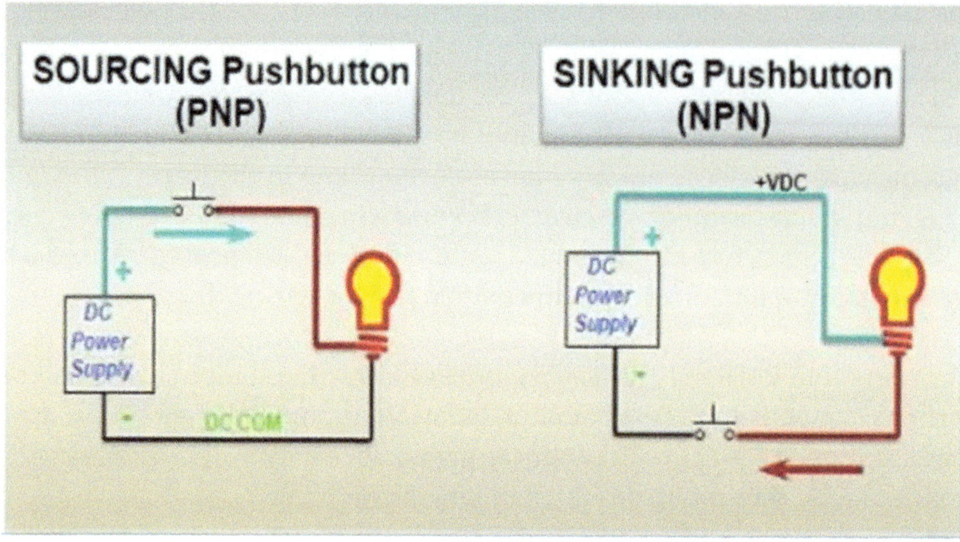

Safety Note: As evident from the above diagram, sourcing input contacts are much safer than sinking input contacts. If the negative of the power supply is grounded and the sinking input contact wire lead shorts or faults to ground, the circuit will operate or start the equipment automatically and the operator will have no control.

22 Overview of Motion Controller Applications

Motion Control Systems

The basic architecture of a motion control system contains:

- A motion controller to move something to a desired position or set and maintain the velocity of a system component or fluid. Motion controllers typically use closed-loop feedback, but not always.
- A drive or amplifier to transform the control signal from the motion controller into a higher power electrical current or voltage that is presented to the actuator.
- An actuator such as a hydraulic pump, air cylinder, linear actuator, or electric motor for output of motion.
- One or more feedback sensors, such as optical encoders, resolvers, or Hall effect devices, to return the position or velocity of the actuator to the motion controller in order to close the position or velocity control loops.
- Mechanical components to transform the motion of the actuator into the desired motion, including gears, shafting, ball screws, belts, linkages, and linear and rotational bearings.

Common control functions include:

- Velocity control.
- Position (point-to-point) control: There are several methods for computing a motion trajectory. These are often based on the velocity profiles of a move, such as a triangular profile, trapezoidal profile, or an S-curve profile.
- Pressure or force control.
- Electronic gearing (or cam profiling).

Stepper Motors

A stepper motor is a brushless, synchronous electric motor that converts digital pulses into mechanical shaft rotation.

Every revolution of the stepper motor is divided into a discrete number of steps, in many cases 200 steps, and the motor must be sent a separate pulse for each step. The stepper motor can only take one step at a time, and each step is the same size. Because each pulse causes the motor to rotate a precise angle, typically 1.8°, the motor's position can be controlled without any feedback mechanism. As the digital pulses increase in frequency, the step movement changes into continuous rotation, with the speed of rotation directly proportional to the frequency of the pulses. Stepper motors are used every day in both industrial and commercial applications because of their low cost, high reliability, high torque at low speeds, and simple, rugged construction that operates in almost any environment.

Closed-Loop Stepper Motor

Modern motion controllers can include the ability to run stepper motors with encoder feedback, resulting in true closed-loop motion control. This mode of motor control is similar to standard three-phase brushless servo motor control, where the three phases are offset 120 electrical degrees. With steppers, the motor has two phases offset by 90 electrical degrees. The benefits of closed-loop control with stepper motors include greatly improved velocity smoothness and reduced power consumption (compared to open-loop steppers) and much higher torque at low velocity (compared with traditional three-phase brushless servomotors).

Stepper Motor Advantages

1. The rotation angle of the motor is proportional to the input pulse.
2. The motor has full torque at standstill (if the windings are energized).
3. Precise positioning and repeatability of movement because good stepper motors have an accuracy of 3%–5% of a step and this error is noncumulative from one step to the next.
4. Excellent response to starting/stopping/reversing of the motor.
5. Very reliable because there are no contact brushes in the motor. Therefore, the life of the stepper motor is simply dependent on the life of the bearing.
6. The stepper motor's response to digital input pulses provides open-loop control, making the motor simpler and less costly to control.
7. It is possible to achieve very low speed synchronous rotation with a load that is directly coupled to the shaft.
8. A wide range of rotational speeds can be realized as the speed is proportional to the frequency of the input pulses.
9. Linear motion control with lead screw or worm gears.

10. The motor provides no hunting to position.

11. The stepper motor controller requires no tuning.

The rotary motion of a stepper motor can be converted to linear motion using a lead screw/worm gear drive system. The lead, or pitch, of the lead screw is the linear distance traveled for one revolution of the screw. If the lead is equal to 1 in per revolution, and there are 200 full steps per revolution, then the resolution of the lead screw system is 0.005 in per step. Even finer resolution is possible by using the stepper motor/drive system in microstepping mode.

Series versus Parallel Connection

There are two ways to connect a stepper motor, in series or in parallel. A series connection provides a high inductance and therefore greater torque at low speeds. A parallel connection will lower the inductance, which results in increased torque at faster speeds.

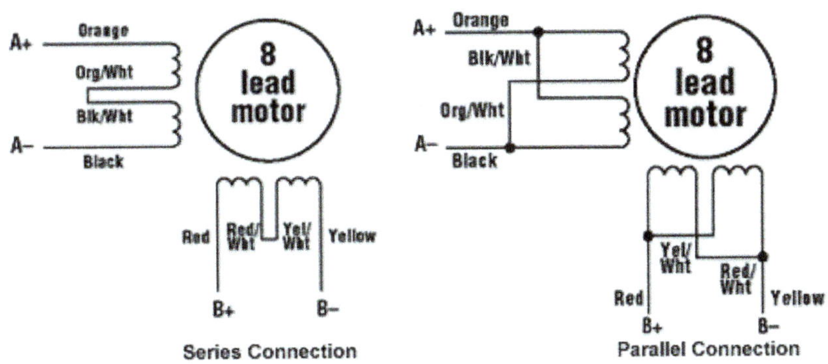

Servomotor Systems

Another option is a servo system. A servo system is typically a low pole count motor that gives high speed but has no inherent positioning capability. To make it a position device, feedback, usually an encoder or resolver, and control loops are required. The servo is essentially turned on and off until the resolver count reaches a certain point. Therefore, the servo works based on error.

For example, the servo is commanded to move 100 revs (revolutions). The resolver count reads zero and the motor is turned on. When the resolver count reaches 100 revs, the motor is turned off. If the position deflects, the motor is turned back on to bring it back to position.

How the servo responds to error depends on a gain setting. If the gain setting is high, the motor will respond very quickly to any changes in error. If the gain setting is low, the motor will not respond as quickly to changes in error. Any time gain settings are involved, though, time delays are introduced into the motion control system.

Advanced Motion Controls

Servo drives and amplifiers are used extensively in motion control systems where precise control of position and/or velocity is required. The drive/amplifier simply translates the low-energy

Benefits of Choosing a Soft Starter

Soft starters are often the more economical choice for applications that only require speed and torque control during motor start-up. Additionally, they are often the ideal solution for applications where space is a concern, as they usually take up less space than compatible variable frequency drives.

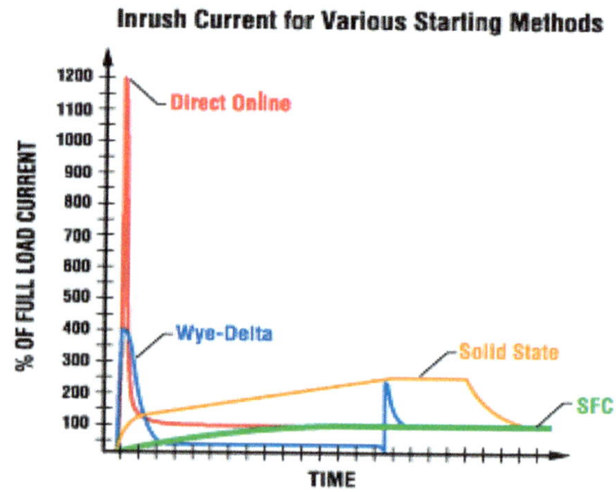

When accelerating an alternating current (AC) motor to full speed using a full voltage connection, a large inrush current may be required. Additionally, the torque of the AC motor is mostly uncontrolled and can shock the connected equipment, potentially causing damage. Variable frequency drives and reduced voltage soft starters and can both be used to reduce inrush currents and limit torque, thereby protecting expensive equipment and extending the life of the motor and coupling devices.

Choosing between a variable frequency drive and soft starter often depends on the type of application, the mechanical system requirements, and cost (for both initial installation and the life cycle of the system).

Variable Frequency Drive

The primary function of a variable-frequency drive (VFD) is to vary the speed of a three-phase AC induction motor. VFDs also provide nonemergency start and stop control, acceleration and deceleration, and overload protection. In addition, VFDs can reduce the amount of motor start-up in-rush current by accelerating the motor gradually. For these reasons, VFDs are suitable for conveyors, fans, and pumps that benefit from reduced and controlled motor operating speed.

One of the little-known benefits of a VFD is improved loop performance when it is used as the final control element of a process control loop. The VFD can replace both a control valve and its electromechanical positioner. Because most VFDs also have internal proportional-integral-derivative (PID) controller function blocks, they can often replace the process loop controller as well. It is well known that stickiness in a process control valve is the single biggest reason for poor control loop performance, and using a VFD to drive the pump motor for liquid flow control can provide better loop performance.

How Does a Variable Frequency Drive Work?

The first stage of a variable frequency AC drive, or VFD, is the converter. The converter is composed of six diodes, which are similar to check valves used in plumbing systems. They cause current to flow in only one direction: the direction shown by the arrow in the diode symbol. For example, whenever A-phase voltage (voltage is similar to pressure in plumbing systems) is more positive than B- or C-phase voltages, that diode will conduct or operate and cause current to flow. When B-phase becomes more positive than A-phase, the B-phase diode will conduct or operate and the A-phase diode will stop conducting or stop (check) the current flow. The same is true for the three diodes on the negative side of the bus. Thus, we get six current "pulses" as each diode conducts and stops conducting. This is called a *six-pulse VFD,* which is the standard configuration for current VFDs.

Note: The VFD and uninterruptible power supply (UPS) operate on the same basic architecture and principles. Images source: VFDs.com

Conversion from AC to DC to AC Pulse Width Modulation

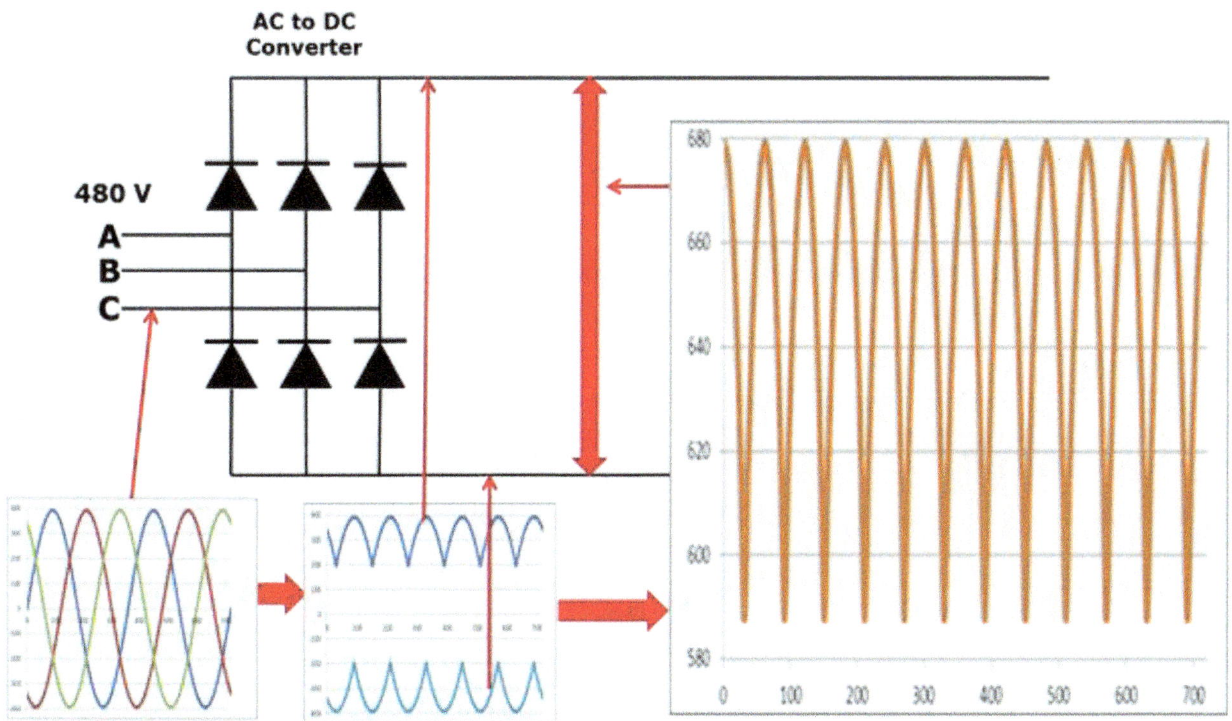

Let us assume that the drive is operating on a 480 V power system. The 480 V rating is root mean square (rms). The peaks on a 480 V system are 679 V. As you can see, the VFD direct current (DC) bus has a DC voltage with an AC ripple. The voltage runs between approximately 580 V and 680 V.

We can get rid of the AC ripple on the DC bus by adding a capacitor. A capacitor operates in a similar fashion to a reservoir or accumulator in a plumbing system. This capacitor absorbs the AC ripple and delivers a smooth DC voltage. The AC ripple on the DC bus is typically less than 3 V. Thus, the voltage on the DC bus becomes "approximately" 650 VDC. The actual voltage will depend on the voltage level of the AC line feeding the drive, the level of voltage unbalance on the

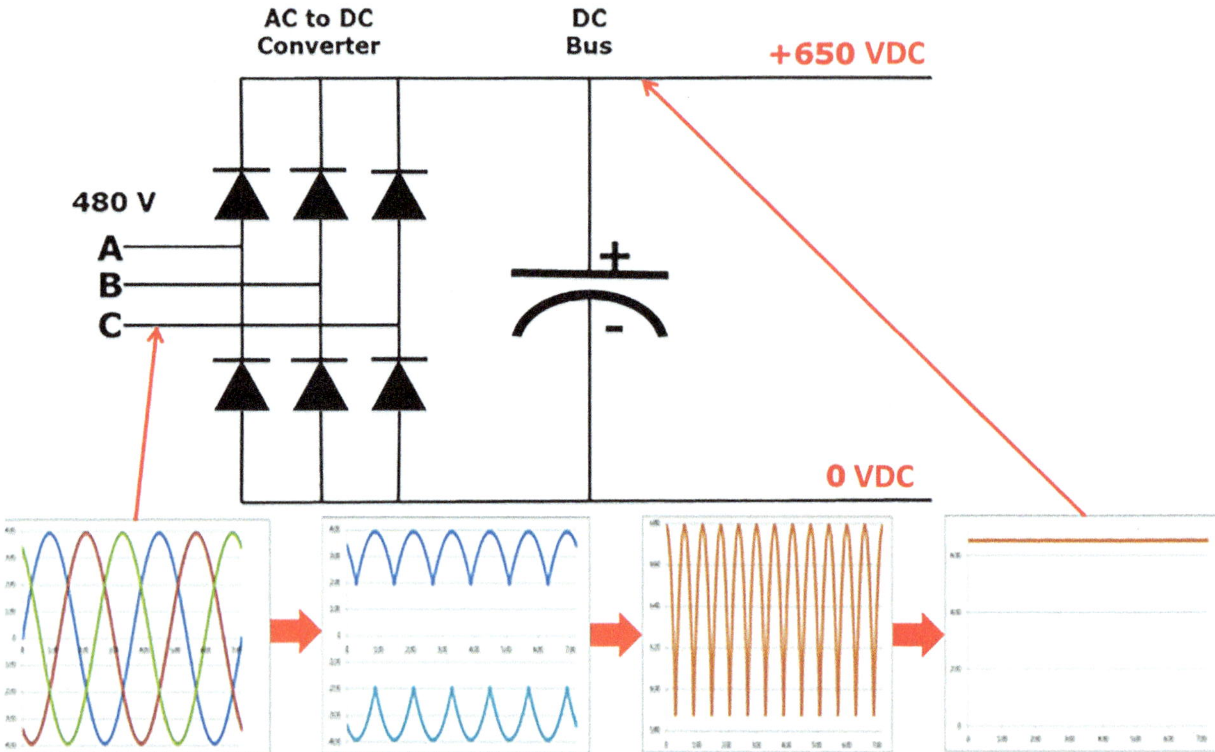

power system, the motor load, the impedance of the power system, and any reactors or harmonic filters on the drive.

The diode bridge rectifier that converts AC to DC is sometimes referred to as a *converter*. The device that converts the DC back to AC is also a converter, but to distinguish it from the diode converter, it is usually referred to as an *inverter*. It has become common in the industry to refer to any DC-to-AC converter as an inverter.

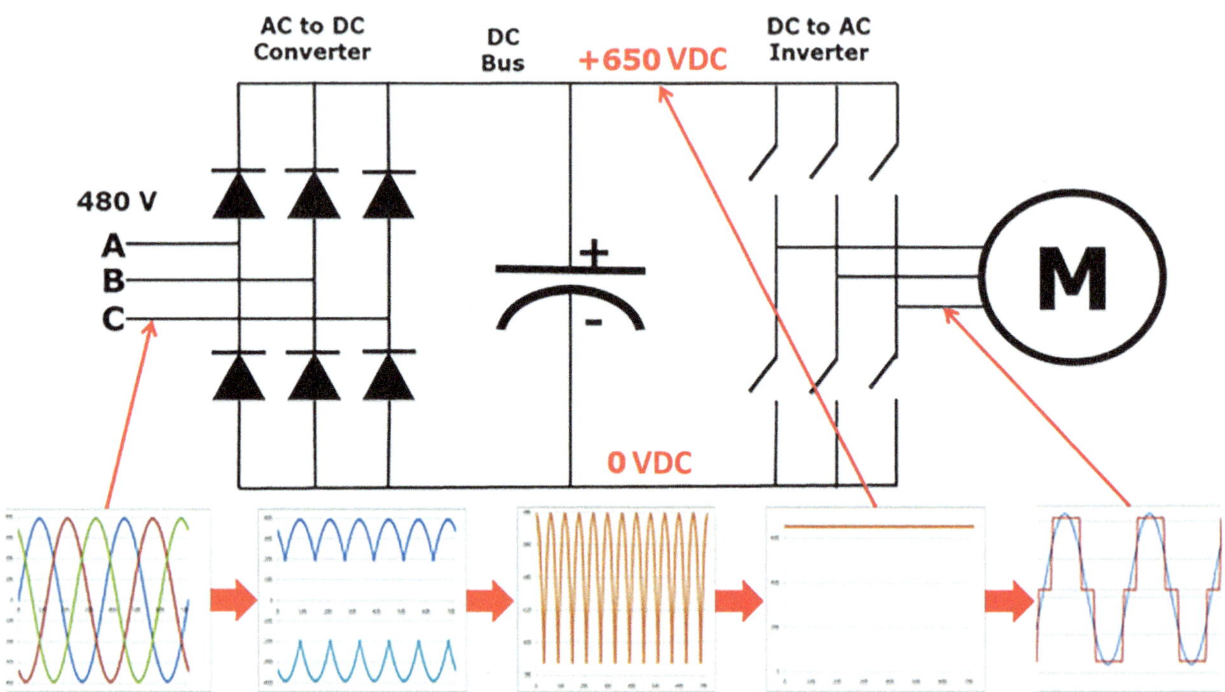

Note that in a real VFD, the switches shown would actually be transistors.

When we close one of the top switches in the inverter, that phase of the motor is connected to the positive DC bus and the voltage on that phase becomes positive. When we close one of the bottom switches in the converter, that phase is connected to the negative DC bus and becomes negative. Thus, we can make any phase on the motor become positive or negative at will and can thus generate any frequency that we want. So, we can make any phase positive, negative, or zero. The following image is a schematic of a generic variable speed drive (VSD).

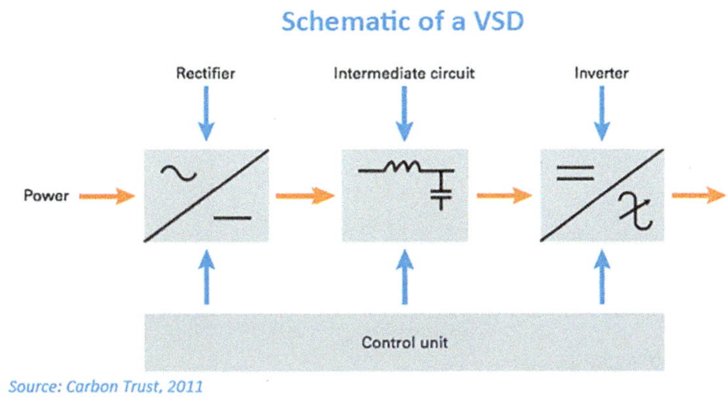

Notice that the output from the VFD is a "rectangular" wave form. VFDs do not produce a sinusoidal output. This rectangular waveform would not be a good choice for a general-purpose distribution system but is perfectly adequate for a motor.

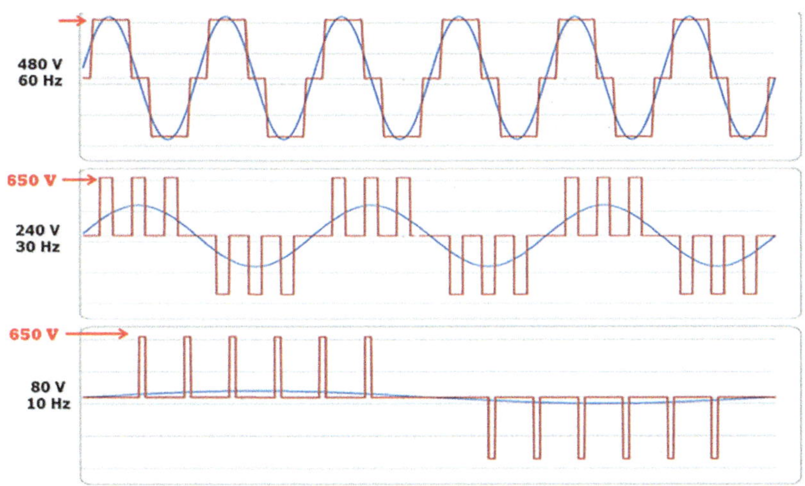

The blue sine waves are shown for comparison purposes only. The drive does not generate this sine wave.

If we want to reduce the motor frequency to 30 Hz, then we simply switch the inverter output transistors more slowly. However, if we reduce the frequency to 30 Hz, we must also reduce the voltage to 240 V to maintain the V/Hz ratio. How are we going to reduce the voltage if the only voltage we have is 650 VDC?

This is called pulse width modulation (PWM). Imagine that we could control the pressure in a water line by turning the valve on and off at a high rate of speed. Although this would not be practical for plumbing systems, it works very well for VFDs. Notice that during the first half cycle, the voltage is on half the time and off half the time. Thus, the average voltage is half of 480 V, or 240 V. By pulsing the output, we can achieve any average voltage on the output of the VFD.

Volts to Hertz Ratio Relationship

The volts to hertz relationship must be constant. The impedance of the motor is set by the frequency. The relationship of inductive impedance (X_L), motor impedance (Z), current (I), and power in kVA (P) is as follows:

$$X_L = 2\pi fL \quad Z = \sqrt{R^2 + X_L{}^2} \quad I = \frac{V}{\sqrt{R^2 + X_L{}^2}} = \frac{V}{Z} \quad P = I^2 \bullet Z$$

The first formula shows that (X_L) is directly proportional to the frequency. If the frequency of the current and voltage into the motor increases, the impedance of the motor increases. The resistance of a motor is almost negligible compared to the inductance, so we will just ignore the resistance for this example.

Work is equal to power per time. We want to keep the horsepower (hp) constant even at low speeds. Assume the motor draws 2 A at a voltage of 240 V. 240 V/2 A = 120 Ω of impedance at 60 Hz. We must keep the Volts to hertz ratio constant, so at 30 Hz the voltage must be 120 V. What are the impedance, current, and power per time at 30 Hz?

$$\text{At} \quad X_L = 2\pi fL \quad \text{so} \quad L = \frac{120\ \Omega}{2\pi f} = \frac{120\ \Omega}{2\pi \bullet 60} = 0.3183\ \text{H}$$

$$\text{At} \quad I = \frac{V}{Z} = \frac{240\ \text{V}}{120\ \Omega} = 2\ \text{A}$$

$$\text{At} \quad X_L = 2\pi fL \quad \text{so} \quad X_L = 2\pi \bullet 30\ \text{Hz} \bullet 0.3183\ \text{H} = 60\ \Omega$$

$$\text{At} \quad I = \frac{V}{Z} = \frac{120\ \text{V}}{60\ \Omega} = 2\ \text{A}$$

Notice that the impedance decreased proportionally to the frequency. It now has half the opposition to the flow of current into the motor. Work equals power, so:

$$\text{At} \quad W = I^2 \bullet Z = 2^2 \bullet 240 = 960\ \text{kVA}$$

$$\text{At} \quad W = I^2 \bullet Z = 2^2 \bullet 120 = 480\ \text{kVA}$$

We are going half the speed at 30 Hz compared to 60 Hz. It takes twice the time to do the same work over distance (60 Hz/30 Hz = 2); this is half the distance the motor shaft was turning in time. The unit Hz is proportional to the rpm (rotations per minute). If we integrate the moment about the motor shaft, we find the work done by a motor:

rpm • 2π • moment = rpm • 2π • radius • force. The work done about the shaft is force multiplied by distance multiplied by time.

Force times distance equals work. Multiply this by time and we have work per unit of time.

The work being done by the motor is now half, but at half the speed. So, if the motor runs twice as long, the work done at twice the time period is the same as that being done at the 60 Hz frequency.

Work equals 906 kVA at 60 Hz = (2 • 480 KVA) at 30 Hz. The horsepower of the motor has not changed with a change in speed as long as the volts to hertz (V/Hz) ratio is kept constant. If the frequency is too high in the volts to hertz setting, the impedance is too high. The motor starves for current to generate magnetic flux and slips. If the frequency is too low in the V/Hz setting, the impedance is too low and too much current is generated, and extra heat will be produced in the motor windings. This current saturates the stator, and the motor's magnetic field is saturated and cannot produce the speed required for the amount of work needed. The slip means less work is being done in time, so the horsepower output decreases.

Important Note About Low Frequency in VFDs

A motor must be capable of doing work at lower speeds (lower frequencies). The internal fan is running more slowly and a lot less air is flowing over the motor to keep it cool inside. This is important in totally enclosed fan cooled (TEFC) motors used in hazardous locations. The motor may exceed the temperature limit of the environment, or the motor insulation breaks down in time due to the increase in temperature and heat dissipation.

There are motors rated just for running at low frequencies, and manufactures also make adaptive fan kits to fit on the end of motors to help keep them cool.

Motors that have P leads may be used to protect the motor and the hazardous environment. P leads are connected to an internal temperature switch or thermistor and are in series with the motor control shutdown circuit. TEFC motors with P leads may be required by the National Electric Code (NEC) for certain operating temperatures in classified locations (hazardous environments) per Article 500.

VFDs Put Noise into the Electrical System

Note that VFDs generate harmonics, and they are reflected back into the electrical system. This is apparent with large speed drive and phase-angle fired speed and heating controllers. Sometimes it is necessary to put harmonic filters on the drive, on programmable logic controller/distributed control system power supplies, and/or on control cabinets. This higher rms voltage from the harmonics can damage equipment and cause constant blowing of fuses and tripping of breakers because of the extra heating effect of the odd harmonics.

Note: For noise protection of data equipment and instrumentation cabinets and enclosures, two grounding paths should be attached as single point ground to the grid from the enclosure, not just one path. This technique is shown in the grounding diagram example. The conductors must be different lengths. This will ensure that one of the bonding conductor paths does not stop draining the voltage noise signal to ground because of resonance at an incoming frequency. This can cause an infinite impedance in the bonding conductor, and the current will not flow to ground. There will then be a standing voltage on the cable or enclosure, and it will act as a transmission antenna, introducing noise into the controller or signal cables.

All grounding and bonding conductor connections in a plant should be exothermally welded. The beam structures, tanks, cable trays, and equipment should be connected to the grounding grid by CadWeld. Mechanical connections are not as reliable as a CadWeld, shown at right.

24 Emergency Standby Systems

NEC Article 700 – Emergency Systems

The requirements of Article 700 apply only to the wiring methods for "emergency systems" that are essential for safety to human life and are required by federal, state, or municipal governments or other agencies having jurisdiction. When normal power is lost, emergency systems shall be able to supply standby power in 10 s or less.

NEC Article 701 – Legally Required Standby Systems

Legally required standby systems provide electric power to aid in firefighting, rescue operations, control of health hazards, and similar operations. When normal power is lost, legally required systems shall be able to supply standby power in 60 s or less, instead of the 10 s or less required of emergency systems.

NEC Article 702 – Optional Standby Systems

Optional standby systems are intended to protect public or private facilities or property where life safety does not depend on the performance of the system. These systems are typically installed to provide an alternate source of electric power for such facilities as industrial and commercial buildings, farms, and residences, and to serve loads that, when stopped during any power outage, could cause discomfort, serious interruption of the process, or damage to the product or process. Optional standby systems are intended to supply on-site generated power to selected loads either automatically or manually.

Uninterruptible Power Supply

An uninterruptible power supply (UPS), also called *uninterruptible power source* or *battery/flywheel backup*, is an electrical apparatus that provides emergency power to a load when the input power source, typically the main power, fails. A UPS differs from an auxiliary or emergency power system or standby generator in that it will provide nearly instantaneous protection from input power interruptions by supplying energy stored in batteries, supercapacitors, or flywheels. The on-battery runtime of most UPSs is relatively short, only a few minutes to a few hours (maybe up

25 Hydraulics and Pneumatics

Fluid Power Systems

Fluid power is the use of fluids under pressure to generate, control, and transmit power. Fluid power is subdivided into hydraulics using a liquid such as mineral oil or water, and pneumatics using a gas such as air or other gases. Compressed-air and water-pressure systems were once used to transmit power from a central source to industrial users over extended geographic areas; fluid power systems today are usually within a single building or mobile machine.

Hydraulic Systems

The basic idea behind any hydraulic system is very simple: Force that is applied at one point is transmitted to another point using an incompressible fluid. The fluid is almost always an oil of some sort. The force is almost always multiplied in the process. The picture below shows the simplest possible hydraulic system:

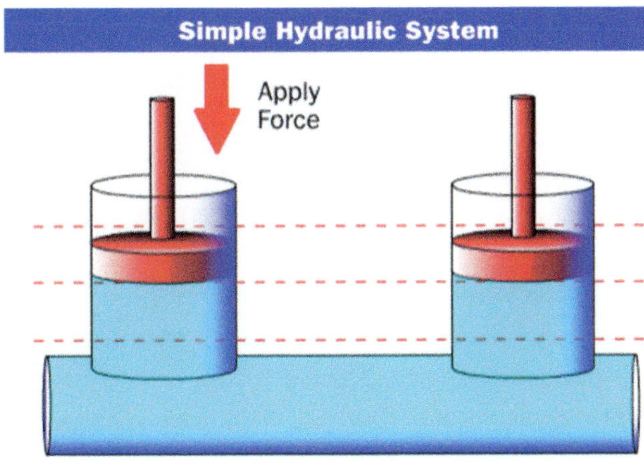

A simple hydraulic system consisting of two pistons and an oil-filled pipe connecting them.

If you apply a downward force to one piston (the left one in this drawing), then the force is transmitted to the second piston through the oil in the pipe. Because oil is incompressible—the efficiency is very good—almost all the applied force appears at the second piston.

The great thing about hydraulic systems is that the pipe connecting the two cylinders can be any length and shape, which enables installers to snake it through all sorts of things separating the two pistons. The pipe can also fork, so that one *master cylinder* can drive more than one **slave** cylinder if desired.

In a hydraulic system, all you do is change the size of one piston and cylinder relative to the other to make a hydraulic amplifier, as shown in the next image. This is called *hydraulic multiplication*. The piston on the right has a surface area nine times greater than the piston on the left. When force is applied to the left piston, it will move nine units for every one unit that the right piston moves, and the force is multiplied by nine on the right-hand piston.

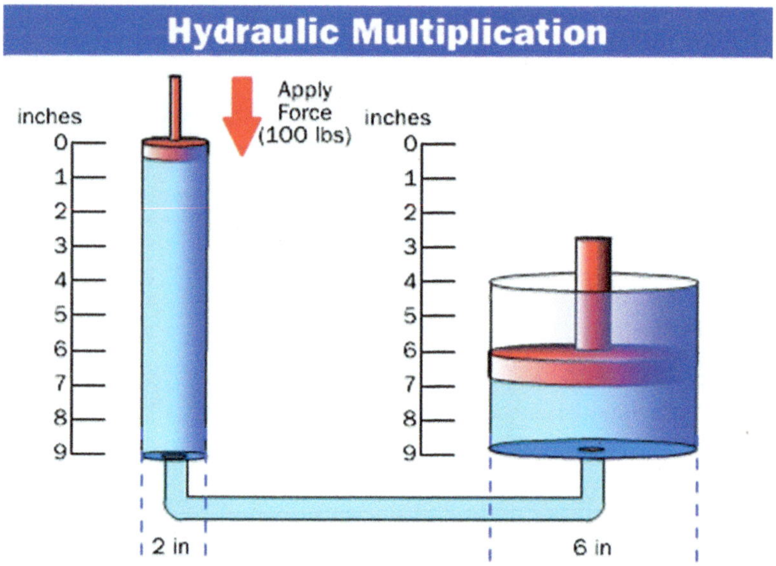

To determine the **multiplication factor**, start by looking at the size of the pistons. Assume that the piston on the left is 2 in in diameter (1-inch radius), while the piston on the right is 6 in in diameter (3-inch radius). The area of the two pistons is Pi multiplied by r^2. The area of the left piston is therefore 3.14, while the area of the piston on the right is 28.26.

The piston on the right is nine times larger than the piston on the left. What that means is that any force applied to the left-hand piston will appear nine times greater on the right-hand piston. So if you apply a 100-pound downward force to the left piston, a 900-pound upward force will appear on the right. The only catch is that you will have to depress the left piston 9 in to raise the right piston 1 in.

Pneumatic Systems

Pneumatics is a branch of technology that deals with the study and application of pressurized gas to produce mechanical motion. Pneumatic systems used extensively in industry are commonly powered by compressed air or compressed inert gases. A centrally located and electrically powered compressor powers cylinders, air motors, and other pneumatic devices. A pneumatic system controlled through manual or automatic solenoid valves is selected when it provides a lower cost, more flexible, or safer alternative to electric motors and actuators.

Mechanical flow diagram of a large compressor

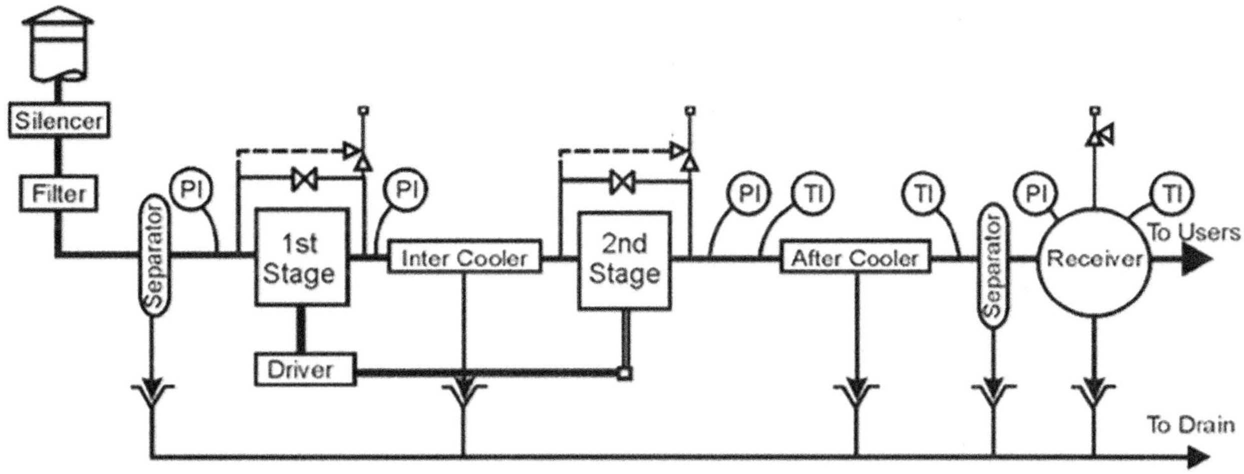

Instrumentation air header (fluid distribution header or manifold)

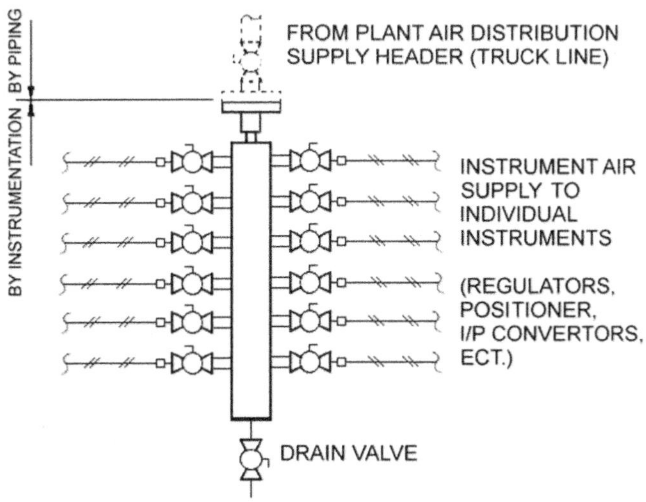

26 Fluid Power Schematic Symbols

Hydraulic and Pneumatic Circuit Design

Lines

Connecting pressure lines (usually representing plastic tubing for pneumatic [air] lines with low pressures, metal piping for hydraulic [fluid] lines with high pressure)

	continuous line for flow line
	dashed line for pilot, drain
	envelope for long and short dashes around two or more component symbols.

Circular

	large circle – pump, motor
	small circle – measuring devices
	semicircle – rotary actuator

Square

	one square – pressure control function two or three adjacent squares – directional control

Diamond

	diamond – fluid conditioner (filter, separator, lubricator, heat exchanger)

Miscellaneous Symbols

	spring
	slow restriction

Triangle

	solid – direction of hydraulic fluid flow
	open – direction of pneumatic flow

Pump and Compressor Symbols

(Turn rotary torque from an AC or DC electric motor into pressure in a hydraulic or pneumatic system)

Fixed Displacement Hydraulic Pump

 unidirectional

 bidirectional

Variable Displacement Hydraulic Pump

 unidirectional

 bidirectional

Compressor

Motor Symbols

Fixed Displacement Hydraulic Motor

 unidirectional

 bidirectional

Variable Displacement Hydraulic Motor

 unidirectional

 bidirectional

Pneumatic Motor

 unidirectional

 bidirectional

Rotary Actuator Symbol

 hydraulic

 pneumatic

Cylinder Symbols

Single-Acting Cylinder

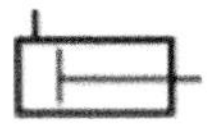

returned by external force

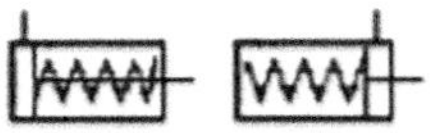

returned by spring or extended by spring force

Double-Acting Cylinder

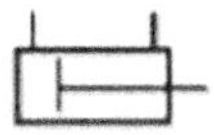

single piston rod (fluid required to extend and retract)

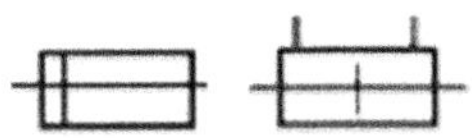

double-ended piston rod

Cylinders with Cushions

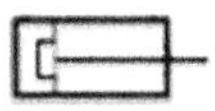

single fixed cushion

double fixed cushion

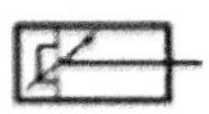

single adjustable cushion

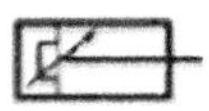

double adjustable cushion

Directional Control Valve Symbols

Directional Control Valve (2 Ports/2 Positions)

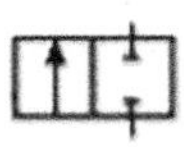

normally closed directional control valve with two ports and two finite positions

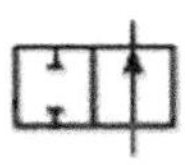

normally open directional control valve with two ports and two finite positions

Directional Control Valve (3 Ports/2 Positions)

normally closed directional control valve with three ports and two finite positions

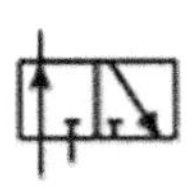

normally open directional control valve with three ports and two finite positions

Directional Control Valve (4 Ports/2 Positions)

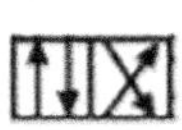

directional control valve with four ports and two finite positions

Directional Control Valve (4 Ports/3 Positions)

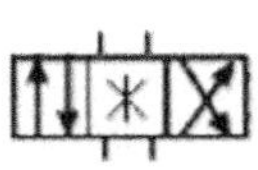

directional control valve with four ports and three finite positions
* – center position can have various flow paths

Directional Control Valve (5 Ports/2 Positions)
(Normally a Pneumatic Valve)

directional control valve with five ports and two finite positions

Directional Control Valve (5 Ports/3 Positions)
(Normally a Pneumatic Valve)

directional control valve with five ports and three finite positions

Proportional Directional Control and
Electrohydraulic Servo Valves
(The spool position on these valves is variable, allowing for variable flow conditions.)

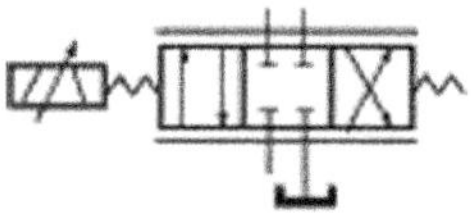

single-stage *direct operation* unit that accepts an analog signal and provides a similar analog fluid power output

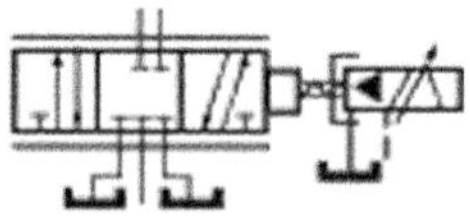

two-stage with mechanical feedback *indirect pilot operation* unit that accepts an analog signal and provides a similar analog fluid power output

Control Method Operator Symbols for Valves

Manual Control

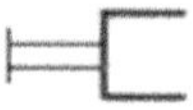

general symbol of a valve's manual operator (without showing the control type)

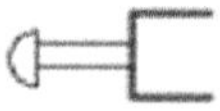

pushbutton

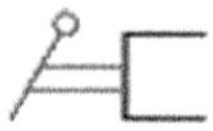

lever

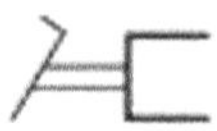

foot pedal

Mechanical Valve Control

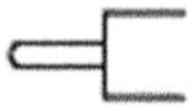

plunger or tracer

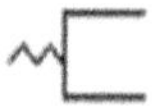

spring (used on one side of a valve to hold it in the normally open or normally closed state)

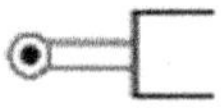

roller

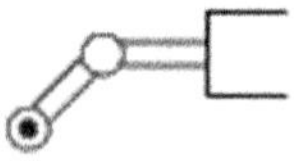

roller (one direction only)

Electrical/Solenoid Valve Control

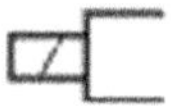

solenoid (one side is winding shown)

Pilot Operation
(Uses Pressure to Actuate Valve)

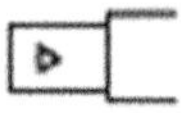

pneumatic actuated pilot

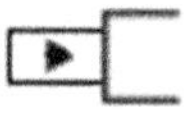

hydraulic actuated pilot

Pilot-Operated Two-Stage Valve
(Uses a Second Lesser Force to Actuate the Pilot Actuation of the Valve)

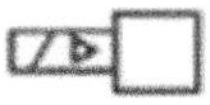

pneumatic: solenoid first stage

pneumatic: air pilot second stage

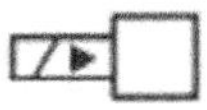

hydraulic: solenoid first stage

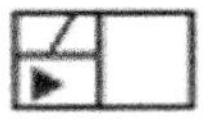

hydraulic: hydraulic pilot second stage

Pressure Control Valve Symbols

Pressure-Relief Valve (Safety Valve) Normally Closed

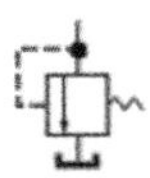

Line pressure is limited to the setting of the valve; secondary part is directed to tank.

Proportional Pressure-Relief Valve

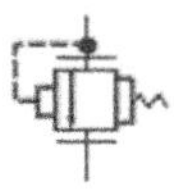

Line pressure is limited to and proportional to an electronic signal.

Sequence Valve

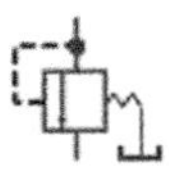

When the line pressure reaches the setting of the valve, the valve opens, permitting flow to the secondary port. The pilot must be externally drained to the tank.

Pressure-Reducing Valve (Hydraulic Pressure Regulator)

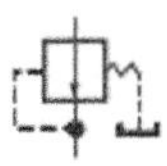

Pressure downstream of valve is limited to the setting of the valve.

Flow Control Valve Symbols

Throttle Valve

adjustable output flow

Flow Control Valves

flow control valve with fixed output (Variations in inlet pressure do not affect rate of flow.)

flow control valve with fixed output and relief port to reservoir with relief for excess flow (Variations in inlet pressure do not affect rate of flow.)

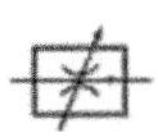

flow control valve with variable output

flow control valve with fixed orifice

flow control valve with metered flow toward right, free flow to left

flow control valve with pressure-compensated flow-control fixed output flow regardless of load

flow control valve with pressure and temperature compensated

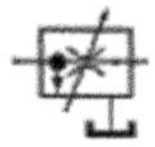
flow control valve with variable output and relief port to reservoir

Flow-Dividing Valve

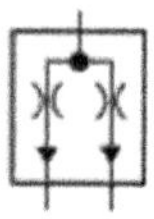
flow is divided equally to two outputs

Shutoff Valve

shutoff valve simplified symbol

Tank Symbols

Accumulator

accumulator symbol (stores pressure)

Reservoir (Tank)

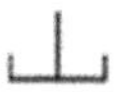
reservoir symbol (holds fluid medium of the system)

Filter, Water Trap, Lubricator, and Miscellaneous Apparatus Symbols

Filter or Strainer

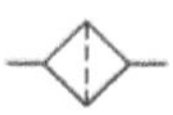
with manual drain

Water Trap

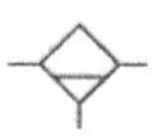
with manual drain

with automatic drain

Filter with Water Trap

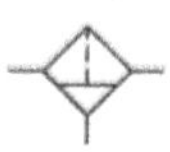
with manual drain

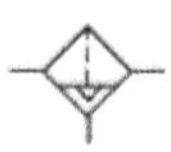
with automatic drain

Air Dryer

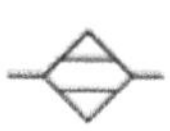
refrigerant, or chemical removal of water from compressed air line

Lubricator

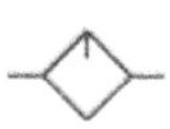
oil vapor is injected into air line

Conditioning Unit (FRL, Pressure Regulator)

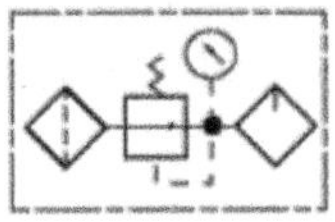

compound symbol of filter, regulator, lubricator unit (FRL symbol)

simplified symbol

Heat Exchanger Symbol

air- or water-cooled unit designed to remove heat from oil returning to reservoir

27 ISA Standards for Documentation

ISA Instrument or Function Symbol—ANSI/ISA-5.1-2009

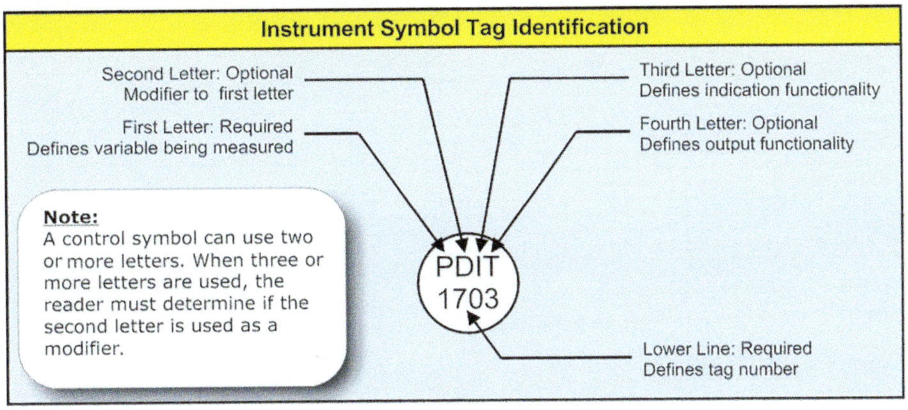

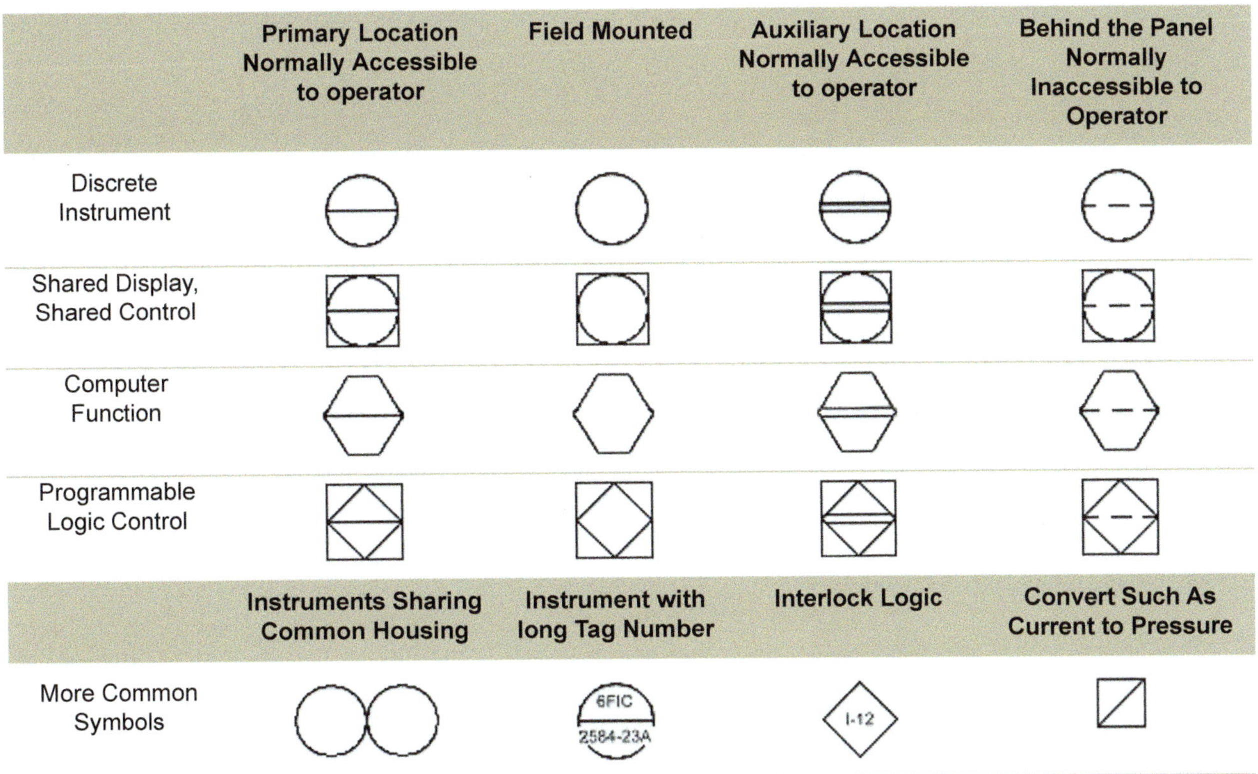

ISA Line-Type Symbols—ANSI/ISA-5.1-2009

B.9.5 Shared display, shared control, and wireless instrumentation:

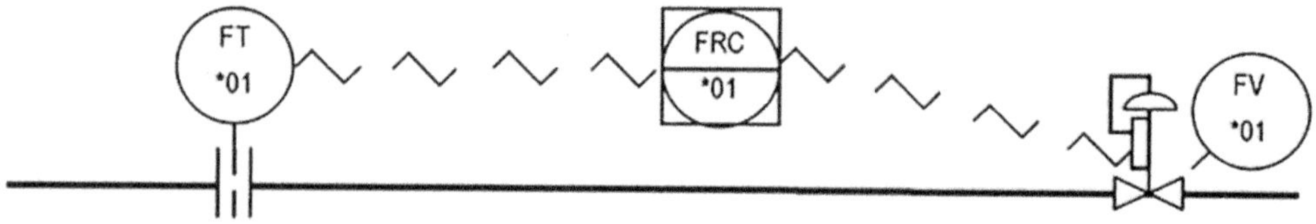

Standard Line Types—ANSI/ISA-5.1-2009

1. Instrument Supply or Connected to Process

2. Undefined Signal 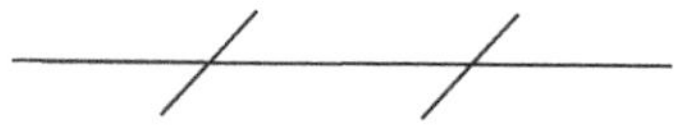

3. Pneumatic Signal 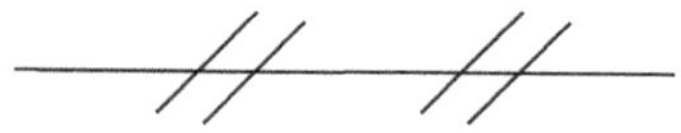

4. Electric Signal 

5. Hydraulic Signal 

6. Capillary Signal 

7. Electromagnetic or Sonic Signal (Guided) 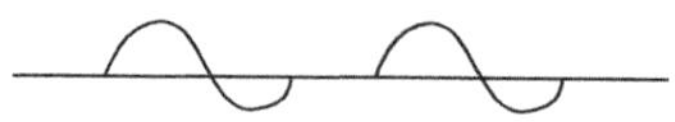

8. Electromagnetic or Sonic Signal (Not Guided) 

9. Internal Systems Link (Software or Data Link) 

10. Mechanical Link 

11. Pneumatic Binary 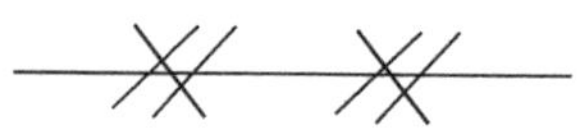

12. Electrical Binary 

ISA Identification Letters

	Initiating or Measured Variable	Modifier	Readout or Passive Function	Output Modifier Function	
A	Analysis (5,19)		Alarm		
B	Burner, Combustion		User's Choice (1)	User's Choice (1)	User's Choice (1)
C	User's Choice (1)			Control (13)	
D	User's Choice (1)	Differential (4)			
E	Voltage		Sensor (Primary Element)		
F	Flow Rate	Ratio (Fraction) (4)			
G	User's Choice (1)		Glass, Viewing Device (9)		
H	Hand				High (7, 15, 16)
I	Current (Electrical)		Indicate (10)		
J	Power	Scan (7)			
K	Time, Time Schedule	Time Rate of Change (4, 21)		Control Station (22)	
L	Level		Light (11)		Low (7, 15, 16)
M	User's Choice (1)	Momentary (4)			Middle, Intermediate (7,15)
N	User's Choice (1)		User's Choice (1)	User's Choice (1)	User's Choice (1)
O	User's Choice (1)		Orifice, Restriction		
P	Pressure, Vacuum		Point (Test) Connection		
Q	Quantity	Integrate, Totalize (4)			
R	Radiation		Record (17)		
S	Speed, Frequency	Safety (8)		Switch (13)	
T	Temperature			Transmit (18)	
U	Multivariable (6)		Multifunction (12)	Multifunction (12)	Multifunction (12)
V	Vibration, Mechanical Analysis (19)			Valve, Damper, Louver (13)	
W	Weight, Force		Well		
X	Unclassified (2)	X-Axis	Unclassified (2)	Unclassified (2)	Unclassified (2)
Y	Event, State, or Presence (20)	Y-Axis		Relay, Compute, Convert (13, 14, 18)	
Z	Position, Dimension	Z-Axis		Driver, Actuator, Unclassified Final Control Element	

ISA P&ID Identification (Controllers and Readouts)

Initiating or Measured Variable		Controllers				Readout Devices	
	First Letter	Recording	Indicating	Blind	Self-Actuated Control Valves	Recording	Indicating
Analysis	A	ARC	AIC	AC		AR	AI
Burner, Combustion	B	BRC	BIC	BC		BR	BI
User's Choice	C						
User's Choice	D						
Voltage	E	ERC	EIC	EC		ER	EI
Flow Rate	F	FRC	FIC	FC	FCV/FICV	FR	FI
Flow Quantity	FQ	FQRC	FQIC			FQR	FQI
Flow Ratio	FF	FFRC	FFIC	FFC		FFR	FFI
User's Choice	G						
Hand	H		HIC	HC		HR	HI
Current (Electrical)	I	IRC	IIC			IR	II
Power	J	JRC	JIC			JR	JI
Time	K	KRC	KIC	KC	KCV	KR	KI
Level	L	LRC	LIC	LC	LCV	LR	LI
User's Choice	M						
User's Choice	N						
User's Choice	O						
Pressure, Vacuum	P	PRC	PIC	PC	PCV	PR	PI
Pressure Differential	PD						
Quantity	Q	QRC	QIC			QR	QI
Radiation	R	RRC	RIC	RC		RR	RI
Speed, Frequency	S	SRC	SIC	SC	SCV	SR	SI
Temperature	T	TRC	TIC	TC	TCV	TR	TI
Temperature Differential	TD	TDRC	TDIC	TDC	TDCV		
Multivariable	U					UR	UI
Vibration, Mechanical Analysis	V					VR	VI
Weight, Force	W	WRC	WIC	WC	WCV	WR	WI
Weight, Force Differential	WD	WDRC	WDIC	WDC	WDCV	WDR	WDI
Unclassified	X						
Event, State, or Presence	Y		YIC	YC		YR	YI
Position, Dimension	Z	ZRC	ZIC	ZC	ZCV	ZR	ZI
Gauging Deviation	ZD	ZDRC	ZDIC	ZDC	ZDCV	ZDR	ZDI

Note: The letter *Z* can be used a second modifier to indicate a safety instrumented system (SIS) function.

Examples: BZIC (burner SIS indicating controller) or PZIT (pressure SIS indicating transmitter)

ISA P&ID Identification (Transmitters, Switches, and Alarms)

Initiating or Measured Variable	Switches and Alarm Devices				Transmitters		
	First Letter	High	Low	Comb	Recording	Indicating	Blind
Analysis	A	ASH	ASL	ASHL	ART	AIT	AT
Burner, Combustion	B	BSH	BSL	BSHL	BRT	BIT	BT
User's Choice	C						
User's Choice	D						
Voltage	E	ESH	ESL	ESHL	ERT	EIT	ET
Flow Rate	F	FSH	FSL	FSHL	FRT	FIT	FT
Flow Quantity	FQ	FQSH	FQSL			FQIT	FQT
Flow Ratio	FF	FFSH	FFSL				
User's Choice	G						
Hand	H			HS	HRT	HIT	HT
Current (Electrical)	I	ISH	ISL	ISHL	IRT	IIT	IT
Power	J	JSH	JSL	JSHL	JRT	JIT	JT
Time	K	KSH	KSL	KSHL	KRT	KIT	KT
Level	L	LSH	LSL	LSHL	LRT	LIT	LT
User's Choice	M						
User's Choice	N						
User's Choice	O						
Pressure, Vacuum	P	PSH	PSL	PSHL	PRT	PIT	PT
Pressure Differential	PD	PDSH	PDSL		PDRT	PDIT	PDT
Quantity	Q	QSH	QSL	QSHL	QRT	QIT	QT
Radiation	R	RSH	RSL	RSHL	RRT	RIT	RT
Speed, Frequency	S	SSH	SSL	SSHL	SRT	SIT	ST
Temperature	T	TSH	TSL	TSHL	TRT	TIT	TT
Temperature Differential	TD	TDSH	TDSL		TDRT	TDIT	TDT
Multivariable	U						
Vibration, Mechanical Analysis	V	VSH	VSL	VSHL	VRT	VIT	VT
Weight, Force	W	WSH	WSL	WSHL	WRT	WIT	WT
Weight, Force Differential	WD	WDSH	WDSL		WDRT	WDIT	WDT
Unclassified	X						
Event, State, or Presence	Y	YSH	YSL				YT
Position, Dimension	Z	ZSH	ZSL	ZSHL	ZRT	ZIT	ZT
Gauging Deviation	ZD	ZDSH	ZDSL		ZDRT	ZDIT	ZDT

ISA P&ID Identification (Compute, Relay, and Elements)

Initiating or Measured Variable	First Letter	Solenoids, Relays, Computing Devices	Primary Element	Test Point	Well of Probe	Viewing Device, Glass	Safety Device	Final Element
Analysis	A	AY	AE	AP	AW			AV
Burner, Combustion	B	BY	BE		BW	BG		BV
User's Choice	C							
User's Choice	D							
Voltage	E	EY	EE					EZ
Flow Rate	F	FY	FE	FP		FG		FV
Flow Quantity	FQ	FQY	FQE					FQV
Flow Ratio	FF		FE					FFV
User's Choice	G							
Hand	H	HY						HV
Current (Electrical)	I	IY	IE					IZ
Power	J	JY	JE					JV
Time	K	KY	KE					KV
Level	L	LY	LE		LW	LG		LV
User's Choice	M							
User's Choice	N							
User's Choice	O							
Pressure, Vacuum	P	PY	PE	PP			PSV, PSE	PV
Pressure Differential	PD	PDY	PE	PP				PDV
Quantity	Q	QY	QE					QZ
Radiation	R	RY	RE		RW			RV
Speed, Frequency	S	SY	SE					SV
Temperature	T	TY	TE	TP	TW	TG	TSE	TV
Temperature Differential	TD	TDY	TE	TP	TW	TDG		TDV
Multivariable	U	UY						UV
Vibration, Mechanical Analysis	V	VY	VE		VW	VG		VZ
Weight, Force	W	WY	WE		WW	WG		WZ
Weight, Force Differential	WD	WDY	WE		WDW	WDG		WDZ
Unclassified	X							
Event, State, or Presence	Y	YY	YE		YW	YG		YZ
Position, Dimension	Z	ZY	ZE		ZW	ZG		ZV
Gauging Deviation	ZD	ZDY	ZDE		ZDW	ZDG		ZDV

Piping and Equipment Symbols

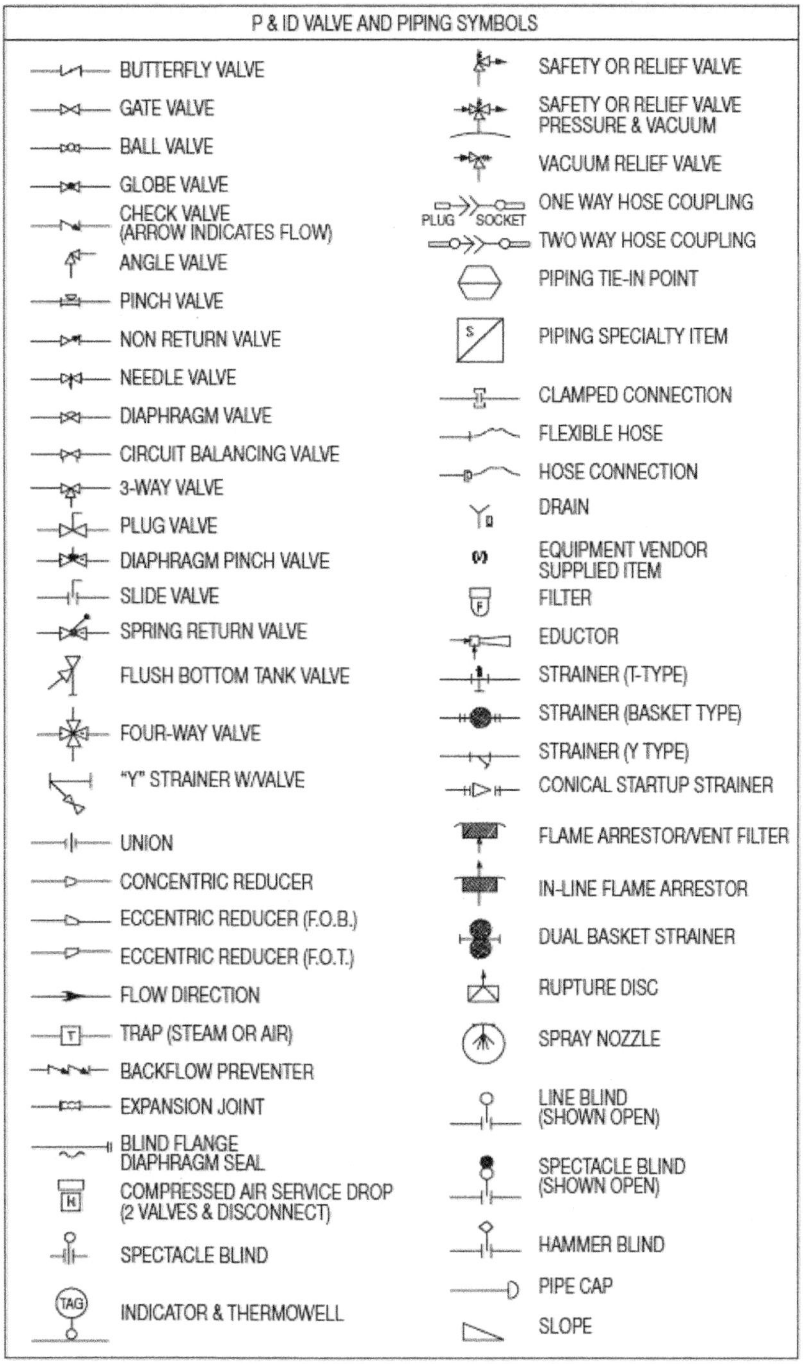

Standard Piping and Instrumentation Diagram

This is a standard ISA piping and instrumentation diagram (P&ID) as might be included on the CSE exam. The exam may have questions related to symbols and connections. Familiarize yourself with ANSI/ISA-5.1-2009, *Instrumentation Symbols and Identification,* for the exam.

ISA Standard P&ID as Might Be Included on the CSE Exam

P&ID Sample 1 (Functions).

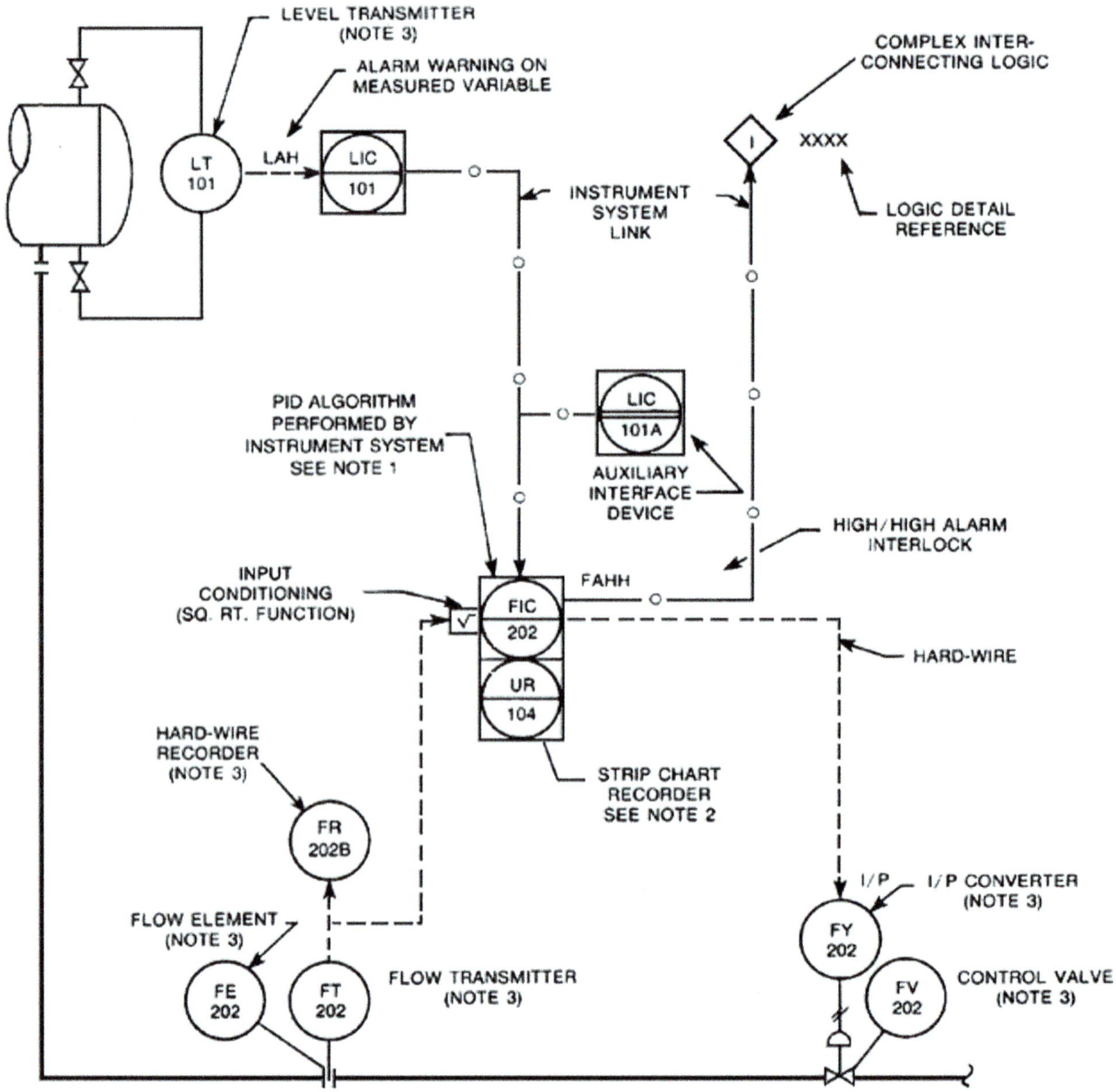

ISA Standard P&ID as Might Be Included on the CSE Exam (continued)

P&ID Sample 2 (Alarms).

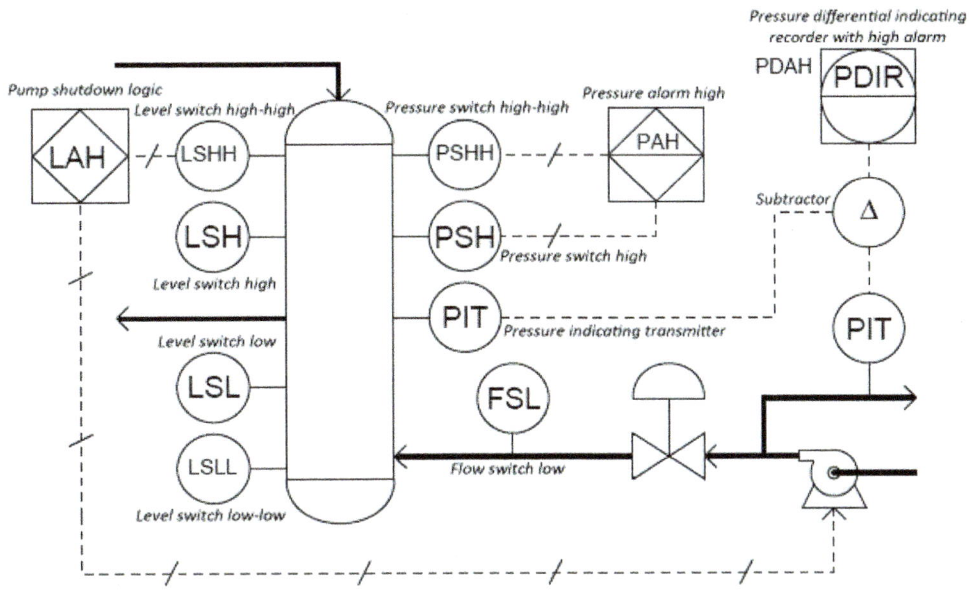

P&ID Sample 3 (Separator).

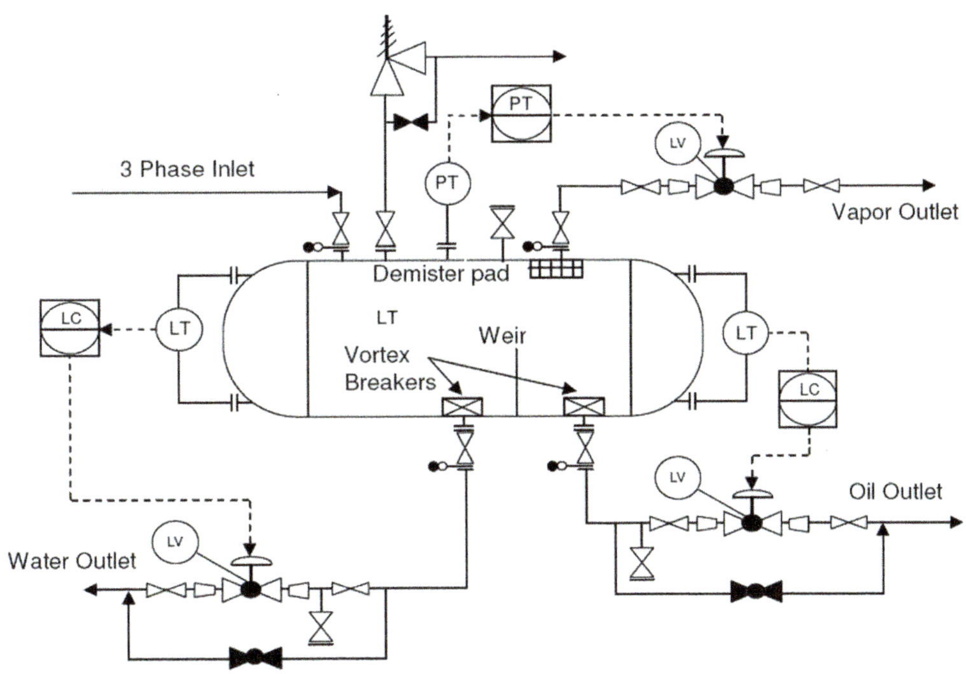

PFD or MFD as Might Be Included on the CSE Exam

PFD Sample 1.

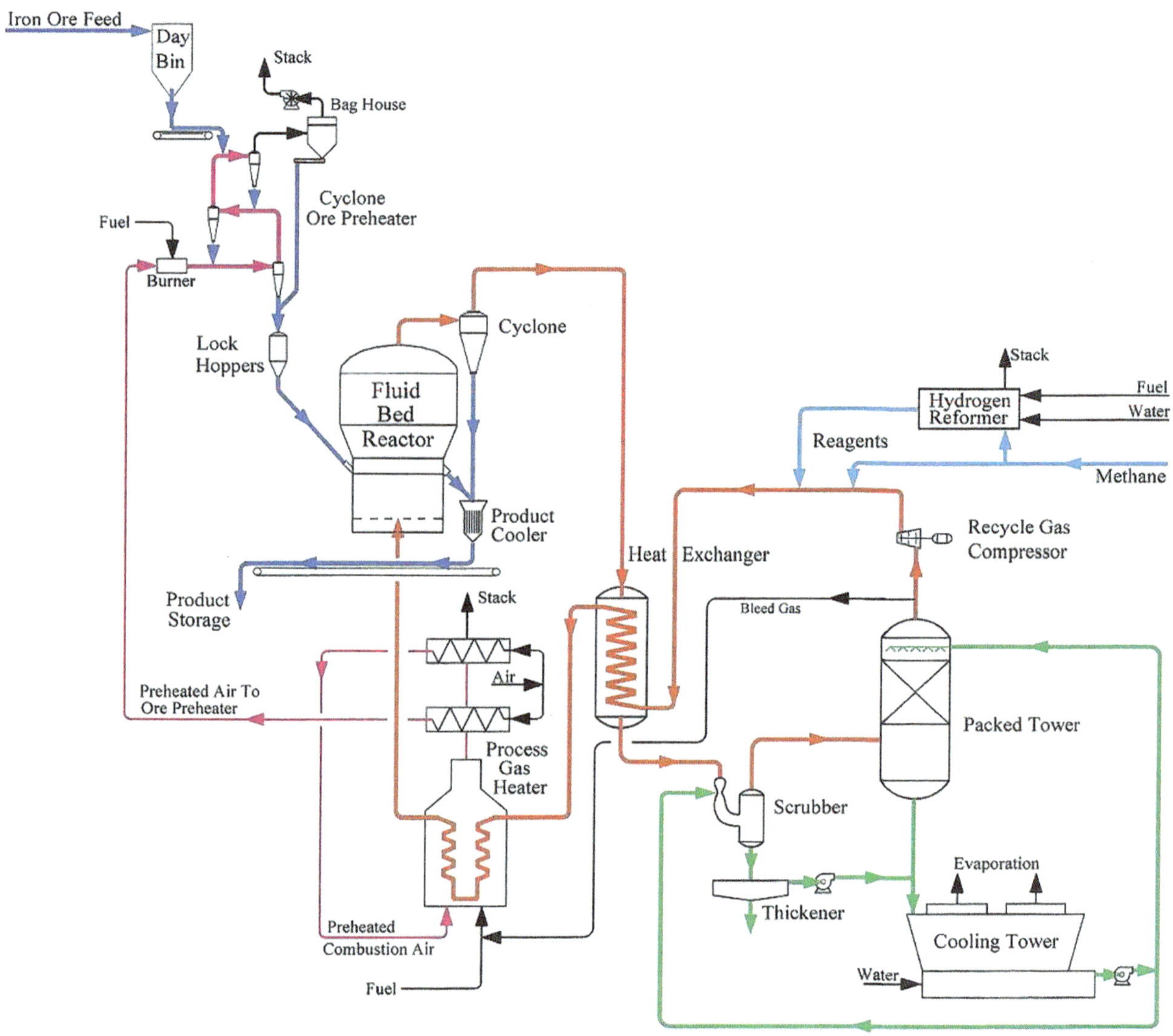

PFD or MFD as Might Be on the CSE Exam (continued)

PFD Sample 2.

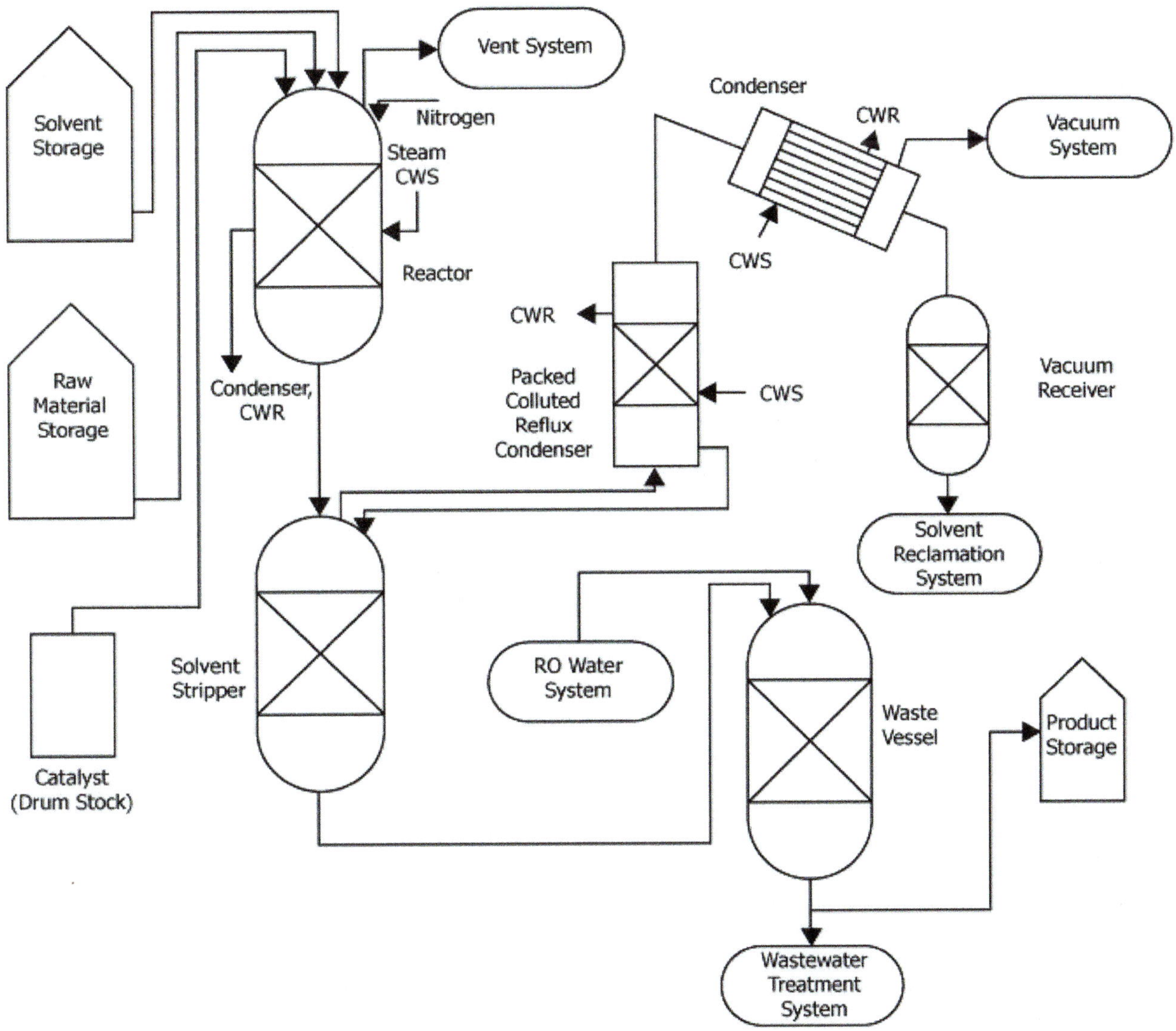

Block Flow Diagram (BFD)

BFD Sample 1.

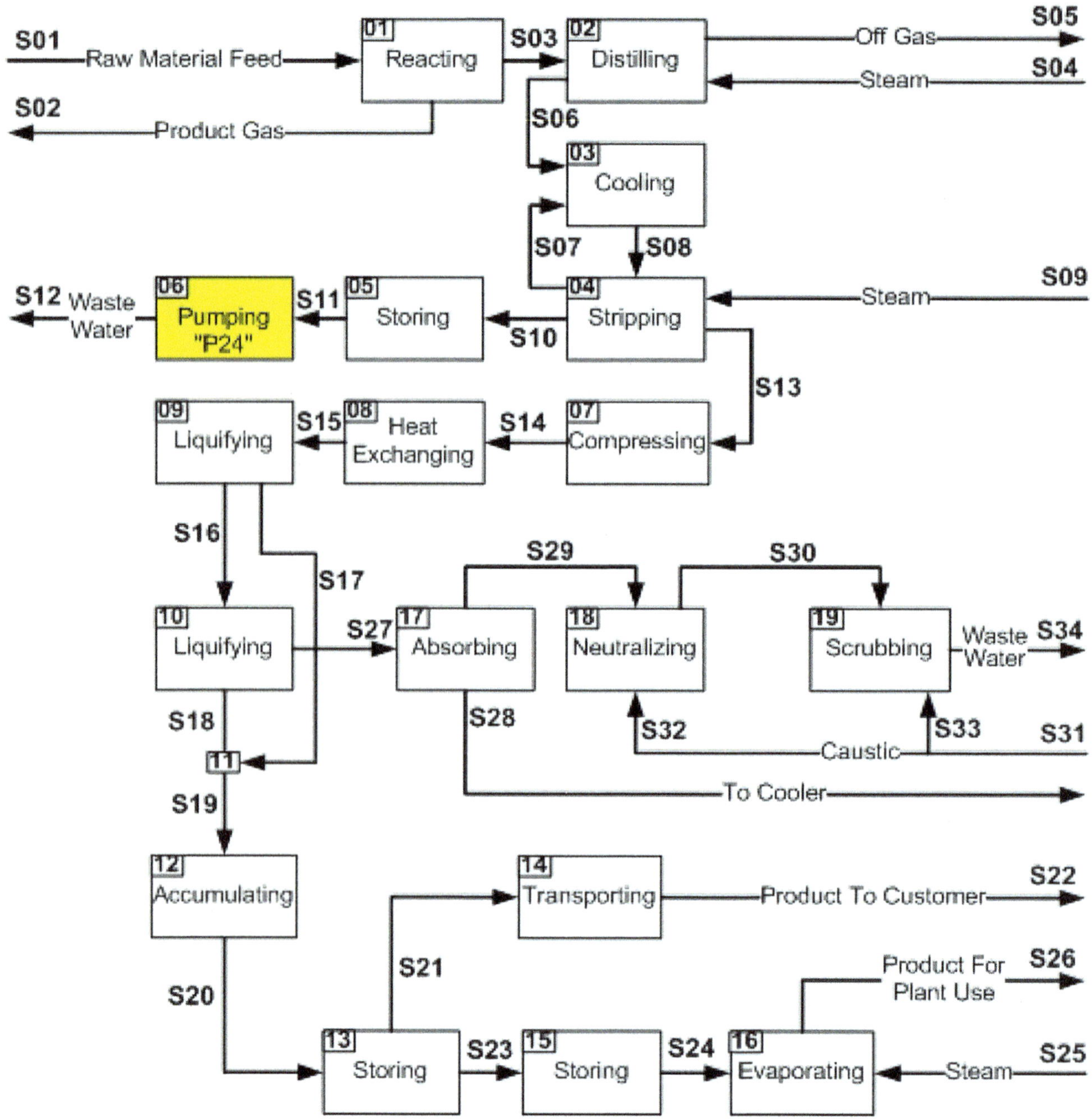

ISA Standard Loop Diagram

The following is a standard ISA instrument loop diagram that might be included on the CSE exam. The exam may have questions related to terminals, symbols, and connections. Familiarize yourself with ISA-5.4-1991, *Standard Instrument Loop Diagrams*, for the exam.

A typical ISA loop diagram of a distributed control system (DCS) controller in a cabinet (it could look the same for a programmable logic controller—PLC).

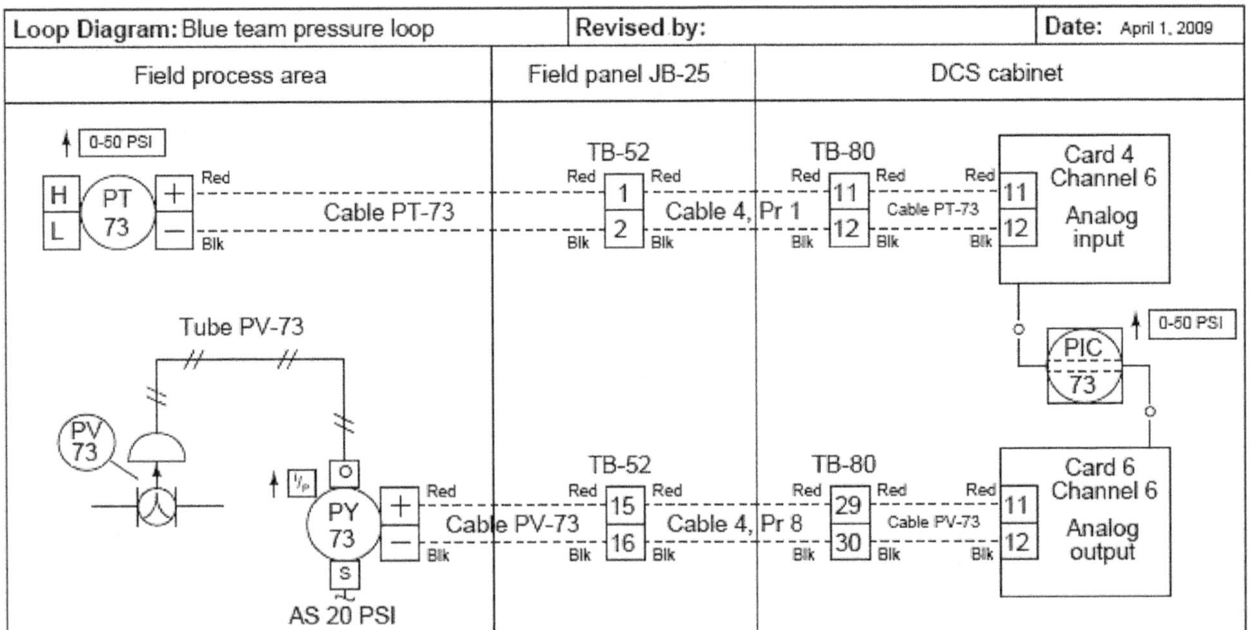

Tag #	Description	Manufacturer	Model	Input range	Output range	Notes
PT-73	Pressure transmitter	Rosemount	3051CD	0-50 PSI	4-20 mA	
PIC-73	Controller	Emerson	DeltaV	4-20 mA	4-20 mA	HART-enabled input Direct-acting control
PY-73	I/P transducer	Fisher	846	4-20 mA	3-15 PSI	
PV-73	Control valve	Fisher	Vee-ball	3-15 PSI	0-100%	Fail-open

A typical ISA loop diagram of a single-loop controller in a control cabinet or panel.

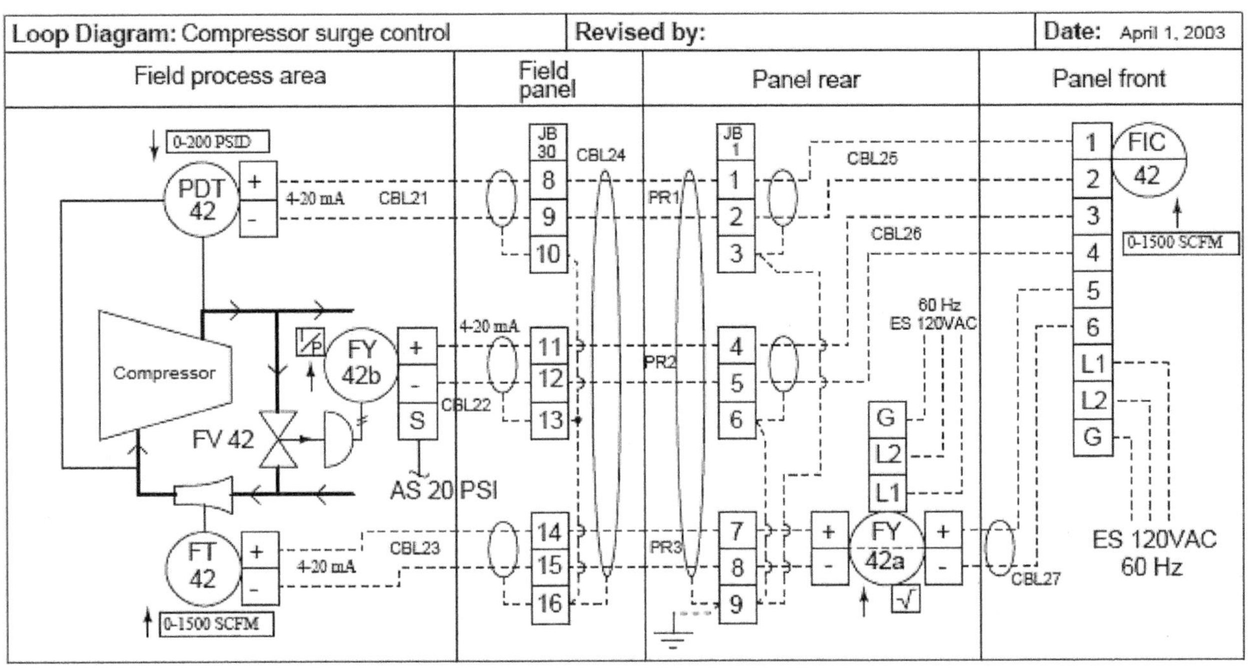

Familiarize yourself with this type of loop diagram. Notice the redundant output models.

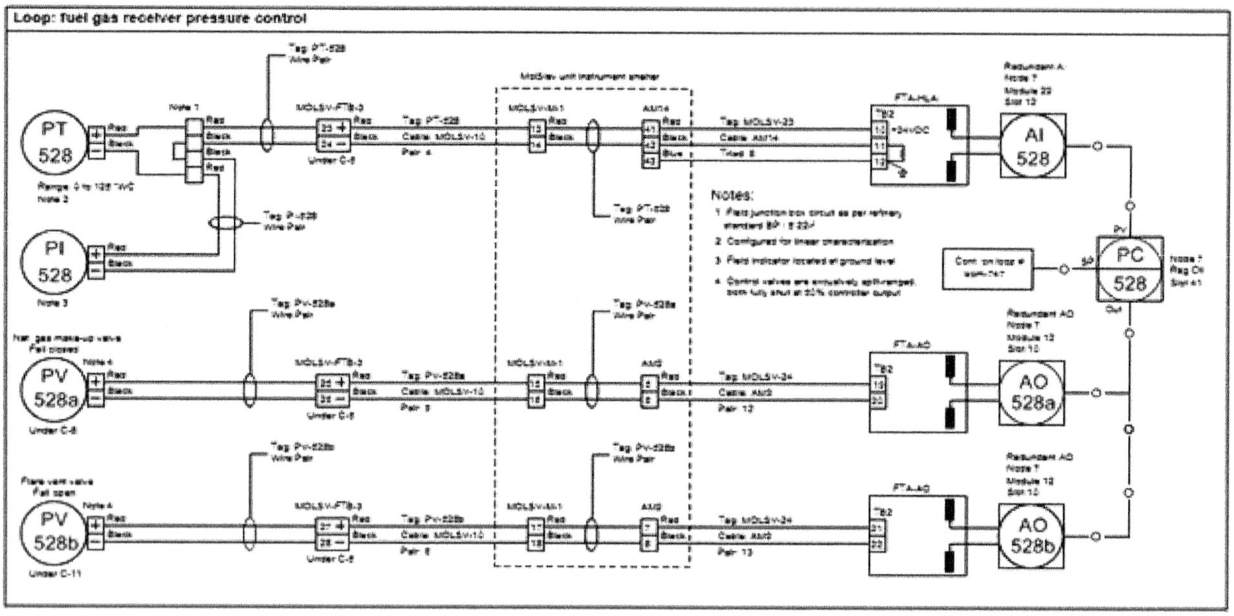

Familiarize yourself with the pneumatic designations for tubes and bulkhead connections.

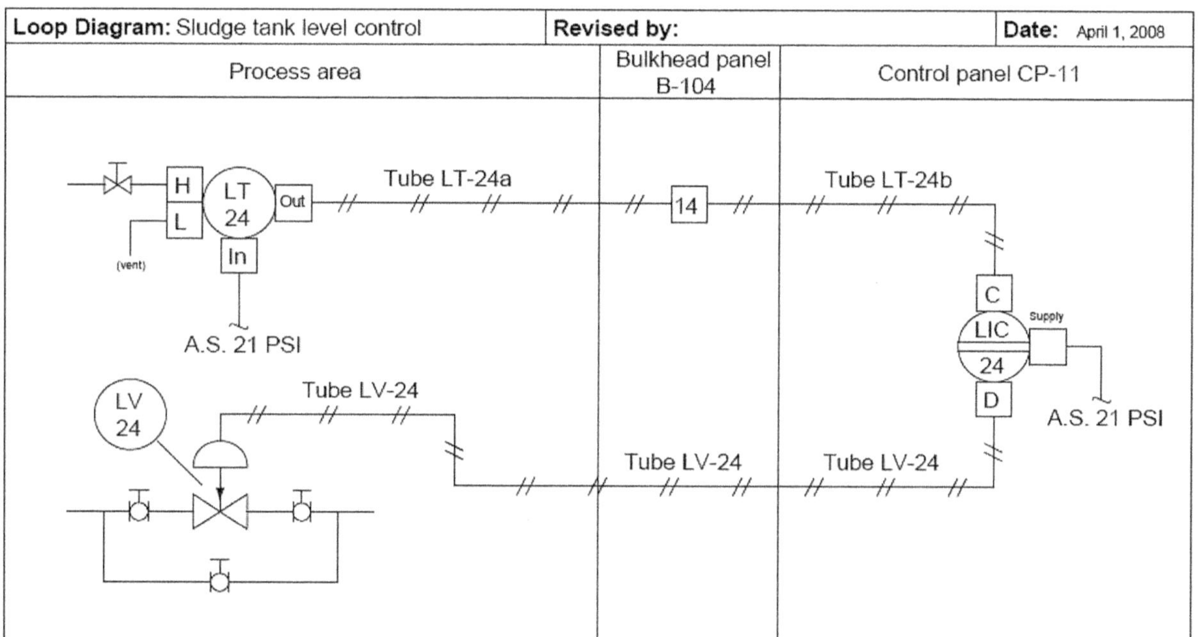

Tag #	Description	Manufacturer	Model	Input range	Output range	Notes
LT-24	Level transmitter	Foxboro	13A	25-150 "H_2O	3-15 PSI	
LIC-24	Controller	Foxboro	130	3-15 PSI	3-15 PSI	
LV-24	Control valve	Fisher	Easy-E / 667	3-15 PSI	0-100%	Fail closed

Instrument Location and Elevation Plan Drawing

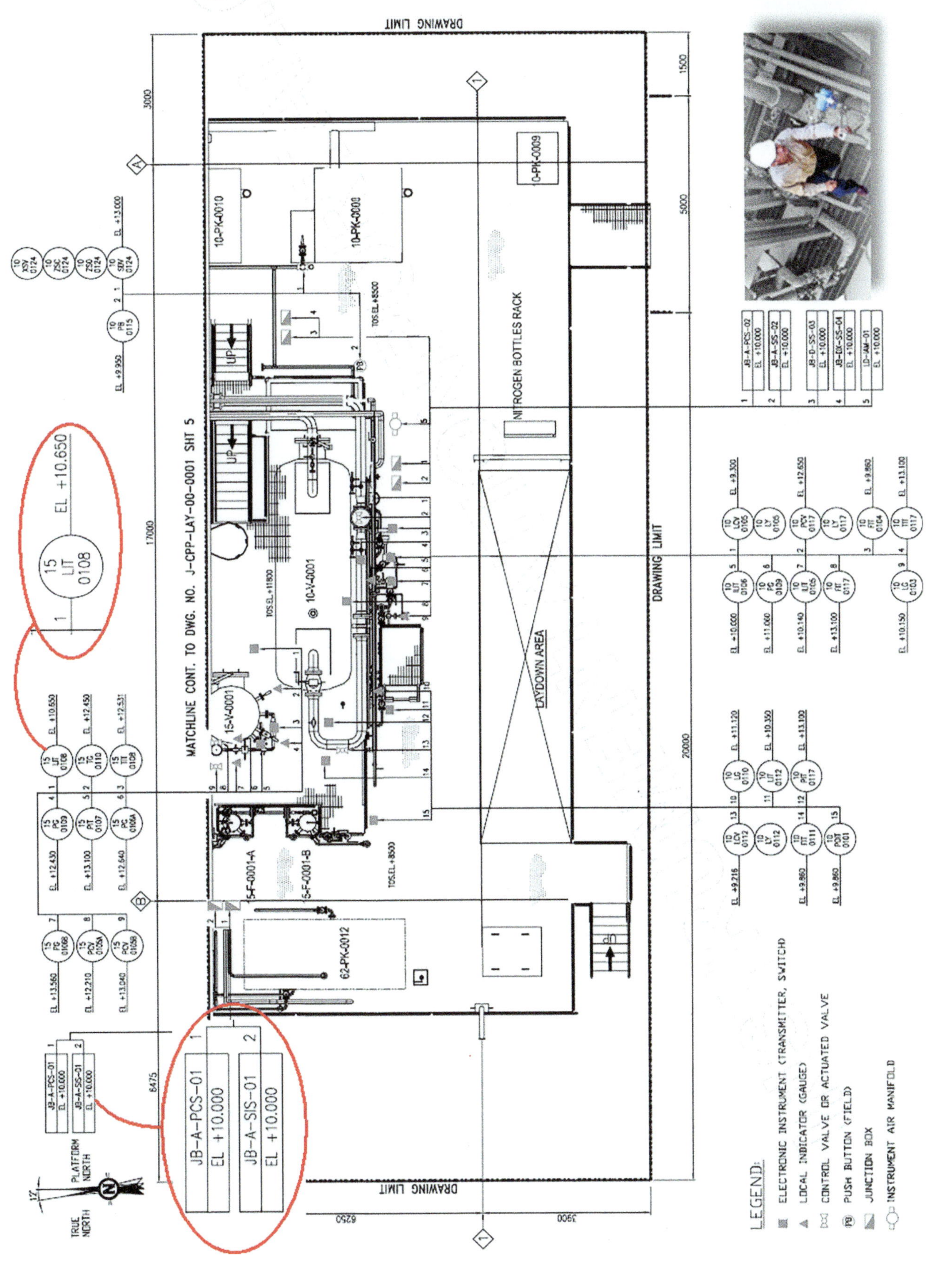

Instrument Index Sheet

A typical instrument index sheet, showing instrument tag number, P&ID, loop diagram, and so on. This is a list of all the instruments installed on a project. It is very useful for cross-referencing prints.

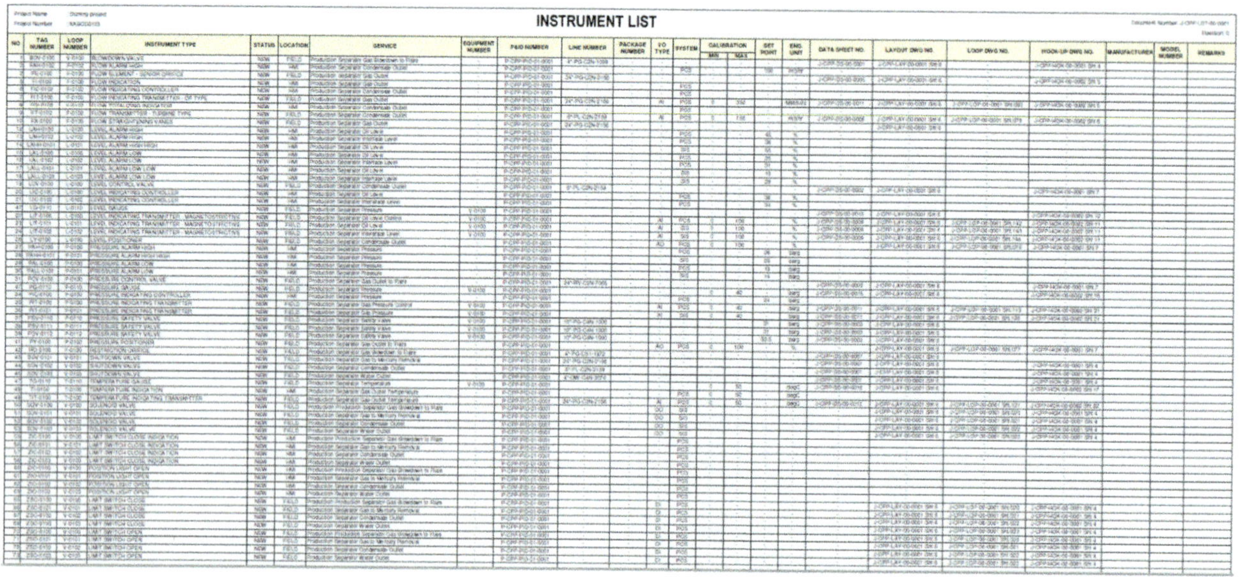

Blowup of left side of the sheet.

NO	TAG NUMBER	LOOP NUMBER	INSTRUMENT TYPE	STATUS	LOCATION	SERVICE
1	BDV-0100	V-0100	BLOWDOWN VALVE	NEW	FIELD	Production Separator Gas Blowdown to Flare
2	FAH-0102	F-0102	FLOW ALARM HIGH	NEW	HMI	Production Separator Condensate Outlet
2	FE-0100	F-0100	FLOW ELEMENT - SENIOR ORIFICE	NEW	FIELD	Production Separator Gas Outlet
5	FI-0100	F-0100	FLOW INDICATION	NEW	HMI	Production Separator Gas Outlet
6	FIC-0102	F-0102	FLOW INDICATING CONTROLLER	NEW	HMI	Production Separator Condensate Outlet
7	FIT-0100	F-0100	FLOW INDICATING TRANSMITTER - DP TYPE	NEW	FIELD	Production Separator Gas Outlet
8	FQI-0102	F-0102	FLOW TOTALIZING INDICATOR	NEW	HMI	Production Separator Condensate Outlet
9	FT-0102	F-0102	FLOW TRANSMITTER - TURBINE TYPE	NEW	FIELD	Production Separator Condensate Outlet
10	FX-0100	F-0100	FLOW STRAIGHTENING VANES	NEW	FIELD	Production Separator Gas Outlet
12	LAH-0100	L-0100	LEVEL ALARM HIGH	NEW	HMI	Production Separator Oil Level
13	LAH-0102	L-0102	LEVEL ALARM HIGH	NEW	HMI	Production Separator Interface Level
14	LAHH-0101	L-0101	LEVEL ALARM HIGH HIGH	NEW	HMI	Production Separator Oil Level
15	LAL-0100	L-0100	LEVEL ALARM LOW	NEW	HMI	Production Separator Oil Level
16	LAL-0102	L-0102	LEVEL ALARM LOW	NEW	HMI	Production Separator Interface Level
17	LALL-0101	L-0101	LEVEL ALARM LOW LOW	NEW	HMI	Production Separator Oil Level
18	LALL-0103	L-0103	LEVEL ALARM LOW LOW	NEW	HMI	Production Separator Interface Level
19	LCV-0100	L-0100	LEVEL CONTROL VALVE	NEW	FIELD	Production Separator Condensate Outlet

Blowup of the middle of the sheet.

P&ID NUMBER	LINE NUMBER	I/O TYPE	SYSTEM	CALIBRATION		SET POINT	ENG. UNIT	DATA SHEET NO.	LAYOUT DWG NO.	LOOP DWG NO.
				MIN	MAX					
P-CPP-PID-01-0001	4"-PG-C2N-1099	-	-	-	-	-	-	J-CPP-DS-00-0001	J-CPP-LAY-00-0001 Sht 6	-
P-CPP-PID-01-0001	-	-	PCS	-	-	100	m3/hr	-	-	-
P-CPP-PID-01-0001	24"-PG-C2N-2156	-	-	-	-	-	-	J-CPP-DS-00-0005	J-CPP-LAY-00-0001 Sht 6	-
P-CPP-PID-01-0001	-	-	PCS	-	-	-	-	-	-	-
P-CPP-PID-01-0001	-	-	PCS	-	-	-	-	-	-	-
P-CPP-PID-01-0001	24"-PG-C2N-2156	AI	PCS	0	350	-	MMScfd	J-CPP-DS-00-0011	J-CPP-LAY-00-0001 Sht 6	J-CPP-LOP-00-0001 Sht.080
P-CPP-PID-01-0001	-	-	PCS	-	-	-	-	-	-	-
P-CPP-PID-01-0001	6"-PL-C2N-2159	AI	PCS	0	150	-	m3/hr	J-CPP-DS-00-0006	J-CPP-LAY-00-0001 Sht 6	J-CPP-LOP-00-0001 Sht.079
P-CPP-PID-01-0001	24"-PG-C2N-2156	-	-	-	-	-	-	-	J-CPP-LAY-00-0001 Sht 6	-
P-CPP-PID-01-0001	-	-	PCS	-	-	45	%	-	-	-
P-CPP-PID-01-0001	-	-	PCS	-	-	36	%	-	-	-
P-CPP-PID-01-0001	-	-	SIS	-	-	55	%	-	-	-
P-CPP-PID-01-0001	-	-	PCS	-	-	25	%	-	-	-
P-CPP-PID-01-0001	-	-	PCS	-	-	31	%	-	-	-
P-CPP-PID-01-0001	-	-	SIS	-	-	16	%	-	-	-
P-CPP-PID-01-0001	-	-	SIS	-	-	29	%	-	-	-
P-CPP-PID-01-0001	6"-PL-C2N-2159	-	-	-	-	-	-	J-CPP-DS-00-0002	J-CPP-LAY-00-0001 Sht 6	-

DCS or PLC I/O List (a List of Inputs and Outputs with Tags and Calibration Data)

A typical instrument I/O list showing instrument tag number, P&ID, loop diagram, and so on.

Project Name : Dummy project
Project Number : XABCD0123

I/O LIST

Document Number: J-CPP-LST-00-0002
Revision: 0

NO	TAG NUMBER	LOOP NUMBER	INSTRUMENT TYPE	SERVICE	P&ID NUMBER	LOCATION	I/O TYPE	SYSTEM	CALIBRATION MIN	CALIBRATION MAX	SET POINT	ENG. UNIT	REMARKS
1	FIT-0100	F-0100	FLOW INDICATING TRANSMITTER - DP TYPE	Production Separator Gas Outlet	P-CPP-PID-01-0001	FIELD	AI	PCS	0	350	-	MMScfd	
2	FT-0102	F-0102	FLOW TRANSMITTER - TURBINE TYPE	Production Separator Condensate Outlet	P-CPP-PID-01-0001	FIELD	AI	PCS	0	150	-	m3/hr	
3	LIT-0100	L-0100	LEVEL INDICATING TRANSMITTER - MAGNETOSTRICTIVE	Production Separator Oil Level Control	P-CPP-PID-01-0001	FIELD	AI	PCS	0	100	-	%	
4	LIT-0101	L-0101	LEVEL INDICATING TRANSMITTER - MAGNETOSTRICTIVE	Production Separator Oil Level	P-CPP-PID-01-0001	FIELD	AI	SIS	0	100	-	%	
5	LIT-0102	L-0102	LEVEL INDICATING TRANSMITTER - MAGNETOSTRICTIVE	Production Separator Interaface Level	P-CPP-PID-01-0001	FIELD	AI	SIS	0	100	-	%	
6	LY-0100	L-0100	LEVEL POSITIONER	Production Separator Condensate Outlet	P-CPP-PID-01-0001	FIELD	AO	PCS	0	100	-	%	
7	PIT-0100	P-0100	PRESSURE INDICATING TRANSMITTER	Production Separator Gas Pressure Control	P-CPP-PID-01-0001	FIELD	AI	PCS	0	40	-	barg	
8	PIT-0101	P-0101	PRESSURE INDICATING TRANSMITTER	Production Separator Gas Pressure	P-CPP-PID-01-0001	FIELD	AI	SIS	0	40	-	barg	
9	PY-0100	P-0100	PRESSURE POSITIONER	Production Separator Gas Outlet to Flare	P-CPP-PID-01-0001	FIELD	AO	PCS	0	100	-	%	
10	TIT-0100	T-0100	TEMPERATURE INDICATING TRANSMITTER	Production Separator Gas Outlet Temperature	P-CPP-PID-01-0001	FIELD	AI	PCS	0	50	-	degC	
11	XSV-0100	V-0100	SOLENOID VALVE	Production Production Separator Gas Blowdown to Flare	P-CPP-PID-01-0001	FIELD	DO	SIS	-	-	-	-	
12	XSV-0101	V-0101	SOLENOID VALVE	Production Separator Gas to Mercury Removal	P-CPP-PID-01-0001	FIELD	DO	SIS	-	-	-	-	
13	XSV-0102	V-0102	SOLENOID VALVE	Production Separator Condensate Outlet	P-CPP-PID-01-0001	FIELD	DO	SIS	-	-	-	-	
14	XSV-0103	V-0103	SOLENOID VALVE	Production Separator Water Outlet	P-CPP-PID-01-0001	FIELD	DO	SIS	-	-	-	-	
15	ZSC-0100	V-0100	LIMIT SWITCH CLOSE	Production Production Separator Gas Blowdown to Flare	P-CPP-PID-01-0001	FIELD	DI	PCS	-	-	-	-	
16	ZSC-0101	V-0101	LIMIT SWITCH CLOSE	Production Separator Gas to Mercury Removal	P-CPP-PID-01-0001	FIELD	DI	PCS	-	-	-	-	
17	ZSC-0102	V-0102	LIMIT SWITCH CLOSE	Production Separator Condensate Outlet	P-CPP-PID-01-0001	FIELD	DI	PCS	-	-	-	-	
18	ZSC-0103	V-0103	LIMIT SWITCH CLOSE	Production Separator Water Outlet	P-CPP-PID-01-0001	FIELD	DI	PCS	-	-	-	-	
19	ZSO-0100	V-0100	LIMIT SWITCH OPEN	Production Production Separator Gas Blowdown to Flare	P-CPP-PID-01-0001	FIELD	DI	PCS	-	-	-	-	
20	ZSO-0101	V-0101	LIMIT SWITCH OPEN	Production Separator Gas to Mercury Removal	P-CPP-PID-01-0001	FIELD	DI	PCS	-	-	-	-	
21	ZSO-0102	V-0102	LIMIT SWITCH OPEN	Production Separator Condensate Outlet	P-CPP-PID-01-0001	FIELD	DI	PCS	-	-	-	-	
22	ZSO-0103	V-0103	LIMIT SWITCH OPEN	Production Separator Water Outlet	P-CPP-PID-01-0001	FIELD	DI	PCS	-	-	-	-	
23	LT-1101	L-1101	LEVEL INDICATING TRANSMITTER - MAGNETOSTRICTIVE	Gas Compressor Suction Scrubber	P-CPP-PID-01-0002	FIELD	AI	PCS	0	100	-	%	
24	LY-1101	L-1101	LEVEL POSITIONER	Suction Scrubber Outlet to Degassing Column	P-CPP-PID-01-0002	FIELD	AO	PCS	0	100	-	%	
25	LT-1201	L-1201	LEVEL INDICATING TRANSMITTER - MAGNETOSTRICTIVE	Compressor Discharge Scrubber	P-CPP-PID-01-0002	FIELD	AI	PCS	0	100	-	%	
26	LY-1201	L-1201	LEVEL POSITIONER	Discharge Scrubber Outlet to Degassing Column	P-CPP-PID-01-0002	FIELD	AO	PCS	0	100	-	%	
27	LT-1202	L-1202	LEVEL INDICATING TRANSMITTER - MAGNETOSTRICTIVE	Compressor Discharge Scrubber	P-CPP-PID-01-0002	FIELD	AI	SIS	0	100	-	%	
28	ZSO-1201	V-1201	LIMIT SWITCH OPEN	Discharge Scrubber Outlet to Degassing Column	P-CPP-PID-01-0002	FIELD	DI	PCS	-	-	-	-	
29	ZSC-1201	V-1201	LIMIT SWITCH CLOSE	Discharge Scrubber Outlet to Degassing Column	P-CPP-PID-01-0002	FIELD	DI	PCS	-	-	-	-	
30	SOV-1201	V-1201	SOLENOID VALVE	Discharge Scrubber Outlet to Degassing Column	P-CPP-PID-01-0002	FIELD	DO	SIS	-	-	-	-	

Blowup of left side of the sheet.

NO	TAG NUMBER	LOOP NUMBER	INSTRUMENT TYPE	SERVICE
1	FIT-0100	F-0100	FLOW INDICATING TRANSMITTER - DP TYPE	Production Separator Gas Outlet
2	FT-0102	F-0102	FLOW TRANSMITTER - TURBINE TYPE	Production Separator Condensate Outlet
3	LIT-0100	L-0100	LEVEL INDICATING TRANSMITTER - MAGNETOSTRICTIVE	Production Separator Oil Level Control
4	LIT-0101	L-0101	LEVEL INDICATING TRANSMITTER - MAGNETOSTRICTIVE	Production Separator Oil Level
5	LIT-0102	L-0102	LEVEL INDICATING TRANSMITTER - MAGNETOSTRICTIVE	Production Separator Interaface Level
6	LY-0100	L-0100	LEVEL POSITIONER	Production Separator Condensate Outlet
7	PIT-0100	P-0100	PRESSURE INDICATING TRANSMITTER	Production Separator Gas Pressure Control
8	PIT-0101	P-0101	PRESSURE INDICATING TRANSMITTER	Production Separator Gas Pressure
9	PY-0100	P-0100	PRESSURE POSITIONER	Production Separator Gas Outlet to Flare

Blowup of right side of the sheet.

P&ID NUMBER	LOCATION	I/O TYPE	SYSTEM	CALIBRATION		SET POINT	ENG. UNIT	REMARKS
				MIN	MAX			
P-CPP-PID-01-0001	FIELD	AI	PCS	0	350	-	MMScfd	
P-CPP-PID-01-0001	FIELD	AI	PCS	0	150	-	m3/hr	
P-CPP-PID-01-0001	FIELD	AI	PCS	0	100	-	%	
P-CPP-PID-01-0001	FIELD	AI	SIS	0	100	-	%	
P-CPP-PID-01-0001	FIELD	AI	SIS	0	100	-	%	
P-CPP-PID-01-0001	FIELD	AO	PCS	0	100	-	%	
P-CPP-PID-01-0001	FIELD	AI	PCS	0	40	-	barg	
P-CPP-PID-01-0001	FIELD	AI	SIS	0	40	-	barg	
P-CPP-PID-01-0001	FIELD	AO	PCS	0	100	-	%	

ISA Standard HMI Graphical Display Symbols and Designations

This shows a standard process plant graphical human machine interface (HMI) display. Questions relating to the colors and functions of the on-screen switches and text may be included on the exam. Familiarize yourself with ISA-5.3-1983, *Graphic Symbols for Distributed Control/Shared Display Instrumentation, Logic, and Computer Systems,* for the exam.

HMI Sample.

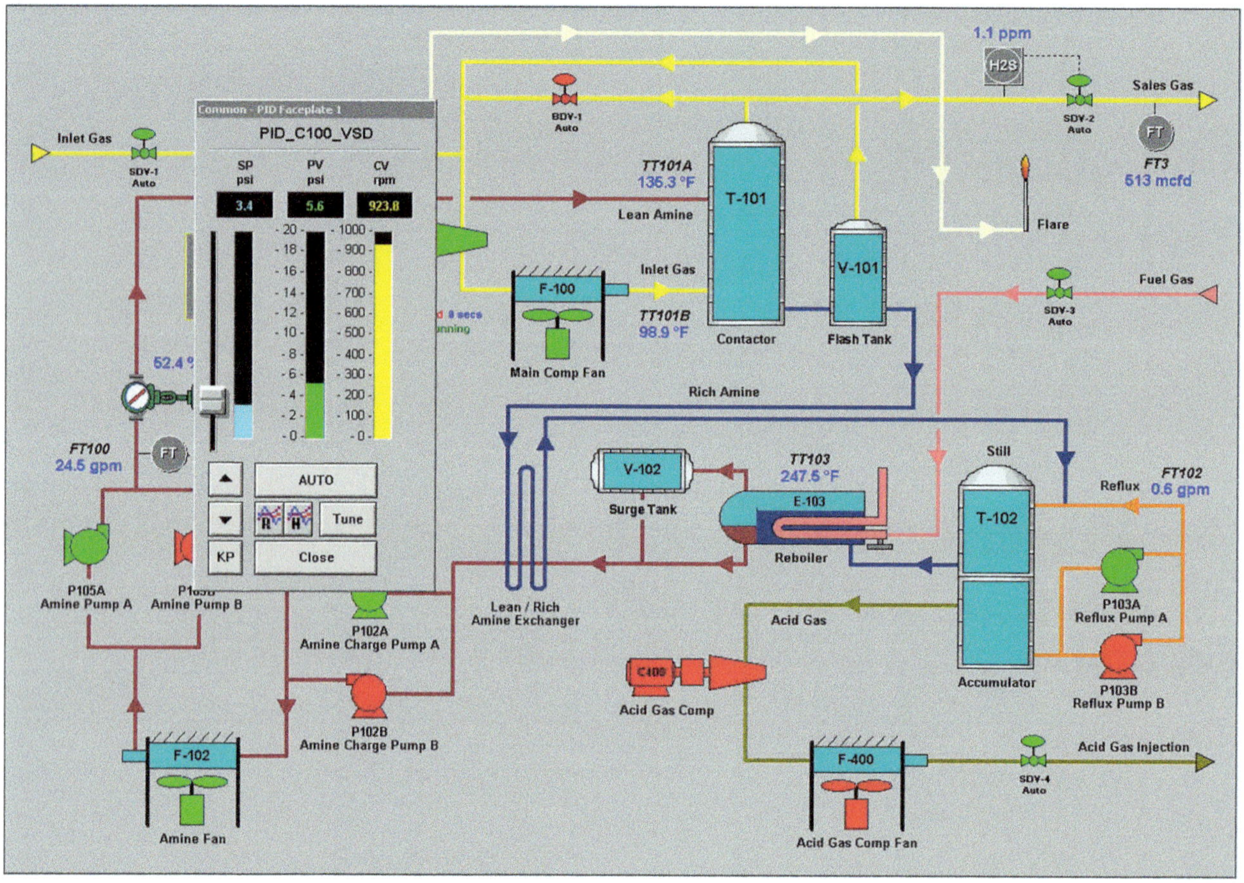

NFPA 79 Colors for Graphical Displays (Industrial Machinery)

Colors	Purposes: Safety of Persons or Environment	Purposes: Condition of Process	Purposes: State of Equipment
Red	Danger	Emergency	Faulty
Yellow (Amber)	Warning/Caution	Abnormal	Abnormal
Green	Safe	Normal	Normal
Blue	Mandatory action		
Clear, White, Gray, Black	Mandatory action		

Battery Limits of the Plant

A battery (naval or military) refers to a grouping of artillery or ships for a common purpose to facilitate command and control. The physical and communication limits became known as *battery limits*. The term was first used in oil refineries to designate a group of plant, equipment, and systems with a common purpose, such as crude distillation or reforming.

Many capital projects are divided into two parts, inside the battery limits (ISBL) and outside the battery limits (OSBL). ISBL typically includes the process units and the focus of the project, whether it is a grassroots facility or a revamp of an existing unit. OSBL consists of all connections necessary to make ISBL function, such as feed and product streams, utilities, and waste streams.

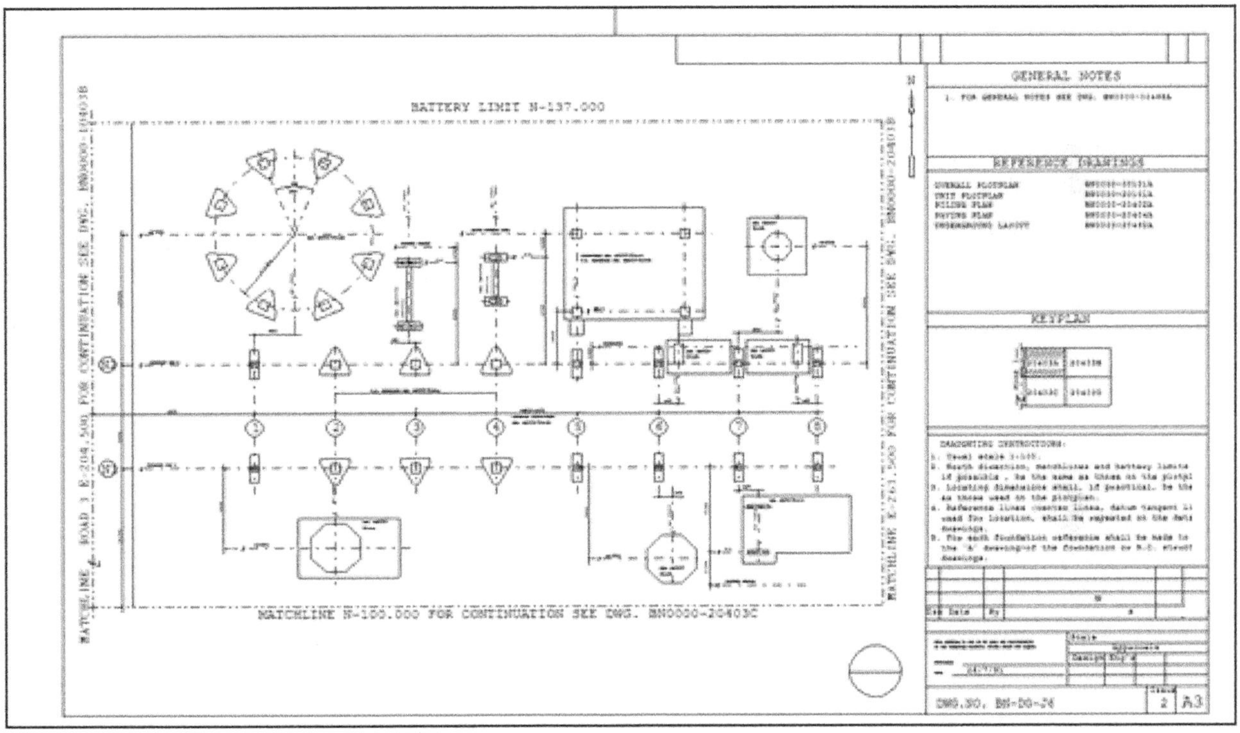

28 Overview of Safety Instrumented Systems

Overview of Process Safety and Shutdown

On the CSE exam, there will be questions on safety instrumented systems (SISs) and safety integrity levels (SILs). We will discuss some of the calculations and data you may encounter on the test.

SISs

US Occupational Safety and Health Administration (OSHA) law incorporates as a guideline that "good engineering practice" will be used in evaluating and engineering SISs. This means that the program follows the codes and standards published by such organizations as the American Society of Mechanical Engineers (ASME), American Petroleum Institute (API), American National Standards Institute (ANSI), National Fire Protection Association (NFPA), American Society for Testing and Materials (ASTM), National Board of Boiler and Pressure Vessel Inspectors (NBBI), and International Society of Automation (ISA). Other countries have similar requirements.

ANSI/ISA-S84.00.01-2003, (IEC 61511 Mod), *Application of Safety Instrumented Systems for the Process Industries*, (referred to as IEC 61511/ISA-84 in this book) addresses the application of SISs to take a process to a safe state when predetermined conditions are violated, such as set points for pressure, temperature, and level. Its objective is to define requirements for SISs. SISs are also called *emergency shutdown systems*, *safety shutdown systems*, and *safety interlock systems*. SISs provide safety control functions and complement the basic process control system (BPCS), which provides normal process control. IEC 61511/ISA-84 address the entire life cycle for SISs, and compliance with the standard requires a significant effort.

Complying with IEC 61511/ISA-84

OSHA endorsed IEC 61511/ISA-84 as a "national consensus standard" in a March 23, 2000, OSHA letter to ISA. This letter states that the standard is considered "a recognized and generally accepted good engineering practice" for SISs. Paragraph (d)(3)(ii) of the process safety management (PSM) standard specifies, "The employer shall document that equipment complies with recognized and generally accepted good engineering practices." The letter states that in evaluating whether an employer's engineering practices with respect to SISs comply with PSM, OSHA would consider, among other factors, whether the employer meets the requirements of IEC 61511/ISA-84.

In the letter, OSHA states that it is also important to note that there are a large percentage of processes that are not covered by PSM, which may include SISs covered by the standard. OSHA states that *"the employer may be in violation of the General Duty Clause, Section 5(a)(1) of the OSH Act, if SISs are utilized that do not conform with IEC 61511/ISA-84, and hazards exist related to the SISs that could seriously harm employees"*. Consequently, this means that companies must comply with IEC 61511/ISA-84, not only for PSM-covered processes, but also for other processes that use SISs where hazards to personnel may be present.

Regardless of regulatory requirements, IEC 61511/ISA-84 is an internationally recognized standard and represents good engineering practice.

Other Codes Related to SIS

- NFPA 85 – *Boiler and Combustion Systems Hazard Code*
- NFPA 86 – *Standard for Ovens and Furnaces*
- IEC 61508 – *Functional Safety: Safety-Related Systems*
- IEC 61511 – *Functional Safety: Safety Instrumented Systems for the Process Industry Sector*
- ANSI/ISA S84.01 – *Application of Safety Instrumented Systems for the Process Industries*

ISA and OSHA Letter Defining the Requirements for Implementation of SIS

- Standard Number: 1910.119

This letter constitutes OSHA's interpretation only of the requirements discussed and may not be applicable to any situation not delineated within the original correspondence.

November 29, 2005

Ms. Lois M. Ferson
Manager of Standards Services
ISA
67 Alexandria Drive
P.O. Box 12277
Research Triangle Park, NC 27709

Dear Ms. Ferson:

Thank you for your October 25, 2004 letter to the Occupational Safety and Health Administration (OSHA), on behalf of the Instrumentation, Systems, and Automation Society (ISA), regarding a consensus standard jointly issued by the American National Standards Institute (ANSI) and ISA that may be applied under OSHA's Process Safety Management (PSM) standard, 29 CFR 1910.119. We apologize for the delay in our response.

Background: ISA is an ANSI accredited organization. In March 2000, OSHA issued a letter to ISA, in which OSHA responded to questions pertaining to ANSI/ISA-84.01-1996, "Application of Safety Instrumented Systems for the Process Industries." Since then, based on experience gained from its use and the availability of newer technology, ISA has updated and modified ANSI/ISA-84.01-1996. This new standard — ANSI/ISA-84.00.01-2004 Parts 1-3 (IEC 61511 Mod), "Functional Safety: Safety Instrumented Systems for the Process Industry Sector," has been approved by ISA and by ANSI. With this background, ISA is again requesting a response to the following questions:

Question 1: Does the modification of corporate standards to comply with ANSI/ISA- S84.00.01-2004 Parts 1-3 (IEC 61511 Mod) ensure compliance with related areas of 29 CFR 1910.119?

Response: As stated in our March 23, 2000 response, the PSM standard contains a number of requirements for equipment associated with a covered process, which may include Safety Instrumented Systems (SIS). OSHA considers the revised ANSI/ISA - S84.00.01-2004 Parts 1-3 (IEC 61511 Mod) to be recognized and generally accepted good engineering practices for SIS. Therefore, if an employer chooses to use S84.00.01-2004 Parts 1-3 as a basis ("code or standard employed") for SIS and meets all S84.00.01-2004 Parts 1-3 requirements and other OSHA PSM requirements related to SIS, the employer will then be considered in compliance with OSHA PSM requirement for SIS.

Question 2: We are interested if your agency views ANSI/ISA-S84.01 as the benchmark OSHA will use to measure compliance with 29 CFR 1910.119?

Response: As stated in our previous response, the PSM standard is a performance-oriented standard. Therefore, employers have the flexibility in complying with the requirements of PSM, including, among other aspects, the use of recognized and generally accepted good engineering practices. With respect to SIS, as stated in our previous response, OSHA does not specify or benchmark S84.00.001-2004, Parts 1-3, as the only recognized and generally accepted good engineering practice. Additionally, it is important to note that an employer not covered under the PSM standard may be in violation of the General Duty Clause, Section 5(a)(1) of the OSH Act, if the employer's use of SIS results in a serious, recognized hazard to employees. In support of a Section 5(a)(1) citation, industry consensus standards, such as ANSI/ISA - S84.00.01-2004, can be used as evidence that a hazard is recognized and can feasibly be abated.

Thank you for your interest in occupational safety and health. We hope you find this information helpful. Please be aware that OSHA's enforcement guidance is subject to periodic review and clarification, amplification, or correction. Such guidance could also be affected by subsequent rulemaking. In the future, should you wish to verify that the guidance provided herein remains current, you may consult OSHA's web-site at www.osha.gov. If you have any further questions, please feel free to contact the Office of General Industry Compliance Assistance at (202) 693-1850.

Sincerely,

Richard E. Fairfax,
Director
Directorate of Enforcement Program

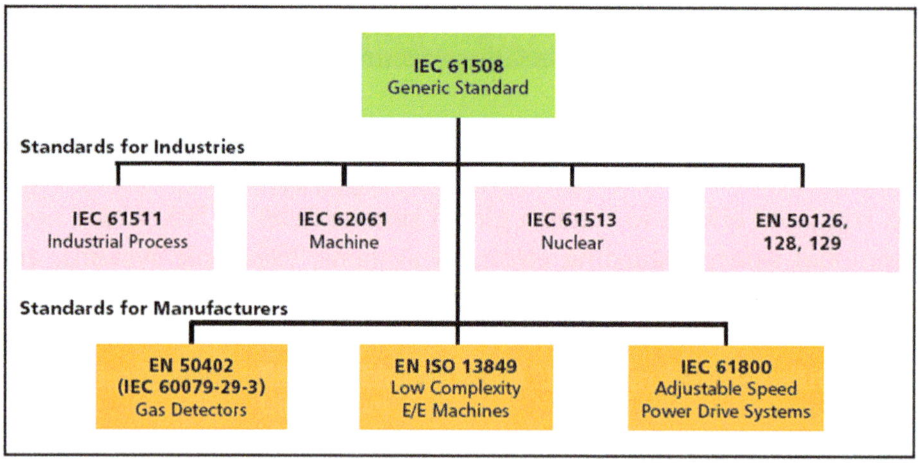

Safety Instrumented Function and SIL

A safety instrumented function (SIF) is designed to minimize process risks to a tolerable level (or as low as reasonably practical—ALARP). Each SIF is assigned an SIL during SIL analysis—risk assessment:

- **SIL 0/none** – lowest risk
- **SIL 1** – 95% of SIFs
- **SIL 2** – 5% of SIFs
- **SIL 3** – <1% of SIFs (not likely in refineries, but possible in off-shore platforms or nuclear)
- **SIL 4** – highest risk (only seen in the nuclear industry)

Risk analysis and protection layers.

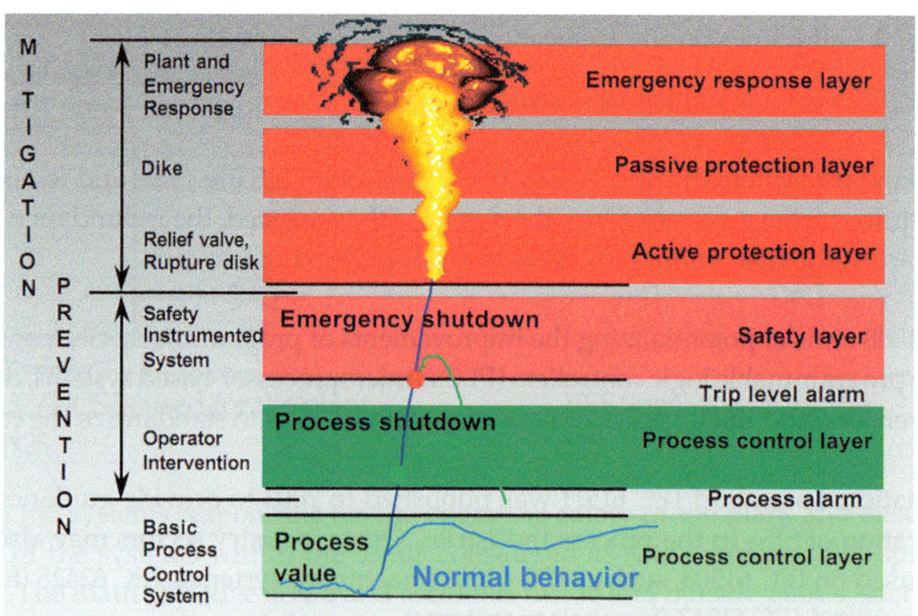

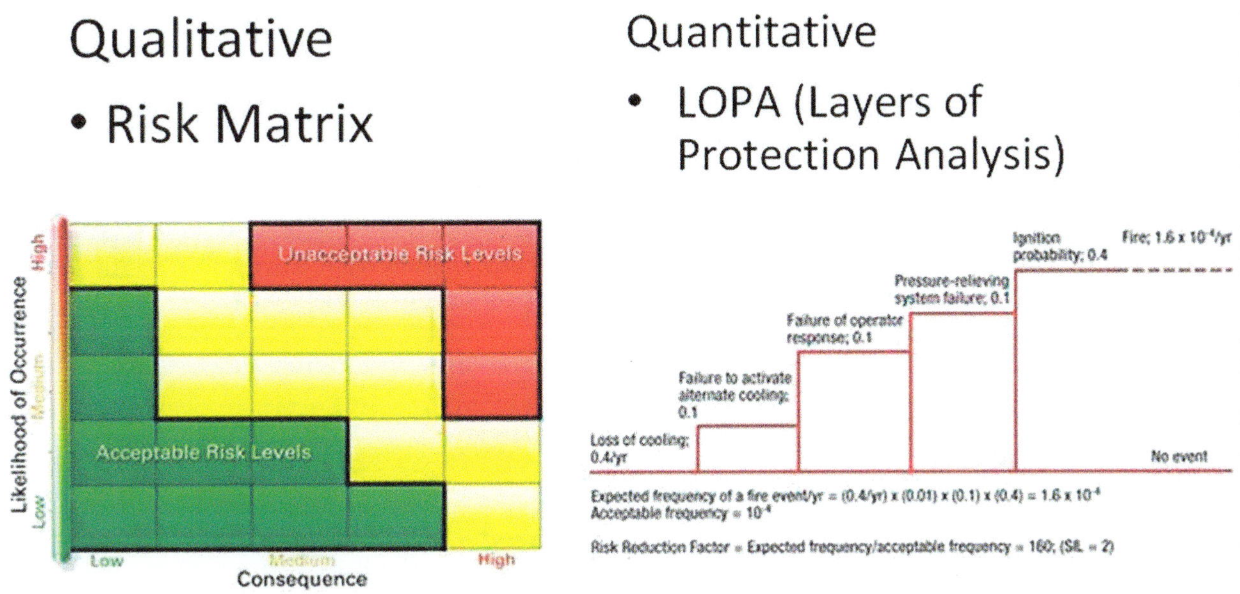

Designing an SIS System

SIL – Unit for functional safety.

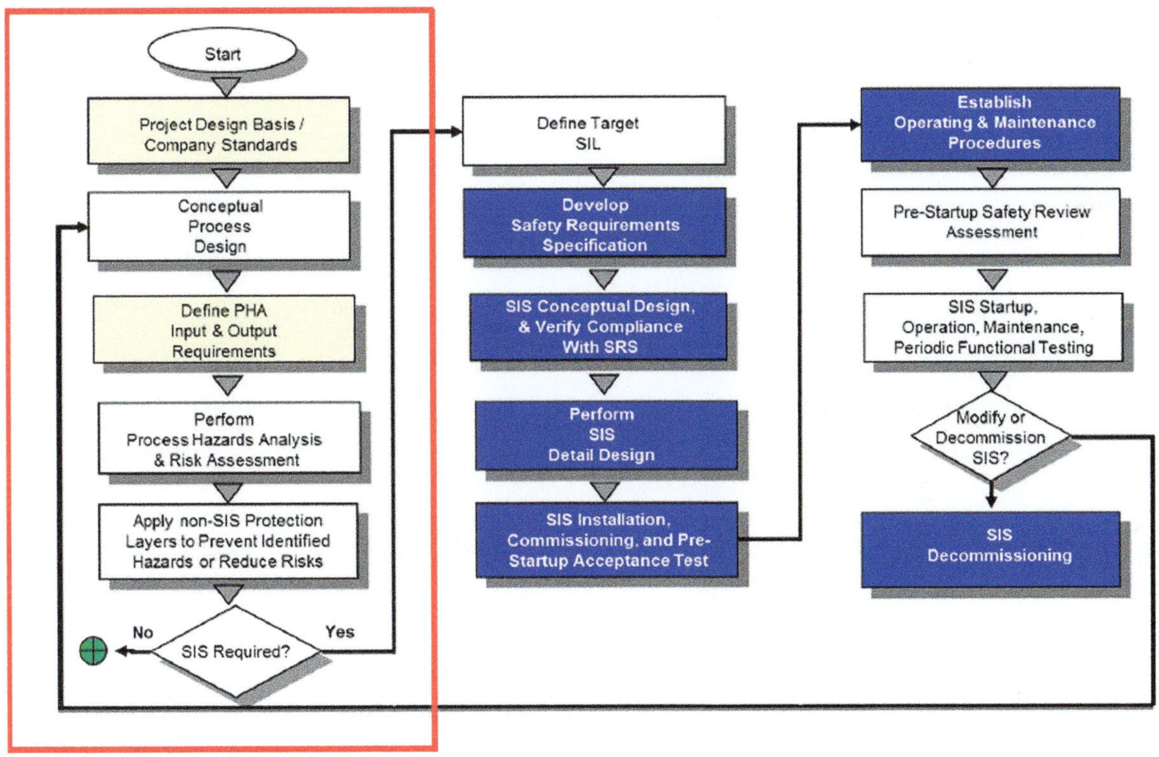

Functional safety is part of the overall safety that depends on an SIS, made up of equipment such as fire and gas detection systems that execute SIFs. A safety function is designed to ensure or maintain a safety state of the SIS when a dangerous event occurs.

Each safety function has an SIL. The SIL is the probability for the system to execute the safety functions required in all specified input conditions within a specified time interval. The IEC 61508 is a functional safety standard and details the requirements necessary to achieve each SIL.

Obtaining the SIL is done by:

- Guaranteeing the integrity of the cycle of development of the system in the fields of specification, design, and testing, with the goal of avoiding and eliminating systematic failures
- Guaranteeing the robustness of the design by measurements allowing the systematic fault tolerances (diagnostics, access control, environment, etc.)
- Respecting the constraints on the equipment architecture for the rate of diagnostic coverage to determine the safe failure fraction (SFF)
- Guaranteeing a probability of failure on demand (PFD), as a function of the failure rate and the test interval, or as probability of failures per hour (PFH)
- If software is included, guaranteeing the integrity and robustness of the design concerning only systematic failures

If it is concluded that an SIS is required, then ANSI/ISA-84.00.01 (IEC 61511 modified) and IEC 61508 require that a target SIL be assigned. The assignment of an SIL is a corporate decision based on risk management and risk tolerance philosophy. Safety regulations require that the assignment of SILs should be carefully performed and documented. A qualitative view of the SIL has slowly developed over the last few years as the concept of SIL has been adopted at many chemical and petrochemical plants. This qualitative view can be expressed in terms of the impact of the SIS failure on plant personnel and the public or community.

- **SIL 4** – Catastrophic Community Impact
- **SIL 3** – Employee and Community Protection
- **SIL 2** – Major Property and Production Protection—possible injury to employees
- **SIL 1** – Minor Property and Production Protection

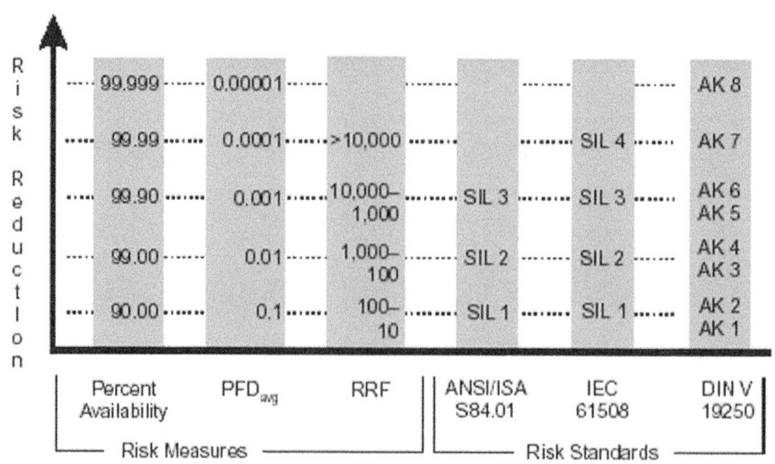

SFF

The SFF is one parameter that is necessary to assess the SIL capability of SIF functions. The SFF is the percentage of safe failures, such as those that are safe or detected.

The calculation is based on the architecture of each SIF and on a functional analysis by carrying out a failure mode effect and diagnostic analysis (FMEDA).

SFF Safe Failure Fraction	Hardware Fault Tolerance		
	0	1	2
< 60%	*Not allowed*	SIL 1	SIL 2
60% - ≤ 90%	SIL 1	SIL 2	SIL 3
90% - ≤ 99%	SIL 2*	SIL 3†	SIL 4
≥ 99%	SIL 3	SIL 4	SIL 4

Example: To be rated SIL 2, a simple nonredundant control unit that will not ensure the safety function in the event of one hardware fault must have an SFF between 90% and 99% (i.e., the percentage of undetected dangerous failures shall not be greater than 10%).

The table above, taken from IEC 61508-1:2010, *Functional safety of electrical/electronic/programmable electronic safety-related systems – Part 1: General requirements,* gives the SIL levels in relation to the SFF and the tolerance for hardware fault.

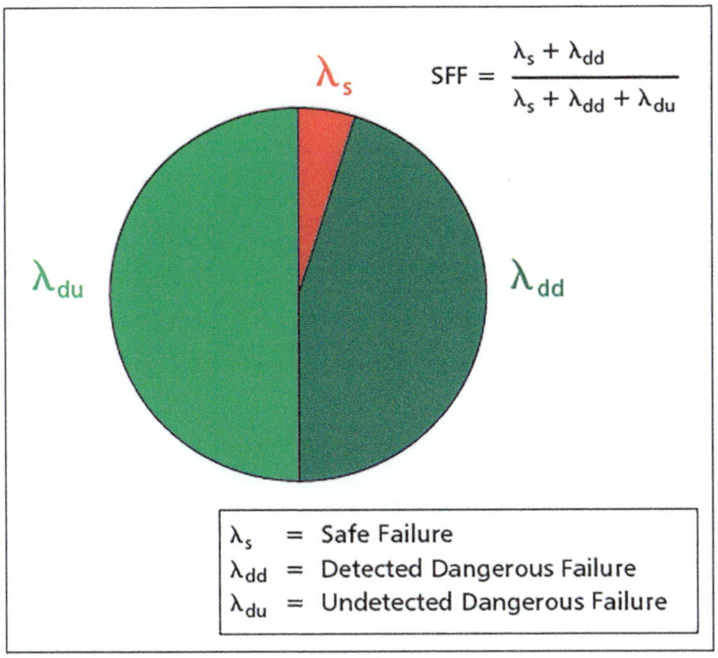

Example: SFF of 94% means that 6% of the failures are dangerous and undetected

The qualitative parameter SFF is not enough. As such undetected dangerous failures exist, their probability to occur during the testing interval should be determined. IEC 61508 describes two modes of operation for a safety function:

1. Low demand mode of operation
2. High demand or continuous mode

A safety function operating in demand mode is only performed when required (i.e., on demand) in order to transfer the equipment under control (EUC) into a specified state. The safety-related system that performs the safety function has no influence on the EUC until there is a demand for the safety function to be performed. This type of system can be as simplistic as a gas detection system in a boiler room that cuts the gas supply in the event of gas leakage.

A safety function operating in continuous mode operates to retain the EUC within its normal safe state. That is, the safety-related system continuously controls the EUC, and a dangerous failure of the safety-related equipment will lead to a hazard. A simple example is a gas concentration measurement by a gas detector system associated with control ventilation and heating to regulate the concentration of gas in a tank.

PFD and PFH

Depending of the timing between the demand and the test proof, IEC 61508 defines the following:

- Low-demand mode is where the frequency of demand for operation made on a safety-related system is no greater than one per year and no greater than twice the proof test frequency.
- High-demand or continuous mode is where the frequency of demand for operation made on a safety-related system is greater than once per year or greater than twice the proof test frequency.

In the context of this definition, continuous is regarded as very high demand. In relation with these two modes of operation, IEC 61508 relates the SIL of a safety function to:

- The PFD, the average probability of failure to perform its design function on demand, in the case of low demand mode.
- The PFH, the probability of a dangerous failure per hour, in the case of high-demand or continuous mode. The PFH is sometimes referred to as the *dangerous failure rate* (i.e., dangerous failures per hour).

SIL	PFD: Low Demand Mode (<1 year and <2 demands between each test/maintenance)	PFH: High Demand or Continuous Mode (>1 year or ≥ 2 demands between each test/maintenance	Risk Reduction
4	$\geq 10^{-5}$ to $< 10^{-4}$	$\geq 10^{-9}$ to $< 10^{-8}$	10000-100000
3	$\geq 10^{-4}$ to $< 10^{-3}$**	$\geq 10^{-8}$ to $< 10^{-7}$	1000-10000
2	$\geq 10^{3}$ to $< 10^{-2}$	$\geq 10^{-7}$ to $< 10^{-6}$	100-1000
1	$\geq 10^{2}$ to $< 10^{-1}$	$\geq 10^{-6}$ to $< 10^{-5}$	10-100

For a simple safety system:

$$PFD_{avg} = \frac{1}{2}\lambda_{du} \bullet \left(T_p + \text{MTTR}\right)$$

This means that statistically, a dangerous failure will remain undetected during half of the proof test interval, T_p.

SIL Capability and Safety System

Each component of an SIS involved in the safety function has a SIL level. The overall SIL of a safety function is determined by calculations based on the failure rate probability of each component. Each component contributes its part to the final SIL level. The weakest link of the chain reflects the maximum achievable SIL level. It is useless to request a SIL 3 controller if the sensor is only SIL 1 and the actuator has no SIL capability. The figure below shows that the weakest component is often the actuator.

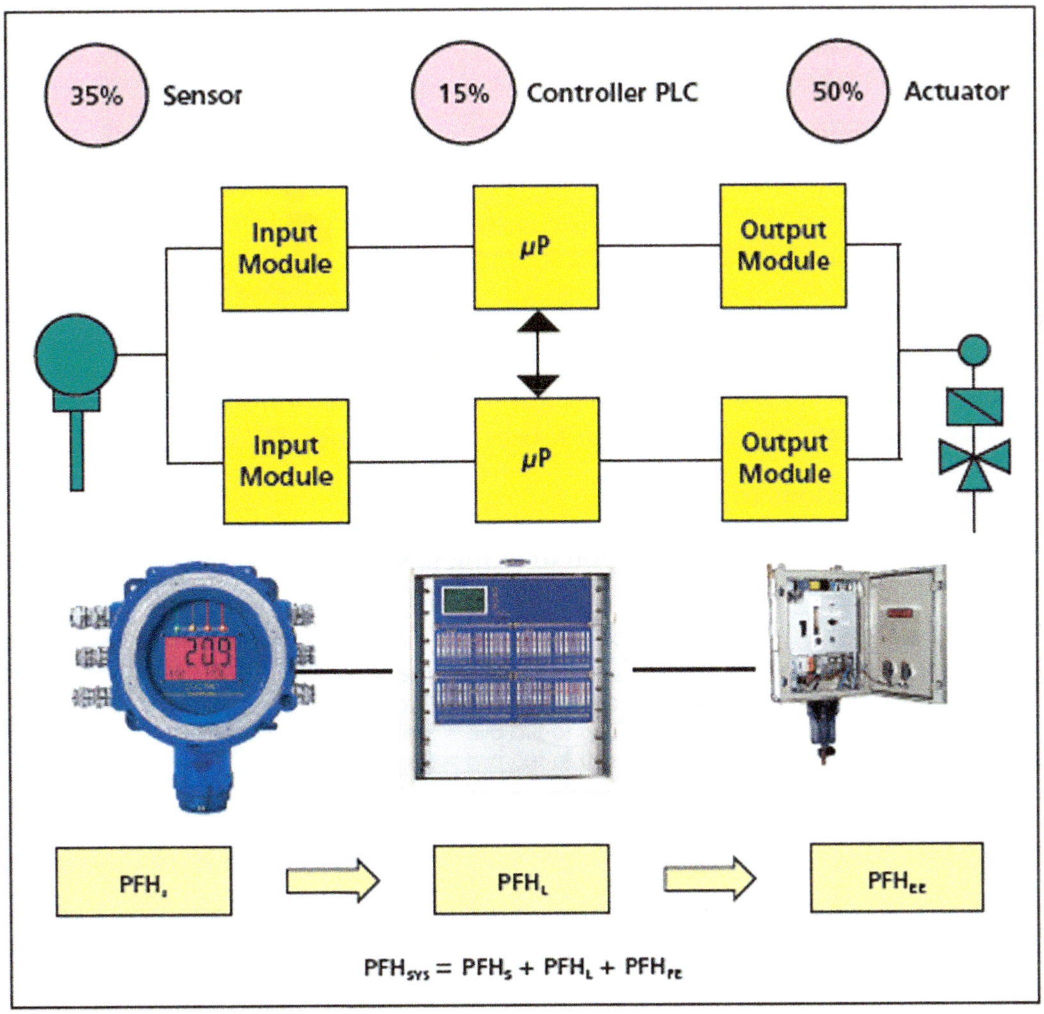

SIF

The SIF sheet includes the following information:

- Input
  - Type
  - Redundancy
  - Voting architecture
  - Testing interval

Voting Probabilities

Types of Voting	Probabilities (Safe)	Probabilities (Dangerous)
1oo1 = one out of one	0.01	0.02
1oo2 = one out of two	0.02	0.0004
2oo2 = two out of two	0.0001	0.04
2oo3 = two out of three	0.0003	0.0012

The SIS Calculations

The correct operation of an SIS requires a series of equipment to function properly. It must have sensors capable of detecting abnormal operating conditions, such as high flow, low level, or incorrect valve positioning. A logic solver is required to receive the sensor input signal(s), make appropriate decisions based on the nature of the signal(s), and change its outputs according to user-defined logic. The logic solver may use electrical, electronic, or programmable electronic equipment, such as relays, trip amplifiers, or PLCs. Next, the change of the logic solver output(s) results in the final element(s) taking action on the process (e.g., closing a valve) to bring it to a safe state. Support systems, such as power, instrument air, and communications, are generally required for SIS operation. The support systems should be designed to provide the required integrity and reliability.

In addition, it is necessary to:

- Understand the common mode failures; know which types of safe and nonsafe failures are possible in a specific system, and how to prevent them; and when, how, where, and which redundancy level is most appropriate for each case.
- Define the preventive maintenance level appropriate for each application.

The simple use of modern, sophisticated, or even certified equipment does not absolutely ensure any improvement on the reliability and safety of operation, when compared with traditional technologies, except when the system is deployed with criteria and knowledge of advantages and limitations inherent to each type of technology available. In addition, the entire SIS life cycle should be kept in mind.

Examples of typical safety systems include:

- Emergency shutdown system
- Safety shutdown system
- Safety interlock system
- Fire and gas system

Quantification of Reliability in Almost Absolute Terms

Reliability essentially has a probability nature and can be quantified in almost absolute terms by mean time between failures (MTBF) and mean time to failure (MTTF). Importantly, those two times are usually the mean values calculated using an identical number of instruments and, therefore, for any particular instrument their values may be different from the average.

The MTBF is a parameter expressing the mean time between failures occurring in an instrument, calculated in a specific period of time. In cases where the equipment has high reliability, in practice it will difficult to count the number of failures. A nonprecise number may be generated for the MTBF, and, in that case, using the manufacturer value is recommended.

The MTTF is an alternative mode to quantify reliability. It is normally used for devices such as thermocouples, as they are discharged when they fail. MTTF expresses the mean time before the failure occurs, calculated in an identical number of devices.

The final associated reliability in terms of importance to the measurement system is expressed by the mean time to repair (MTTR), that is, the mean time to repair an instrument or even the mean time to replace a piece of equipment. The combination of MTBF and MTTR shows the availability:

$$\text{availability} = \text{MTBF}/(\text{MTBF} + \text{MTTR})$$

The availability measures the proportion of time in which the instrument works without failures. The objective with measurement systems is to maximize the MTBF, minimize the MTTR, and, consequently, maximize the availability.

Failure Models – The Bathtub Curve

The failure mode in a device may change throughout its life cycle. It may remain unchanged, decrease, or, at least, increase. In electronic devices, it is common to have a behavior such as that shown in Figure 28-1, also known as the *bathtub curve.*

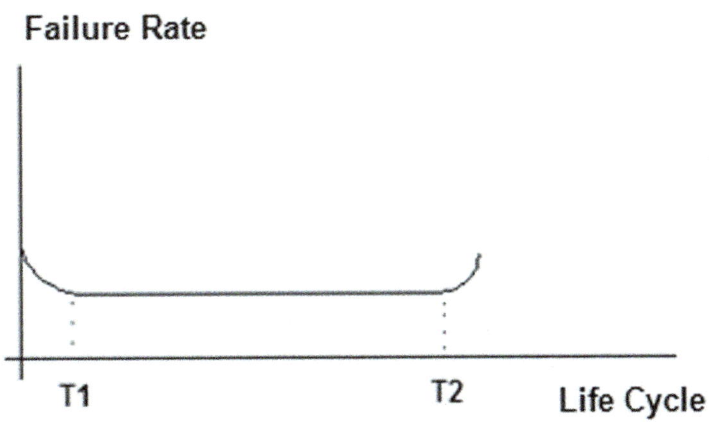

Figure 28-1. Typical Curve of reliability variation of an electronic component.

Manufacturers usually apply burn-in tests in such a way to eliminate the phase until T1, until products are placed in the market. But the mechanical components will have a higher failure rate in the end of their life cycle, as shown in Figure 28-2.

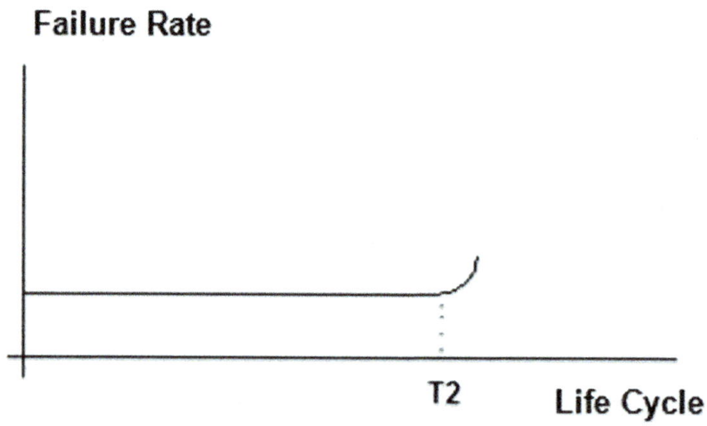

Figure 28-2. Typical curve of reliability variation of a mechanical component.

In practice, where systems are electronic and mechanical compositions, the failure models are complex. The more components, the higher the incidents and probabilities of failures.

Reliability Laws

In the practice of controls engineering, we usually have several components, and the measurement system is complex. We may have components in series or in parallel. Reliability of components in series should take into consideration the probability of individual failures in a time period. For a measurement system with *n* components in series, reliability (*Rs*) is the product of individual reliabilities:

$$R_s = R_1 \bullet R_2 \ldots R_n$$

Imagine we have a measurement system composed of a sensor, a conversion element, and a circuit of signal processing, where we have the following reliability: 0.9, 0.95, and 0.099, respectively. In such case, the system reliability will be:

$$0.9 \bullet 0.95 \bullet 0.009 = 0.85$$

The reliability can be increased by placing components in parallel; this means that the system fails if all components fail. In such case, R_s is demonstrated by:

$R_s = 1 - F_s$, where *Fs* is the nonreliability of the system.
The nonreliability is $F_s = F_1 \bullet F_2 \ldots F_3$

For example, in a safe measurement system, there are three identical instruments in parallel. The reliability of each one is 0.95, and that of the system is:

$$R_s = 1 - [(1 - 0.95) \bullet (1 - 0.95) \bullet (1 - 0.95)] = 0.999875$$

Improving the Reliability of a Measurement System

What we look for, in practice, is to minimize the level of failures. An important requirement is to ensure action to prevent failures and act before T2 (see Figures 28-1 and 28-2), when the statistical frequency of failures increases. The idea is to make T (time period or life cycle) equal to T2 and then maximize the period without failures.

There are several ways to increase the reliability of a measurement system:

- **Choice of instruments** – One should always be aware of the instrument specified, its influences regarding the process, materials, environment, and so on.
- **Protection of instruments** – Providing the instruments with appropriate protections may help to improve and ensure a higher level of reliability. For example, thermocouples should be protected in adverse operation conditions.
- **Regular calibration** – Most failures may be caused by drifts that may change and generate incorrect outputs. Therefore, according to the good instrumentation practices, we recommend that the instruments are periodically checked and calibrated.
- **Redundancy** – In these cases, there is more than one piece of equipment working in parallel and locked with a key, sometimes automatically. Here the reliability is significantly improved.

SIL and Availability

SIL is a statistical representation of the safety availability of an SIS at the time of process demand. It is at the heart of acceptable SIS design and includes the following factors:

- Device integrity
- Diagnostics
- Systematic and common cause failures
- Testing
- Operation
- Maintenance

Sample of SIL Evaluation

Acronyms

- **EUC** – equipment under control
- **F_T** – tolerable risk level
- **F_{NP}** – present risk level
- **MTBF** – mean time between failures

- **MTTF** – mean time to failure
- **PFD** – Probability of failure on demand
- **RRF** – risk reduction factor
- **RRF** – F_{NP}/F_T
- **PFD** – 1/RRF

Example SIL Determination Steps

IEC 61508 contains guidance on using both qualitative and quantitative methods to determine the SIL for a system based on risk frequency and consequence tables and graphs. The following steps illustrate application of the general guidelines contained in IEC 61508:

1. Set the target tolerable risk level (F_T), where F_T is the risk frequency, often determined as the hazardous event frequency multiplied by the consequence of hazardous event expressed numerically.
2. Calculate the present risk level (F_{NP}) for the EUC, which is the risk frequency with no protective functions present (or unprotected risk).
3. The ratio F_{NP}/F_T gives the risk reduction factor (RRF) required to achieve the target tolerable risk.
4. Determine the amount of RRF to be assigned to the SIS (RRF). The reciprocal of RRF gives the target average PFD the SIS must achieve.
5. Translate the PFD value into an SIL value (using guidance tables).

Metrics Used in the Reliability Engineering Field Involving SIS

Reliability R(t)

Reliability is a metric developed to determine the probability of success of an operation in a specified period of time.

$$\text{MTTF} = \frac{1}{\lambda}$$

$$R(t) = \exp(-\lambda t)$$

When λ (failure rate) is too low, the nonreliability function (*F(t)*) or the probability of failure (PF) is shown by: $PF(t) = \lambda t$

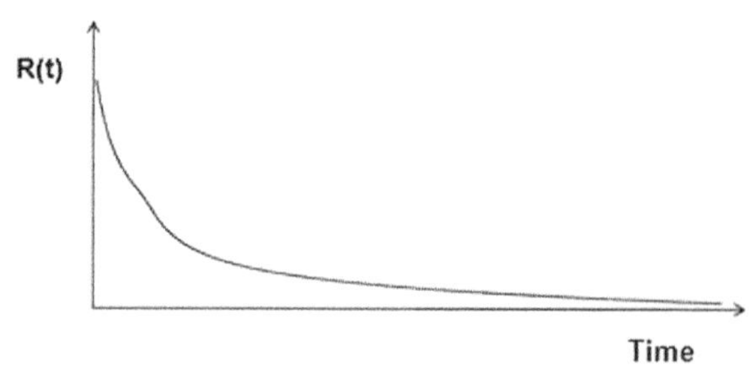

MTTR

The reliability measurement requires that a system has success in an operation during a time interval. In this sense, appears the MTTR metrics, which is the time in which a failure is detected and its repair occurs (or the reestablishment of operating success).

The reestablishment of the operating success is shown by:

$$\mu = 1/\text{MTTR}$$

In practice, it is not simple to estimate that rate, mainly when periodic inspection activities occur, as the failure may occur just after an inspection.

MTBF

The MTBF is a basic measure of the reliability in repairable items of a piece of equipment. It may be expressed in hours or years. It is commonly used in systems reliability and sustainability analysis and can be calculated by the following formula:

$$\text{MTBF} = \text{MTTR} + \text{MTTF}$$

where

- MTTR = mean time to repair
- MTTF = mean time to failure, which is the inverse of the sum of all failure rates

As the MTTR is too low in practice, it is common to assume the MTBF is equal to the MTTF.

Availability A(t) and Unavailability U(t)

Another useful metric is the availability. It is defined as the probability of a device being available (without failures) when a time (t) requires it to operate within the operating conditions to which it was designed.

Unavailability is given by: $U(t) = 1 - A(t)$

Availability is not only a reliability function, but it is also a maintenance function. The table below shows the relationship between reliability, maintenance, and availability. Note in this table that an increase in the maintenance ability implies a decrease in the time necessary to conduct the maintenance actions.

Reliability	Maintenance	Availability
Continual	Decreases	Decreases
Continual	Increases	Increases
Increases	Continual	Increases
Decreases	Continual	Decreases

Relationship between reliability, maintenance, and availability.

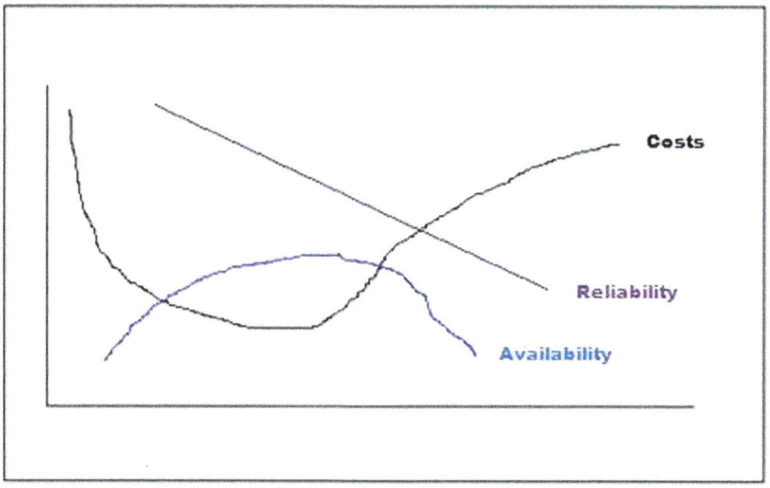

Reliability, availability, and costs.

Probability of Failure on Demand (PFD_{avg}) and Periodic Test and Inspection

PFD_{avg} is the probability of failure that a system (for failure prevention) has when a failure occurs. The SIL level is related to this probability of failure on demand and to the factor of risk reduction (how much must be protected to ensure an acceptable risk when a failure event occurs).

PFD is the indicator of reliability that is appropriate for the safety systems.

If it is not tested, the failure probability tends to move toward a factor of 1.0 with the time. Periodic tests maintain the probability of failure within the desired limit.

The figure below shows the architecture details versus the voting and PFD, and the formula below shows the correlation between the PFD and factor of risk reduction.

Voting	(PFD_{avg})	Architecture
1oo1	$\lambda_{du} * T/2$	1oo1
1oo2	$\frac{(\lambda_{du})^2 * T^2}{3}$	1oo2
2oo2	$\lambda_{du} * T$	2oo2
2oo3	$(\lambda_{du})^2 * T^2$	2oo3

SIF - Safety Instrumented Function   PFD: Probability of Failure on Demand

$$PFD_{avg} = \frac{tolerable\ frequency\ of\ an\ accident}{accident\ frequency\ with\ no\ protection} = \frac{1}{risk\ reduction\ factor\ (RRF)}$$

Correlation between PFD_{avg} and the Factor of Risk Reduction

The failure probability may be calculated using the following equation:

$$PF_{avg} = (Cpt \bullet \lambda_{du} \bullet TI/2) + ((1 - Cpt) \bullet \lambda_{du} \bullet MT/2)$$

where

λ_{du} = Failure rate, which is dangerous when undetected Cpt = percentage of failures detected by a test (proof test)

MT = mission time—the period in which a set of equipment will be operated

TI = test interval period

Example:

Suppose that a valve is used in an SIS and has an annual failure rate of 0.002. Every year a verification and inspection test is conducted. It is estimated that 70% of failures are detected in such tests. The valve will be used for 25 years, and its usage demand is estimated as once every 100 years. What is the average probability of failure?

Using the previous equation, we have:

$\lambda_{du} = 0.002$
$Cpt = 0.7$
$MT = 25$ years
$TI = 1$ year

$$PFD_{avg} = (0.7) \bullet 0.002 \bullet 1/2 + (1 - 0.7) \bullet 0.002 \bullet 25/2 = 0.0082$$

SIS Calculations – Worked Example

Consider a system with EUC that has an unprotected risk frequency (F_{NP}) of one hazardous event per 5 years ($F_{NP} = 0.2$/year), [0.2 = 1/5], with a consequence classified as "Critical." Tables B1 and B2 show examples of guidance tables used for risk classification and class interpretation of accidents from IEC 61508-5.

Risk Classification of Accidents (F_T): Table B1 of IEC 61508-5

Frequency	Catastrophic	Critical	Marginal	Negligible
	> 1 Death	1 Death or Injuries	Minor Injury	Production Loss
1 per year	I	I	I	II
1 per 5 years	I	I	II	III
1 per 50 years	I	II	III	III
1 per 500 years	II	III	III	IV
1 per 5000 years	III	III	IV	IV
1 per 50,000 years	IV	IV	IV	IV

Risk Classification of Accidents: Table B2 of IEC 61508-5

Risk Class	Interpretation
I	Intolerable risk
II	Undesirable risk, tolerable only if risk reduction is impracticable or if costs are grossly disproportionate to the improvement gained
III	Tolerable risk if the cost of risk reduction would exceed the improvement gained
IV	Negligible risk

Using Tables B1 and B2, the unprotected risk is determined as Class I. The target is to reduce this risk to a tolerable risk of Class III, that is, one hazardous event per 500 to 5000 years.

If we consider the safest target, wherein F_T is equal to 1 hazardous event in 5,000 years, this represents a frequency of 0.0002 events/year.

This gives a target RRF of F_{NP}/F_T or 0.2/0.0002, which is equal to 1000.

If there are no non-SIS protective layers assigned to the system, the SIS must fulfill the total RRF of 1000.

Now PFD is equal to 1/RRF, or 1/1000, which is 0.001 or $1 \cdot 10^{-3}$.

Using the SIL assignments in the following table, this gives a SIL target 2 for a PFD equal to 10^{-3}.

SIL	Availability	PDF_{avg}	MTBF
4	>99.99%	10^{-5} to $<10^{-4}$	100,000–10,000
3	99.9%	10^{-4} to $<10^{-3}$	10,000–1,000
2	99–99.9%	10^{-3} to $<10^{-2}$	1,000–100
1	90–99%	10^{-2} to $<10^{-1}$	100–10

Calculating PFD

$$PFD = \frac{1}{RRF} \text{ or } \frac{(system\ FR)(test\ interval)}{2}$$

Note: This will probably be useful on the CSE exam.

Calculating MTTF Based on Failure Rates

$$failure\ rate\ (FR) = \frac{number\ of\ failures}{total\ time\ \text{(hours or years)}}$$

Note: 1 year = 8,760 h

MTTF is normally expressed in years. On the CSE exam, it will probably be only 1 year.

1 failure in 1 year is:

$$\textit{failure rate }(FR) = \frac{1\text{ failure}}{(1\text{ year})\bullet(8{,}760\text{ h})} = \frac{1}{8{,}760\text{ h}} = 1.14\times10^{-4}/\text{h}$$

1 failure in 10 years is:

$$\textit{failure rate }(FR) = \frac{1\text{ failure}}{(10\text{ years})\bullet(8{,}760\text{ h})} = \frac{1}{87{,}600\text{ h}} = 1.14\times10^{-5}/\text{h}$$

$$\text{MTTF} = \frac{1}{FR_1 + FR_2 + FR_3 + \dots FRn}$$

Calculating MTBF Based on Failures

$$MTBF = \frac{(\textit{number of pieces of equipment} \bullet \textit{time period})}{(\textit{number of failures during that time})}$$

Example: Calculate MTTF and MTBF

Suppose 10 devices are tested for 500 h. During the test, two failures occur.

The estimate of the MTBF is:

$$MTBF = (10 \bullet 500)/2 = 2{,}500\text{ h/failure}$$

Whereas for MTTF:

$$MTTF = 2/500 = 4 \bullet 10^{-3}/\text{h}$$

SIS and SIL – Worked Examples

Sample SIS Problems That Might be on the CSE Exam

Example 1: Pump Failure Rate (*FR*)

1200 pumps fail 387 times over a 1-year period. What is the MTBF? (1200 • 12 months)/387 = 37 months/failure

The MTBF is 37 months/failure

For a single item, it is just the (time period/number of failures).

Example: If a pump failed twice in 1 year, the MTBF would be (12 months/2 failures) = 6 months MTBF

What is the *FR* of the previous example problem per year?

$$FR = \frac{387 \text{ failures}}{1 \text{ year}} = 387/\text{year}$$

What is the failure rate in 10 years?

$$FR = \frac{387 \text{ failures}}{\left(\frac{1}{10}\right) \text{years}} = \frac{387 \text{ failures} \cdot 10}{1 \text{ year}} = 3870/10 \text{ years}$$

What is the failure rate in hours for the above failure rate of 1 year?
Note: 1 year = 8760 h (24 h • 365 days)

$$FR = \frac{387 \text{ failures}}{1 \text{ years}} \cdot \frac{1 \text{ year}}{8760 \text{ h}} = \frac{387}{8760 \text{ h}} = 1.9517 \times 10^{-2}/\text{h}$$

What is the failure rate in hours for the above failure rate of 10 years?
Note: 1 year = 8760 h (24 h • 365 days)

$$FR = \frac{3870 \text{ failures}}{10 \text{ years}} \cdot \frac{1 \text{ year}}{8760 \text{ h}} = \frac{3870}{87600 \text{ h}} = 4.418 \times 10^{-2}/\text{h}$$

Example 2: MTBF over 10 Years

What is the MTBF for the previous example over a 10-year period?

The failure rate was 3870 in 10 years:

$$MTBF = \frac{1}{FR} = \frac{1}{3870} = 2.58 \cdot 10^{-4} \text{ years}$$

Example 3: Test Interval

An SIL 3 interlock with a RRF equal to 1175 is required to mitigate a risk class I hazard to class III. If the covert failure rates of the SIS loop components are as follows, recommend a test frequency:

Inputs = $1.2 \cdot 10^{-5}$/per hour
Logic solver = $7.0 \cdot 10^{-10}$/per hour
Valves = $2.75 \cdot 10^{-5}$/per hour

$$\textit{Failure Rate}\ (FR) = \frac{\text{number of failures}}{\text{total time (hours or years)}}$$

$$PFD = \frac{1}{\text{RRF}} \text{ or } \frac{(\textit{system FR})\,(\text{Test interval})}{2}$$

where

FR = failure rate (dangerous)

T_I = proof test interval

The PFD_{avg} can be calculated for each component of the system and then summed together. (e.g., S = sensor, LS = logic solver, and FE = final element)

$$PFD_{avg} = \frac{1}{\text{RRF}_{avg}} \quad \text{or} \quad \frac{FR_S(T_I)}{2} + \frac{FR_{LS}(T_I)}{2} + \frac{FR_{FE}(T_I)}{2}$$

$$PFD_{avg} = \frac{1}{\text{RRF}_{avg}} \quad \text{or} \quad \frac{(FR_S + FR_{LS} + FR_{FE}) \bullet T_I}{2}$$

$$8.51 \bullet 10^{-4}(PFD_{avg}) = \frac{1}{1175(\text{RRF}_{avg})} \quad \text{or} \quad \frac{(1.2 \bullet 10^{-5} + 7.0 \bullet 10^{-10} + 2.75 \bullet 10^{-5}) \bullet T_I}{2}$$

$$(2)8.51 \bullet 10^{-4} = \frac{(2)}{1175} \quad \text{or} \quad \frac{(1.2 \bullet 10^{-5} + 0.00007 \bullet 10^{-5} + 2.75 \bullet 10^{-5}) \bullet T_I}{1} \bullet \frac{(2)}{(2)}$$

$$1.702 \bullet 10^{-3} = (1.0 \bullet 10^{-5} + 0.00007 \bullet 10^{-5} + 3.0 \bullet 10^{-5}) \bullet T_I$$

$$T_I = \frac{170.2 \bullet 10^{-5}}{(1.2 \bullet 10^{-5} + 0.00007 \bullet 10^{-5} + 2.75 \bullet 10^{-5})} = \frac{170.2 \bullet 10^{-5}}{3.95007 \bullet 10^{-5}} = 43.09 \text{ h}$$

29 Overview of NEC/NFPA and Other Codes

NFPA 30	2003	National Fire Protection Association	Flammable and Combustible Liquids Code	49 CFR 192
NFPA 54	2002	National Fire Protection Association	National Fuel and Gas Code	24 CFR 3280
NFPA 58	2001	National Fire Protection Association	Standard for Liquefied Petroleum Gases	49 CFR 173
NFPA 58	2004	National Fire Protection Association	Standard for the Storage and Handling of Liquefied Petroleum Gases	49 CFR 192.11(b)
NFPA 59	2004	National Fire Protection Association	Standard for the Storage and Handling of Liquefied Petroleum Gases at Utility Gas Plants	49 CFR 192.11(b)
NFPA 70	2005	National Fire Protection Association	National Electrical Code	49 CFR 192.189(c)
NFPA 101	2000	National Fire Protection Association	Life Safety Code	59 CFR 130
UL 50	1995	Underwriters Laboratories	Standard for Enclosures for Electrical Equipment	46 CFR 111.81-1(d)

List of NFPA Codes (Be Familiar with These Codes)

The CSE exam will include code questions. We covered American Society of Mechanical Engineers (ASME) in Chapter 14 pressure-relief valves and safety rupture disks. We will now talk about the codes for the installation, maintenance, and operation of control systems in manufacturing and process plants. Our focus will be on the installation of electrical systems.

Here are the major codes the CSE exam may cover:

NFPA 30 – *Flammable and Combustible Liquids Code*

NFPA 70 NEC – *National Electrical Code* (be extremely familiar with this code)

NFPA 70E – *Standards for Electrical Safety in the Workplace*

NFPA 77 – *Static Electricity* (very important in the application of loading stations)

NFPA 79 – *Industrial Machinery* (electrical code specifically focusing on industrial controls for machinery)

NFPA 407 – *Aircraft Fuel Servicing*

NFPA 780 – *Lightning Protection* (protection of equipment and tanks from damage and explosion)

NFPA 496 – *Purged and Pressurized Systems* (used extensively in petrochemical production)

Voltage Drop Calculations – Worked Examples

Transmitter Sample Problem: What is the voltage drop for AWG 18 stranded wire in a steel conduit 565 ft long, if the wire is a coated cable? Wire carries 20 mA of DC current.

Note: Coated (wires are jacketed); uncoated (wires are not jacketed).

Find the resistance for AWG 18 stranded wire (coated) in NEC Table 8.

Refer to Table A17, "NEC Conductor Properties and Impedance" for a reproduction of NEC Table 8.

Look under the DC Resistance column.

Resistance per 1,000 ft = 8.45 Ω

$$V_d = \left(\frac{2 \bullet L}{1000}\right) \bullet I \bullet R; \quad \left(\frac{2 \bullet 565}{1000}\right) \bullet 0.020 \text{ mA} \bullet 8.45\ \Omega = 0.191 \text{ V or } 191 \text{ mV}$$

Motor Sample Problem: A 480 V, three-phase 50 hp motor draws 65 A and is 600 ft away.

a) What is the allowable voltage dropped across the motor cable?

b) What size wire or cable should we use for a 3% voltage drop per NEC?

a) Find the allowable voltage drop for the motor cable first.

$$\%_{drop} = \frac{V_d}{V_{source}}; \quad V_d = V_{source} \bullet \%_{drop} = 480 \bullet 0.03 = 14.4 \text{ volts drop maximun}$$

b) Find the wire size for the maximum allowable voltage drop across the motor cable.

$$\text{cm} = \left(\frac{\sqrt{3} \bullet L \bullet I \bullet k}{V_d}\right) = \left(\frac{\sqrt{3} \bullet 600 \bullet 65 \bullet 12.9}{14.4}\right) = 60{,}514 \text{ cm}$$

Find the cm (area circular mils) of stranded wire (uncoated) in NEC Table 8.

Refer to Table A17, "NEC Conductor Properties and Impedance" for a reproduction of NEC Table 8

Look under the Circular Mill (Area) column.

AWG 3 = 52,620 cm
AWG 2 = 66,360 cm
We need 60,514 cm, so use AWG 2.

Proof of voltage drop, using the resistance for AWG 2 stranded wire (uncoated) listed in NEC Table 8.

Resistance per 1000 ft = 0.194 Ω

$$V_d = \left(\frac{\sqrt{3} \bullet L}{1000}\right) \bullet I \bullet Z_e = \left(\frac{\sqrt{3} \bullet 600}{1000}\right) \bullet 65 \bullet 0.194 = 13.1 \text{ or } 13 \text{ V dropped along the wire.}$$

The wire size gives less than the required maximum of 3% voltage drop.

Sample Problem: A fieldbus cabinet is being installed for a critical system 1,000 ft from the 24 VDC supply. The load requirement is 10 A at a minimum operating voltage of 22.5 VDC. What size wire or cable will be used to power the fieldbus cabinet?

Note: This is not a typical branch circuit or feeder. We will not be using the standard 2% for feeders and 3% for branch circuits voltage drop allowed by the NEC. This would be minimum operating voltage required by the manufacturer.

The allowable voltage drop is as follows:

24 – 22.5 = 1.5 VDC The wire size in circular mils will be:

$$\text{cm} = \frac{2 \bullet L \bullet I \bullet k}{VD} = \frac{2 \bullet 1000 \bullet 10 \bullet 12}{1.5} = 160{,}000 \text{ cm}$$

(**Note:** $k = 12$ for 131°F)

AWG 3/0 is required. (3/0 AWG = 167,800, 2/0 AWG = 133,100 cm)

Sample Problem: A 125 VDC panelboard for critical systems is being installed 1,200 ft from the battery rack. The load requirement is 60 A at a minimum operating voltage of 2% voltage drop on the feeder per NEC. What size wire or cable will be used to power the panelboard?

The allowable voltage drop is as follows:

125 • 0.02 = 2.5 VDC drop.

The wire size in circular mils will be:

$$\text{cm} = \frac{2 \bullet L \bullet I \bullet k}{VD} = \frac{2 \bullet 1200 \bullet 60 \bullet 12}{2.5} = 691{,}200 \text{ cm}$$

(**Note:** $k = 12$ for 131°F)

The wire is so large we will make parallel runs of cable.

250 kcmil • 3 = 750 kcmil, or 300 kcmil • 2 = 600 kcmil, or 400 kcmil • 2 = 800 kcmil

Three sets of 250 kcmil cables is the best solution.

Note: Two sets of 300 kcmil cables are too small to meet the NEC 2% feeder requirement for effective operation of the electrical system.

NEC Article 500 Explosion-Proof Installations

(Hazardous Locations per NFPA 70 – NEC)

What Does Explosion-Proof Mean?

To understand what is meant by *explosion-proof,* we must look at the context of the term and the organization that defined it. The National Fire Protection Association (NFPA) began publishing the NEC in 1897. The NEC *national electrical code* is also known as NFPA 70 and as ANSI/NFPA 70 from its inclusion in the body of NFPA codes.

The NEC includes definitions for several types of protection techniques that are acceptable when designing products for use in hazardous (classified) locations: explosion-proof, dust-ignition-proof, dust-tight, purged/pressurized, intrinsically safe, and hermetically sealed. These definitions set the criteria that must be met by all components installed in hazardous (classified) locations, referred to as simply *HAZLOC*.

To meet the criteria for the explosion-proof rating, an enclosure must be able to contain any explosion originating within its housing and prevent sparks from within its housing from igniting vapors, gases, dust, or fibers in the air surrounding it. Therefore, explosion-proof, when referring to electrical enclosures, does not mean the enclosure is able to withstand an exterior explosion. Instead, it means the enclosure can prevent an internal spark or explosion from causing a much larger blast.

Additionally, the NEC states that equipment must meet the temperature requirements of the specific application in which it is to be installed. This means that the operating temperature of the motor (and its enclosure) or other component cannot be greater than the lowest ignition/combustion temperature of the gases or dusts in the atmosphere where the component is to be installed.

Whenever electrical equipment is installed within a hazardous area, one or more protection techniques must be employed to ensure that life and property are not endangered. The principal protection techniques include:

- Explosion-proof equipment
- Dust-ignition-proof equipment
- Dust-tight systems
- Purged and pressurized equipment
- Intrinsic safety systems
- Nonincendive circuit
- Oil immersion
- Combustible gas detection system

Class I Hazardous Location NEC Article 501

Class I Location Definition

According to the NEC, there are three types of hazardous locations. The first type of hazard is one is created by the presence of *flammable gases or vapors* in the air, such as natural gas or gasoline vapor. When these materials are found in the atmosphere, a potential for explosion exists, as the materials could be ignited if an electrical or other source of ignition is present. The NEC Code writers have referred to this first type of hazard as *Class I*. So, a *Class I Hazardous Location* is one in which *flammable gases or vapors* may be present in the air in sufficient quantities to be explosive or ignitable. Some typical Class I locations include:

- Petroleum refineries, and gasoline storage and dispensing areas
- Dry cleaning plants where vapors from cleaning fluids can be present
- Spray finishing areas

- Aircraft hangars and fuel servicing areas
- Utility gas plants and operations involving storage and handling of liquefied petroleum gas or natural gas

All of these are Class I, gas or vapor, hazardous locations. All require special Class I hazardous location equipment.

Class I Division Definitions

Class I locations are further subdivided into two divisions, Division 1 and Division 2. The division defines the likelihood of the hazardous material being present in a flammable concentration.

Division	Definitions
Division 1	In which an ignitable concentration of flammable gases or vapors: • Exists under normal operating conditions • Exists frequently because of repair operations, maintenance operations, or leakage • Is released through breakdown or faulty operation of equipment or processes in which the breakdown causes electrical equipment to become a source of ignition
Division 2	In which an ignitable concentration of flammable gases or vapors: • Is normally confined within closed containers when handled, processed, or used • Is normally prevented by positive mechanical ventilation • Is adjacent to a Class I, Division 1 location

Class I Group Definitions

Combustible and flammable gases and vapors are divided into four groups. The classification is based on maximum explosion pressures, and maximum safe clearance between parts of a clamped joint in an enclosure per NEC Article 500.

Class I Groups

Class	Division	Group	Flammable Material	
Class I	Division 1 and 2	A	Acetylene	
Class I	Division 1 and 2	B	Hydrogen Butadiene	Ethylene oxide Propylene oxide
Class I	Division 1 and 2	C	Ethylene Cyclopropane Ethyl ether	
Class I	Division 1 and 2	D	Propane Acetone Ammonia Benzene Butane	Ethanol Gasoline Methanol Natural gas

Class I Temperature Definition

The temperature marking specified shall not exceed the ignition temperature of the specific gas or vapor to be encountered.														
Temp Code	**T1**	**T2**	**T2A**	**T2B**	**T2C**	**T2D**	**T3**	**T3A**	**T3B**	**T3C**	**T4**	**T4A**	**T5**	**T6**
°C	450	300	280	260	230	215	200	180	165	160	135	120	100	85
°F	842	572	536	500	446	419	392	356	329	320	275	248	212	185

Class II Hazardous Location NEC Article 502

Class II Location Definition

The second type of hazard listed by the NEC are those areas made hazardous by the presence of combustible dust. These are referred to in the code as *Class II Locations*. Finely pulverized material suspended in the atmosphere can cause a powerful explosion such as might occur at a processing or manufacturing facility. Some typical Class II locations are:

- Grain elevators
- Flour and feed mills
- Plants that manufacture, use, or store magnesium or aluminum powders
- Producers of plastics, medicines, and fireworks
- Producers of starch or candies
- Spice-grinding plants, sugar plants, and cocoa plants
- Coal preparation plants and other carbon handling or processing areas

Class II Division Definitions

Class II locations are further subdivided into two divisions, Division 1 and Division 2. The division defines the likelihood of the combustible dust being present in an ignitable concentration.

Class	Division	Group	Combustible Dust	
Class II	Division 1 and 2	E	Aluminum Magnesium	Commercial alloys
Class II	Division 1 and 2	F	Coal Black carbon	Charcoal Coal dust
Class II	Division 1 and 2	G	Flour Grain Wood	Plastic Chemical

Division	Definitions
Division 1	In which combustible dusts: • Exist under normal conditions. • Exist because of mechanical failure or abnormal operation of equipment and machinery. This might also provide a source of ignition through simultaneous failure of electric equipment, operation of protection devices, or from other causes. • Are present in hazardous quantities of electrically conductive nature.
Division 2	In which combustible dusts: • Are not normally in the air • Accumulations are not sufficient to interfere with normal operation • Are suspended in the air as a result of infrequent malfunctioning of handling equipment and processing equipment • Accumulations may be sufficient to interfere with the safe dissipation of heat from electrical equipment • Accumulation may be ignitable by abnormal operation or failure of electrical equipment

Class II Group Definitions

Class II Temperature Class

The temperature marking specified shall not exceed the ignition temperature of the specific gas or vapor to be encountered. For organic dusts that may dehydrate or carbonize, the temperature marking shall not exceed the lower of either the ignition temperature or 165°C (329°F).

Class III Hazardous Location NEC Article 503

Class III Location Definition

Class III hazardous locations, according to the NEC, are areas where there are *easily ignitable fibers* or *flyings* present, due to the types of materials being handled, stored, or processed. The fibers and flyings are not likely to be suspended in the air, but can collect around machinery or on lighting fixtures and where heat, a spark, or hot metal can ignite them. Typical Class III locations include the following:

- Textile mills, cotton gins
- Cotton seed mills and flax processing plants
- Plants that shape, pulverize, or cut wood and create sawdust or flyings

Class III Division Definitions

Class III locations are further subdivided into two divisions, Division 1 and Division 2. The division defines the likelihood of the combustible dust being present in an ignitable concentration.

Division	Definitions
Division 1	In which easily ignitable fibers or materials producing combustible flyings are: • Handled • Manufactured • Used
Division 2	In which easily ignitable fibers or materials producing combustible flyings are: • Stored • Handled • Processed other than manufactured

Class III Group Definitions

There are no specific groups for Class III	
Combustible Dust	
Industry	*Type of Materials*
Textile mills Combustible fiber Manufacturing and processing plants Cotton plants Clothing manufacturing plants Woodworking plants Similar hazardous industry	Rayon Cotton Sisal or henequen Hemp Cocoa fiber Oakum Spanish moss Other materials of similar nature

Use of IEC Zone Classifications

Classification Comparison (Zone/Division) for a Class I Location

Zone 0	Zone 1	Zone 2
Where ignitable concentrations of flammable gases, vapors, or liquids are present continuously or for long periods of time under normal operating conditions.	Where ignitable concentrations of flammable gases, vapors, or liquids: • Are likely to exist under normal operating conditions • May exist frequently because of repair, maintenance operations, or leakage	Where ignitable concentrations of flammable gases, vapors, or liquids: • Are not likely to exist under normal operating conditions • Occur for only a short period of time • Become hazardous only in case of an accident or some unusual operating condition

Division 1	Division 2
Where ignitable concentrations of flammable gases, vapors, or liquids: • Are likely to exist under normal operating conditions • Exist frequently because of maintenance/repair work or frequent equipment failure	Where ignitable concentrations of flammable gases, vapors, or liquids: • Are not likely to exist under normal operating conditions • Are normally in closed containers where the hazard can only escape through accidental rupture or breakdown of such containers or in case of abnormal operation of equipment

Note: Per NEC Article 505-10(b)(1), a division classified product may be installed in a zone classified location, but the reverse is not true. Typically, a zone classified product provides protection utilizing a protection method not available in the class/division scheme.

Group Comparison (Zone/Division) for a Class I Location

Zone	Class/Division
IIC – Acetylene and hydrogen	A – Acetylene
	B – Hydrogen
IIB – Ethylene	C – Ethylene
IIA – Propane	D – Propane

Protection Methods Comparison Class

Zone 0	Zone 1	Zone 2
• Explosion-proof • Intrinsically safe (allows two faults) • Purged/pressurized • (Type X or Y)1 (US only)	• Encapsulation, "m" • Flame-proof, "d" • Increased safety, "e" • Intrinsically safe, "ib" (allows only one fault) • Powder-filled, "q" • Purged/pressurized, "p" • Any Class I, Zone 0 method • Any Class I, Division I method (US only)	• Energy limited, "nC" • Hermetically sealed, "nC" • Nonincendive, "nC" • Nonsparking, "nA" • Restricted breathing, "nR" • Sealed device, "nC" • Any Class I, Zone 0 or 1 method • Any Class I, Division 1 or 2 method (US only)

Division 1	Division 2
• Explosion proof • Intrinsically safe (allows two faults) • Purged/pressurized (Type X or Y)	• Hermetically sealed • Nonincendive • Nonsparking • Oil immersion • Sealed device • Purged/pressurized (Type L) • Any Class I, Zone 1 or 2 method (US only) • Any Class I, Division 1 method

Designation of NEC/CEC Classification

The following chart explains the HAZLOC classification and rating HAZLOC on the equipment label.

Hazardous Location Classifications per NEC Article 500

Class I Division 2 Group D T4

Type of Flammable Substance

Class I — Approved for the strictest class, therefore can be used for all classes

Class I — Flammable gas, vapors, and liquids

Class II — Combustible dusts

Class III — Ignitable fibers and flyings

Area Classification

Division 1 — Approved for the strictest division, therefore can be used for both divisions

Division 1 — Flammable substances are continually present or are likely to exist under normal operating conditions.

Division 2 — Flammable substances are not likely to exist under normal operating conditions.

Gas Group

Group B — Approved for Group B; therefore, also approved for Groups C and D, but not Group A. If no groups are listed, the equipment is approved for all groups. The gases are grouped according to certain physical characteristics on their explosive behavior.

Temperature Code

If no temperature code is listed, the equipment must meet the strictest temperature code, (T6).

This is the maximum temperature that the equipment is allowed to emit without causing an explosion/fire.

Example:

This will probably be on the CSE exam.

Let us illustrate our NEC Code "translation" with an example. How would we classify a storage area where LP (liquid propane) gas is contained in closed tanks? LP gas is a Class I substance (gas or vapor). It is Division 2 because it would only be in the atmosphere if an accidental rupture or leakage occurred, and it is Group D material.

Note: This is for a storage system, separate from the process unit location. If the electrical equipment were in the area of processing vessels and process piping systems (the process unit), the equipment would be rated for a Division 1 location.

Summary of the Various Hazardous (Classified) Locations

Summary of Class I, II, III Hazardous Locations

Class	Groups	Division 1	Division 2
I Gases, vapors, and liquids (Article 501)	A: Acetylene B: Hydrogen, etc. C: Ether, etc. D: Hydrocarbons, fuels, solvents, etc.	Normally explosive and hazardous.	Not normally present in an explosive concentration (but may accidentally exist).
II Dusts (Article 502)	E: Metal dusts (conductive and explosive) F: Carbon dusts (some are conductive, and all are explosive) G: Flour, starch, grain, combustible plastic, or chemical dust (explosive)	Ignitable quantities of dust normally are or may be in suspension, or conductive dust may be present.	Dust not normally suspended in an ignitable concentration (but may accidentally exist). Dust layers are present.
III Fibers and flyings (Article 503)	Textiles, woodworking, etc. (easily ignitable, but not likely to be explosive)	Handled or used in manufacturing.	Stored or handled in storage (exclusive of manufacturing).

Hazardous Location Wiring Methods

Class	Division	Methods Permitted by the NEC
I (Article 501)	1	• Any suitable type of wire or cable if installed in threaded metallic conduit with approved termination fittings (end seals). • Type MI (mineral insulated) cable installed with approved termination fittings. • Any suitable type of wire or cable if installed in nonmetallic conduit encased in concrete and buried under at least 2 ft of earth. • In certain industrial establishments, Type MC-HL or Type ITC-HL continuously corrugated welded armor (CCW) sheathed cable with approved termination fittings. • Optical fiber cable types (OFN, OFC, etc.) installed in raceways in accordance with NEC 501.10 and 501.15.
	2	• All wiring methods permitted for Class I Division 1. • Type PLTC or PLTC-ER cables installed in accordance with Article 725 including cable trays. Cables must be installed with approved termination fittings. • Type ITC or ITC-ER cables installed as permitted in Article 727.4 and terminated with approved fittings. • Type MC, MV, TC, or TC-ER cables installed in cable trays with approved termination fittings (end seals). • Optical fiber cable types (OFN, OFC, etc.) installed in raceways in accordance with NEC 501.10(A) and 501.15.

Class	Division	Methods Permitted by the NEC
II (Article 502)	1	• Any suitable type of wire or cable if installed in threaded metallic conduit with approved termination fittings (end seals). • Type MI (mineral insulated) cable installed with approved termination fittings. • In certain industrial establishments, Type MC-HL continuously corrugated aluminum cable having an overall jacket and listed for use in hazardous locations with approved termination fittings. • Optical fiber cable types (OFN, OFC, etc.) installed in raceways in accordance with NEC 501.10(A) and 501.15.
	2	• All wiring methods permitted for Class II Division 1. • Any suitable type of wire or cable if installed in metallic conduit with approved termination fittings (end seals). • Type MC or MI cables if installed with approved termination fittings (end seals). • Type PLTC or PLTC-ER cables installed in accordance with Article 725 including cable trays. Cables must be installed with approved termination fittings. • Type ITC or ITC-ER cables installed as permitted in Article 727.4 and terminated with approved fittings. • Type MC, MI, TC, or TC-ER cables installed in cable trays. • Optical fiber cable types (OFN, OFC, etc.) installed in raceways in accordance with NEC 501.10(A) and 501.15.
III (Article 503)	1 and 2	• Any suitable type of wire or cable if installed in metallic or PVC conduit with approved termination fittings (end seals). • Type PLTC or PLTC-ER cables installed in accordance with Article 725 including cable trays. Cables must be installed with approved termination fittings. • Type ITC or ITC-ER installed as permitted in Article 727.4 and terminated with approved fittings. • Type MC, MI, TC, or TC-ER cables installed in cable trays with approved termination fittings (end seals).

Purged and Pressurized Systems

Purged and pressurized enclosures are referred to in NEC Article 500, but for purging buildings you must refer to NFPA 496, *Standard for Purged and Pressurized Enclosures for Electrical Equipment*, for installation details.

See NFPA 496 for more details.

Intrinsically Safe Systems

Zener diode barrier (configurations)

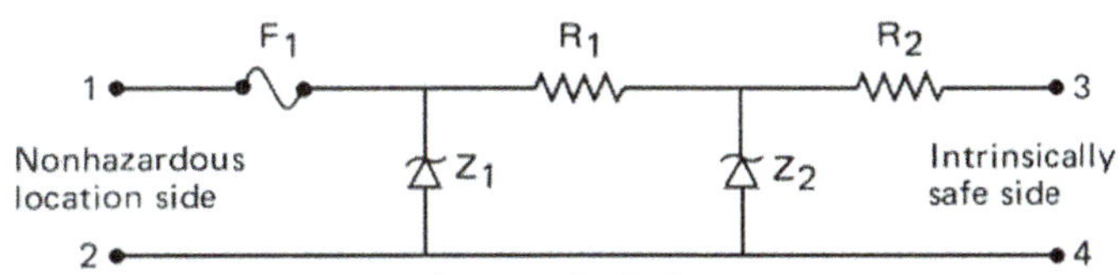

Fused zener diode barrier.

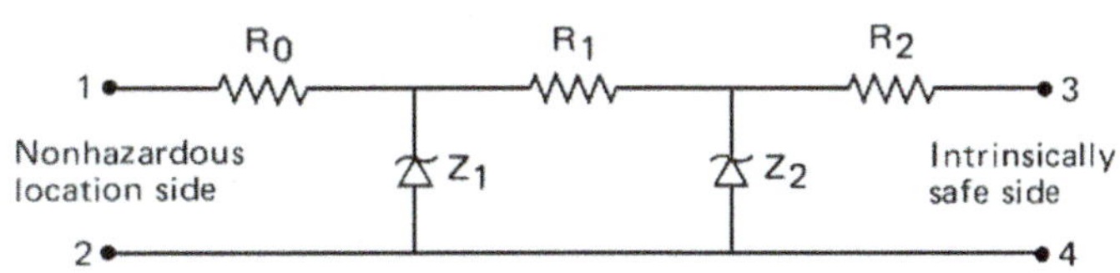

Unfused zener diode barrier.

Isolated Circuit Barriers

The grounding requirement, maintenance, and testing can be considerably reduced by using isolated barriers that do not require a maintained IS (isolated) ground.

The circuits are floating ground systems and are usually equipped with three transformers (input, output, and power). A comparison of wiring methods using isolated barriers follows.

Conventional Passive IS Zener Barriers

1. A good ground connection must be provided and maintained.
2. Field devices must be isolated from ground.
3. Voltage drop across the barriers can make some applications difficult.
4. Improper connection or voltage surges could blow the fuse.
5. Poor common mode rejection values may result.

Active (powered) IS isolation Barriers

1. A ground connection is not required.
2. Field devices can be grounded.
3. Full voltage is available to field devices.
4. Reverse polarity is protected and surge arrestors are incorporated.
5. This method tolerates high common mode voltage.
6. Signal conditioning and circuit protection are combined.
7. Installation is simple with the elimination of ground loops.

NEMA Electrical Enclosures Types and Uses

Definitions from NEMA 250-2003

This topic will probably be on the CSE exam. This is an overview of NEMA enclosure designations. The following summary provides the essential information needed to choose the appropriate enclosure type for a specific application. It is also recommended to double-check with the AHJ for each installation.

Nonhazardous Location NEMA Enclosure Types

- **Type 1 –** General purpose, indoor use. Provides a degree of protection against human contact with electrically charged, live parts and against ingress of solid foreign objects such as falling dirt.
- **Type 2 –** Drip-proof, indoor use. Same protection as Type 1 but adds protection against dripping and light splashing of water.

- **Types 3R, 3RX –** Rain-tight, sleet-resistant, rust-resistant. Indoor or outdoor use. Same protection as Type 1, but adds a degree of protection against ingress of falling dirt, rain, sleet, and snow; also protects against damage due to external ice formation. The *X* designation indicates corrosion resistance.
- **Types 3, 3X –** Dust-tight, rain-tight, sleet-tight. Indoor or outdoor use. Same protection as Type 3R but adds protection against windblown dust. The *X* designation indicates corrosion resistance.
- **Types 3S, 3SX –** Dust-tight, rain-tight, sleet-tight. Indoor or outdoor use. Same protection as Type 3 but includes external mechanisms that remains operable when ice-laden. The *X* designation indicates corrosion resistance.
- **Types 4, 4X –** Water-tight, dust-tight, sleet-resistant. Indoor or outdoor use. Same protection as Type 3 but adds protection against splashing and hose-directed water. The *X* designation indicates corrosion resistance.
- **Type 5 –** Dust-tight. Indoor use. Same protection as Type 2 but adds gaskets to prevent ingress of settling dust, lint, fibers, and flyings.
- **Types 6, 6P –** Submersible, indoor or outdoor use. Same protection as Type 4, but adds protection against occasional temporary submersion (Type 6) or prolonged submersion (Type 6P) at a limited depth.
- **Types 12, 12K –** General purpose, indoor use. Protects against falling dirt and circulating dust, lint, fibers, and flyings. Protects against ingress of dripping and splashing water. Rust-resistant. Type 12 enclosures *do not* include knockouts; Type 12K enclosures *do* include knockouts.
- **Type 13 –** General purpose, indoor use. Same protection as Type 12, but adds protection against ingress of spraying, splashing, or seeping oil and noncorrosive coolants.

Table 29-1 – Indoor Nonhazardous Locations

From NEMA 250-2003 – Comparison of Specific Applications of Enclosures

Provides a Degree of Protection Against the Following Conditions	Type of Enclosure									
	1*	2*	4	4X	5	6	6P	12	12K	13
Access to hazardous parts	X	X	X	X	X	X	X	X	X	X
Ingress of solid foreign objects (falling dirt)	X	X	X	X	X	X	X	X	X	X
Ingress of water (dripping and light splashing)	–	X	X	X	X	X	X	X	X	X
Ingress of solid foreign objects (circulating dust, lint, fibers, and flyings**)	–	–	X	X	–	X	X	X	X	X
Ingress of solid foreign objects (settling airborne dust, lint, fibers, and flyings**)	–	–	X	X	X	X	X	X	X	X
Ingress of water (hose down and splashing water)	–	–	X	X	–	X	X	–	–	–
Oil and coolant seepage	–	–	–	–	–	–	–	X	X	X
Oil or coolant spraying and splashing	–	–	–	–	–	–	–	–	–	X
Corrosive agents	–	–	–	X	–	–	X	–	–	–
Ingress of water (occasional temporary submersion)	–	–	–	–	–	X	X	–	–	–
Ingress of water (occasional prolonged submersion)	–	–	–	–	–	–	X	–	–	–

* These enclosures may be ventilated.

** These fibers and flyings are nonhazardous materials and are not considered Class III type ignitable fibers or combustible flyings. For Class III type ignitable fibers or combustible flyings, see NEC Article 500.

Table 29-2 – Outdoor Nonhazardous Locations

From NEMA 250-2003 – Comparison of Specific Applications of Enclosures

Provides a Degree of Protection Against the Following Conditions	Type of Enclosure									
	3	3X	3R*	3RX*	3S	3SX	4	4X	6	6P
Access to hazardous parts	X	X	X	X	X	X	X	X	X	X
Ingress of water (rain, snow, and sleet**)	X	X	X	X	X	X	X	X	X	X
Sleet***	–	–	–	–	X	X	–	–	–	–
Ingress of solid foreign objects (windblown dust, lint, fibers, and flyings)	X	X	–	–	X	X	X	X	X	X
Ingress of water (hose down)	–	–	–	–	–	–	X	X	X	X
Corrosive agents	–	X	–	X	–	X	–	X	–	X
Ingress of water (occasional temporary submersion)	–	–	–	–	–	–	–	–	X	X
Ingress of water (occasional prolonged submersion)	–	–	–	–	–	–	–	–	–	X

* These enclosures may be ventilated.

** External operating mechanisms are not required to be operable when the enclosure is ice covered.

*** External operating mechanisms are operable when the enclosure is ice covered.

Hazardous Location NEMA Enclosure Types

Note that all equipment designed for use in hazardous locations must be certified by a nationally recognized testing laboratory, such as Underwriters Laboratories (UL). In addition to the NEMA type, look for the appropriate hazardous location equipment markings.

- **Type 7** – Explosion-proof, indoor use. Class I, Division 1 hazardous locations, Groups A, B, C, and D.
- **Type 8** – Explosion-proof, indoor or outdoor use. Class I, Division 1 hazardous locations, Groups A, B, C, and D.
- **Type 9** – Dust ignition proof, indoor use. Class II, Division 1 hazardous locations, Groups E, F, and G.
- **Type 10** – MSHA. Meets the requirements of the Mine Safety and Health Administration, 30 CFR Part 18.

Table 29-3 – Hazardous Locations

From NEMA 250-2003 – Comparison of Specific Applications of Enclosures

Note: NEMA Types 7 and 9 are for indoor use only; NEMA Type 8 is for indoor and outdoor use.

Provides a Degree of Protection Against Atmospheres Typically Containing		Enclosure Types 7 and 8, Class I Groups**				Enclosure Type 9, Class II Groups			
(See NFPA 497M for Complete Listing)	Class	A	B	C	D	E	F	G	10
Acetylene	I	X	–	–	–	–	–	–	–
Hydrogen, manufactured gas	I	–	X	–	–	–	–	–	–
Diethyl ether, ethylene, cyclopropane	I	–	–	X	–	–	–	–	–
Gasoline, hexane, butane, naphtha, propane, acetone, toluene, isoprene	I	–	–	–	X	–	–	–	–
Metal dust	II	–	–	–	–	X	–	–	–
Carbon black, coal dust, coke dust	II	–	–	–	–	–	X	–	–
Flour, starch, grain dust	II	–	–	–	–	–	–	X	–
Fibers, flyings*	III	–	–	–	–	–	–	X	–
Methane with or without coal dust	MSHA	–	–	–	–	–	–	–	X

* For Class III type ignitable fibers or combustible flyings, see NEC Article 500.

** Due to the characteristics of the gas, vapor, or dust, a product suitable for one class or group may not be suitable for another class or group unless marked on the product.

Temperature Rise Calculation

First calculate the surface area of the enclosure (*exclude displays and human machine interfaces—HMIs*) and, from the expected heat load and the surface area, determine the heat input power in W/ft^2. Then the expected temperature rise can be read from the Sealed Enclosure Temperature Rise graph. The enclosure can be cooled by a fan, air conditioner, or vortex cooler.

Example:

What temperature rise can be expected from a 48 in by 36 in by 16 in painted steel enclosure with 300 W of heat dissipated within it?

Surface area = 2 [(48 • 36) + (48 • 16) + (36 • 16)] / 144 = 42.67 ft^2

Input power = 300 / 42.67 = 7.03 W/ft^2

From the heat transfer graph below:

Temperature rise = approximately 30°F (16.7°C)

Note: If outside temperature = 100°F, inside temperature = 130°F.

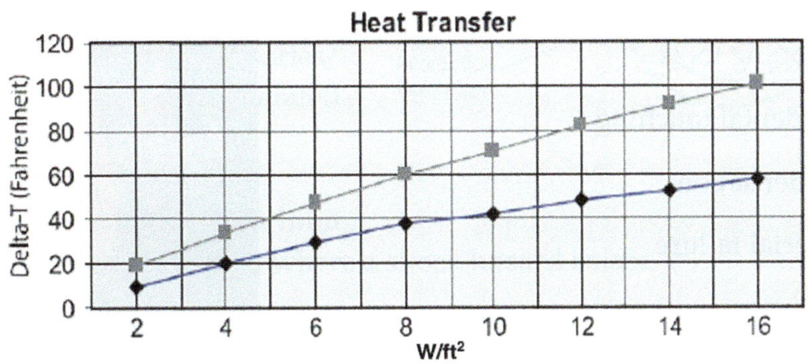

Prohibited Approach (inner boundary) – A distance from an exposed part that is considered the same as making contact with the live part.

This distance is not common between equipment. Some equipment will have a greater flash protection boundary, while other equipment will have a lesser boundary.

Arc Flash Analysis

According to NFPA 70E, what are the available methods for performing an arc flash hazard analysis? There are two methods that the technical committee to NFPA 70E considers applicable (i.e., equally valid). They are the incident energy (calculation) method and the hazard/risk category (table) method.

Required Arc Flash Warning Label

What information is required to be included on an arc flash boundary hazard warning label?

Items 1 a, b, and c apply when an incident energy analysis is performed. Item 1 d applies when the hazard/risk category method is used. In accordance with 130.5(C), all the following information should be included:

1. At least one of the following:
   a. Available incident energy in cal/cm^2 and the working distance
   b. Minimum arc rating of clothing
   c. Required level of PPE
   d. Highest hazard/risk category (HRC) for the equipment
2. Applicable nominal system voltage
3. Applicable arc flash boundary

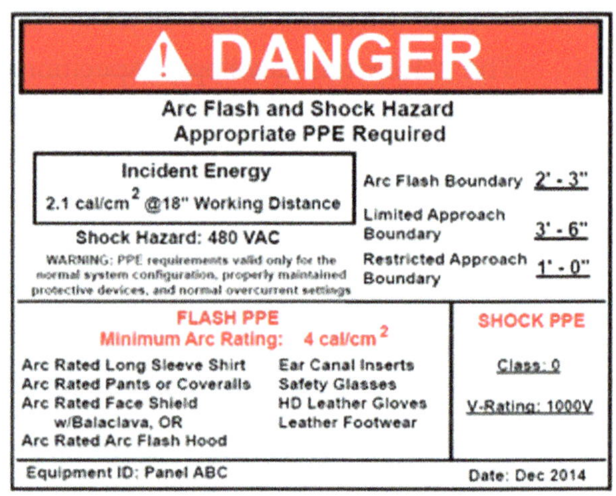

NFPA 77 Static Electricity

The buildup of static electricity in flowing applications is a major concern. It is important that proper grounding be implemented to protect personnel from shock and possible explosions due to sparks.

NFPA 30, *Flammable and Combustible Liquids Code,* is required by federal regulation 29 CFR 1910.106. NFPA 30 refers the user to NFPA 77 under these articles: A.4.3.2.5.4, A.4.5.3.4, A.7.6.10.4, A.7.6.11.2, and A.7.9.4

NFPA 30 A.7.9.4 The prevention of electrostatic ignition in equipment is a complex subject. Refer to NFPA 77, *Recommended Practice on Static Electricity,* for guidance.

NFPA 77 covers proper grounding techniques for loading stations, where these hazards may occur.

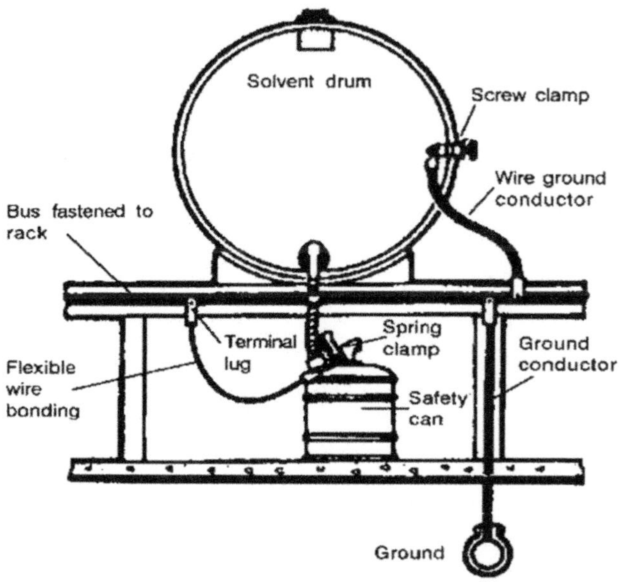

Proper protection grounding.

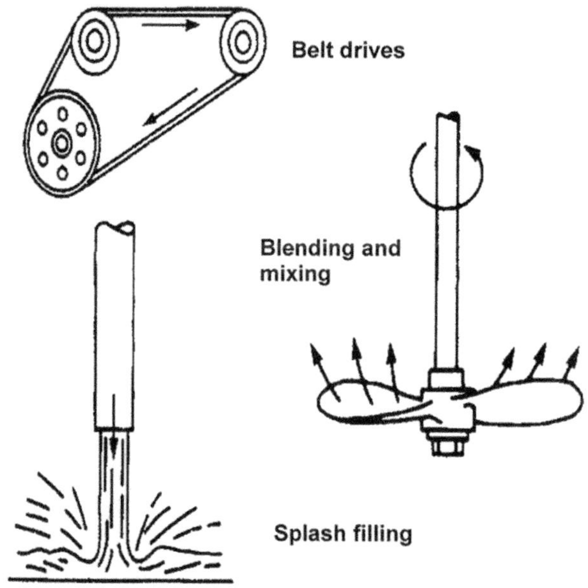

Static electric generators.

Important Articles of NFPA 77

1.2 Purpose

The purpose of this recommended practice is to assist the user in controlling the hazards associated with the generation, accumulation, and discharge of static electricity by providing the following:

- Basic understanding of the nature of static electricity
- Guidelines for identifying and assessing the hazards of static electricity
- Techniques for controlling the hazards of static electricity
- Guidelines for controlling static electricity in selected industrial applications

8.1 General Overview

This chapter discusses the assessment and control of static electricity hazards involved with the storage, handling, and use of flammable and combustible liquids and their vapors and mists. While focused on flammable and combustible liquids, the principles of this chapter also apply to noncombustible liquids and vapors (e.g., wet steam) where their storage, use, and handling can cause a static electricity ignition hazard. The chapter begins with a discussion of the combustion characteristics of liquids and their vapors and mists, followed by a discussion of charge generation and dissipation in liquids.

Emphasis is then given to processes involving the following:

- Flow in pipe, hose, and tubing
- Storage tanks

- Loading of tank vehicles
- Vacuum trucks
- Railroad tank cars
- Marine vessel and barge cargo tanks
- Process vessels
- Gauging and sampling
- Tank cleaning
- Portable tanks and containers
- Vacuum cleaning

8.3.1 Charge Generation

Charge separation occurs where liquids flow through pipes, hose, and filters; where splashing occurs during transfer operations; or where liquids are stirred or agitated. The greater the area of the interface between the liquid and the surfaces and the higher the flow velocity, the greater is the rate of charging. The charges become mixed with the liquid and are carried to receiving vessels, where they can accumulate. The charge is often characterized by its bulk charge density and its flow as a streaming current to the vessel (see the figure below.)

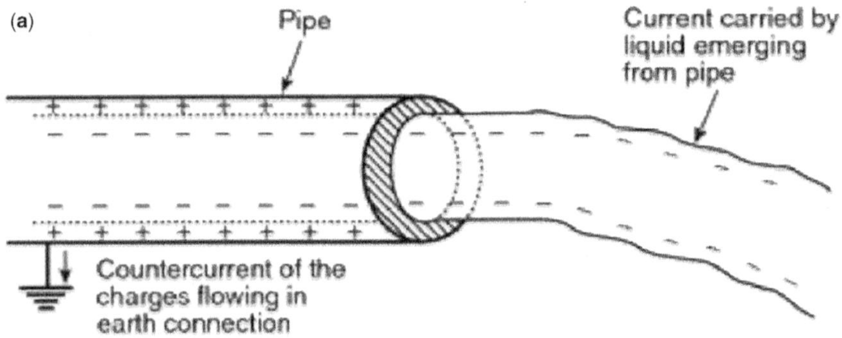

Flow (pipe charging)

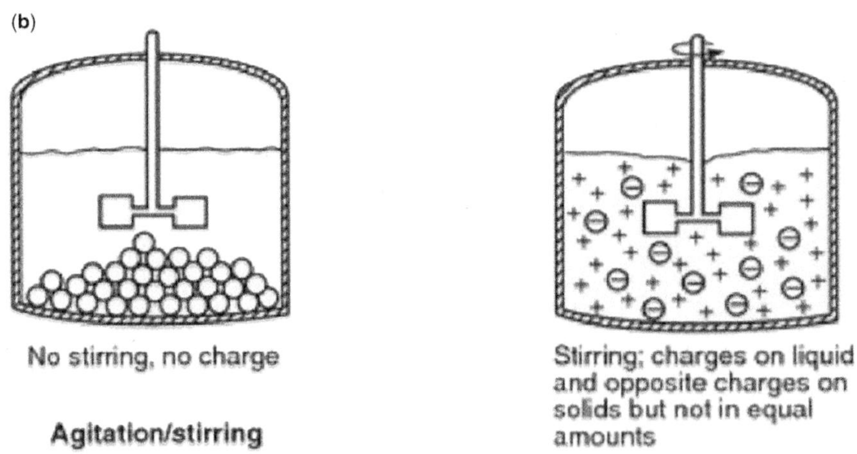

Agitation/stirring

NFPA 780 Lightning Protection (Formerly NFPA 78)

This may not be mentioned on the CSE exam, but you should be familiar with surge protection devices and applications for plant electrical systems and equipment. A lightning strike can generate up to 300,000 V and shoot through a concrete wall 2 ft thick. A direct lightning strike can cause an enormous amount of physical damage. Lightning strikes that hit equipment and storage or process vessels containing flammable materials can cause devastating accidents at refineries, bulk plants, processing sites, and other facilities.

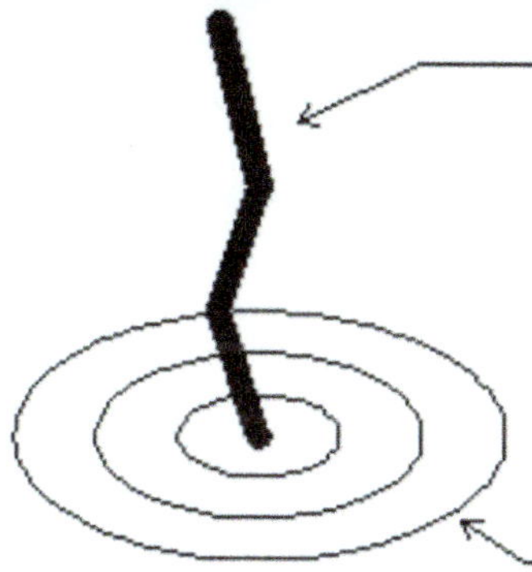

However, the indirect effects from a nearby strike can also cause damage by inducing voltage surges onto electrical system main lines, feeders, and data cables. Lightning-induced voltage surges are often described as a "secondary effect" of lightning, and there are three recognized means by which these surges are induced in electrical system main lines, feeders, or data/telecommunications cables:

- Resistive coupling
- Inductive coupling
- Capacitive coupling

NFPA 780 covers proper grounding techniques for lightning protection. Lightning surge arrestors and lightning protection equipment should be used to protect the process control system and ensure it continues to function correctly.

NFPA 780 and NFPA 70 (NEC)

The NEC does not require a lightning protection system to be installed. However, if lightning protection is installed, NEC article 250.60 and the fine print notes (FPNs), in addition to NEC article 250.106 and associated FPNs, refer users to NFPA 780. Some state, county, or city government (AHJ) and/or facility owners usually insist on installing a lightning protection system. In Article 100 of the 2014 NEC, the term *authority having jurisdiction* is defined as "an organization, office, or individual responsible for enforcing the requirements of a code or standard, or for approving equipment, materials, an installation, or a procedure."

Strike-Termination Devices

A lightning protection system consists of five main components: air strike terminations on the roof, roof conductors and down conductors, connecting means of conductors to devices and electrodes, earth-grounding electrodes, and surge protection systems, where needed. When installed correctly, these components will provide a safe, continuous path to ground for dissipating lightning current.

Starting at the roof and based on the type of roof, the number of strike-termination devices can be determined. In Figure 4.1.2 of NFPA 780, the roof types are described and the minimum number of devices are recommended. The following roof styles and roof-based supplementary items should be considered: flat or gently sloping roofs; dormers; domed roofs; and roofs with ridges, wells, chimneys, or vents. Most standards strongly recommend that these striking devices be sized and positioned in accordance with the design concept and theory of a 150-ft-radius that is known in the industry as a *rolling ball* (sphere).

Connecting Conductors to Electrodes

In many installations, properly sized lightning system conductors are used to connect strike-termination devices on the roof to the earth-grounding electrodes. However, it is not unusual to connect the strike terminals to the structural steel columns or to concrete-reinforcing rebar. These metal items are then used instead of conductors to provide the low-impedance path for dissipating the lightning current to the earth electrodes. Designers and installers must remember that strike-termination devices can be constructed using air terminals, metal masts, and permanent metal parts of structures and overhead ground wires as well as any combination of such.

Routing Down Conductors

To ensure a lightning strike current is safely carried to the earth-grounding electrodes, the conductors must be run without shape bends and installed in a horizontal or sloping-downward direction. They must be sized at least as large as required by Tables 4.1.1.1(A) and (B) of NFPA 780 based on the minimum class of material used for construction purposes. The NEC and NFPA 780 require the appropriate bonding of the down conductors and structural steel and other metal bodies to prevent side flashes.

Conductor and Electrode Connection

The down conductors must be attached permanently to the earth-grounding electrodes by bolding, brazing, welding or high-compression connectors listed for the purpose, and clamps that are suitable for direct burial. If these connections are not performed efficiently, the integrity of the lightning protection system is jeopardized. When selecting connectors and fittings, it is important that they be compatible for terminating the conductors to the devices and electrodes and for meeting the condition in which they are used.

Earth-Grounding Electrodes

Down conductors must be connected to earth-grounding electrodes that are dedicated to the lightning protection system. In other words, the grounding-electrode system for the service equipment for the facility, as well as that used for grounding telecommunications and cable television systems, must not be used in lieu of lightning protection electrodes.

Electrodes that are suitable for lightning protection systems are ground rods, concrete-encased electrodes, ground-ring electrodes, radial electrodes, or plate electrodes. Installing multiple rods or driving longer rods deeper into the soil of the earth can decrease resistance; long grounding conductors buried in the earth can also be used. Note that other methods—such as salt, chemical, and other commercial type techniques—are available and can be used to obtain a lower earth (soil) resistance.

Summary of Lightning Protection Components

Air Terminal Height

The tip of an air terminal shall be not less than 254 mm (10 in) above the object or area it is to protect.

Conductor Bends

No bend of a conductor shall form an included angle of less than 90°, nor shall it have a radius of bend less than 203 mm (8 in), as shown in the figure below.

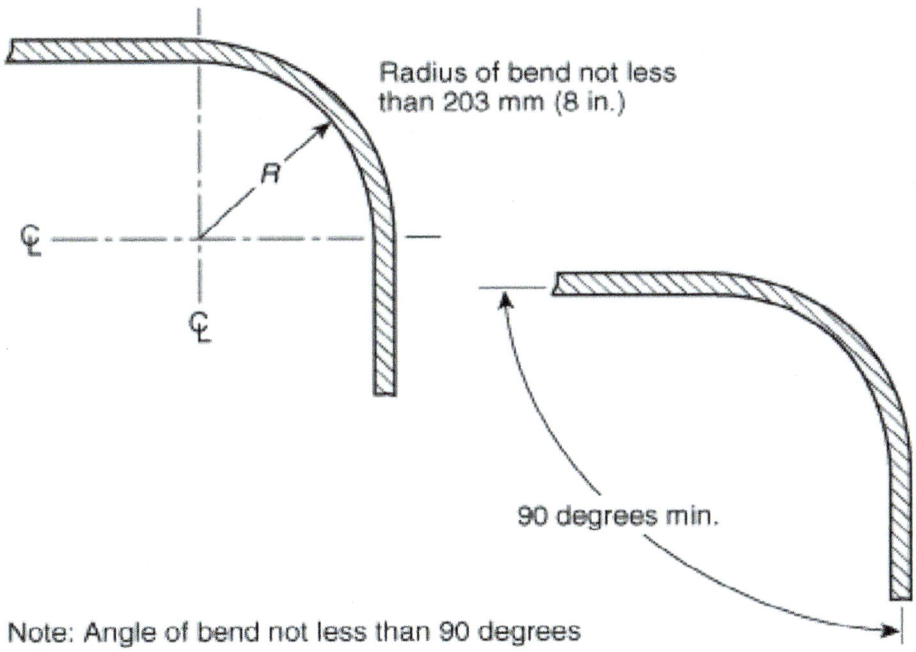

Conductor Size and Material for Lightning Protection

4.1.1.1 Ordinary structures shall be protected according to 4.1.1.1(A) or 4.1.1.1(B).

Ordinary structures not exceeding 23 m (75 ft) in height shall be protected with Class I materials as shown in Table 4.1.1.1(A).

Table 4.1.1.1(A) Minimum Class I Material Requirements

Minimum Class I Material Requirements			
Type of Conductor	**Parameter**	**SI**	**Copper US**
Air terminal, solid	Diameter	9.5 mm	3/8 in
Air terminal, tubular	Diameter Wall thickness	15.9 mm 0.8 mm	5/8 in 0.033 in
Main conductor, cable	Size each strand Weight per length Cross-section area	278 g/m 29 mm^2	17 AWG 187 lb/1000 ft 57,400 circular mils
Bonding conductor, cable (solid or stranded)	Size each strand Cross-section area		17 AWG 26,240 circular mils
Bonding conductor, solid strip	Thickness Width	1.30 mm 12.7 mm	1/2 in
Main conductor, solid strip	Thickness Cross-section area	1.30 mm 29 mm^2	0.051 in 57,400 circular mils

Neither NFPA 780 nor NFPA 70 implies that lightning strikes can be prevented, but injury to personnel and damage to facilities and installation materials can certainly be reduced with a properly designed and installed lightning protection system.

Transient Protection from Lightning Strikes

Definitions Used in Transient and Surge Protection

- Suppressed voltage rating (SVR)
- Surge protective devices (SPDs)
- Transient voltage surge suppressor (TVSS)

Transient Protection Is Required by NFPA 780

According to NFPA 780 A.4.18.2.5: "Most services to facilities will require discrete surge suppression devices installed to protect against damaging surges. Occasionally, services will be located in an area or manner where the threat from lightning-induced surges and overvoltage transients may be negligible. For example, the requirements in 4.18.2.3 (also see A.4.18.6.1) exempt services less than 30 m (100 ft) in length that are run in grounded metal conduit between buildings requiring surge protection. These are examples of acceptable exceptions where SPDs may not be required on each service entrance. The standard recognizes that there can be acceptable exceptions and consequently allows for such exceptions to the requirements for surge suppression on electrical utility, data, and other signal lines, provided a competent engineering authority has determined that the threat is negligible or that the system is protected in a manner equivalent to surge suppression."

Allowance for the exemption of surge suppression at specific locations in this standard is not intended as a means to provide a broad exemption simply because surge suppression may be considered inconvenient to install. Rather, it recognizes that all possible circumstances and configurations, particularly those in specialized industries, cannot be covered by this standard.

Determinations made by an engineering authority for exempting installation of SPDs should focus on the likelihood of lightning activity in the region, the level of damage that may be incurred, and the potential loss to human life or essential services due to inadequate overvoltage protection.

NFPA 79 Industrial Machinery

The wire sizing and color codes for wires and buttons are covered in NFPA 79 *industrial machinery*.

Conductor Sizing

Conductors shall not be smaller than:

- Power circuits.. #14 AWG
- Lighting and control circuits on machine and in raceways .... #16 AWG
- Exception: in jacket multiconductor cable assembly......................... #18 AWG
- Control circuits .... #18 AWG
- Electronic, control conductors in raceways....................................... #24 AWG

Conductor Colors

- **Black** – Line, load, and control circuits at line voltage
- **Red** – AC control circuits less than line voltage

- **Blue** – DC control circuits
- **Yellow** – Interlock control circuits powered from external power supply
- **Green** – Equipment ground conductor where insulated or covered

Pushbutton Functions for Color

- **Red** – Emergency stop, stop, off
- **Yellow** – Return, emergency return, intervention—suppress abnormal conditions
- **Green** – Start, on
- **Black** – No specified function assigned
- **White** – Any function not covered above
- **Clear** – Any function not covered above
- **Blue**– Any function not covered above
- **Grey** – Any function not covered above

NFPA 79 Table 10.3.2. Colors for Machine Indicator Lights and Icons

NFPA 79 Code Excerpt Article 10.3.2	Color	Purposes		
		Safety of Persons or Equipment	Condition of Process	State of Equipment
Indicator lights and icons on a color graphic interface device shall be color coded with respect to the condition (status) of the machine in accordance with Table 10.3.2. Alternate purposes shall be permitted to indicate machine or process status.	**Red**	Danger	Emergency	Faulty
	Yellow/(amber)	Warning/Caution	Abnormal	Abnormal
	Green	Safe	Normal	Normal
	Blue	Mandatory Action		
	Clear, White, Grey, Black	No Specific Meaning Assigned		

NFPA 496 Purged and Pressurized Systems

Questions related to this standard may be included on the CSE exam. This standard shall apply to all purged and pressurized enclosures. The standard's intent is to provide information on the methods for purging and pressurizing enclosures to prevent ignition of flammable atmospheres.

Refineries and similar facilities often have electrical control equipment and instrumentation located in rooms that are within or near Class I hazardous locations. If you install a purging and pressurizing system in these rooms, the NEC allows you to install general-purpose equipment (intended for use in an unclassified location) in such locations.

Purging for Class I Hazardous Locations (NEC/NFPA):

- **Type X purging** – Reduces the classification from Division 1 to nonhazardous
- **Type Y purging** – Reduces the classification from Division 1 to Division 2
- **Type Z purging** – Reduces the classification from Division 2 to nonhazardous

Note: At least *four volumes* of purge gas must pass through the enclosure, while maintaining a minimum pressure of *0.1 in of water,* before operation of the equipment inside. A *minimum* of 0.1 in of water pressure must be maintained in the enclosure when operating.

A warning label shall be mounted on the enclosure. On Type Y and Type Z purge failure, an alarm or pressure switch can be used to remove power from the enclosure. With Type X purge, this power must be removed with an explosion-proof switch.

Overview of the NFPA 496 Articles

Refineries and similar facilities often have electrical control equipment and instrumentation located in rooms that are within or near Class I hazardous locations. If you install a purging and pressurizing system in these rooms, the NEC allows you to install general-purpose equipment (intended for use in an unclassified location) in such locations.

NEC Section 500-4 suggests you can reduce, limit, or even eliminate hazards by adequate positive pressure ventilation from a source of clear air, coupled with effective safeguards against ventilation failure. The FPN in Section 500-4 in the NEC refers to NFPA 496 (*Purged and Pressurized Enclosures for Electrical Equipment*) for requirements pertaining to the design of a purged and pressurized room.

Requirements included in Chapter 5, NFPA 496, provide guidelines for preventing the entry of flammable vapors or gases into the room housing electrical-related equipment.

Factors to Consider (NFPA 496, Section 5-3)

As a designer or installer, you must consider many factors in the design and layout of the control room. First, the number of people in the room is important when calculating the volume of air required and the access requirements. An appendix in NFPA 496 states that a control room located in a hazardous (classified) location should have as few doors as possible so you can maintain positive pressure within the room, while maintaining the need for egress of personnel per Appendix A-5-4.1.

A control room typically contains data processing; communications; heating, ventilation, and air conditioning (HVAC); lighting; power; and electrical equipment, as well as process control instruments and panels. It is the designer and installer's job to understand the varied needs for protection from flammable atmospheres. The volume of air introduced must satisfy the need for cooling the electrical equipment and preventing heating problems as well.

Location of the Control Room (NFPA 496, Sections 5-3.1(c) and 5-3.2)

You must also consider the location of the control room in relation to the source of flammable gases or vapors. Pay particular attention to the direction of the prevailing wind. One side of the room may face a location generally free from trace amounts of flammable vapors or gases, or the height of the fan intake may be sufficient to provide a clean source of air. If you need ducting to reach an uncontaminated source, it must be noncombustible material, free of leaks, and protected against damage or corrosion.

Positive Pressure Air Systems (NFPA 496, Section 5-4.1)

You must maintain positive pressure of at least 0.1 in wc (25 Pa) in the control room with all openings closed. Sensitive pressure switches and other devices are available to measure these low values. The code permits this minimum air pressure to drop to a lower level when doors and other apertures remain open, if a minimum air velocity of 60 ft/min (0.3 m/s) is maintained through the openings.

Type X Equipment (NFPA 496, Section 5-4.4)

If you locate a control room in a Class I, Division 1 (or Zone 1) location, containing equipment that can only function safely in an unclassified location, you must use a Type X purging system. Type X purging systems reduce the hazards from Division I (or Zone 1) to unclassified. You must cut off power immediately when the positive pressure air system fails. You also must detect failure of the system at the discharge end of the fan.

The code does not consider an electrical interlock that indicates when the pressurizing fan motor is running to be reliable for this purpose because of the possibility of a broken belt or other equipment failure. The sensing device must start an audible or visual alarm located in a constantly attended position.

You must take off the electrical power circuit for the positive pressure air system equipment ahead of any service disconnects feeding the control room. The airflow-monitoring switch, electrical disconnect, and motor for the air system fan must be suitable for the area in question (as it would be classified if there was no positive ventilation system). This provision allows for the re-pressurization of the room after the air system fails.

One method to determine the degree of safety for such a situation involves the use of combustible gas detectors. You can use these detectors to be sure the atmosphere around the electrical equipment is nonflammable. As an alternative, you could also employ a purge timer to prevent reapplying power too soon after the pressurizing air system restarts. The period should be sufficient to allow at least four air changes within the room.

Type Y Equipment (NFPA 496, Section 5-4.5)

Type Y purging systems reduce the classification within a room from Division 1 (or Zone 1) to Division 2 (or Zone 2). If the control room location and/or equipment is suitable for Type Y purges, then it is not necessary to de-energize the power supply circuit to the control room equipment immediately upon a positive pressure air system failure. However, for safety's sake, you should de-energize the equipment as soon as possible after you detect air failure, or begin some means of monitoring the atmosphere within the room.

Type Z Equipment (NFPA 496, Section 5-4.5)

Type Z purging systems reduce the classification from Division 2 (Zone 2) to unclassified. The design conditions and requirements are the same as for Type Y equipment.

Examples of Purged and Pressurized Systems

Basic design of purged enclosures.

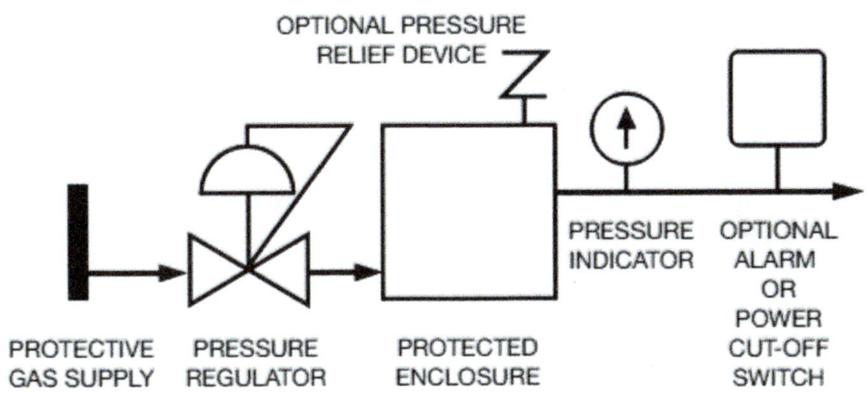

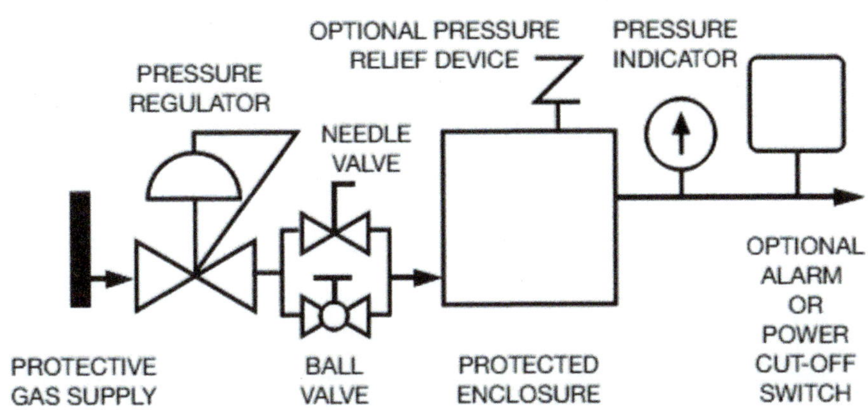

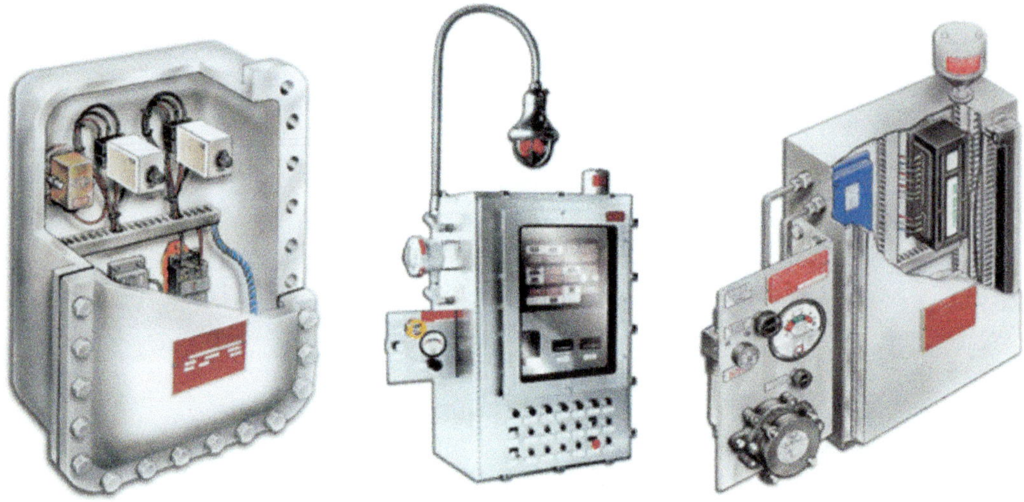

Standard explosion-proof enclosure.
Example of purged electrical enclosures (designed per NFPA 496).

40 CFR and the Environmental Protection Agency – Leak Detection and Repair

Title 40 is a part of the US Code of Federal Regulations. Title 40 arranges mainly environmental regulations that were promulgated by the US Environmental Protection Agency (EPA), based on the provisions of US laws (statutes of the US Federal Code). Parts of the regulation may be updated annually on July 1.

The Clean Air Act

The Clean Air Act (CAA) requires refineries to develop and implement a leak detection and repair (LDAR) program to control fugitive emissions. Fugitive emissions occur from valves, pumps, compressors, pressure-relief valves, flanges, connectors, and other piping components. Comparison monitoring conducted by the EPA National Enforcement Investigation Center (NEIC) shows that the number of leaking valves and components is up to 10 times greater than had been reported by certain refineries. The EPA believes this great disparity between what refineries are reporting and what the EPA is finding may be attributable to refineries not monitoring in the manner prescribed in 40 CFR Part 60, Appendix A, Method 21.

Federal regulations require refiners to routinely monitor for leaks and to fix any equipment found leaking. Failure to identify leaking equipment results in necessary repairs not being made and continuing fugitive emissions of volatile organic compounds (VOCs) and other hazardous chemicals. The EPA estimates that the failure to identify and repair leaks at petroleum refineries could be resulting in additional VOC emissions of 80 million pounds annually. VOCs contribute to ground-level ozone, a principal component of smog, which can cause significant health and environmental problems.

What the Law Requires

Specific requirements for refinery fugitive emissions are identified in 40 CFR Part 60, *New Source Performance Standards* (NSPS), and 40 CFR Parts 61 and 63, *National Emission Standards for Hazardous Air Pollutants* (NESHAP). Many state and local air agencies incorporate federal requirements, but some have established more stringent requirements as authorized by law. The various regulations require refineries to implement an LDAR program to reduce fugitive emissions from valves, pumps, compressors, pressure-relief valves, flanges, connectors, and other piping components.

Valves are usually the single largest source of fugitive emissions. Emissions from any single piece of equipment are usually small. Based on the large number of equipment components that can leak and are subject to LDAR requirements, however, cumulative emissions can be very large. To obtain a proper reading of emissions from leaking components, the monitoring equipment must be calibrated correctly and held at the component interface where leakage could occur (e.g., at the seal between the valve stem and housing) for a sufficient length of time to obtain a valid measurement.

For example, if a closed venting system is being used to reduce emissions and to comply with the requirements of the standard, it must be capable of recovering (condensers, absorbers, etc.) or reducing (flares, enclosed combustion devices, etc.) the VOC emissions at some defined level of efficiency. The closed systems must be monitored to verify that they meet or exceed all the requirements of their design.

Any leaks, as indicated by an instrument reading greater than *500 parts per million (ppm)* above background, must be repaired as soon as practicable. The first attempt at repair must be made no later than 5 calendar days after the leak is detected as per federal regulation 40 CFR 60.482-10.

Note: This may be on the CSE exam: Valves will not be allowed to leak at gasket seals more than 500 ppm!

See the "EPA Regulations" section in Chapter 14 for more detailed information on the 40 CFR pressure-relief valve and rupture disk requirements.

30 Equations for Pumping, Piping, and Sizing Motors

Find Pipe Diameter with Velocity of Flow Known

$$ID\,(\text{in}) = \sqrt{\frac{\text{gpm} \bullet 0.4085}{velocity\,(\text{ft/s})}}$$

$$ID\,(\text{in}) = \sqrt{\frac{\text{acfm} \bullet 3.0558}{velocity\,(\text{ft/s})}}$$

Find Flow Velocity with Pipe Diameter Known

$$velocity\,(\text{ft/s}) = \frac{\text{gpm} \bullet 0.4085}{ID^2\,(\text{in})}$$

$$velocity\,(\text{ft/s}) = \frac{\text{acfm} \bullet 3.0558}{ID^2\,(\text{in})}$$

Change acfm to scfm with Standard Temperature and Pressure Correction

$$scfm = acfm\left(\frac{14.7 + psig}{14.7}\right)\left(\frac{520}{460 + T_{deg\ F}}\right) = acfm\left(\frac{P_f}{P_s}\right)\left(\frac{T_s}{T_f}\right)$$

Note: scfh equals scfm times 60 (acfm equals scfm when both are at 60°F and 14.696 psia)

Change scfm to acfm with Standard Temperature and Pressure Correction

$$acfm = scfm\left(\frac{14.7}{14.7 + \text{psig}}\right)\left(\frac{460 + T_{\text{deg F}}}{520}\right)$$

where

T = temperature

Find the Reynolds Number for the Flow

$$R_e = \frac{3160 \bullet flow\ rate\,(\text{gpm}) \bullet SG}{Pipe\ ID\,(\text{in}) \bullet Viscosity\,(cP)}$$ **Note:** for liquids

Centipoise equals centistokes times specific gravity where specific gravity is assumed to be 0.8 (except for water). To find the exact cp of your fluid: $cp = cSt \bullet$ (weight per gallon • 0.120).

$$R_e = \frac{7740 \bullet Velocity\,(\text{ft/s}) \bullet Pipe\ ID\,(\text{in})}{Viscosity\,(\text{cSt})} \quad \textbf{Note: } \text{for liquids}$$

$$R_e = \frac{6.316 \bullet Flow\ Rate\,(\text{lb/h})}{Pipe\ ID\,(\text{in}) \bullet Viscosity\,(\text{cSt})} \quad \textbf{Note: } \text{for gases and steam}$$

Calculate the Piping Head Losses to Size a Control Valve

Calculate the pump head needed for a piping system.

Note: Feet of head in system (h_{sys}) equals height plus head loss in the pipe plus head loss in the fittings plus head across the meter element plus head across the control valve plus head for pressure in the vessel.

The head for the pump is all in feet of head. This head must be in feet of head at the maximum operating flow rate.

$$h_{sys} = h_{elevation} + h_{pipe} + h_{fitting} + h_{meter} + h_{valve} + h_{vessel}$$

Head Losses for Piping System (h_{pipe})

The Darcy–Weisbach equation for piping head loss in feet of head loss across the piping system is:

$$h_{pipe} = f\left(\frac{Length\,(\text{ft})}{Pipe\ ID\,(\text{ft})}\right) \bullet \frac{V^2\,(\text{ft/s})}{2 \bullet g} = f\left(\frac{Length\,(\text{ft}) \bullet 12}{Pipe\ ID\,(\text{in})}\right) \bullet \frac{V^2\,(\text{ft/s})}{64.34}$$

$$\frac{Moody\ friction\ factor\ formula\ for\ Darcy\text{-}Weisbach\ equation}{Note\text{: } e = 0.00015\ for\ steel\ pipes}$$

$$f = 0.0055 + 0.0055\left[20{,}000\left(\frac{e \bullet 12}{Pipe\ ID\,(\text{in})}\right) + \frac{10^6}{Re}\right]^{\frac{1}{3}}$$

where

h_{pipe} = head drop across pipe system
f = friction factor of pipe
V = velocity in feet per second
e = relative roughness of pipe

Head Losses for Hand Valves and Fittings ($h_{fitting}$)

Use this formula to obtain head loss of fittings and hand valves. Then add this head loss to the piping head loss obtained by the Darcy–Weisbach equation for the total piping system head loss as:

$$h_{fitting} = K \bullet \frac{V^2\,(\text{ft/s})}{2 * g} = K \bullet \frac{V^2\,(\text{ft/s})}{64.34}$$

where K is computed as: **Note:** Le and D are in feet, so D (ft) = d (in)/12.

$$K = f \bullet (Le/D)$$

L_e = equivalent length (length of pipe with same resistance as the fitting/valve)
D = pipe diameter in feet
f = friction factor

See Appendix Table A28, "Head Loss in Piping Systems" for K factors for head loss in various piping fittings.

Calculating the Motor Horsepower of Process Pumps

Note: Feet of head in system (h_{sys}) equals height plus head loss in the pipe plus head loss in the fittings plus head across the meter element plus head across the control valve plus head for pressure in the vessel.

Once the total dynamic head is found, the power required to operate the pump can be determined. The *theoretical horsepower* or *brake horsepower* is defined as:

$$\text{hp} = \frac{h_{sys}(\text{feet of head in the system}) \bullet \text{gpm} \bullet 8.33 \bullet SG}{33{,}000 \bullet \textit{efficiency of pump}}$$

OR

$$\text{hp} = \frac{h_{sys}(\text{feet of head in the system}) \bullet \text{gpm} \bullet SG}{3{,}960 \bullet \textit{efficiency of pump}}$$

where

$$1 \text{ hp} = 746 \text{ W}$$

Single-phase motor amps

$\text{A} = (\text{hp} \bullet 746)/(V \bullet PF \bullet EFF)$

Three-phase motor amps

$\text{A} = (\text{hp} \bullet 746)/(V \bullet \text{sqrt } 3 \bullet PF \bullet EFF)$

Calculating the Motor Horsepower of Hydraulic Fluid Pumps

This will be the size of the hydraulic system pump motor. Verify the actual size with the manufacturer. This is useful for sizing motor starters in the motor control center (MCC).

$$\text{hp} = \frac{\text{psi} \bullet \text{gpm}}{1714 \bullet \textit{efficiency of pump}}$$

Calculating the Motor Horsepower of Fans

$$\text{hp} = \frac{\text{cfm} \bullet \text{psf}}{33{,}000 \bullet \textit{efficiency of fan}}$$

$$\text{hp} = \frac{\text{cfm} \bullet \text{piw}}{6{,}356 \bullet \textit{efficiency of fan}}$$

$$\text{hp} = \frac{\text{cfm} \bullet \text{psi}}{229 \bullet \textit{efficiency of fan}}$$

where

cfm = cubic feet per minute
psf = pounds per square foot
piw = inches of water gauge
psi = pounds per square inch
Efficiency of fan = (%/100)

Calculating the Motor Horsepower of Conveyors

hp (vertical):

$$\text{hp} = \frac{F \bullet V}{33{,}000}$$

hp (horizontal):

$$\text{hp} = \frac{F \bullet V \bullet \text{coefficient of friction}}{33{,}000}$$

where

F = force (in pounds)
V = velocity (in feet per minute)

Coefficient of friction

Ball or roller slide = 0.02
Dovetail slide = 0.20
Hydrostatic ways = 0.01
Rectangle ways with Gib = 0.01 to 0.25

Piping Absolute Roughness Values

Absolute Roughness and C-Factors of Common Pipe Materials

Pipe Material Surface		Roughness e [ft]			C-Factor			
		Low	High	Design	Low	High	Clean	Design
Steel	Welded and seamless	0.0001	0.008	0.0002	80	150	140	100
	Corroded	0.0005	0.0133					
	Riveted	0.003	0.003				139	100
	Stainless			0.00005				
	Galvanized, plain	0.0002	0.0008	0.0005				
	Corrugated						60	60
Mineral	Concrete	0.001	0.01	0.004	85	152	120	100
	Cement	0.0013	0.004	0.003	140	160	150	140
	Vitrified clays							110
	Brick sewer							110
Iron	Cast, plain	0.0004	0.0027	0.00085	80	150	130	100
	Cast, tar (asphalt) coated	0.0002	0.0008	0.0004	50	145	130	100
	Cast, cement lined	0.000008		0.000008			150	140
	Cast, bituminous lined	0.000008		0.000008	130	160	148	140
	Cast, centrifugally spun	0.00001		0.00001				
	Wrought, plain	0.0001	0.0003	0.0002	80	150	130	100
Misc.	Fiber						150	140
	Copper and brass	0.000005	0.003	0.000005	120	150	140	130
	Wood stave	0.0006	0.003	0.002	110	145	120	110
	Transite	0.000008		0.000008				
	Lead, tin, glass			0.000005	120	150	140	130
	Plastic			0.000005	120	150	140	130

31 Calculating Volume in Tanks

The following calculations are for obtaining the volume of cylindrical and irregularly shaped tanks.

See Chapter 8, "Level Measurement and Calibration," for the level transmitter calibration.

This section is for information only. All units are in inches. This topic will not be on the CSE exam.

Note: All measurements for calculations are in units of inches. Diameter (D) = h in these examples.

(Hint: Multiply the tank diameter (h) by the percentage level signal to get y, and then calculate the formula.)

Cylindrical Tanks Upright

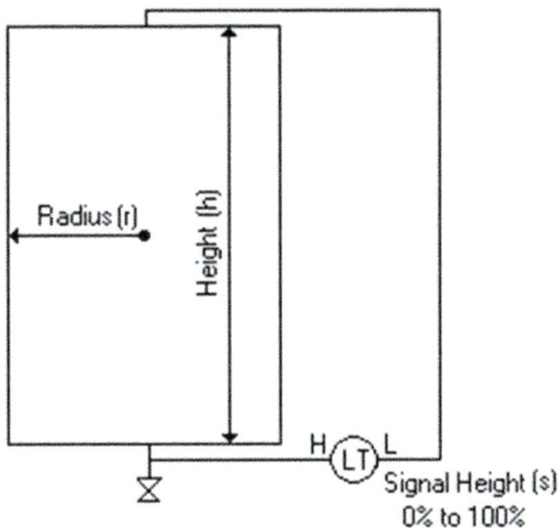

$$y = h \bullet s = height \bullet 0.00 \text{ to } 1.00 \text{ (\% of signal from transmitter)}$$

$$V_{gal} = (\pi \bullet r^2 \bullet y)/231 \text{ in}^3$$

where

y = height of process fluid in inches
h = tank height in inches
V = volume in gallons
r = radius of tank in inches
s = percent height of process variable

Cylindrical Tanks on Side

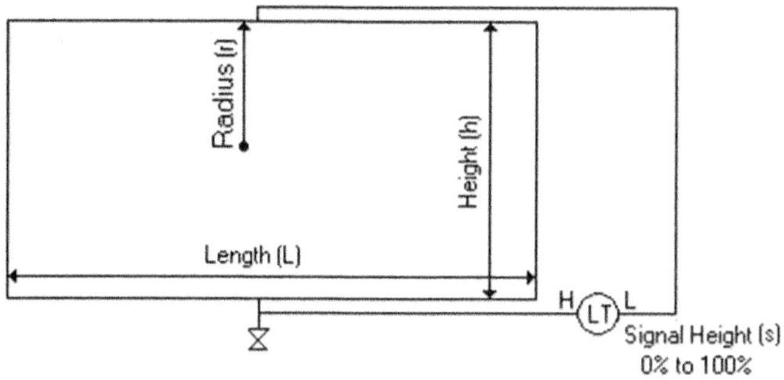

$$y = h \bullet s = height \bullet 0.00 \text{ to } 1.00 \text{ (\% of signal from transmitter)}$$

$$V_{gal} = length \bullet \left[r^2 \bullet \cos^{-1}\left(1 - \frac{y}{r}\right) + \sqrt{(2r - y)y} \bullet (y - r) \right] / 231 \text{ in}^3$$

Important Note: The ($\cos^{-1}$) or (arccos) or (acos) function must return radians, *not* degrees. All measurements for the calculations are in units of inches.

Spherical Tanks

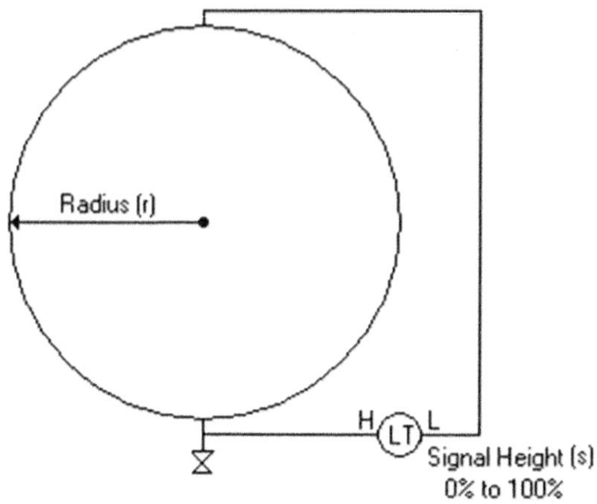

$$y = h \bullet s = height \bullet 0.00 \text{ to } 1.00 \text{ (\% of signal from transmitter)}$$

$$V_{gal} = \frac{1}{3}\pi(3r - y)y^2 / 231 \text{ in}^3$$

Bullet Tanks

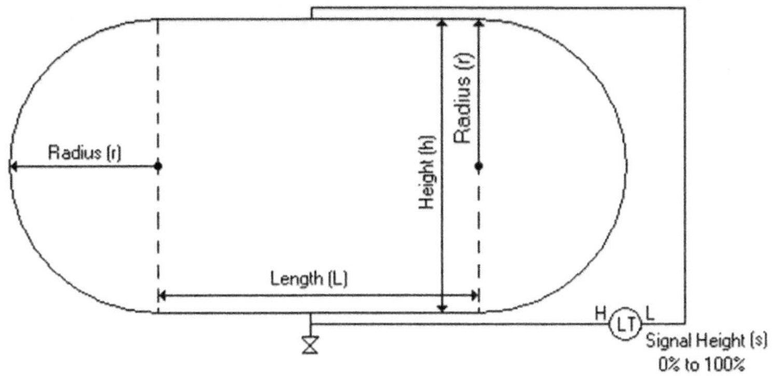

$$y = h \bullet s = \text{height} \bullet 0.00 \text{ to } 1.00 \text{ (\% of signal from transmitter)}$$

$$V_{\text{gal}} = \left\{ \left[\left[length \bullet \left\{ \cos^{-1}\left(1 - \frac{y}{r}\right) r^2 + \sqrt{(2r - y)y} \bullet (y - r) \right\} \right] + \left[\frac{1}{3} \pi (3r - y) y^2 \right] \right] \right\} / 231 \text{ in}^3$$

Note: All measurements are in inches. For volume in cubic feet (ft^3), divide V_{gal} by 7.4805. See Chapter 8, "Level Measurement and Calibration," in this manual for different end cap (tank heads) calculations.

Important Note: The ($\cos^{-1}$) or (arccos) or (acos) function must return radians, NOT degrees.

32 Exam Sample Questions

Sample Questions

1. At 433°F, a type J thermocouple with a 32°F reference junction (ice bath) will produce an output in millivolts (mV) that is most nearly equal to:

   A. 9.04 mV
   B. 10.51 mV
   C. 12.05 mV
   D. 17.79 mV

2. The flow of water in a 6-inch pipe is measured with an orifice plate and differential pressure transmitter. At a flow rate of 200 gpm, the differential pressure is 35 in of water. At a flow rate of 312 gpm, the differential pressure will be approximately equal to:

   A. 16.4 in wc
   B. 32.5 in wc
   C. 85.4 in wc
   D. 100 in wc

3. A tank level is measured using a differential pressure transmitter and a bubbler tube. The tank is vented to atmosphere. The bubbler tube is 1 ft from the bottom of the tank, and the tank wall is 20 ft high. A 0–10 psig differential pressure gauge, accurate to 0.25% of full scale, is connected to the bubbler tube connection at the high side of the transmitter. The low-pressure side is connected to the tank top. With the tank containing liquid with a specific gravity (SG) of 1.1 and the level in the tank at 16 ft, the gauge reading in pounds per square inch (psi) is most nearly equal to:

   A. 4.80
   B. 9.35
   C. 13.00
   D. 7.10

4. Which of the following practices is important in routing optic cable?

   A. Laying cable in trays with high-horsepower motor wiring should be avoided.
   B. Conduit fittings that require small radius bends should be avoided.
   C. Overhead runs on messenger wires should be limited to 75 ft.
   D. Underground fiber-optic runs must be covered with concrete.

5. Compared to a control loop with no dead time (pure time delay), a control loop with an appreciable dead time tends to require:

   A. Less proportional gain and less integral action
   B. More proportional gain and less integral action
   C. More proportional gain and more integral action
   D. Less proportional gain and more integral action

6. The definition and classification of hazardous areas for the purpose of wiring and electrical equipment is found in codes published by:

   A. National Fire Protection Association
   B. International Society of Automation
   C. Electric Power Research Institute
   D. Occupational Safety and Health Administration

7. Given the following data for liquid flow:
   Flow rate: 0–200 gpm
   Water at: 125°F and 75 psia
   Pipe size: 4-inch, schedule 40

   The orifice bore for a pressure differential range of 100 in of water is most nearly equal to:

   A. 2.33 in
   B. 3.50 in
   C. 1.50 in
   D. 0.75 in

8. A control valve is to be sized for the following conditions:
   Liquid flow: 50 gpm
   Specific gravity: 0.81
   Inlet pressure: 240 psig
   Delta pressure drop of across the valve: 10 psi

   The required flow coefficient (C_v) for the valve will most nearly be:

   A. 10.4
   B. 14.2
   C. 22.0
   D. 35.5

9. A control valve is to be sized for the following service conditions, saturated steam:
   Maximum flow rate: 30,000 lb/h
   P_1 (upstream pressure): 40 psia
   P_2 (downstream pressure): 30 psia

The required flow coefficient (C_v) for the valve will most nearly be:

A. 260
B. 540
C. 760
D. 198

10. The control algorithm for a flow control loop is under consideration. It is determined that the flow must be maintained near set point with little or no offset and that the signal will be rapid response and noisy. The best choice of control modes for this loop will be:

    A. Proportional mode
    B. Integral plus derivative
    C. Proportional plus integral
    D. Proportional plus integral plus derivative

11. According to ISA S5.1, *Instrumentation Symbols and Identification*, the terms *record* or *recording* can apply to which of the following?

    I Graphical data in a strip or circular chart
    II Table of numerical data in a computer memory
    III Listing of alarms by a control computer

    A. I and II
    B. II and III
    C. I and III
    D. I, II, and III

12. An orifice plate with an opening diameter of 2.324 in is to be used to measure the flow of water in a 4-inch, schedule 40 line. The flow rate is specified as 0–200 gpm at a pressure of 75 psia and a temperature of 125°F. What is the differential pressure in inches of water for the transmitter measurement across the primary element, the head in inches of water column most nearly equal to:

    A. 98 in wc
    B. 100 in wc
    C. 110 in wc
    D. 108 in wc

13. An SIL 1 interlock has a risk reduction factor (RRF) of 42.76. The target RRF is 75. How can you increase the RRF to meet or exceed the target RRF?

    A. Add more field sensors.
    B. Add dual solenoids to the one—and only one—block valve.
    C. Double the testing frequency.
    D. None of the above.

14. The plant has three pumps that fail in 7 years. What is the failure rate (FR) of the pumps in hours?

    A. $4.89 \bullet 10^{-5}$
    B. $4.29 \bullet 10^{-1}$
    C. $4.89 \bullet 10^{-4}$
    D. $1.14 \bullet 10^{-3}$

15. Which of the following types of valves has the highest gain when the valve is nearly closed?

    A. Quick-opening
    B. Equal percentage
    C. Fail open
    D. Linear

16. Which of the following types of control systems is normally programmed in ladder logic?

    A. Programmable logic controllers (PLCs)
    B. Distributed control systems (DCSs)
    C. Single-loop digital controllers
    D. Supervisory control and data acquisition (SCADA)

17. What is the resistance of 2000 ft of copper wire (specific resistance = 10.37) given a cross-sectional area of 10,370 cmil and a wire temperature of 20°C?

    A. 1
    B. 2
    C. 10
    D. 20

18. Which of the following protection techniques is acceptable for equipment located in a Class I, Division 1 area of an industrial facility?

    A. Explosion-proof apparatus and nonincendive equipment
    B. Explosion-proof apparatus and intrinsically safe equipment
    C. Dust ignition proof and nonincendive equipment
    D. Hermetically sealed and intrinsically safe equipment

19. To minimize electrical interference when AC power and DC signal wiring meet in a control panel, it is BEST to:

    A. Use a different size wire
    B. Cross the wires at 90°
    C. Run the wires parallel to each other
    D. Twist the AC wires around the DC wires

20. In Figure 32-1, if only the open flow area (X) of the feedwater control valve increased, which of the following best describes how the mass flow (F) would change?

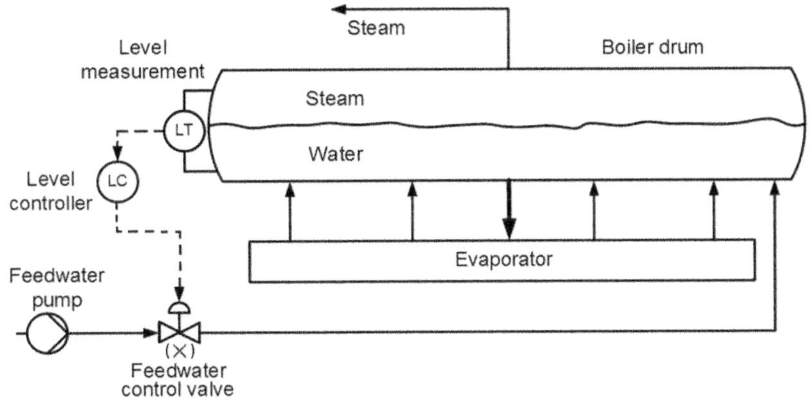

Figure 32-1

A. $F2 = F1(X1/X2)^{0.5}$
B. $F2 = F1(X2/X1)^{0.5}$
C. $F2 = F1(X2/X1)$
D. $F2 = F1(X2/X1)^2$

The following illustration is used for questions 21 through 25.

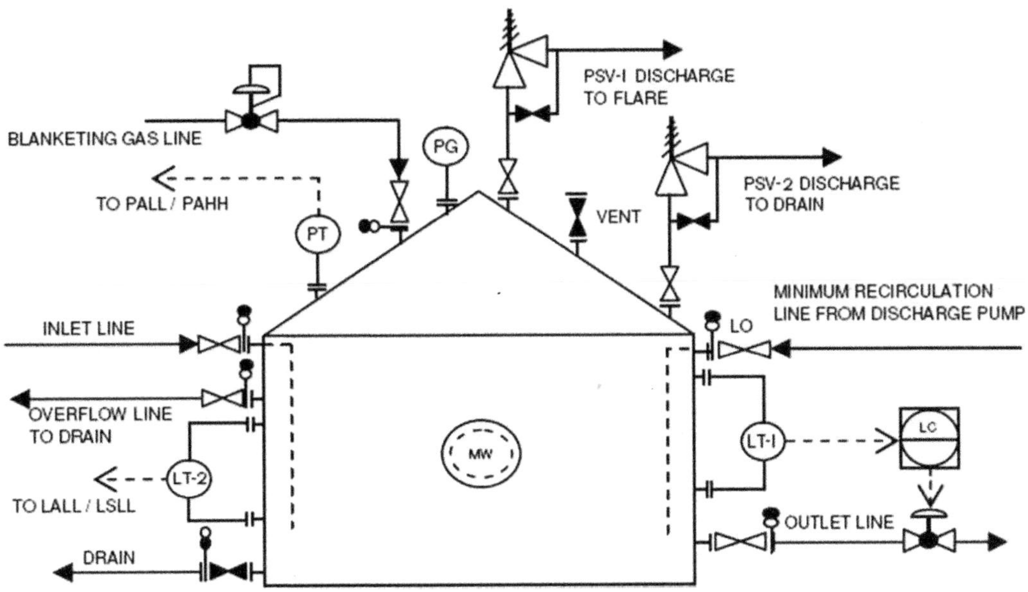

Figure 32-2

The following data is used for questions 21–25; see Figure 32-2.

Vessel Data:

Maximum allowable working pressure (MAWP): 100 psig
Back pressure on relief valve: 5 psig
Flow rate: 150 gpm

Level Transmitter LT-1 Data:
Maximum fluid level: 10 ft (from bottom of tank)
Minimum fluid level: 1 ft (from bottom of tank)
Transmitter level: 2 ft (below minimum level)
Wet leg level: 12 ft (from bottom of tank)
Wet leg (fill fluid) SG: 1.1
Fluid: water.
Operating pressure: 80 psig
Operating temperature: 100°F

21. In Figure 32-2, what is the maximum vessel pressure (in psig) allowed by ASME Code Section VIII when only PSV-1 is in service and relieving?

    A. 100 psig
    B. 110 psig
    C. 120 psig
    D. 130 psig

22. In Figure 32-2, what is the maximum vessel pressure (in psig) allowed by ASME Code Section VIII when both PSV-1 and PSV-2 are in service and relieving?

    A. 100 psig
    B. 116 psig
    C. 110 psig
    D. 120 psig

23. In Figure 32-2, what is the calibration range for level transmitter LT-1? Give the LVR (lower range value) to URV (upper range value) in inches of water.

    A. −145.2 to −37.78
    B. −145.2 to −37.2
    C. −147.73 to −40.31
    D. −147.6 to −39.6

24. In Figure 32-2, what is the orifice size of PSV-1 when only PSV-1 is in service and relieving?

    A. G
    B. E
    C. F
    D. H

25. In Figure 32-2, what is the span for level transmitter LT-1 in inches of water?

    A. 108 in
    B. 107.42 in
    C. 132 in
    D. 131.29 in

26. An SIL 3 interlock with an RRF of 1250 is required to mitigate a Category I hazard to Category III. If the failure rates of the SIS loop components are as follows, recommend a test frequency:

    Inputs = $1.0 \cdot 10^{-5}$/per h
    Logic solver = $7.0 \cdot 10^{-10}$/per h
    Valves = $3.0 \cdot 10^{-5}$/per h

    A. Once every 40 h
    B. Once every 80 h
    C. Once every 336 h
    D. Once every 600 h

Answers to Exam Sample Questions

1. C.
2. C.
3. D.
4. B.
5. D.
6. A.
7. A.
8. B.
9. B.
10. C.
11. D.
12. B.
13. C.
14. A.
15. A.
16. A.
17. B.
18. B.
19. B.
20. C.
21. B.
22. B.
23. C.
24. D.
25. B.
26. A.

Explanations and Proofs of Exam Sample Questions

1. The correct answer is C: 12.05.

   Find the nearest temperature for 433°F in Table A1, "Thermocouple Table (Type J)," in the appendix of this manual.

   The nearest temperature in the first column is 430. Look at the column headers at the bottom of the chart. Find the column header labeled 3. Follow the column up to the row with the 430 value. Where they meet is a total of 430°F + 3°F = (433°F).

   Locate the mV value; the answer is 12.044 mV.

2. The correct answer is C: 85.4 in wc.

   The change in head measurement can be defined as a ratio of the change in the flow rate squared:

$$h_1 F_2^2 = h_2 F_1^2 \qquad h_2 = \left(\frac{F_2}{F_1}\right)^2 h_1$$

$$85.176\ in = \left(\frac{312\ \text{gpm}}{200\ \text{gpm}}\right)^2 35\ in$$

   The new head measurement for a flow rate of 312 gpm is 85.175 in of water.

3. The correct answer is D: 7.10

   The bubbler measures the water column exerting pressure back on it. The level in the tank is 16 ft. The end of the bubbler tube is 1 ft from the bottom of the tank. The head being measured is 16 ft – 1 ft = 15 ft. The head is multiplied by the SG and then divided by 2.3 ft per psi:

$$7.1739\ \text{psi} = \frac{15\ \text{ft} \bullet 1.1\ \text{SG}}{\left(2.3\ \dfrac{\text{ft}}{\text{psi}}\right)}$$

4. The correct answer is B: Conduit fittings that require small radius bends should be avoided.

Acceptable	A. Electromagnetic fields from cables in trays do not affect optical cables.
Not Acceptable	B. Small radius bends can cause the cable to crack or break.
Acceptable	C. Overhead runs on messenger wires should be limited to 75 ft.
Acceptable	D. Underground fiber optic runs must be covered with concrete.

5. The correct answer is D: Less proportional gain and more integral action.

   Less proportional gain (K_C) and more integral action, or less integral time (T_I), should be chosen for the examination. It was stated that the control loop has a large dead time. It can be seen from the equations below that as the deadtime (θ) becomes larger, (T_I) becomes smaller. This means more repeats per minute of the proportion action for error correction. Note that more integral action means a smaller integral time.

Note: Looking at the Cohen-Coon tuning methods for a PI controller, a larger dead time in the equation gives a smaller controller gain and a smaller integral time.

$$K_C = \frac{1}{K_P}\frac{\tau}{\theta}\left(\frac{9}{10}+\frac{\theta}{12\tau}\right) \quad T_I = \frac{\theta[30+3(\theta/\tau)]}{9+20(\theta/\tau)}$$

Note: Looking at the Ziegler-Nichols tuning methods for a PI controller, a larger dead time in the equation gives a smaller controller gain and a smaller integral time.

$$K_C = \frac{0.9\tau}{K_P\theta} \quad T_I = \frac{\theta}{0.3}$$

where

K_C = controller gain
K_P = process gain
T_D = controller derivative time
T_I = controller integral time
t = process time constant
θ = process dead time

6. The correct answer is A: National Fire Protection Association.

NFPA 70 – NEC (National Electrical Code) is adopted into state law by most cities.
NFPA 496 covers the installation of electrical systems in hazardous locations.
NFPA 79 covers the installation of electrical systems for industrial machinery.

7. The correct answer is A: 2.33 in.

First, find the data needed for the calculation.

From Table A6, "Properties of Water Specific Gravity and lb/h to gpm," the SG of water at 125°F = ?

Temperature
120°F = 0.9901
125°F = ?
130°F = 0.9872

$$0.98865 = (SG) = \left[\left(\frac{125-120}{130-120}\right)(0.9872-0.9901)\right] + 0.9901$$

From Table A15, "Standard Pipe Dimensions and Data," the internal diameter of 4-inch, schedule 40 pipe is 4.026 in.

The equation for liquid flow through an orifice is:

$$Q\,(\text{gpm}) = 5.667\ SD^2\sqrt{\frac{h}{G_f}}$$

where

Q = gallons per minute of flow of liquid

Solve for S, the Spink factor, to find the orifice beta from Table 3, "The Spink Factor (S)."

$$S\,(spink) = \frac{\sqrt{SG} \bullet gpm}{5.667 \bullet D^2 \bullet \sqrt{h}}$$

$$0.2165 = \frac{\sqrt{0.98865} \bullet 200}{5.667 \bullet 4.026^2 \bullet \sqrt{100}}$$

From Table 3, "The Spink Factor (S)," the beta is as follows:

Beta

0.575 = 0.2144

? = 0.2165

0.600 = 0.2369

$$0.5773 = (beta) = \left[\left(\frac{0.2165 - 0.2144}{0.2369 - 0.2144}\right)(0.600 - 0.575)\right] + 0.575$$

Find the orifice hole diameter (d):

d = beta • pipe ID = hole size

d = 0.5773 • 4.026 = 2.324 inches

8. The correct answer is B: 14.2.

   The equation for flow through a valve for liquid is:

$$\text{gpm} = C_V\sqrt{\frac{DP}{SG}}$$

where

C_V = flow coefficient of the valve

DP = delta or differential pressure across the valve

Solve for C_v in the equation:

$$C_V = \frac{\sqrt{SG} \bullet gpm}{\sqrt{DP}}$$

$$14.23 = \frac{\sqrt{0.81} \bullet 50}{\sqrt{10}}$$

9. The correct answer is B: 540.

   The equation for flow through a valve for steam is:
   Note: N_6 is always equal to 63.3 for psia.

$$w = \left(N_6 F_p C_v Y\right)\sqrt{x P_1 \gamma_1}$$

Set $F_P = 1$ and $Y = 1$; the pipe size was not given.

From Table A9, "Properties of Saturated Steam," we can find the specific volume of the steam at a pressure of 40 psia equals 10.498 ft³/lb.

$$\gamma_1 = \text{specific weight is the reciprocal of specific volume } \frac{1}{(\text{ft}^3/\text{lb})} = \frac{\text{lb}}{\text{ft}^3}$$

$$\gamma_1 = \text{specific weight is the reciprocal of specific volume } \frac{1}{(\text{ft}^3/\text{lb})} = \frac{1}{10.498} = 0.09526\frac{\text{lb}}{\text{ft}^3}$$

$$0.25 = x = \frac{\Delta P}{P} = \frac{10}{40}$$

Find C_v:

$$C_v = \frac{w\ (in\ lb/h)}{63.3F_pY\sqrt{xP_1\gamma_1}} = \frac{30{,}000}{63.3(1)(1)\sqrt{(0.25)(50)(0.09526)}} = 485.58$$

10. The correct answer is C: Proportional plus integral.

    The control loop is noisy. This means that there may be quick changes in the process variable (*PV*). The derivative mode will provide quick changes in the manipulated variable (*MV*), due to extreme changes in the error signal (*e*). This may cause the system to amplify the noise causing unnecessary oscillations.

$$MV = e \bullet Kc$$

$$e = SP - PV$$

    where

$$Kc = \text{controller gain}$$
$$MV = \text{manipulated variable}$$
$$SP = \text{set point}$$
$$e = \text{error}$$

    The best mode for this type of control is proportional and integral modes only (PI).

11. The correct answer is D I, II, and III.

    All the following types of data can be recorded by the process control system, a DCS, PLC, SCADA, or DDC system.

    I. Graphical data in a strip or circular chart
    II. A table of numerical data in a computer memory
    III. A listing of alarms by a control computer

12. The correct answer is B: 100 in wc.

    First, find the data needed for the calculation.

    From Table A6, "Properties of Water Specific Gravity and lb/h to gpm," the SG of water at 125°F.

    125 = ?

    Temperature
    120°F = 0.9901
    125°F = ?
    130°F = 0.9872

$$0.98865 = (\text{SG}) = \left[\left(\frac{125-120}{130-120}\right)(0.9872-0.9901)\right] + 0.9901$$

    From Table A15, "Standard Pipe Dimensions and Data," the internal diameter of 4-inch, schedule 40 pipe is 4.026 in.

    The equation for liquid flow through an orifice is:

$$Q\,(\text{gpm}) = 5.667\ SD^2\sqrt{\frac{h}{G_f}}$$

    where
    Q = gallons per minute of flow of liquid

    Find the beta ratio of the orifice:

$$0.5772 = \frac{2.324}{4.026}$$

    Solve for (S) the Spink factor, to find the orifice beta from Table 3, "The Spink Factor (S)."

    From Table 3, "The Spink Factor (S)," the beta is as follows:

    Beta
    0.575 = 0.2144
    0.5772 = ?
    0.600 = 0.2369

$$0.2163 = (spink) = \left[\left(\frac{0.5772-0.575}{0.600-0.575}\right)(0.2369-0.2144)\right] + 0.2144$$

    Solve for the head in inches water column:

$$\sqrt{h} = \frac{\sqrt{\text{SG}} \bullet Q\,(\text{gpm})}{5.667SD^2}$$

$$h = \left(\frac{\sqrt{\text{SG}} \bullet Q\,(\text{gpm})}{5.667SD^2}\right)^2$$

$$100.19 = \left(\frac{\sqrt{0.9887} \bullet 200}{5.667\ (0.2163)\ 4.026^2}\right)^2$$

13. The correct answer is C: Double the testing frequency.

The risk reduction factor (RRF) is inversely related to the probability of failure on demand (PFD). We will substitute the formula for PFD to derive the RRF.

Calculating PFD:

$$PFD = \frac{1}{\text{RRF}} \text{ or } \frac{(system\ FR)(test\ interval)}{2}$$

where
FR = failure rate

By increasing the testing frequency or testing interval, the time between tests becomes smaller.

14. The correct answer is A: $4.89 \bullet 10^{-5}$/h.

$$Failure\ Rate\ (FR) = \frac{number\ of\ failures}{total\ time\ \text{(hours or years)}}$$

where
FR = failure rate

Note: 1 year = 8,760 h

MTTF: 3 failures in 7 years is

$$Failure\ Rate\ (FR) = \frac{\text{3 failures}}{\text{7 years} \bullet \text{8,760 h}} = \frac{3}{61{,}320\ \text{h}} = 4.89 \times 10^{-5}/\text{h}$$

15. The correct answer is A: Quick-opening.

The graph below shows that the quick-opening valve has the largest gain with flow versus stem (spindle) travel for percentage of open signal. The same is true for C_v versus stem (spindle) travel for percentage of open signal.

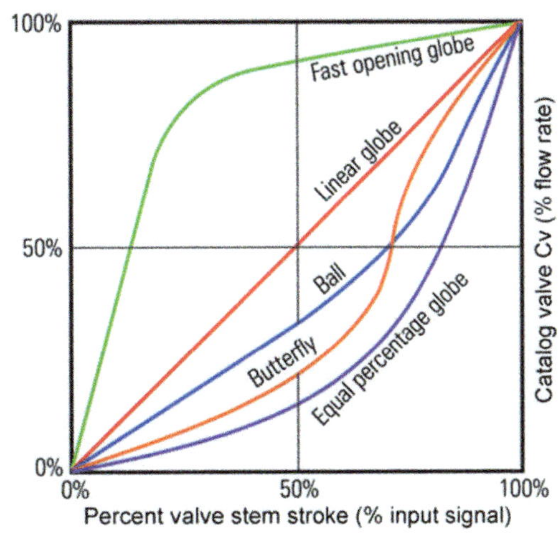

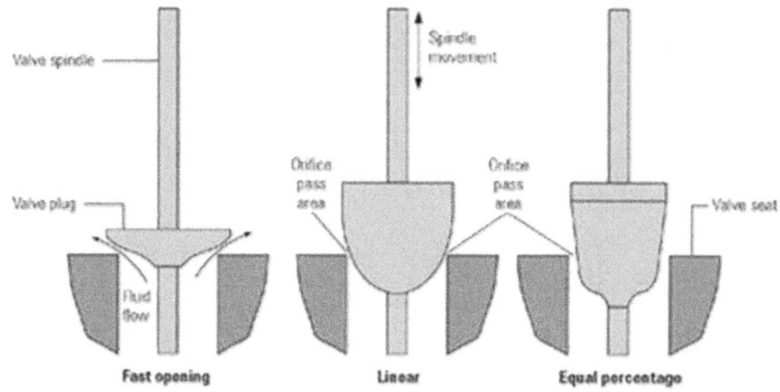

The shape of the trim determines the valve characteristic

16. The correct answer is A: Programmable logic controllers (PLCs).

    The PLC (programmable logic controller) is typically the only controller programmed in RLL (relay ladder logic).

17. The correct answer is B: 2.

    $k = 10.37$; the specific resistance of copper for 1 cm of 1 ft in length (for 20°C)
    cm = circular mils of copper

$$R = \left(\frac{k \bullet L}{cm}\right)$$

$$2 = \left(\frac{10.37 \bullet 2000 \text{ ft}}{10370 \text{ cm}}\right)$$

18. The correct answer is B: Explosion-proof apparatus and intrinsically safe equipment.

    Although explosion-proof apparatus and nonincendive equipment is listed in NEC Article 500, it only applies to Class I, Division 2 installations. Intrinsically safe equipment can be used in any class and division. The industrial standard for this application is intrinsically safe equipment.

19. The correct answer is B: Cross the wires at 90°.

    To reduce the effect of electromagnetic induction, also known as *transformer action,* in wires, it is necessary to separate the wires with a steel barrier. This is not always possible. By crossing the wires at 90° angles when intersecting other wires, the magnetic field around the conductor cannot induce a voltage into the other conductor.

20. The correct answer is C: $F_2 = F_1(X_2/X_1)$.

    The flow of the mass is proportional to the area multiplied by the velocity.

$$F_1 = V_1 A_1 \quad \text{and} \quad F_2 = V_2 A_2$$

$$\frac{F_1}{A_1} = V_1 \quad \text{and} \quad \frac{F_2}{A_2} = V_2$$

If the velocity was constant:

$$\frac{F_1}{A_1} = \frac{F_2}{A_2}$$

$$F_2 = F_1\left(\frac{A_2}{A_1}\right)$$

21. The correct answer is B: 110.

The ASME VIII Code requires that when a pressure-relieving device is used as the primary relief device, it must be sized to prevent the pressure from rising above 110% of the MAWP (UG-125(c)).

$$110\,(\text{psig}) = 100\,(\text{psig}) \bullet 1.10\,(\text{allowable over pressure})$$

22. The correct answer is B: 116.

The ASME VIII pressure relief Code requires that when a pressure-relieving device is used as the secondary relief device or as multiple relief devices, the size must prevent the pressure from rising above 116% of the MAWP in section (UG-125(c)(1)).

$$116\text{ psig} = 100\text{ psig} \bullet 1.16\text{ (allowable over pressure)}$$

23. The correct answer is C: –147.73 to –40.31.

From Table A6, "Properties of Water Specific Gravity and lb/h to gpm," the SG of water at 100°F is 0.9946.

The transmitter maximum high side will be maximum vessel level height minus the lowest vessel level height plus the transmitter height to the lower vessel line tap, multiplied by the specific gravity of the water at 100°.

$$\{(10\text{ ft} - 1\text{ ft})\text{ tank} + (2\text{ ft})\text{ transmitter}\} \bullet 0.9946$$

$$\text{h (in)} = [\{9\text{ ft (vessel)} + 2\text{ ft (transmitter)}\} \bullet 12\text{ in}] \bullet 0.9946\text{ (SG)}$$

$$131.287\text{ (in)} = [11\text{ ft} \bullet 2\text{ in}] \bullet 0.9946$$

The transmitter minimum high side will be lowest vessel level height plus the transmitter height to the vessel line tap, multiplied by the specific gravity of the water at 220°.

$$\text{head (in)} = [\{0\text{ ft (vessel)} + 2\text{ ft (transmitter)}\} \bullet 12\text{ in}] \bullet 0.9946\text{ (SG)}$$

$$23.87\text{ (in)} = [2\text{ ft} \bullet 12\text{ in}] \bullet 0.9946$$

The transmitter low side will be the lowest vessel level height, plus the wet leg height, plus the transmitter height to the upper vessel line tap, multiplied by the specific gravity of the wet-leg fluid with an SG of 1.1.

$$\text{h (in)} = [\{12 - 1\text{ ft (vessel)} + 2\text{ ft (transmitter)}\} \bullet 12\text{ in}] \bullet 1.1\text{ (SG)}$$

$$171.60\text{ (in)} = [13\text{ ft} \bullet 12\text{ in}] \bullet 1.1$$

The lower range value (LRV) for the transmitter is:

High side – Low side

−147.73 (LRV inches) = 23.87 (high side inches) − 171.6 (low side inches)

The upper range value (URV) for the transmitter is:

High side – Low side

−40.313 (LRV inches) = 131.287 (high side inches) − 171.6 (low side inches)

The transmitter calibration is –147.73 to –40.313 in of water.
The span of the transmitter is abs (–147.73) – abs (–40.31) = 107.42 in or (8.95 ft).

Remember to set the elevate switch or variable in the transmitter.

24. The correct answer is D: H.

We will size a pressure-relief valve for liquid service. Size the orifice for the following criteria.

Application: Primary Relief.

From Table A6, "Properties of Water Specific Gravity and lb/h to gpm," the SG of water at 100° is 0.9946.

Use 10% over pressure as permitted by ASME code. $P_1 = (1.10)$ MAWP + 14.7
$K = 0.67$
$K_V = 1$ (except for very viscous applications)
$K_U = 38$ for gallons per minute application
$P_1 = (1.1)\ (100) + 14.7 = 124.7$
$P_2 = 5 + 14.7 =$ atmospheric pressure
$Q = 150$ gallons per minute of liquid flow

$$A = \frac{Q\sqrt{SG}}{K_u K K_v \sqrt{P_1 - P_2}} = \frac{150\sqrt{0.9946}}{(38)(0.67)(1)\sqrt{124.7 - 19.7}} = 0.573\ in^2$$

Use Table 13-1, "ASME Standard Nozzle Orifice Data," located in Chapter 13, *Pressure relief valves and rupture disks*, to find the orifice size for the relief valve.

$G = 0.503$ in^2
$H = 0.785$ in^2

25. The correct answer is B: 107.42.

From Table A6, "Properties of Water Specific Gravity and lb/h to gpm," the SG of water at 100°F is 0.9946.

The span of the vessel is:

$$span\,(in) = [(10 \text{ ft maximum level} - 1 \text{ foot minimum level}) \bullet 12\ in] \bullet \text{SG}$$
$$107.42\,(in) = [9\ ft \bullet 12\ in] \bullet 0.9946$$

The span of the level measurement in the vessel is: 103.31 in or (8.61 ft).

26. The correct answer is A: Once every 40 h.

The equations used are as follows:

$$Failure\ Rate\ (FR) = \frac{number\ of\ failures}{total\ time\ \text{(hours or years)}}$$

$$PFD = \frac{1}{\text{RRF}} \quad \text{or} \quad \frac{(system\ FR)(test\ interval)}{2}$$

where

FR = failure rate (dangerous)
T_I = proof test interval

The PFD_{avg} can be calculated for each component of the system (e.g., S = sensor, LS = logic solver, and FE = final element) and then summed together.

$$PFD_{avg} = \frac{1}{\text{RRF}_{\text{avg}}} \quad \text{or} \quad \frac{FR_{\text{S}}(T_I)}{2} + \frac{FR_{\text{LS}}(T_I)}{2} + \frac{FR_{\text{FE}}(T_I)}{2}$$

$$PFD_{avg} = \frac{1}{\text{RRF}_{\text{avg}}} \quad \text{or} \quad \frac{(FR_{\text{S}} + FR_{\text{LS}} + FR_{\text{FE}}) \bullet T_I}{2}$$

$$8.0 \bullet 10^{-4}(PFD_{avg}) = \frac{1}{1250\ (\text{RRF}_{\text{avg}})} \quad \text{or} \quad \frac{(1.0 \bullet 10^{-5} + 7.0 \bullet 10^{-10} + 3.0 \bullet 10^{-5}) \bullet T_I}{2}$$

$$(2)8.0 \bullet 10^{-4} = \frac{(2)}{1250} \quad \text{or} \quad \frac{(1.0 \bullet 10^{-5} + 0.00007 \bullet 10^{-5} + 3.0 \bullet 10^{-5}) \bullet T_I}{1} \bullet \frac{(2)}{(2)}$$

$$1.6 \bullet 10^{-3} = (1.0 \bullet 10^{-5} + 0.00007 \bullet 10^{-5} + 3.0 \bullet 10^{-5}) \bullet T_I$$

$$\frac{160 \bullet 10^{-5}}{(1.0 \bullet 10^{-5} + 0.00007 \bullet 10^{-5} + 3.0 \bullet 10^{-5})} = T_I$$

$$\frac{160 \bullet 10^{-5}}{4.00007 \bullet 10^{-5}} = 40 \text{ h}$$

33 Guide to the Fisher Control Valve Handbook

The *Fisher Control Valve Handbook* from Emerson is a supplement book with many worked examples.

Purchasing this book or printing it out is highly recommended.

The handbook can help aid in study for the CSE exam. The information and tables in the *Fisher Control Valve Handbook* will be constantly referenced. I have repeated the most common data from the book needed for the CSE exam. The book is not required, but it is recommended and may be downloaded in PDF format from the Fisher Controls public website at the following address:

https://www.emerson.com/documents/automation/control-valve-handbook-en-3661206.pdf

If you wish to obtain a hard copy of the handbook, the *Fisher Control Valve Handbook* can be acquired for free, or for about $20, from your local instrumentation supplier. The book is also available from Brown's Technical Book Shop, 1517 San Jacinto, Houston, Texas, 77002. http://www.browntechnical.org.

I suggest tabbing the handbook for quick reference.

Important Sections to Review

- Chapter 3 – Valve and Actuator Types
- Chapter 5 – USA Regulatory Requirement for Fugitive Emissions
- Chapter 5 – Control Valve Selection (and sizing)

Important Pages to Possibly Tab

Valve and Materials Selection

- Chart for Test Frequency for Valves Leaking 500 ppm
- Sliding-Stem Valve Environmental Packing Selection
- Rotary Valve Environmental Packing Selection
- Designations for the High Nickel Alloys

Pressure-Temperature Ratings for Standard Class

- Cast Carbon Steel (ASTM A216 Grade WCC)
- Cast Chromium-Molybdenum Steel (ASTM A217 Grade WC9)
- Cast Chromium-Molybdenum Steel (ASTM A217 Grade C5)
- Cast Type 304L Stainless Steel (ASTM A351 Grade CF3)
- Cast Type 316 Stainless Steel (ASTM A351 Grade CF8M)
- Cast Type 317 Stainless Steel (ASTM A479 Grade UNS S31700)
- Cast Iron (ASTM A126)
- Cast Bronze Valves (ASTM B61 and B62)

Valve Trim Material Temperature Limits

Ambient Temperature Corrosion information

Elastomer Information

Application of Valves for Liquid Level Systems

Application of Valves for Flow Control Processes

Actuator Sizing Methods

- Typical Packing Friction Values
- Actuator Force Calculations
- Typical Rotary Shaft Valve Torque Factors (V–Notch Ball Valve)
- Typical Rotary Shaft Valve Torque Factors (High-Performance Butterfly Valve)
- Packing Selection Guidelines for Sliding-Stem Valves (100 ppm)
- Packing Selection Guidelines for Rotary Valves (500 ppm)

Valve Sizing Methods

- Valve Sizing Procedure
- Abbreviations and Terminology
- Equation Constants
- Determining q_{MAX} (the Maximum Flow Rate)
- Determining P_{MAX} (the Allowable Sizing Pressure Drop)
- Liquid Sizing Sample Problem
- Sizing Coefficients (C_v, X_T) for Single-Ported Globe Valve Bodies
- Sizing Coefficients (C_v, X_T) for Rotary Shaft Valve Bodies

Electrical Apparatus

- NEMA Enclosure Ratings and Definitions
- Hazardous Location Classification
- IEC Temperature Codes (Temperature Ratings)
- NEMA and IEC Enclosure Rating Comparison

Engineering Data

- Standard Specifications for Valve Materials (Temperature Range and Harness)
- Valve Materials Properties for Pressure
- Physical Constants of Hydrocarbons
- Specific Heat Ratio (k)
- Physical Constants of Various Fluids
- Properties of Water
- Properties of Saturated Steam
- Properties of Superheated Steam

Piping System Applications

- Velocity of Liquids in Pipe
- Flow of Water through Schedule 40 pipe
- Flow of Air through Schedule 40 pipe
- Calculations for Pipe Other Than Schedule 40
- Flow Correction Formulas for Steam, Vapor, Temperature and Pressure
- Pipe Data – Carbon and Alloy Steel – Stainless Steel

Conversions and Equivalents

- Length Equivalents
- Whole Inch–Millimeter Equivalents
- Fractional Inches to Millimeters
- Area Equivalents
- Volume Equivalents
- Volume Rate Equivalents
- Mass Conversion—Pounds to Kilograms
- Pressure Equivalents
- Temperature Conversions

Table A1 – Thermocouple Table: Type J (continued)

°F	0	1	2	3	4	5	6	7	8	9	10	°F
950	27.953	27.985	28.016	28.047	28.078	28.109	28.141	28.172	28.203	28.234	28.266	950
960	28.266	28.297	28.328	28.359	28.391	28.422	28.453	28.485	28.516	28.547	28.579	960
970	28.579	28.610	28.641	28.672	28.704	28.735	28.767	28.798	28.829	28.861	28.892	970
980	28.892	28.923	28.955	28.986	29.018	29.049	29.080	29.112	29.143	29.175	29.206	980
990	29.206	29.238	29.269	29.301	29.332	29.363	29.395	29.426	29.458	29.489	29.521	990
1000	29.521	29.552	29.584	29.616	29.647	29.679	29.710	29.742	29.773	29.805	29.836	1000
1010	29.836	29.868	29.900	29.931	29.963	29.995	30.026	30.058	30.089	30.121	30.153	1010
1020	30.153	30.184	30.216	30.248	30.279	30.311	30.343	30.375	30.406	30.438	30.470	1020
1030	30.470	30.502	30.533	30.565	30.597	30.629	30.660	30.692	30.724	30.756	30.788	1030
1040	30.788	30.819	30.851	30.883	30.915	30.947	30.979	31.011	31.043	31.074	31.106	1040
1050	31.106	31.138	31.170	31.202	31.234	31.266	31.298	31.330	31.362	31.394	31.426	1050
1060	31.426	31.458	31.490	31.522	31.554	31.586	31.618	31.650	31.682	31.714	31.746	1060
1070	31.746	31.778	31.811	31.843	31.875	31.907	31.939	31.971	32.003	32.035	32.068	1070
1080	32.068	32.100	32.132	32.164	32.196	32.229	32.261	32.293	32.325	32.358	32.390	1080
1090	32.390	32.422	32.455	32.487	32.519	32.551	32.584	32.616	32.648	32.681	32.713	1090
1100	32.713	32.746	32.778	32.810	32.843	32.875	32.908	32.940	32.973	33.005	33.037	1100
1110	33.037	33.070	33.102	33.135	33.167	33.200	33.232	33.265	33.298	33.330	33.363	1110
1120	33.363	33.395	33.428	33.460	33.493	33.526	33.558	33.591	33.624	33.656	33.689	1120
1130	33.689	33.722	33.754	33.787	33.820	33.853	33.885	33.918	33.951	33.984	34.016	1130
1140	34.016	34.049	34.082	34.115	34.148	34.180	34.213	34.246	34.279	34.312	34.345	1140
1150	34.345	34.378	34.411	34.444	34.476	34.509	34.542	34.575	34.608	34.641	34.674	1150
1160	34.674	34.707	34.740	34.773	34.806	34.840	34.873	34.906	34.939	34.972	35.005	1160
1170	35.005	35.038	35.071	35.104	35.138	35.171	35.204	35.237	35.270	35.304	35.337	1170
1180	35.337	35.370	35.403	35.437	35.470	35.503	35.536	35.570	35.603	35.636	35.670	1180
1190	35.670	35.703	35.736	35.770	35.803	35.837	35.870	35.903	35.937	35.970	36.004	1190
1200	36.004	36.037	36.071	36.104	36.138	36.171	36.205	36.238	36.272	36.305	36.339	1200
1210	36.339	36.373	36.406	36.440	36.473	36.507	36.541	36.574	36.608	36.642	36.675	1210
1220	36.675	36.709	36.743	36.777	36.810	36.844	36.878	36.912	36.945	36.979	37.013	1220
1230	37.013	37.047	37.081	37.114	37.148	37.182	37.216	37.250	37.284	37.318	37.352	1230
1240	37.352	37.386	37.420	37.454	37.488	37.522	37.556	37.590	37.624	37.658	37.692	1240
1250	37.692	37.726	37.760	37.794	37.828	37.862	37.896	37.930	37.964	37.999	38.033	1250
1260	38.033	38.067	38.101	38.135	38.169	38.204	38.238	38.272	38.306	38.341	38.375	1260
1270	38.375	38.409	38.444	38.478	38.512	38.546	38.581	38.615	38.650	38.684	38.718	1270
1280	38.718	38.753	38.787	38.822	38.856	38.890	38.925	38.959	38.994	39.028	39.063	1280
1290	39.063	39.097	39.132	39.166	39.201	39.235	39.270	39.305	39.339	39.374	39.408	1290
1300	39.408	39.443	39.478	39.512	39.547	39.582	39.616	39.651	39.686	39.720	39.755	1300
1310	39.755	39.790	39.825	39.859	39.894	39.929	39.964	39.998	40.033	40.068	40.103	1310
1320	40.103	40.138	40.173	40.207	40.242	40.277	40.312	40.347	40.382	40.417	40.452	1320
1330	40.452	40.487	40.522	40.556	40.591	40.626	40.661	40.696	40.731	40.766	40.801	1330
1340	40.801	40.836	40.872	40.907	40.942	40.977	41.012	41.047	41.082	41.117	41.152	1340
1350	41.152	41.187	41.222	41.258	41.293	41.328	41.363	41.398	41.433	41.469	41.504	1350
1360	41.504	41.539	41.574	41.610	41.645	41.680	41.715	41.751	41.786	41.821	41.856	1360
1370	41.856	41.892	41.927	41.962	41.998	42.033	42.068	42.104	42.139	42.174	42.210	1370
1380	42.210	42.245	42.281	42.316	42.351	42.387	42.422	42.458	42.493	42.528	42.564	1380
1390	42.564	42.599	42.635	42.670	42.706	42.741	42.777	42.812	42.848	42.883	42.919	1390
1400	42.919	42.954	42.990	43.025	43.061	43.096	43.132	43.167	43.203	43.239	43.274	1400
1410	43.274	43.310	43.346	43.381	43.417	43.452	43.488	43.524	43.559	43.595	43.631	1410
1420	43.631	43.667	43.702	43.738	43.774	43.809	43.845	43.881	43.917	43.953	43.988	1420
1430	43.988	44.024	44.060	44.096	44.131	44.167	44.203	44.239	44.275	44.310	44.346	1430
1440	44.346	44.382	44.418	44.454	44.490	44.525	44.561	44.597	44.633	44.669	44.705	1440
1450	44.705	44.741	44.777	44.812	44.848	44.884	44.920	44.956	44.992	45.028	45.064	1450
1460	45.064	45.099	45.135	45.171	45.207	45.243	45.279	45.315	45.351	45.387	45.423	1460
1470	45.423	45.458	45.494	45.530	45.566	45.602	45.638	45.674	45.710	45.746	45.782	1470
1480	45.782	45.818	45.853	45.889	45.925	45.961	45.997	46.033	46.069	46.105	46.141	1480
1490	46.141	46.177	46.212	46.248	46.284	46.320	46.356	46.392	46.428	46.464	46.500	1490
1500	46.500	46.535	46.571	46.607	46.643	46.679	46.715	46.751	46.786	46.822	46.858	1500
1510	46.858	46.894	46.930	46.966	47.001	47.037	47.073	47.109	47.145	47.181	47.216	1510
1520	47.216	47.252	47.288	47.324	47.359	47.395	47.431	47.467	47.503	47.538	47.574	1520
1530	47.574	47.610	47.646	47.681	47.717	47.753	47.788	47.824	47.860	47.896	47.931	1530
1540	47.931	47.967	48.003	48.038	48.074	48.110	48.145	48.181	48.217	48.252	48.288	1540
1550	48.288	48.324	48.359	48.395	48.430	48.466	48.502	48.537	48.573	48.608	48.644	1550
1560	48.644	48.679	48.715	48.750	48.786	48.822	48.857	48.893	48.928	48.964	48.999	1560
1570	48.999	49.034	49.070	49.105	49.141	49.176	49.212	49.247	49.283	49.318	49.353	1570
1580	49.353	49.389	49.424	49.460	49.495	49.530	49.566	49.601	49.636	49.672	49.707	1580
1590	49.707	49.742	49.778	49.813	49.848	49.883	49.919	49.954	49.989	50.024	50.060	1590
°F	0	1	2	3	4	5	6	7	8	9	10	°F

°F	0	1	2	3	4	5	6	7	8	9	10	°F
1600	50.060	50.095	50.130	50.165	50.200	50.235	50.271	50.306	50.341	50.376	50.411	1600
1610	50.411	50.446	50.481	50.517	50.552	50.587	50.622	50.657	50.692	50.727	50.762	1610
1620	50.762	50.797	50.832	50.867	50.902	50.937	50.972	51.007	51.042	51.077	51.112	1620
1630	51.112	51.147	51.181	51.216	51.251	51.286	51.321	51.356	51.391	51.425	51.460	1630
1640	51.460	51.495	51.530	51.565	51.599	51.634	51.669	51.704	51.738	51.773	51.808	1640
1650	51.808	51.843	51.877	51.912	51.947	51.981	52.016	52.051	52.085	52.120	52.154	1650
1660	52.154	52.189	52.224	52.258	52.293	52.327	52.362	52.396	52.431	52.465	52.500	1660
1670	52.500	52.534	52.569	52.603	52.638	52.672	52.707	52.741	52.776	52.810	52.844	1670
1680	52.844	52.879	52.913	52.947	52.982	53.016	53.050	53.085	53.119	53.153	53.188	1680
1690	53.188	53.222	53.256	53.290	53.325	53.359	53.393	53.427	53.462	53.496	53.530	1690
1700	53.530	53.564	53.598	53.632	53.667	53.701	53.735	53.769	53.803	53.837	53.871	1700
1710	53.871	53.905	53.939	53.973	54.007	54.041	54.075	54.109	54.143	54.177	54.211	1710
1720	54.211	54.245	54.279	54.313	54.347	54.381	54.415	54.449	54.483	54.516	54.550	1720
1730	54.550	54.584	54.618	54.652	54.686	54.719	54.753	54.787	54.821	54.855	54.888	1730
1740	54.888	54.922	54.956	54.990	55.023	55.057	55.091	55.124	55.158	55.192	55.225	1740
1750	55.225	55.259	55.293	55.326	55.360	55.393	55.427	55.461	55.494	55.528	55.561	1750
1760	55.561	55.595	55.628	55.662	55.695	55.729	55.762	55.796	55.829	55.863	55.896	1760
1770	55.896	55.930	55.963	55.997	56.030	56.063	56.097	56.130	56.164	56.197	56.230	1770
1780	56.230	56.264	56.297	56.330	56.364	56.397	56.430	56.464	56.497	56.530	56.564	1780
1790	56.564	56.597	56.630	56.663	56.697	56.730	56.763	56.796	56.829	56.863	56.896	1790
1800	56.896	56.929	56.962	56.995	57.028	57.062	57.095	57.128	57.161	57.194	57.227	1800
1810	57.227	57.260	57.293	57.326	57.360	57.393	57.426	57.459	57.492	57.525	57.558	1810
1820	57.558	57.591	57.624	57.657	57.690	57.723	57.756	57.789	57.822	57.855	57.888	1820
1830	57.888	57.920	57.953	57.986	58.019	58.052	58.085	58.118	58.151	58.184	58.217	1830
1840	58.217	58.249	58.282	58.315	58.348	58.381	58.414	58.446	58.479	58.512	58.545	1840
1850	58.545	58.578	58.610	58.643	58.676	58.709	58.741	58.774	58.807	58.840	58.872	1850
1860	58.872	58.905	58.938	58.971	59.003	59.036	59.069	59.101	59.134	59.167	59.199	1860
1870	59.199	59.232	59.265	59.297	59.330	59.363	59.395	59.428	59.460	59.493	59.526	1870
1880	59.526	59.558	59.591	59.623	59.656	59.689	59.721	59.754	59.786	59.819	59.851	1880
1890	59.851	59.884	59.916	59.949	59.982	60.014	60.047	60.079	60.112	60.144	60.177	1890
1900	60.177	60.209	60.242	60.274	60.307	60.339	60.371	60.404	60.436	60.469	60.501	1900
1910	60.501	60.534	60.566	60.599	60.631	60.663	60.696	60.728	60.761	60.793	60.826	1910
1920	60.826	60.858	60.890	60.923	60.955	60.987	61.020	61.052	61.085	61.117	61.149	1920
1930	61.149	61.182	61.214	61.246	61.279	61.311	61.343	61.376	61.408	61.440	61.473	1930
1940	61.473	61.505	61.537	61.570	61.602	61.634	61.667	61.699	61.731	61.763	61.796	1940
1950	61.796	61.828	61.860	61.893	61.925	61.957	61.989	62.022	62.054	62.086	62.118	1950
1960	62.118	62.151	62.183	62.215	62.247	62.280	62.312	62.344	62.376	62.409	62.441	1960
1970	62.441	62.473	62.505	62.537	62.570	62.602	62.634	62.666	62.699	62.731	62.763	1970
1980	62.763	62.795	62.827	62.860	62.892	62.924	62.956	62.988	63.020	63.053	63.085	1980
1990	63.085	63.117	63.149	63.181	63.214	63.246	63.278	63.310	63.342	63.374	63.406	1990
2000	63.406	63.439	63.471	63.503	63.535	63.567	63.599	63.632	63.664	63.696	63.728	2000
2010	63.728	63.760	63.792	63.824	63.856	63.889	63.921	63.953	63.985	64.017	64.049	2010
2020	64.049	64.081	64.113	64.146	64.178	64.210	64.242	64.274	64.306	64.338	64.370	2020
2030	64.370	64.402	64.435	64.467	64.499	64.531	64.563	64.595	64.627	64.659	64.691	2030
2040	64.691	64.723	64.756	64.788	64.820	64.852	64.884	64.916	64.948	64.980	65.012	2040
2050	65.012	65.044	65.076	65.109	65.141	65.173	65.205	65.237	65.269	65.301	65.333	2050
2060	65.333	65.365	65.397	65.429	65.461	65.493	65.525	65.557	65.590	65.622	65.654	2060
2070	65.654	65.686	65.718	65.750	65.782	65.814	65.846	65.878	65.910	65.942	65.974	2070
2080	65.974	66.006	66.038	66.070	66.102	66.134	66.166	66.199	66.231	66.263	66.295	2080
2090	66.295	66.327	66.359	66.391	66.423	66.455	66.487	66.519	66.551	66.583	66.615	2090
2100	66.615	66.647	66.679	66.711	66.743	66.775	66.807	66.839	66.871	66.903	66.935	2100
2110	66.935	66.967	66.999	67.031	67.063	67.095	67.127	67.159	67.191	67.223	67.255	2110
2120	67.255	67.287	67.319	67.351	67.383	67.415	67.447	67.479	67.511	67.543	67.575	2120
2130	67.575	67.607	67.639	67.671	67.703	67.735	67.767	67.799	67.831	67.863	67.895	2130
2140	67.895	67.927	67.959	67.991	68.023	68.055	68.087	68.119	68.150	68.182	68.214	2140
2150	68.214	68.246	68.278	68.310	68.342	68.374	68.406	68.438	68.470	68.502	68.534	2150
2160	68.534	68.566	68.597	68.629	68.661	68.693	68.725	68.757	68.789	68.821	68.853	2160
2170	68.853	68.884	68.916	68.948	68.980	69.012	69.044	69.076	69.108	69.139	69.171	2170
2180	69.171	69.203	69.235	69.267	69.299	69.330	69.362	69.394	69.426	69.458	69.490	2180
2190	69.490	69.521	69.553									2190
°F	0	1	2	3	4	5	6	7	8	9	10	°F

Table A2 – Thermocouple Table: Type K

Thermoelectric Voltage in Millivolts

°F	-10	-9	-8	-7	-6	-5	-4	-3	-2	-1	0	°F
-450							-6.458	-6.457	-6.457	-6.456	-6.456	-450
-440	-6.456	-6.455	-6.454	-6.454	-6.453	-6.452	-6.451	-6.450	-6.449	-6.448	-6.446	-440
-430	-6.446	-6.445	-6.444	-6.443	-6.441	-6.440	-6.438	-6.436	-6.435	-6.433	-6.431	-430
-420	-6.431	-6.429	-6.427	-6.425	-6.423	-6.421	-6.419	-6.416	-6.414	-6.411	-6.409	-420
-410	-6.409	-6.406	-6.404	-6.401	-6.398	-6.395	-6.392	-6.389	-6.386	-6.383	-6.380	-410
-400	-6.380	-6.377	-6.373	-6.370	-6.366	-6.363	-6.359	-6.355	-6.352	-6.348	-6.344	-400
-390	-6.344	-6.340	-6.336	-6.332	-6.328	-6.323	-6.319	-6.315	-6.310	-6.306	-6.301	-390
-380	-6.301	-6.296	-6.292	-6.287	-6.282	-6.277	-6.272	-6.267	-6.262	-6.257	-6.251	-380
-370	-6.251	-6.246	-6.241	-6.235	-6.230	-6.224	-6.218	-6.213	-6.207	-6.201	-6.195	-370
-360	-6.195	-6.189	-6.183	-6.177	-6.171	-6.165	-6.158	-6.152	-6.146	-6.139	-6.133	-360
-350	-6.133	-6.126	-6.119	-6.113	-6.106	-6.099	-6.092	-6.085	-6.078	-6.071	-6.064	-350
-340	-6.064	-6.057	-6.049	-6.042	-6.035	-6.027	-6.020	-6.012	-6.004	-5.997	-5.989	-340
-330	-5.989	-5.981	-5.973	-5.965	-5.957	-5.949	-5.941	-5.933	-5.925	-5.917	-5.908	-330
-320	-5.908	-5.900	-5.891	-5.883	-5.874	-5.866	-5.857	-5.848	-5.840	-5.831	-5.822	-320
-310	-5.822	-5.813	-5.804	-5.795	-5.786	-5.776	-5.767	-5.758	-5.749	-5.739	-5.730	-310
-300	-5.730	-5.720	-5.711	-5.701	-5.691	-5.682	-5.672	-5.662	-5.652	-5.642	-5.632	-300
-290	-5.632	-5.622	-5.612	-5.602	-5.592	-5.581	-5.571	-5.561	-5.550	-5.540	-5.529	-290
-280	-5.529	-5.519	-5.508	-5.497	-5.487	-5.476	-5.465	-5.454	-5.443	-5.432	-5.421	-280
-270	-5.421	-5.410	-5.399	-5.388	-5.377	-5.365	-5.354	-5.343	-5.331	-5.320	-5.308	-270
-260	-5.308	-5.296	-5.285	-5.273	-5.261	-5.250	-5.238	-5.226	-5.214	-5.202	-5.190	-260
-250	-5.190	-5.178	-5.166	-5.153	-5.141	-5.129	-5.117	-5.104	-5.092	-5.079	-5.067	-250
-240	-5.067	-5.054	-5.042	-5.029	-5.016	-5.003	-4.991	-4.978	-4.965	-4.952	-4.939	-240
-230	-4.939	-4.926	-4.913	-4.900	-4.886	-4.873	-4.860	-4.847	-4.833	-4.820	-4.806	-230
-220	-4.806	-4.793	-4.779	-4.766	-4.752	-4.738	-4.724	-4.711	-4.697	-4.683	-4.669	-220
-210	-4.669	-4.655	-4.641	-4.627	-4.613	-4.599	-4.584	-4.570	-4.556	-4.542	-4.527	-210
-200	-4.527	-4.513	-4.498	-4.484	-4.469	-4.455	-4.440	-4.425	-4.411	-4.396	-4.381	-200
-190	-4.381	-4.366	-4.351	-4.336	-4.321	-4.306	-4.291	-4.276	-4.261	-4.246	-4.231	-190
-180	-4.231	-4.215	-4.200	-4.185	-4.169	-4.154	-4.138	-4.123	-4.107	-4.091	-4.076	-180
-170	-4.076	-4.060	-4.044	-4.029	-4.013	-3.997	-3.981	-3.965	-3.949	-3.933	-3.917	-170
-160	-3.917	-3.901	-3.885	-3.869	-3.852	-3.836	-3.820	-3.803	-3.787	-3.771	-3.754	-160
-150	-3.754	-3.738	-3.721	-3.705	-3.688	-3.671	-3.655	-3.638	-3.621	-3.604	-3.587	-150
-140	-3.587	-3.571	-3.554	-3.537	-3.520	-3.503	-3.486	-3.468	-3.451	-3.434	-3.417	-140
-130	-3.417	-3.400	-3.382	-3.365	-3.348	-3.330	-3.313	-3.295	-3.278	-3.260	-3.243	-130
-120	-3.243	-3.225	-3.207	-3.190	-3.172	-3.154	-3.136	-3.119	-3.101	-3.083	-3.065	-120
-110	-3.065	-3.047	-3.029	-3.011	-2.993	-2.975	-2.957	-2.938	-2.920	-2.902	-2.884	-110
-100	-2.884	-2.865	-2.847	-2.829	-2.810	-2.792	-2.773	-2.755	-2.736	-2.718	-2.699	-100
-90	-2.699	-2.680	-2.662	-2.643	-2.624	-2.605	-2.587	-2.568	-2.549	-2.530	-2.511	-90
-80	-2.511	-2.492	-2.473	-2.454	-2.435	-2.416	-2.397	-2.378	-2.359	-2.339	-2.320	-80
-70	-2.320	-2.301	-2.282	-2.262	-2.243	-2.223	-2.204	-2.185	-2.165	-2.146	-2.126	-70
-60	-2.126	-2.106	-2.087	-2.067	-2.048	-2.028	-2.008	-1.988	-1.969	-1.949	-1.929	-60
-50	-1.929	-1.909	-1.889	-1.869	-1.850	-1.830	-1.810	-1.790	-1.770	-1.749	-1.729	-50
-40	-1.729	-1.709	-1.689	-1.669	-1.649	-1.628	-1.608	-1.588	-1.568	-1.547	-1.527	-40
-30	-1.527	-1.507	-1.486	-1.466	-1.445	-1.425	-1.404	-1.384	-1.363	-1.343	-1.322	-30
-20	-1.322	-1.301	-1.281	-1.260	-1.239	-1.218	-1.198	-1.177	-1.156	-1.135	-1.114	-20
-10	-1.114	-1.094	-1.073	-1.052	-1.031	-1.010	-0.989	-0.968	-0.947	-0.926	-0.905	-10
0	-0.905	-0.883	-0.862	-0.841	-0.820	-0.799	-0.778	-0.756	-0.735	-0.714	-0.692	0

°F	0	1	2	3	4	5	6	7	8	9	10	°F
0	-0.692	-0.671	-0.650	-0.628	-0.607	-0.586	-0.564	-0.543	-0.521	-0.500	-0.478	0
10	-0.478	-0.457	-0.435	-0.413	-0.392	-0.370	-0.349	-0.327	-0.305	-0.284	-0.262	10
20	-0.262	-0.240	-0.218	-0.197	-0.175	-0.153	-0.131	-0.109	-0.088	-0.066	-0.044	20
30	-0.044	-0.022	0.000	0.022	0.044	0.066	0.088	0.110	0.132	0.154	0.176	30
40	0.176	0.198	0.220	0.242	0.264	0.286	0.308	0.330	0.353	0.375	0.397	40
50	0.397	0.419	0.441	0.463	0.486	0.508	0.530	0.552	0.575	0.597	0.619	50
60	0.619	0.642	0.664	0.686	0.709	0.731	0.753	0.776	0.798	0.821	0.843	60
70	0.843	0.865	0.888	0.910	0.933	0.955	0.978	1.000	1.023	1.045	1.068	70
80	1.068	1.090	1.113	1.136	1.158	1.181	1.203	1.226	1.249	1.271	1.294	80
90	1.294	1.316	1.339	1.362	1.384	1.407	1.430	1.453	1.475	1.498	1.521	90

°F	0	1	2	3	4	5	6	7	8	9	10	°F
100	1.521	1.543	1.566	1.589	1.612	1.635	1.657	1.680	1.703	1.726	1.749	100
110	1.749	1.771	1.794	1.817	1.840	1.863	1.886	1.909	1.931	1.954	1.977	110
120	1.977	2.000	2.023	2.046	2.069	2.092	2.115	2.138	2.161	2.184	2.207	120
130	2.207	2.230	2.253	2.276	2.298	2.321	2.344	2.367	2.390	2.413	2.436	130
140	2.436	2.459	2.483	2.506	2.529	2.552	2.575	2.598	2.621	2.644	2.667	140
150	2.667	2.690	2.713	2.736	2.759	2.782	2.805	2.828	2.851	2.874	2.897	150
160	2.897	2.920	2.944	2.967	2.990	3.013	3.036	3.059	3.082	3.105	3.128	160
170	3.128	3.151	3.174	3.197	3.220	3.244	3.267	3.290	3.313	3.336	3.359	170
180	3.359	3.382	3.405	3.428	3.451	3.474	3.497	3.520	3.544	3.567	3.590	180
190	3.590	3.613	3.636	3.659	3.682	3.705	3.728	3.751	3.774	3.797	3.820	190
200	3.820	3.843	3.866	3.889	3.912	3.935	3.958	3.981	4.004	4.027	4.050	200
210	4.050	4.073	4.096	4.119	4.142	4.165	4.188	4.211	4.234	4.257	4.280	210
220	4.280	4.303	4.326	4.349	4.372	4.395	4.417	4.440	4.463	4.486	4.509	220
230	4.509	4.532	4.555	4.578	4.601	4.623	4.646	4.669	4.692	4.715	4.738	230
240	4.738	4.760	4.783	4.806	4.829	4.852	4.874	4.897	4.920	4.943	4.965	240
250	4.965	4.988	5.011	5.034	5.056	5.079	5.102	5.124	5.147	5.170	5.192	250
260	5.192	5.215	5.238	5.260	5.283	5.306	5.328	5.351	5.374	5.396	5.419	260
270	5.419	5.441	5.464	5.487	5.509	5.532	5.554	5.577	5.599	5.622	5.644	270
280	5.644	5.667	5.690	5.712	5.735	5.757	5.779	5.802	5.824	5.847	5.869	280
290	5.869	5.892	5.914	5.937	5.959	5.982	6.004	6.026	6.049	6.071	6.094	290
300	6.094	6.116	6.138	6.161	6.183	6.205	6.228	6.250	6.272	6.295	6.317	300
310	6.317	6.339	6.362	6.384	6.406	6.429	6.451	6.473	6.496	6.518	6.540	310
320	6.540	6.562	6.585	6.607	6.629	6.652	6.674	6.696	6.718	6.741	6.763	320
330	6.763	6.785	6.807	6.829	6.852	6.874	6.896	6.918	6.941	6.963	6.985	330
340	6.985	7.007	7.029	7.052	7.074	7.096	7.118	7.140	7.163	7.185	7.207	340
350	7.207	7.229	7.251	7.273	7.296	7.318	7.340	7.362	7.384	7.407	7.429	350
360	7.429	7.451	7.473	7.495	7.517	7.540	7.562	7.584	7.606	7.628	7.650	360
370	7.650	7.673	7.695	7.717	7.739	7.761	7.783	7.806	7.828	7.850	7.872	370
380	7.872	7.894	7.917	7.939	7.961	7.983	8.005	8.027	8.050	8.072	8.094	380
390	8.094	8.116	8.138	8.161	8.183	8.205	8.227	8.250	8.272	8.294	8.316	390
400	8.316	8.338	8.361	8.383	8.405	8.427	8.450	8.472	8.494	8.516	8.539	400
410	8.539	8.561	8.583	8.605	8.628	8.650	8.672	8.694	8.717	8.739	8.761	410
420	8.761	8.784	8.806	8.828	8.851	8.873	8.895	8.918	8.940	8.962	8.985	420
430	8.985	9.007	9.029	9.052	9.074	9.096	9.119	9.141	9.163	9.186	9.208	430
440	9.208	9.231	9.253	9.275	9.298	9.320	9.343	9.365	9.388	9.410	9.432	440
450	9.432	9.455	9.477	9.500	9.522	9.545	9.567	9.590	9.612	9.635	9.657	450
460	9.657	9.680	9.702	9.725	9.747	9.770	9.792	9.815	9.837	9.860	9.882	460
470	9.882	9.905	9.927	9.950	9.973	9.995	10.018	10.040	10.063	10.086	10.108	470
480	10.108	10.131	10.153	10.176	10.199	10.221	10.244	10.267	10.289	10.312	10.334	480
490	10.334	10.357	10.380	10.402	10.425	10.448	10.471	10.493	10.516	10.539	10.561	490
500	10.561	10.584	10.607	10.629	10.652	10.675	10.698	10.720	10.743	10.766	10.789	500
510	10.789	10.811	10.834	10.857	10.880	10.903	10.925	10.948	10.971	10.994	11.017	510
520	11.017	11.039	11.062	11.085	11.108	11.131	11.154	11.176	11.199	11.222	11.245	520
530	11.245	11.268	11.291	11.313	11.336	11.359	11.382	11.405	11.428	11.451	11.474	530
540	11.474	11.497	11.519	11.542	11.565	11.588	11.611	11.634	11.657	11.680	11.703	540
550	11.703	11.726	11.749	11.772	11.795	11.818	11.841	11.864	11.887	11.910	11.933	550
560	11.933	11.956	11.978	12.001	12.024	12.047	12.070	12.093	12.116	12.140	12.163	560
570	12.163	12.186	12.209	12.232	12.255	12.278	12.301	12.324	12.347	12.370	12.393	570
580	12.393	12.416	12.439	12.462	12.485	12.508	12.531	12.554	12.577	12.600	12.624	580
590	12.624	12.647	12.670	12.693	12.716	12.739	12.762	12.785	12.808	12.831	12.855	590
600	12.855	12.878	12.901	12.924	12.947	12.970	12.993	13.016	13.040	13.063	13.086	600
610	13.086	13.109	13.132	13.155	13.179	13.202	13.225	13.248	13.271	13.294	13.318	610
620	13.318	13.341	13.364	13.387	13.410	13.433	13.457	13.480	13.503	13.526	13.549	620
630	13.549	13.573	13.596	13.619	13.642	13.665	13.689	13.712	13.735	13.758	13.782	630
640	13.782	13.805	13.828	13.851	13.874	13.898	13.921	13.944	13.967	13.991	14.014	640
650	14.014	14.037	14.060	14.084	14.107	14.130	14.154	14.177	14.200	14.223	14.247	650
660	14.247	14.270	14.293	14.316	14.340	14.363	14.386	14.410	14.433	14.456	14.479	660
670	14.479	14.503	14.526	14.549	14.573	14.596	14.619	14.643	14.666	14.689	14.713	670
680	14.713	14.736	14.759	14.783	14.806	14.829	14.853	14.876	14.899	14.923	14.946	680
690	14.946	14.969	14.993	15.016	15.039	15.063	15.086	15.109	15.133	15.156	15.179	690
°F	0	1	2	3	4	5	6	7	8	9	10	°F

Table A2 – Thermocouple Table: Type K (continued)

°F	0	1	2	3	4	5	6	7	8	9	10	°F
700	15.179	15.203	15.226	15.250	15.273	15.296	15.320	15.343	15.366	15.390	15.413	700
710	15.413	15.437	15.460	15.483	15.507	15.530	15.554	15.577	15.600	15.624	15.647	710
720	15.647	15.671	15.694	15.717	15.741	15.764	15.788	15.811	15.834	15.858	15.881	720
730	15.881	15.905	15.928	15.952	15.975	15.998	16.022	16.045	16.069	16.092	16.116	730
740	16.116	16.139	16.163	16.186	16.209	16.233	16.256	16.280	16.303	16.327	16.350	740
750	16.350	16.374	16.397	16.421	16.444	16.468	16.491	16.514	16.538	16.561	16.585	750
760	16.585	16.608	16.632	16.655	16.679	16.702	16.726	16.749	16.773	16.796	16.820	760
770	16.820	16.843	16.867	16.890	16.914	16.937	16.961	16.984	17.008	17.031	17.055	770
780	17.055	17.078	17.102	17.125	17.149	17.173	17.196	17.220	17.243	17.267	17.290	780
790	17.290	17.314	17.337	17.361	17.384	17.408	17.431	17.455	17.478	17.502	17.526	790
800	17.526	17.549	17.573	17.596	17.620	17.643	17.667	17.690	17.714	17.738	17.761	800
810	17.761	17.785	17.808	17.832	17.855	17.879	17.902	17.926	17.950	17.973	17.997	810
820	17.997	18.020	18.044	18.068	18.091	18.115	18.138	18.162	18.185	18.209	18.233	820
830	18.233	18.256	18.280	18.303	18.327	18.351	18.374	18.398	18.421	18.445	18.469	830
840	18.469	18.492	18.516	18.539	18.563	18.587	18.610	18.634	18.657	18.681	18.705	840
850	18.705	18.728	18.752	18.776	18.799	18.823	18.846	18.870	18.894	18.917	18.941	850
860	18.941	18.965	18.988	19.012	19.035	19.059	19.083	19.106	19.130	19.154	19.177	860
870	19.177	19.201	19.224	19.248	19.272	19.295	19.319	19.343	19.366	19.390	19.414	870
880	19.414	19.437	19.461	19.485	19.508	19.532	19.556	19.579	19.603	19.626	19.650	880
890	19.650	19.674	19.697	19.721	19.745	19.768	19.792	19.816	19.839	19.863	19.887	890
900	19.887	19.910	19.934	19.958	19.981	20.005	20.029	20.052	20.076	20.100	20.123	900
910	20.123	20.147	20.171	20.194	20.218	20.242	20.265	20.289	20.313	20.336	20.360	910
920	20.360	20.384	20.407	20.431	20.455	20.479	20.502	20.526	20.550	20.573	20.597	920
930	20.597	20.621	20.644	20.668	20.692	20.715	20.739	20.763	20.786	20.810	20.834	930
940	20.834	20.857	20.881	20.905	20.929	20.952	20.976	21.000	21.023	21.047	21.071	940
950	21.071	21.094	21.118	21.142	21.165	21.189	21.213	21.236	21.260	21.284	21.308	950
960	21.308	21.331	21.355	21.379	21.402	21.426	21.450	21.473	21.497	21.521	21.544	960
970	21.544	21.568	21.592	21.616	21.639	21.663	21.687	21.710	21.734	21.758	21.781	970
980	21.781	21.805	21.829	21.852	21.876	21.900	21.924	21.947	21.971	21.995	22.018	980
990	22.018	22.042	22.066	22.089	22.113	22.137	22.160	22.184	22.208	22.232	22.255	990
1000	22.255	22.279	22.303	22.326	22.350	22.374	22.397	22.421	22.445	22.468	22.492	1000
1010	22.492	22.516	22.540	22.563	22.587	22.611	22.634	22.658	22.682	22.705	22.729	1010
1020	22.729	22.753	22.776	22.800	22.824	22.847	22.871	22.895	22.919	22.942	22.966	1020
1030	22.966	22.990	23.013	23.037	23.061	23.084	23.108	23.132	23.155	23.179	23.203	1030
1040	23.203	23.226	23.250	23.274	23.297	23.321	23.345	23.368	23.392	23.416	23.439	1040
1050	23.439	23.463	23.487	23.510	23.534	23.558	23.581	23.605	23.629	23.652	23.676	1050
1060	23.676	23.700	23.723	23.747	23.771	23.794	23.818	23.842	23.865	23.889	23.913	1060
1070	23.913	23.936	23.960	23.984	24.007	24.031	24.055	24.078	24.102	24.126	24.149	1070
1080	24.149	24.173	24.197	24.220	24.244	24.267	24.291	24.315	24.338	24.362	24.386	1080
1090	24.386	24.409	24.433	24.457	24.480	24.504	24.527	24.551	24.575	24.598	24.622	1090
1100	24.622	24.646	24.669	24.693	24.717	24.740	24.764	24.787	24.811	24.835	24.858	1100
1110	24.858	24.882	24.905	24.929	24.953	24.976	25.000	25.024	25.047	25.071	25.094	1110
1120	25.094	25.118	25.142	25.165	25.189	25.212	25.236	25.260	25.283	25.307	25.330	1120
1130	25.330	25.354	25.377	25.401	25.425	25.448	25.472	25.495	25.519	25.543	25.566	1130
1140	25.566	25.590	25.613	25.637	25.660	25.684	25.708	25.731	25.755	25.778	25.802	1140
1150	25.802	25.825	25.849	25.873	25.896	25.920	25.943	25.967	25.990	26.014	26.037	1150
1160	26.037	26.061	26.084	26.108	26.132	26.155	26.179	26.202	26.226	26.249	26.273	1160
1170	26.273	26.296	26.320	26.343	26.367	26.390	26.414	26.437	26.461	26.484	26.508	1170
1180	26.508	26.532	26.555	26.579	26.602	26.626	26.649	26.673	26.696	26.720	26.743	1180
1190	26.743	26.767	26.790	26.814	26.837	26.861	26.884	26.907	26.931	26.954	26.978	1190
1200	26.978	27.001	27.025	27.048	27.072	27.095	27.119	27.142	27.166	27.189	27.213	1200
1210	27.213	27.236	27.259	27.283	27.306	27.330	27.353	27.377	27.400	27.424	27.447	1210
1220	27.447	27.471	27.494	27.517	27.541	27.564	27.588	27.611	27.635	27.658	27.681	1220
1230	27.681	27.705	27.728	27.752	27.775	27.798	27.822	27.845	27.869	27.892	27.915	1230
1240	27.915	27.939	27.962	27.986	28.009	28.032	28.056	28.079	28.103	28.126	28.149	1240
1250	28.149	28.173	28.196	28.219	28.243	28.266	28.289	28.313	28.336	28.360	28.383	1250
1260	28.383	28.406	28.430	28.453	28.476	28.500	28.523	28.546	28.570	28.593	28.616	1260
1270	28.616	28.640	28.663	28.686	28.710	28.733	28.756	28.780	28.803	28.826	28.849	1270
1280	28.849	28.873	28.896	28.919	28.943	28.966	28.989	29.013	29.036	29.059	29.082	1280
1290	29.082	29.106	29.129	29.152	29.176	29.199	29.222	29.245	29.269	29.292	29.315	1290
°F	0	1	2	3	4	5	6	7	8	9	10	°F

°F	0	1	2	3	4	5	6	7	8	9	10	°F
1300	29.315	29.338	29.362	29.385	29.408	29.431	29.455	29.478	29.501	29.524	29.548	1300
1310	29.548	29.571	29.594	29.617	29.640	29.664	29.687	29.710	29.733	29.757	29.780	1310
1320	29.780	29.803	29.826	29.849	29.873	29.896	29.919	29.942	29.965	29.989	30.012	1320
1330	30.012	30.035	30.058	30.081	30.104	30.128	30.151	30.174	30.197	30.220	30.243	1330
1340	30.243	30.267	30.290	30.313	30.336	30.359	30.382	30.405	30.429	30.452	30.475	1340
1350	30.475	30.498	30.521	30.544	30.567	30.590	30.613	30.637	30.660	30.683	30.706	1350
1360	30.706	30.729	30.752	30.775	30.798	30.821	30.844	30.868	30.891	30.914	30.937	1360
1370	30.937	30.960	30.983	31.006	31.029	31.052	31.075	31.098	31.121	31.144	31.167	1370
1380	31.167	31.190	31.213	31.236	31.260	31.283	31.306	31.329	31.352	31.375	31.398	1380
1390	31.398	31.421	31.444	31.467	31.490	31.513	31.536	31.559	31.582	31.605	31.628	1390
1400	31.628	31.651	31.674	31.697	31.720	31.743	31.766	31.789	31.812	31.834	31.857	1400
1410	31.857	31.880	31.903	31.926	31.949	31.972	31.995	32.018	32.041	32.064	32.087	1410
1420	32.087	32.110	32.133	32.156	32.179	32.202	32.224	32.247	32.270	32.293	32.316	1420
1430	32.316	32.339	32.362	32.385	32.408	32.431	32.453	32.476	32.499	32.522	32.545	1430
1440	32.545	32.568	32.591	32.614	32.636	32.659	32.682	32.705	32.728	32.751	32.774	1440
1450	32.774	32.796	32.819	32.842	32.865	32.888	32.911	32.933	32.956	32.979	33.002	1450
1460	33.002	33.025	33.047	33.070	33.093	33.116	33.139	33.161	33.184	33.207	33.230	1460
1470	33.230	33.253	33.275	33.298	33.321	33.344	33.366	33.389	33.412	33.435	33.458	1470
1480	33.458	33.480	33.503	33.526	33.548	33.571	33.594	33.617	33.639	33.662	33.685	1480
1490	33.685	33.708	33.730	33.753	33.776	33.798	33.821	33.844	33.867	33.889	33.912	1490
1500	33.912	33.935	33.957	33.980	34.003	34.025	34.048	34.071	34.093	34.116	34.139	1500
1510	34.139	34.161	34.184	34.207	34.229	34.252	34.275	34.297	34.320	34.343	34.365	1510
1520	34.365	34.388	34.410	34.433	34.456	34.478	34.501	34.524	34.546	34.569	34.591	1520
1530	34.591	34.614	34.637	34.659	34.682	34.704	34.727	34.750	34.772	34.795	34.817	1530
1540	34.817	34.840	34.862	34.885	34.908	34.930	34.953	34.975	34.998	35.020	35.043	1540
1550	35.043	35.065	35.088	35.110	35.133	35.156	35.178	35.201	35.223	35.246	35.268	1550
1560	35.268	35.291	35.313	35.336	35.358	35.381	35.403	35.426	35.448	35.471	35.493	1560
1570	35.493	35.516	35.538	35.560	35.583	35.605	35.628	35.650	35.673	35.695	35.718	1570
1580	35.718	35.740	35.763	35.785	35.807	35.830	35.852	35.875	35.897	35.920	35.942	1580
1590	35.942	35.964	35.987	36.009	36.032	36.054	36.076	36.099	36.121	36.144	36.166	1590
1600	36.166	36.188	36.211	36.233	36.256	36.278	36.300	36.323	36.345	36.367	36.390	1600
1610	36.390	36.412	36.434	36.457	36.479	36.501	36.524	36.546	36.568	36.591	36.613	1610
1620	36.613	36.635	36.658	36.680	36.702	36.725	36.747	36.769	36.792	36.814	36.836	1620
1630	36.836	36.859	36.881	36.903	36.925	36.948	36.970	36.992	37.014	37.037	37.059	1630
1640	37.059	37.081	37.104	37.126	37.148	37.170	37.193	37.215	37.237	37.259	37.281	1640
1650	37.281	37.304	37.326	37.348	37.370	37.393	37.415	37.437	37.459	37.481	37.504	1650
1660	37.504	37.526	37.548	37.570	37.592	37.615	37.637	37.659	37.681	37.703	37.725	1660
1670	37.725	37.748	37.770	37.792	37.814	37.836	37.858	37.881	37.903	37.925	37.947	1670
1680	37.947	37.969	37.991	38.013	38.036	38.058	38.080	38.102	38.124	38.146	38.168	1680
1690	38.168	38.190	38.212	38.235	38.257	38.279	38.301	38.323	38.345	38.367	38.389	1690
1700	38.389	38.411	38.433	38.455	38.477	38.499	38.522	38.544	38.566	38.588	38.610	1700
1710	38.610	38.632	38.654	38.676	38.698	38.720	38.742	38.764	38.786	38.808	38.830	1710
1720	38.830	38.852	38.874	38.896	38.918	38.940	38.962	38.984	39.006	39.028	39.050	1720
1730	39.050	39.072	39.094	39.116	39.138	39.160	39.182	39.204	39.226	39.248	39.270	1730
1740	39.270	39.292	39.314	39.335	39.357	39.379	39.401	39.423	39.445	39.467	39.489	1740
1750	39.489	39.511	39.533	39.555	39.577	39.599	39.620	39.642	39.664	39.686	39.708	1750
1760	39.708	39.730	39.752	39.774	39.796	39.817	39.839	39.861	39.883	39.905	39.927	1760
1770	39.927	39.949	39.970	39.992	40.014	40.036	40.058	40.080	40.101	40.123	40.145	1770
1780	40.145	40.167	40.189	40.211	40.232	40.254	40.276	40.298	40.320	40.341	40.363	1780
1790	40.363	40.385	40.407	40.429	40.450	40.472	40.494	40.516	40.537	40.559	40.581	1790
1800	40.581	40.603	40.624	40.646	40.668	40.690	40.711	40.733	40.755	40.777	40.798	1800
1810	40.798	40.820	40.842	40.864	40.885	40.907	40.929	40.950	40.972	40.994	41.015	1810
1820	41.015	41.037	41.059	41.081	41.102	41.124	41.146	41.167	41.189	41.211	41.232	1820
1830	41.232	41.254	41.276	41.297	41.319	41.341	41.362	41.384	41.405	41.427	41.449	1830
1840	41.449	41.470	41.492	41.514	41.535	41.557	41.578	41.600	41.622	41.643	41.665	1840
1850	41.665	41.686	41.708	41.730	41.751	41.773	41.794	41.816	41.838	41.859	41.881	1850
1860	41.881	41.902	41.924	41.945	41.967	41.988	42.010	42.032	42.053	42.075	42.096	1860
1870	42.096	42.118	42.139	42.161	42.182	42.204	42.225	42.247	42.268	42.290	42.311	1870
1880	42.311	42.333	42.354	42.376	42.397	42.419	42.440	42.462	42.483	42.505	42.526	1880
1890	42.526	42.548	42.569	42.591	42.612	42.633	42.655	42.676	42.698	42.719	42.741	1890
°F	0	1	2	3	4	5	6	7	8	9	10	°F

Table A2 – Thermocouple Table: Type K (continued)

°F	0	1	2	3	4	5	6	7	8	9	10	°F
1900	42.741	42.762	42.783	42.805	42.826	42.848	42.869	42.891	42.912	42.933	42.955	1900
1910	42.955	42.976	42.998	43.019	43.040	43.062	43.083	43.104	43.126	43.147	43.169	1910
1920	43.169	43.190	43.211	43.233	43.254	43.275	43.297	43.318	43.339	43.361	43.382	1920
1930	43.382	43.403	43.425	43.446	43.467	43.489	43.510	43.531	43.552	43.574	43.595	1930
1940	43.595	43.616	43.638	43.659	43.680	43.701	43.723	43.744	43.765	43.787	43.808	1940
1950	43.808	43.829	43.850	43.872	43.893	43.914	43.935	43.957	43.978	43.999	44.020	1950
1960	44.020	44.041	44.063	44.084	44.105	44.126	44.147	44.169	44.190	44.211	44.232	1960
1970	44.232	44.253	44.275	44.296	44.317	44.338	44.359	44.380	44.402	44.423	44.444	1970
1980	44.444	44.465	44.486	44.507	44.528	44.550	44.571	44.592	44.613	44.634	44.655	1980
1990	44.655	44.676	44.697	44.719	44.740	44.761	44.782	44.803	44.824	44.845	44.866	1990
2000	44.866	44.887	44.908	44.929	44.950	44.971	44.992	45.014	45.035	45.056	45.077	2000
2010	45.077	45.098	45.119	45.140	45.161	45.182	45.203	45.224	45.245	45.266	45.287	2010
2020	45.287	45.308	45.329	45.350	45.371	45.392	45.413	45.434	45.455	45.476	45.497	2020
2030	45.497	45.518	45.539	45.560	45.580	45.601	45.622	45.643	45.664	45.685	45.706	2030
2040	45.706	45.727	45.748	45.769	45.790	45.811	45.832	45.852	45.873	45.894	45.915	2040
2050	45.915	45.936	45.957	45.978	45.999	46.019	46.040	46.061	46.082	46.103	46.124	2050
2060	46.124	46.145	46.165	46.186	46.207	46.228	46.249	46.269	46.290	46.311	46.332	2060
2070	46.332	46.353	46.373	46.394	46.415	46.436	46.457	46.477	46.498	46.519	46.540	2070
2080	46.540	46.560	46.581	46.602	46.623	46.643	46.664	46.685	46.706	46.726	46.747	2080
2090	46.747	46.768	46.789	46.809	46.830	46.851	46.871	46.892	46.913	46.933	46.954	2090
2100	46.954	46.975	46.995	47.016	47.037	47.057	47.078	47.099	47.119	47.140	47.161	2100
2110	47.161	47.181	47.202	47.223	47.243	47.264	47.284	47.305	47.326	47.346	47.367	2110
2120	47.367	47.387	47.408	47.429	47.449	47.470	47.490	47.511	47.531	47.552	47.573	2120
2130	47.573	47.593	47.614	47.634	47.655	47.675	47.696	47.716	47.737	47.757	47.778	2130
2140	47.778	47.798	47.819	47.839	47.860	47.880	47.901	47.921	47.942	47.962	47.983	2140
2150	47.983	48.003	48.024	48.044	48.065	48.085	48.105	48.126	48.146	48.167	48.187	2150
2160	48.187	48.208	48.228	48.248	48.269	48.289	48.310	48.330	48.350	48.371	48.391	2160
2170	48.391	48.411	48.432	48.452	48.473	48.493	48.513	48.534	48.554	48.574	48.595	2170
2180	48.595	48.615	48.635	48.656	48.676	48.696	48.717	48.737	48.757	48.777	48.798	2180
2190	48.798	48.818	48.838	48.859	48.879	48.899	48.919	48.940	48.960	48.980	49.000	2190
2200	49.000	49.021	49.041	49.061	49.081	49.101	49.122	49.142	49.162	49.182	49.202	2200
2210	49.202	49.223	49.243	49.263	49.283	49.303	49.323	49.344	49.364	49.384	49.404	2210
2220	49.404	49.424	49.444	49.465	49.485	49.505	49.525	49.545	49.565	49.585	49.605	2220
2230	49.605	49.625	49.645	49.666	49.686	49.706	49.726	49.746	49.766	49.786	49.806	2230
2240	49.806	49.826	49.846	49.866	49.886	49.906	49.926	49.946	49.966	49.986	50.006	2240
2250	50.006	50.026	50.046	50.066	50.086	50.106	50.126	50.146	50.166	50.186	50.206	2250
2260	50.206	50.226	50.246	50.266	50.286	50.306	50.326	50.346	50.366	50.385	50.405	2260
2270	50.405	50.425	50.445	50.465	50.485	50.505	50.525	50.545	50.564	50.584	50.604	2270
2280	50.604	50.624	50.644	50.664	50.684	50.703	50.723	50.743	50.763	50.783	50.802	2280
2290	50.802	50.822	50.842	50.862	50.882	50.901	50.921	50.941	50.961	50.981	51.000	2290
2300	51.000	51.020	51.040	51.060	51.079	51.099	51.119	51.139	51.158	51.178	51.198	2300
2310	51.198	51.217	51.237	51.257	51.276	51.296	51.316	51.336	51.355	51.375	51.395	2310
2320	51.395	51.414	51.434	51.453	51.473	51.493	51.512	51.532	51.552	51.571	51.591	2320
2330	51.591	51.611	51.630	51.650	51.669	51.689	51.708	51.728	51.748	51.767	51.787	2330
2340	51.787	51.806	51.826	51.845	51.865	51.885	51.904	51.924	51.943	51.963	51.982	2340
2350	51.982	52.002	52.021	52.041	52.060	52.080	52.099	52.119	52.138	52.158	52.177	2350
2360	52.177	52.197	52.216	52.235	52.255	52.274	52.294	52.313	52.333	52.352	52.371	2360
2370	52.371	52.391	52.410	52.430	52.449	52.468	52.488	52.507	52.527	52.546	52.565	2370
2380	52.565	52.585	52.604	52.623	52.643	52.662	52.681	52.701	52.720	52.739	52.759	2380
239052.759		52.778	52.797	52.817	52.836	52.855	52.875	52.894	52.913	52.932	52.952	2390
2400	52.952	52.971	52.990	53.010	53.029	53.048	53.067	53.087	53.106	53.125	53.144	2400
2410	53.144	53.163	53.183	53.202	53.221	53.240	53.260	53.279	53.298	53.317	53.336	2410
2420	53.336	53.355	53.375	53.394	53.413	53.432	53.451	53.470	53.490	53.509	53.528	2420
2430	53.528	53.547	53.566	53.585	53.604	53.623	53.643	53.662	53.681	53.700	53.719	2430
2440	53.719	53.738	53.757	53.776	53.795	53.814	53.833	53.852	53.871	53.890	53.910	2440
2450	53.910	53.929	53.948	53.967	53.986	54.005	54.024	54.043	54.062	54.081	54.100	2450
2460	54.100	54.119	54.138	54.157	54.176	54.195	54.214	54.233	54.252	54.271	54.289	2460
2470	54.289	54.308	54.327	54.346	54.365	54.384	54.403	54.422	54.441	54.460	54.479	2470
2480	54.479	54.498	54.517	54.536	54.554	54.573	54.592	54.611	54.630	54.649	54.668	2480
2490	54.668	54.687	54.705	54.724	54.743	54.762	54.781	54.800	54.819	54.837	54.856	2490
2500	54.856	54.875	54.894									2500
°F	0	1	2	3	4	5	6	7	8	9	10	°F

Table A3 – Thermocouple Table: Type E

Thermoelectric Voltage in Millivolts

°F	-10	-9	-8	-7	-6	-5	-4	-3	-2	-1	0	-10
-450							-9.835	-9.834	-9.833	-9.832	-9.830	-450
-440	-9.830	-9.829	-9.827	-9.825	-9.823	-9.821	-9.819	-9.817	-9.814	-9.812	-9.809	-440
-430	-9.809	-9.806	-9.803	-9.800	-9.797	-9.793	-9.790	-9.786	-9.782	-9.779	-9.775	-430
-420	-9.775	-9.771	-9.766	-9.762	-9.758	-9.753	-9.749	-9.744	-9.739	-9.734	-9.729	-420
-410	-9.729	-9.724	-9.718	-9.713	-9.707	-9.702	-9.696	-9.690	-9.684	-9.678	-9.672	-410
-400	-9.672	-9.666	-9.659	-9.653	-9.646	-9.639	-9.632	-9.625	-9.618	-9.611	-9.604	-400
-390	-9.604	-9.597	-9.589	-9.581	-9.574	-9.566	-9.558	-9.550	-9.542	-9.534	-9.525	-390
-380	-9.525	-9.517	-9.508	-9.500	-9.491	-9.482	-9.473	-9.464	-9.455	-9.446	-9.436	-380
-370	-9.436	-9.427	-9.417	-9.408	-9.398	-9.388	-9.378	-9.368	-9.358	-9.348	-9.338	-370
-360	-9.338	-9.327	-9.317	-9.306	-9.295	-9.285	-9.274	-9.263	-9.252	-9.241	-9.229	-360
-350	-9.229	-9.218	-9.207	-9.195	-9.184	-9.172	-9.160	-9.148	-9.136	-9.124	-9.112	-350
-340	-9.112	-9.100	-9.088	-9.075	-9.063	-9.050	-9.038	-9.025	-9.012	-8.999	-8.986	-340
-330	-8.986	-8.973	-8.960	-8.947	-8.934	-8.920	-8.907	-8.893	-8.880	-8.866	-8.852	-330
-320	-8.852	-8.839	-8.825	-8.811	-8.797	-8.782	-8.768	-8.754	-8.739	-8.725	-8.710	-320
-310	-8.710	-8.696	-8.681	-8.666	-8.652	-8.637	-8.622	-8.607	-8.591	-8.576	-8.561	-310
-300	-8.561	-8.546	-8.530	-8.515	-8.499	-8.483	-8.468	-8.452	-8.436	-8.420	-8.404	-300
-290	-8.404	-8.388	-8.372	-8.356	-8.339	-8.323	-8.307	-8.290	-8.273	-8.257	-8.240	-290
-280	-8.240	-8.223	-8.206	-8.189	-8.173	-8.155	-8.138	-8.121	-8.104	-8.087	-8.069	-280
-270	-8.069	-8.052	-8.034	-8.017	-7.999	-7.981	-7.963	-7.945	-7.928	-7.910	-7.891	-270
-260	-7.891	-7.873	-7.855	-7.837	-7.819	-7.800	-7.782	-7.763	-7.745	-7.726	-7.707	-260
-250	-7.707	-7.688	-7.670	-7.651	-7.632	-7.613	-7.593	-7.574	-7.555	-7.536	-7.516	-250
-240	-7.516	-7.497	-7.478	-7.458	-7.438	-7.419	-7.399	-7.379	-7.359	-7.339	-7.319	-240
-230	-7.319	-7.299	-7.279	-7.259	-7.239	-7.219	-7.198	-7.178	-7.157	-7.137	-7.116	-230
-220	-7.116	-7.096	-7.075	-7.054	-7.033	-7.013	-6.992	-6.971	-6.950	-6.928	-6.907	-220
-210	-6.907	-6.886	-6.865	-6.843	-6.822	-6.801	-6.779	-6.757	-6.736	-6.714	-6.692	-210
-200	-6.692	-6.671	-6.649	-6.627	-6.605	-6.583	-6.561	-6.539	-6.516	-6.494	-6.472	-200
-190	-6.472	-6.449	-6.427	-6.405	-6.382	-6.359	-6.337	-6.314	-6.291	-6.269	-6.246	-190
-180	-6.246	-6.223	-6.200	-6.177	-6.154	-6.130	-6.107	-6.084	-6.061	-6.037	-6.014	-180
-170	-6.014	-5.991	-5.967	-5.943	-5.920	-5.896	-5.872	-5.849	-5.825	-5.801	-5.777	-170
-160	-5.777	-5.753	-5.729	-5.705	-5.681	-5.656	-5.632	-5.608	-5.584	-5.559	-5.535	-160
-150	-5.535	-5.510	-5.486	-5.461	-5.436	-5.412	-5.387	-5.362	-5.337	-5.312	-5.287	-150
-140	-5.287	-5.262	-5.237	-5.212	-5.187	-5.162	-5.136	-5.111	-5.086	-5.060	-5.035	-140
-130	-5.035	-5.009	-4.984	-4.958	-4.932	-4.907	-4.881	-4.855	-4.829	-4.803	-4.777	-130
-120	-4.777	-4.751	-4.725	-4.699	-4.673	-4.647	-4.621	-4.594	-4.568	-4.542	-4.515	-120
-110	-4.515	-4.489	-4.462	-4.436	-4.409	-4.382	-4.355	-4.329	-4.302	-4.275	-4.248	-110
-100	-4.248	-4.221	-4.194	-4.167	-4.140	-4.113	-4.086	-4.058	-4.031	-4.004	-3.976	-100
-90	-3.976	-3.949	-3.922	-3.894	-3.867	-3.839	-3.811	-3.784	-3.756	-3.728	-3.700	-90
-80	-3.700	-3.672	-3.645	-3.617	-3.589	-3.561	-3.532	-3.504	-3.476	-3.448	-3.420	-80
-70	-3.420	-3.391	-3.363	-3.335	-3.306	-3.278	-3.249	-3.221	-3.192	-3.163	-3.135	-70
-60	-3.135	-3.106	-3.077	-3.048	-3.020	-2.991	-2.962	-2.933	-2.904	-2.875	-2.846	-60
-50	-2.846	-2.816	-2.787	-2.758	-2.729	-2.699	-2.670	-2.641	-2.611	-2.582	-2.552	-50
-40	-2.552	-2.523	-2.493	-2.463	-2.434	-2.404	-2.374	-2.344	-2.315	-2.285	-2.255	-40
-30	-2.255	-2.225	-2.195	-2.165	-2.135	-2.105	-2.074	-2.044	-2.014	-1.984	-1.953	-30
-20	-1.953	-1.923	-1.893	-1.862	-1.832	-1.801	-1.771	-1.740	-1.709	-1.679	-1.648	-20
-10	-1.648	-1.617	-1.587	-1.556	-1.525	-1.494	-1.463	-1.432	-1.401	-1.370	-1.339	-10
0	-1.339	-1.308	-1.277	-1.245	-1.214	-1.183	-1.152	-1.120	-1.089	-1.057	-1.026	0
0	-1.026	-0.994	-0.963	-0.931	-0.900	-0.868	-0.836	-0.805	-0.773	-0.741	-0.709	0
10	-0.709	-0.677	-0.645	-0.614	-0.582	-0.550	-0.517	-0.485	-0.453	-0.421	-0.389	10
20	-0.389	-0.357	-0.324	-0.292	-0.260	-0.227	-0.195	-0.163	-0.130	-0.098	-0.065	20
30	-0.065	-0.033	0.000	0.033	0.065	0.098	0.131	0.163	0.196	0.229	0.262	30
40	0.262	0.294	0.327	0.360	0.393	0.426	0.459	0.492	0.525	0.558	0.591	40
50	0.591	0.624	0.657	0.691	0.724	0.757	0.790	0.824	0.857	0.890	0.924	50
60	0.924	0.957	0.990	1.024	1.057	1.091	1.124	1.158	1.192	1.225	1.259	60
70	1.259	1.292	1.326	1.360	1.394	1.427	1.461	1.495	1.529	1.563	1.597	70
80	1.597	1.631	1.665	1.699	1.733	1.767	1.801	1.835	1.869	1.904	1.938	80
90	1.938	1.972	2.006	2.041	2.075	2.109	2.144	2.178	2.212	2.247	2.281	90
°F	0	1	2	3	4	5	6	7	8	9	10	-10

°F	0	1	2	3	4	5	6	7	8	9	10	°F
100	2.281	2.316	2.351	2.385	2.420	2.454	2.489	2.524	2.558	2.593	2.628	100
110	2.628	2.663	2.698	2.733	2.767	2.802	2.837	2.872	2.907	2.942	2.977	110
120	2.977	3.012	3.048	3.083	3.118	3.153	3.188	3.224	3.259	3.294	3.330	120
130	3.330	3.365	3.400	3.436	3.471	3.507	3.542	3.578	3.613	3.649	3.685	130
140	3.685	3.720	3.756	3.792	3.827	3.863	3.899	3.935	3.970	4.006	4.042	140
150	4.042	4.078	4.114	4.150	4.186	4.222	4.258	4.294	4.330	4.366	4.403	150
160	4.403	4.439	4.475	4.511	4.547	4.584	4.620	4.656	4.693	4.729	4.766	160
170	4.766	4.802	4.839	4.875	4.912	4.948	4.985	5.021	5.058	5.095	5.131	170
180	5.131	5.168	5.205	5.242	5.278	5.315	5.352	5.389	5.426	5.463	5.500	180
190	5.500	5.537	5.574	5.611	5.648	5.685	5.722	5.759	5.796	5.833	5.871	190
200	5.871	5.908	5.945	5.982	6.020	6.057	6.094	6.132	6.169	6.207	6.244	200
210	6.244	6.281	6.319	6.356	6.394	6.432	6.469	6.507	6.544	6.582	6.620	210
220	6.620	6.658	6.695	6.733	6.771	6.809	6.847	6.884	6.922	6.960	6.998	220
230	6.998	7.036	7.074	7.112	7.150	7.188	7.226	7.264	7.302	7.341	7.379	230
240	7.379	7.417	7.455	7.493	7.532	7.570	7.608	7.647	7.685	7.723	7.762	240
250	7.762	7.800	7.839	7.877	7.916	7.954	7.993	8.031	8.070	8.108	8.147	250
260	8.147	8.186	8.224	8.263	8.302	8.340	8.379	8.418	8.457	8.496	8.535	260
270	8.535	8.573	8.612	8.651	8.690	8.729	8.768	8.807	8.846	8.885	8.924	270
280	8.924	8.963	9.002	9.041	9.081	9.120	9.159	9.198	9.237	9.277	9.316	280
290	9.316	9.355	9.395	9.434	9.473	9.513	9.552	9.591	9.631	9.670	9.710	290
300	9.710	9.749	9.789	9.828	9.868	9.907	9.947	9.987	10.026	10.066	10.106	300
310	10.106	10.145	10.185	10.225	10.265	10.304	10.344	10.384	10.424	10.464	10.503	310
320	10.503	10.543	10.583	10.623	10.663	10.703	10.743	10.783	10.823	10.863	10.903	320
330	10.903	10.943	10.983	11.024	11.064	11.104	11.144	11.184	11.224	11.265	11.305	330
340	11.305	11.345	11.385	11.426	11.466	11.506	11.547	11.587	11.627	11.668	11.708	340
350	11.708	11.749	11.789	11.830	11.870	11.911	11.951	11.992	12.032	12.073	12.113	350
360	12.113	12.154	12.195	12.235	12.276	12.317	12.357	12.398	12.439	12.480	12.520	360
370	12.520	12.561	12.602	12.643	12.684	12.724	12.765	12.806	12.847	12.888	12.929	370
380	12.929	12.970	13.011	13.052	13.093	13.134	13.175	13.216	13.257	13.298	13.339	380
390	13.339	13.380	13.421	13.462	13.504	13.545	13.586	13.627	13.668	13.710	13.751	390
400	13.751	13.792	13.833	13.875	13.916	13.957	13.999	14.040	14.081	14.123	14.164	400
410	14.164	14.205	14.247	14.288	14.330	14.371	14.413	14.454	14.496	14.537	14.579	410
420	14.579	14.620	14.662	14.704	14.745	14.787	14.828	14.870	14.912	14.953	14.995	420
430	14.995	15.037	15.078	15.120	15.162	15.204	15.245	15.287	15.329	15.371	15.413	430
440	15.413	15.454	15.496	15.538	15.580	15.622	15.664	15.706	15.748	15.790	15.831	440
450	15.831	15.873	15.915	15.957	15.999	16.041	16.083	16.125	16.168	16.210	16.252	450
460	16.252	16.294	16.336	16.378	16.420	16.462	16.504	16.547	16.589	16.631	16.673	460
470	16.673	16.715	16.758	16.800	16.842	16.884	16.927	16.969	17.011	17.054	17.096	470
480	17.096	17.138	17.181	17.223	17.265	17.308	17.350	17.392	17.435	17.477	17.520	480
490	17.520	17.562	17.605	17.647	17.690	17.732	17.775	17.817	17.860	17.902	17.945	490
500	17.945	17.987	18.030	18.073	18.115	18.158	18.200	18.243	18.286	18.328	18.371	500
510	18.371	18.414	18.456	18.499	18.542	18.585	18.627	18.670	18.713	18.756	18.798	510
520	18.798	18.841	18.884	18.927	18.969	19.012	19.055	19.098	19.141	19.184	19.227	520
530	19.227	19.269	19.312	19.355	19.398	19.441	19.484	19.527	19.570	19.613	19.656	530
540	19.656	19.699	19.742	19.785	19.828	19.871	19.914	19.957	20.000	20.043	20.086	540
550	20.086	20.129	20.172	20.216	20.259	20.302	20.345	20.388	20.431	20.474	20.517	550
560	20.517	20.561	20.604	20.647	20.690	20.733	20.777	20.820	20.863	20.906	20.950	560
570	20.950	20.993	21.036	21.080	21.123	21.166	21.209	21.253	21.296	21.339	21.383	570
580	21.383	21.426	21.470	21.513	21.556	21.600	21.643	21.686	21.730	21.773	21.817	580
590	21.817	21.860	21.904	21.947	21.991	22.034	22.078	22.121	22.165	22.208	22.252	590
600	22.252	22.295	22.339	22.382	22.426	22.469	22.513	22.556	22.600	22.644	22.687	600
610	22.687	22.731	22.774	22.818	22.862	22.905	22.949	22.993	23.036	23.080	23.124	610
620	23.124	23.167	23.211	23.255	23.298	23.342	23.386	23.429	23.473	23.517	23.561	620
630	23.561	23.604	23.648	23.692	23.736	23.780	23.823	23.867	23.911	23.955	23.999	630
640	23.999	24.042	24.086	24.130	24.174	24.218	24.262	24.305	24.349	24.393	24.437	640
650	24.437	24.481	24.525	24.569	24.613	24.657	24.701	24.745	24.789	24.832	24.876	650
660	24.876	24.920	24.964	25.008	25.052	25.096	25.140	25.184	25.228	25.272	25.316	660
670	25.316	25.360	25.404	25.448	25.493	25.537	25.581	25.625	25.669	25.713	25.757	670
680	25.757	25.801	25.845	25.889	25.933	25.977	26.022	26.066	26.110	26.154	26.198	680
690	26.198	26.242	26.286	26.331	26.375	26.419	26.463	26.507	26.552	26.596	26.640	690
°F	0	1	2	3	4	5	6	7	8	9	10	°F

Table A3 – Thermocouple Table: Type E (continued)

°F	0	1	2	3	4	5	6	7	8	9	10	°F
700	26.640	26.684	26.728	26.773	26.817	26.861	26.905	26.950	26.994	27.038	27.082	700
710	27.082	27.127	27.171	27.215	27.259	27.304	27.348	27.392	27.437	27.481	27.525	710
720	27.525	27.570	27.614	27.658	27.703	27.747	27.791	27.836	27.880	27.924	27.969	720
730	27.969	28.013	28.057	28.102	28.146	28.191	28.235	28.279	28.324	28.368	28.413	730
740	28.413	28.457	28.501	28.546	28.590	28.635	28.679	28.724	28.768	28.813	28.857	740
750	28.857	28.901	28.946	28.990	29.035	29.079	29.124	29.168	29.213	29.257	29.302	750
760	29.302	29.346	29.391	29.435	29.480	29.525	29.569	29.614	29.658	29.703	29.747	760
770	29.747	29.792	29.836	29.881	29.925	29.970	30.015	30.059	30.104	30.148	30.193	770
780	30.193	30.238	30.282	30.327	30.371	30.416	30.461	30.505	30.550	30.595	30.639	780
790	30.639	30.684	30.728	30.773	30.818	30.862	30.907	30.952	30.996	31.041	31.086	790
800	31.086	31.130	31.175	31.220	31.264	31.309	31.354	31.398	31.443	31.488	31.533	800
810	31.533	31.577	31.622	31.667	31.711	31.756	31.801	31.846	31.890	31.935	31.980	810
820	31.980	32.025	32.069	32.114	32.159	32.204	32.248	32.293	32.338	32.383	32.427	820
830	32.427	32.472	32.517	32.562	32.606	32.651	32.696	32.741	32.786	32.830	32.875	830
840	32.875	32.920	32.965	33.010	33.054	33.099	33.144	33.189	33.234	33.278	33.323	840
850	33.323	33.368	33.413	33.458	33.503	33.547	33.592	33.637	33.682	33.727	33.772	850
860	33.772	33.816	33.861	33.906	33.951	33.996	34.041	34.086	34.130	34.175	34.220	860
870	34.220	34.265	34.310	34.355	34.400	34.445	34.489	34.534	34.579	34.624	34.669	870
880	34.669	34.714	34.759	34.804	34.849	34.893	34.938	34.983	35.028	35.073	35.118	880
890	35.118	35.163	35.208	35.253	35.298	35.343	35.387	35.432	35.477	35.522	35.567	890
900	35.567	35.612	35.657	35.702	35.747	35.792	35.837	35.882	35.927	35.972	36.016	900
910	36.016	36.061	36.106	36.151	36.196	36.241	36.286	36.331	36.376	36.421	36.466	910
920	36.466	36.511	36.556	36.601	36.646	36.691	36.736	36.781	36.826	36.870	36.915	920
930	36.915	36.960	37.005	37.050	37.095	37.140	37.185	37.230	37.275	37.320	37.365	930
940	37.365	37.410	37.455	37.500	37.545	37.590	37.635	37.680	37.725	37.770	37.815	940
950	37.815	37.860	37.905	37.950	37.995	38.040	38.085	38.130	38.175	38.220	38.265	950
960	38.265	38.309	38.354	38.399	38.444	38.489	38.534	38.579	38.624	38.669	38.714	960
970	38.714	38.759	38.804	38.849	38.894	38.939	38.984	39.029	39.074	39.119	39.164	970
980	39.164	39.209	39.254	39.299	39.344	39.389	39.434	39.479	39.524	39.569	39.614	980
990	39.614	39.659	39.704	39.749	39.794	39.839	39.884	39.929	39.974	40.019	40.064	990
1000	40.064	40.109	40.154	40.199	40.243	40.288	40.333	40.378	40.423	40.468	40.513	1000
1010	40.513	40.558	40.603	40.648	40.693	40.738	40.783	40.828	40.873	40.918	40.963	1010
1020	40.963	41.008	41.053	41.098	41.143	41.188	41.233	41.278	41.323	41.368	41.412	1020
1030	41.412	41.457	41.502	41.547	41.592	41.637	41.682	41.727	41.772	41.817	41.862	1030
1040	41.862	41.907	41.952	41.997	42.042	42.087	42.132	42.176	42.221	42.266	42.311	1040
1050	42.311	42.356	42.401	42.446	42.491	42.536	42.581	42.626	42.671	42.715	42.760	1050
1060	42.760	42.805	42.850	42.895	42.940	42.985	43.030	43.075	43.120	43.165	43.209	1060
1070	43.209	43.254	43.299	43.344	43.389	43.434	43.479	43.524	43.569	43.613	43.658	1070
1080	43.658	43.703	43.748	43.793	43.838	43.883	43.928	43.972	44.017	44.062	44.107	1080
1090	44.107	44.152	44.197	44.242	44.286	44.331	44.376	44.421	44.466	44.511	44.555	1090
1100	44.555	44.600	44.645	44.690	44.735	44.780	44.824	44.869	44.914	44.959	45.004	1100
1110	45.004	45.049	45.093	45.138	45.183	45.228	45.273	45.317	45.362	45.407	45.452	1110
1120	45.452	45.497	45.541	45.586	45.631	45.676	45.720	45.765	45.810	45.855	45.900	1120
1130	45.900	45.944	45.989	46.034	46.079	46.123	46.168	46.213	46.258	46.302	46.347	1130
1140	46.347	46.392	46.437	46.481	46.526	46.571	46.616	46.660	46.705	46.750	46.794	1140
1150	46.794	46.839	46.884	46.929	46.973	47.018	47.063	47.107	47.152	47.197	47.241	1150
1160	47.241	47.286	47.331	47.375	47.420	47.465	47.509	47.554	47.599	47.643	47.688	1160
1170	47.688	47.733	47.777	47.822	47.867	47.911	47.956	48.001	48.045	48.090	48.135	1170
1180	48.135	48.179	48.224	48.268	48.313	48.358	48.402	48.447	48.492	48.536	48.581	1180
1190	48.581	48.625	48.670	48.715	48.759	48.804	48.848	48.893	48.937	48.982	49.027	1190
1200	49.027	49.071	49.116	49.160	49.205	49.249	49.294	49.338	49.383	49.428	49.472	1200
1210	49.472	49.517	49.561	49.606	49.650	49.695	49.739	49.784	49.828	49.873	49.917	1210
1220	49.917	49.962	50.006	50.051	50.095	50.140	50.184	50.229	50.273	50.318	50.362	1220
1230	50.362	50.407	50.451	50.495	50.540	50.584	50.629	50.673	50.718	50.762	50.807	1230
1240	50.807	50.851	50.895	50.940	50.984	51.029	51.073	51.118	51.162	51.206	51.251	1240
1250	51.251	51.295	51.340	51.384	51.428	51.473	51.517	51.561	51.606	51.650	51.695	1250
1260	51.695	51.739	51.783	51.828	51.872	51.916	51.961	52.005	52.049	52.094	52.138	1260
1270	52.138	52.182	52.227	52.271	52.315	52.360	52.404	52.448	52.493	52.537	52.581	1270
1280	52.581	52.625	52.670	52.714	52.758	52.803	52.847	52.891	52.935	52.980	53.024	1280
1290	53.024	53.068	53.112	53.157	53.201	53.245	53.289	53.334	53.378	53.422	53.466	1290
1300	53.466	53.510	53.555	53.599	53.643	53.687	53.732	53.776	53.820	53.864	53.908	1300
1310	53.908	53.952	53.997	54.041	54.085	54.129	54.173	54.218	54.262	54.306	54.350	1310
1320	54.350	54.394	54.438	54.482	54.527	54.571	54.615	54.659	54.703	54.747	54.791	1320
1330	54.791	54.835	54.879	54.924	54.968	55.012	55.056	55.100	55.144	55.188	55.232	1330
1340	55.232	55.276	55.320	55.364	55.408	55.453	55.497	55.541	55.585	55.629	55.673	1340
1350	55.673	55.717	55.761	55.805	55.849	55.893	55.937	55.981	56.025	56.069	56.113	1350
1360	56.113	56.157	56.201	56.245	56.289	56.333	56.377	56.421	56.465	56.509	56.553	1360
1370	56.553	56.597	56.641	56.685	56.729	56.773	56.816	56.860	56.904	56.948	56.992	1370
1380	56.992	57.036	57.080	57.124	57.168	57.212	57.256	57.300	57.344	57.387	57.431	1380
1390	57.431	57.475	57.519	57.563	57.607	57.651	57.695	57.738	57.782	57.826	57.870	1390
1400	57.870	57.914	57.958	58.002	58.045	58.089	58.133	58.177	58.221	58.265	58.308	1400
1410	58.308	58.352	58.396	58.440	58.484	58.527	58.571	58.615	58.659	58.702	58.746	1410
1420	58.746	58.790	58.834	58.878	58.921	58.965	59.009	59.053	59.096	59.140	59.184	1420
1430	59.184	59.228	59.271	59.315	59.359	59.402	59.446	59.490	59.534	59.577	59.621	1430
1440	59.621	59.665	59.708	59.752	59.796	59.839	59.883	59.927	59.970	60.014	60.058	1440
1450	60.058	60.101	60.145	60.189	60.232	60.276	60.320	60.363	60.407	60.451	60.494	1450
1460	60.494	60.538	60.581	60.625	60.669	60.712	60.756	60.799	60.843	60.887	60.930	1460
1470	60.930	60.974	61.017	61.061	61.105	61.148	61.192	61.235	61.279	61.322	61.366	1470
1480	61.366	61.409	61.453	61.496	61.540	61.583	61.627	61.671	61.714	61.758	61.801	1480
1490	61.801	61.845	61.888	61.932	61.975	62.018	62.062	62.105	62.149	62.192	62.236	1490
1500	62.236	62.279	62.323	62.366	62.410	62.453	62.496	62.540	62.583	62.627	62.670	1500
1510	62.670	62.714	62.757	62.800	62.844	62.887	62.931	62.974	63.017	63.061	63.104	1510
1520	63.104	63.148	63.191	63.234	63.278	63.321	63.364	63.408	63.451	63.494	63.538	1520
1530	63.538	63.581	63.624	63.668	63.711	63.754	63.798	63.841	63.884	63.927	63.971	1530
1540	63.971	64.014	64.057	64.101	64.144	64.187	64.230	64.274	64.317	64.360	64.403	1540
1550	64.403	64.447	64.490	64.533	64.576	64.619	64.663	64.706	64.749	64.792	64.835	1550
1560	64.835	64.879	64.922	64.965	65.008	65.051	65.094	65.138	65.181	65.224	65.267	1560
1570	65.267	65.310	65.353	65.396	65.440	65.483	65.526	65.569	65.612	65.655	65.698	1570
1580	65.698	65.741	65.784	65.827	65.871	65.914	65.957	66.000	66.043	66.086	66.129	1580
1590	66.129	66.172	66.215	66.258	66.301	66.344	66.387	66.430	66.473	66.516	66.559	1590
1600	66.559	66.602	66.645	66.688	66.731	66.774	66.817	66.860	66.903	66.946	66.989	1600
1610	66.989	67.031	67.074	67.117	67.160	67.203	67.246	67.289	67.332	67.375	67.418	1610
1620	67.418	67.460	67.503	67.546	67.589	67.632	67.675	67.718	67.760	67.803	67.846	1620
1630	67.846	67.889	67.932	67.974	68.017	68.060	68.103	68.146	68.188	68.231	68.274	1630
1640	68.274	68.317	68.359	68.402	68.445	68.488	68.530	68.573	68.616	68.659	68.701	1640
1650	68.701	68.744	68.787	68.829	68.872	68.915	68.957	69.000	69.043	69.085	69.128	1650
1660	69.128	69.171	69.213	69.256	69.298	69.341	69.384	69.426	69.469	69.511	69.554	1660
1670	69.554	69.597	69.639	69.682	69.724	69.767	69.809	69.852	69.894	69.937	69.979	1670
1680	69.979	70.022	70.064	70.107	70.149	70.192	70.234	70.277	70.319	70.362	70.404	1680
1690	70.404	70.447	70.489	70.531	70.574	70.616	70.659	70.701	70.744	70.786	70.828	1690
1700	70.828	70.871	70.913	70.955	70.998	71.040	71.082	71.125	71.167	71.209	71.252	1700
1710	71.252	71.294	71.336	71.379	71.421	71.463	71.506	71.548	71.590	71.632	71.675	1710
1720	71.675	71.717	71.759	71.801	71.844	71.886	71.928	71.970	72.012	72.055	72.097	1720
1730	72.097	72.139	72.181	72.223	72.266	72.308	72.350	72.392	72.434	72.476	72.518	1730
1740	72.518	72.561	72.603	72.645	72.687	72.729	72.771	72.813	72.855	72.897	72.939	1740
1750	72.939	72.981	73.023	73.066	73.108	73.150	73.192	73.234	73.276	73.318	73.360	1750
1760	73.360	73.402	73.444	73.486	73.528	73.570	73.612	73.654	73.696	73.738	73.780	1760
1770	73.780	73.821	73.863	73.905	73.947	73.989	74.031	74.073	74.115	74.157	74.199	1770
1780	74.199	74.241	74.283	74.324	74.366	74.408	74.450	74.492	74.534	74.576	74.618	1780
1790	74.618	74.659	74.701	74.743	74.785	74.827	74.869	74.910	74.952	74.994	75.036	1790
1800	75.036	75.078	75.120	75.161	75.203	75.245	75.287	75.329	75.370	75.412	75.454	1800
1810	75.454	75.496	75.538	75.579	75.621	75.663	75.705	75.746	75.788	75.830	75.872	1810
1820	75.872	75.913	75.955	75.997	76.039	76.081	76.122	76.164	76.206	76.248	76.289	1820
1830	76.289	76.331	76.373									1830
°F	0	1	2	3	4	5	6	7	8	9	10	°F

Table A4 – Thermocouple Table: Type T

Thermoelectric Voltage in Millivolts

°F	-10	-9	-8	-7	-6	-5	-4	-3	-2	-1	0	°F
-450							-6.258	-6.257	-6.256	-6.255	-6.254	-450
-440	-6.254	-6.253	-6.252	-6.251	-6.250	-6.248	-6.247	-6.245	-6.243	-6.242	-6.240	-440
-430	-6.240	-6.238	-6.236	-6.234	-6.232	-6.230	-6.227	-6.225	-6.222	-6.220	-6.217	-430
-420	-6.217	-6.215	-6.212	-6.209	-6.206	-6.203	-6.200	-6.197	-6.194	-6.191	-6.187	-420
-410	-6.187	-6.184	-6.180	-6.177	-6.173	-6.170	-6.166	-6.162	-6.158	-6.154	-6.150	-410
-400	-6.150	-6.146	-6.141	-6.137	-6.133	-6.128	-6.124	-6.119	-6.115	-6.110	-6.105	-400
-390	-6.105	-6.100	-6.095	-6.090	-6.085	-6.080	-6.075	-6.069	-6.064	-6.059	-6.053	-390
-380	-6.053	-6.047	-6.042	-6.036	-6.030	-6.025	-6.019	-6.013	-6.007	-6.001	-5.994	-380
-370	-5.994	-5.988	-5.982	-5.976	-5.969	-5.963	-5.956	-5.950	-5.943	-5.937	-5.930	-370
-360	-5.930	-5.923	-5.916	-5.909	-5.902	-5.896	-5.888	-5.881	-5.874	-5.867	-5.860	-360
-350	-5.860	-5.853	-5.845	-5.838	-5.830	-5.823	-5.815	-5.808	-5.800	-5.792	-5.785	-350
-340	-5.785	-5.777	-5.769	-5.761	-5.753	-5.745	-5.737	-5.729	-5.721	-5.713	-5.705	-340
-330	-5.705	-5.697	-5.688	-5.680	-5.672	-5.663	-5.655	-5.646	-5.638	-5.629	-5.620	-330
-320	-5.620	-5.612	-5.603	-5.594	-5.585	-5.577	-5.568	-5.559	-5.550	-5.541	-5.532	-320
-310	-5.532	-5.523	-5.513	-5.504	-5.495	-5.486	-5.476	-5.467	-5.458	-5.448	-5.439	-310
-300	-5.439	-5.429	-5.420	-5.410	-5.400	-5.391	-5.381	-5.371	-5.361	-5.351	-5.341	-300
-290	-5.341	-5.332	-5.322	-5.312	-5.301	-5.291	-5.281	-5.271	-5.261	-5.250	-5.240	-290
-280	-5.240	-5.230	-5.219	-5.209	-5.198	-5.188	-5.177	-5.167	-5.156	-5.145	-5.135	-280
-270	-5.135	-5.124	-5.113	-5.102	-5.091	-5.081	-5.070	-5.059	-5.048	-5.036	-5.025	-270
-260	-5.025	-5.014	-5.003	-4.992	-4.980	-4.969	-4.958	-4.946	-4.935	-4.923	-4.912	-260
-250	-4.912	-4.900	-4.889	-4.877	-4.865	-4.854	-4.842	-4.830	-4.818	-4.806	-4.794	-250
-240	-4.794	-4.783	-4.771	-4.759	-4.746	-4.734	-4.722	-4.710	-4.698	-4.685	-4.673	-240
-230	-4.673	-4.661	-4.648	-4.636	-4.624	-4.611	-4.599	-4.586	-4.573	-4.561	-4.548	-230
-220	-4.548	-4.535	-4.523	-4.510	-4.497	-4.484	-4.471	-4.458	-4.445	-4.432	-4.419	-220
-210	-4.419	-4.406	-4.393	-4.380	-4.366	-4.353	-4.340	-4.326	-4.313	-4.300	-4.286	-210
-200	-4.286	-4.273	-4.259	-4.246	-4.232	-4.218	-4.205	-4.191	-4.177	-4.163	-4.149	-200
-190	-4.149	-4.136	-4.122	-4.108	-4.094	-4.080	-4.066	-4.052	-4.037	-4.023	-4.009	-190
-180	-4.009	-3.995	-3.980	-3.966	-3.952	-3.937	-3.923	-3.908	-3.894	-3.879	-3.865	-180
-170	-3.865	-3.850	-3.836	-3.821	-3.806	-3.791	-3.777	-3.762	-3.747	-3.732	-3.717	-170
-160	-3.717	-3.702	-3.687	-3.672	-3.657	-3.642	-3.626	-3.611	-3.596	-3.581	-3.565	-160
-150	-3.565	-3.550	-3.535	-3.519	-3.504	-3.488	-3.473	-3.457	-3.441	-3.426	-3.410	-150
-140	-3.410	-3.394	-3.379	-3.363	-3.347	-3.331	-3.315	-3.299	-3.283	-3.267	-3.251	-140
-130	-3.251	-3.235	-3.219	-3.203	-3.187	-3.171	-3.154	-3.138	-3.122	-3.105	-3.089	-130
-120	-3.089	-3.072	-3.056	-3.040	-3.023	-3.006	-2.990	-2.973	-2.956	-2.940	-2.923	-120
-110	-2.923	-2.906	-2.889	-2.873	-2.856	-2.839	-2.822	-2.805	-2.788	-2.771	-2.754	-110
-100	-2.754	-2.737	-2.719	-2.702	-2.685	-2.668	-2.651	-2.633	-2.616	-2.598	-2.581	-100
-90	-2.581	-2.564	-2.546	-2.529	-2.511	-2.493	-2.476	-2.458	-2.440	-2.423	-2.405	-90
-80	-2.405	-2.387	-2.369	-2.351	-2.334	-2.316	-2.298	-2.280	-2.262	-2.244	-2.225	-80
-70	-2.225	-2.207	-2.189	-2.171	-2.153	-2.134	-2.116	-2.098	-2.079	-2.061	-2.043	-70
-60	-2.043	-2.024	-2.006	-1.987	-1.969	-1.950	-1.931	-1.913	-1.894	-1.875	-1.857	-60
-50	-1.857	-1.838	-1.819	-1.800	-1.781	-1.762	-1.743	-1.724	-1.705	-1.686	-1.667	-50
-40	-1.667	-1.648	-1.629	-1.610	-1.591	-1.572	-1.552	-1.533	-1.514	-1.494	-1.475	-40
-30	-1.475	-1.456	-1.436	-1.417	-1.397	-1.378	-1.358	-1.338	-1.319	-1.299	-1.279	-30
-20	-1.279	-1.260	-1.240	-1.220	-1.200	-1.181	-1.161	-1.141	-1.121	-1.101	-1.081	-20
-10	-1.081	-1.061	-1.041	-1.021	-1.001	-0.980	-0.960	-0.940	-0.920	-0.900	-0.879	-10
0	-0.879	-0.859	-0.839	-0.818	-0.798	-0.777	-0.757	-0.736	-0.716	-0.695	-0.675	0
0	-0.675	-0.654	-0.633	-0.613	-0.592	-0.571	-0.550	-0.530	-0.509	-0.488	-0.467	0
10	-0.467	-0.446	-0.425	-0.404	-0.383	-0.362	-0.341	-0.320	-0.299	-0.278	-0.256	10
20	-0.256	-0.235	-0.214	-0.193	-0.171	-0.150	-0.129	-0.107	-0.086	-0.064	-0.043	20
30	-0.043	-0.022	0.000	0.022	0.043	0.065	0.086	0.108	0.130	0.151	0.173	30
40	0.173	0.195	0.216	0.238	0.260	0.282	0.303	0.325	0.347	0.369	0.391	40
50	0.391	0.413	0.435	0.457	0.479	0.501	0.523	0.545	0.567	0.589	0.611	50
60	0.611	0.634	0.656	0.678	0.700	0.723	0.745	0.767	0.790	0.812	0.834	60
70	0.834	0.857	0.879	0.902	0.924	0.947	0.969	0.992	1.015	1.037	1.060	70
80	1.060	1.083	1.105	1.128	1.151	1.174	1.196	1.219	1.242	1.265	1.288	80
90	1.288	1.311	1.334	1.357	1.380	1.403	1.426	1.449	1.472	1.496	1.519	90
100	1.519	1.542	1.565	1.588	1.612	1.635	1.658	1.682	1.705	1.729	1.752	100
110	1.752	1.776	1.799	1.823	1.846	1.870	1.893	1.917	1.941	1.964	1.988	110
120	1.988	2.012	2.036	2.060	2.083	2.107	2.131	2.155	2.179	2.203	2.227	120
130	2.227	2.251	2.275	2.299	2.323	2.347	2.371	2.395	2.420	2.444	2.468	130
140	2.468	2.492	2.517	2.541	2.565	2.590	2.614	2.639	2.663	2.687	2.712	140
°F	0	1	2	3	4	5	6	7	8	9	10	

°F	0	1	2	3	4	5	6	7	8	9	10	°F
150	2.712	2.737	2.761	2.786	2.810	2.835	2.860	2.884	2.909	2.934	2.958	150
160	2.958	2.983	3.008	3.033	3.058	3.082	3.107	3.132	3.157	3.182	3.207	160
170	3.207	3.232	3.257	3.282	3.307	3.333	3.358	3.383	3.408	3.433	3.459	170
180	3.459	3.484	3.509	3.534	3.560	3.585	3.610	3.636	3.661	3.687	3.712	180
190	3.712	3.738	3.763	3.789	3.814	3.840	3.866	3.891	3.917	3.943	3.968	190
200	3.968	3.994	4.020	4.046	4.071	4.097	4.123	4.149	4.175	4.201	4.227	200
210	4.227	4.253	4.279	4.305	4.331	4.357	4.383	4.409	4.435	4.461	4.487	210
220	4.487	4.513	4.540	4.566	4.592	4.618	4.645	4.671	4.697	4.724	4.750	220
230	4.750	4.776	4.803	4.829	4.856	4.882	4.909	4.935	4.962	4.988	5.015	230
240	5.015	5.042	5.068	5.095	5.122	5.148	5.175	5.202	5.228	5.255	5.282	240
250	5.282	5.309	5.336	5.363	5.389	5.416	5.443	5.470	5.497	5.524	5.551	250
260	5.551	5.578	5.605	5.632	5.660	5.687	5.714	5.741	5.768	5.795	5.823	260
270	5.823	5.850	5.877	5.904	5.932	5.959	5.986	6.014	6.041	6.068	6.096	270
280	6.096	6.123	6.151	6.178	6.206	6.233	6.261	6.288	6.316	6.343	6.371	280
290	6.371	6.399	6.426	6.454	6.482	6.510	6.537	6.565	6.593	6.621	6.648	290
300	6.648	6.676	6.704	6.732	6.760	6.788	6.816	6.844	6.872	6.900	6.928	300
310	6.928	6.956	6.984	7.012	7.040	7.068	7.096	7.124	7.152	7.181	7.209	310
320	7.209	7.237	7.265	7.294	7.322	7.350	7.378	7.407	7.435	7.463	7.492	320
330	7.492	7.520	7.549	7.577	7.606	7.634	7.663	7.691	7.720	7.748	7.777	330
340	7.777	7.805	7.834	7.863	7.891	7.920	7.949	7.977	8.006	8.035	8.064	340
350	8.064	8.092	8.121	8.150	8.179	8.208	8.237	8.266	8.294	8.323	8.352	350
360	8.352	8.381	8.410	8.439	8.468	8.497	8.526	8.555	8.585	8.614	8.643	360
370	8.643	8.672	8.701	8.730	8.759	8.789	8.818	8.847	8.876	8.906	8.935	370
380	8.935	8.964	8.994	9.023	9.052	9.082	9.111	9.141	9.170	9.200	9.229	380
390	9.229	9.259	9.288	9.318	9.347	9.377	9.406	9.436	9.466	9.495	9.525	390
400	9.525	9.555	9.584	9.614	9.644	9.673	9.703	9.733	9.763	9.793	9.822	400
410	9.822	9.852	9.882	9.912	9.942	9.972	10.002	10.032	10.062	10.092	10.122	410
420	10.122	10.152	10.182	10.212	10.242	10.272	10.302	10.332	10.362	10.392	10.423	420
430	10.423	10.453	10.483	10.513	10.543	10.574	10.604	10.634	10.664	10.695	10.725	430
440	10.725	10.755	10.786	10.816	10.847	10.877	10.907	10.938	10.968	10.999	11.029	440
450	11.029	11.060	11.090	11.121	11.151	11.182	11.213	11.243	11.274	11.304	11.335	450
460	11.335	11.366	11.396	11.427	11.458	11.489	11.519	11.550	11.581	11.612	11.643	460
470	11.643	11.673	11.704	11.735	11.766	11.797	11.828	11.859	11.890	11.920	11.951	470
480	11.951	11.982	12.013	12.044	12.075	12.106	12.138	12.169	12.200	12.231	12.262	480
490	12.262	12.293	12.324	12.355	12.386	12.418	12.449	12.480	12.511	12.543	12.574	490
500	12.574	12.605	12.636	12.668	12.699	12.730	12.762	12.793	12.824	12.856	12.887	500
510	12.887	12.919	12.950	12.982	13.013	13.045	13.076	13.108	13.139	13.171	13.202	510
520	13.202	13.234	13.265	13.297	13.328	13.360	13.392	13.423	13.455	13.487	13.518	520
530	13.518	13.550	13.582	13.614	13.645	13.677	13.709	13.741	13.772	13.804	13.836	530
540	13.836	13.868	13.900	13.932	13.964	13.995	14.027	14.059	14.091	14.123	14.155	540
550	14.155	14.187	14.219	14.251	14.283	14.315	14.347	14.379	14.411	14.444	14.476	550
560	14.476	14.508	14.540	14.572	14.604	14.636	14.669	14.701	14.733	14.765	14.797	560
570	14.797	14.830	14.862	14.894	14.926	14.959	14.991	15.023	15.056	15.088	15.121	570
580	15.121	15.153	15.185	15.218	15.250	15.283	15.315	15.347	15.380	15.412	15.445	580
590	15.445	15.477	15.510	15.543	15.575	15.608	15.640	15.673	15.705	15.738	15.771	590
600	15.771	15.803	15.836	15.869	15.901	15.934	15.967	15.999	16.032	16.065	16.098	600
610	16.098	16.130	16.163	16.196	16.229	16.262	16.295	16.327	16.360	16.393	16.426	610
620	16.426	16.459	16.492	16.525	16.558	16.591	16.624	16.657	16.690	16.723	16.756	620
630	16.756	16.789	16.822	16.855	16.888	16.921	16.954	16.987	17.020	17.053	17.086	630
640	17.086	17.120	17.153	17.186	17.219	17.252	17.286	17.319	17.352	17.385	17.418	640
650	17.418	17.452	17.485	17.518	17.552	17.585	17.618	17.652	17.685	17.718	17.752	650
660	17.752	17.785	17.819	17.852	17.886	17.919	17.952	17.986	18.019	18.053	18.086	660
670	18.086	18.120	18.153	18.187	18.221	18.254	18.288	18.321	18.355	18.389	18.422	670
680	18.422	18.456	18.490	18.523	18.557	18.591	18.624	18.658	18.692	18.725	18.759	680
690	18.759	18.793	18.827	18.861	18.894	18.928	18.962	18.996	19.030	19.064	19.097	690
700	19.097	19.131	19.165	19.199	19.233	19.267	19.301	19.335	19.369	19.403	19.437	700
710	19.437	19.471	19.505	19.539	19.573	19.607	19.641	19.675	19.709	19.743	19.777	710
720	19.777	19.811	19.845	19.879	19.913	19.947	19.982	20.016	20.050	20.084	20.118	720
730	20.118	20.152	20.187	20.221	20.255	20.289	20.323	20.358	20.392	20.426	20.460	730
740	20.460	20.495	20.529	20.563	20.597	20.632	20.666	20.700	20.735	20.769	20.803	740
750	20.803	20.838	20.872									750
°F	0	1	2	3	4	5	6	7	8	9	10	°F

Table A5 – Platinum 100 Ohm Resistance Temperature Detector Table in Ohms

°F	°C	Ω
0	−17.78	93.04
5	−15.00	94.12
10	−12.22	95.21
15	−9.44	96.31
20	−6.67	97.39
25	−3.89	98.48
30	−1.11	99.57
32	0.00	100.00
35	1.67	100.65
40	4.44	101.73
45	7.22	102.82
50	10.00	103.90
55	12.78	104.98
60	15.56	106.07
65	18.33	107.15
70	21.11	108.22
75	23.89	109.31
80	26.67	110.38
85	29.44	111.45
90	32.22	112.53
95	35.00	113.61
100	37.78	114.68
105	40.56	115.76
110	43.33	116.83
115	46.11	117.90
120	48.89	118.97
125	51.67	120.04
130	54.44	121.11
135	57.22	122.17
140	60.00	123.24
145	62.78	124.31
150	65.56	125.37
155	68.33	126.44
160	71.11	127.50
165	73.89	128.56
170	76.67	129.62

Table A5 – Platinum 100 Ohm Resistance Temperature Detector Table in Ohms (continued)

°F	°C	Ω
175	79.44	130.68
180	82.22	131.74
185	85.00	132.80
190	87.78	133.86
195	90.56	134.91
200	93.33	135.97
212	100.00	138.50
225	107.22	141.24
250	121.11	146.48
275	135.00	151.70
300	148.89	156.90

Table A6 – Properties of Water Specific Gravity and Pounds per Hour to Gallons per Minute Conversion

Temperature (°F)	Saturation Pressure (psia)	Weight (lb/gal)	Specific Gravity (60/60 °F)	Conversion Factor (lb/h to gpm)
32	0.0885	8.345	1.0013	0.00199
40	0.1217	8.345	1.0013	0.00199
50	0.1781	8.340	1.0007	0.00199
60	0.2653	8.334	1.0000	0.00199
70	0.3631	8.325	0.9989	0.00200
80	0.5069	8.314	0.9976	0.00200
90	0.6982	8.303	0.9963	0.00200
100	0.9492	8.289	0.9946	0.00201
110	1.2748	8.267	0.9919	0.00201
120	1.6924	8.253	0.9901	0.00201
130	2.2225	8.227	0.9872	0.00202
140	2.8886	8.207	0.9848	0.00203
150	3.718	8.182	0.9818	0.00203
160	4.741	8.156	0.9786	0.00204
170	5.992	8.127	0.9752	0.00205
180	7.510	8.098	0.9717	0.00205
190	9.339	8.068	0.9681	0.00206
200	11.526	8.039	0.9646	0.00207
210	14.123	8.005	0.9605	0.00208
212	14.696	7.996	0.9594	0.00208
220	17.186	7.972	0.9566	0.00209
240	24.969	7.901	0.9480	0.00210
260	35.429	7.822	0.9386	0.00211
280	49.203	7.746	0.9294	0.00215

Table A7 – Properties of Water: Specific Volume and Density

Temperature, *t* (°F)	Specific Volume, *v* (ft^3/lb)	Weight Density, *ρ*	
		(lb/ft^3)	(lb/gal)
32	0.01602	62.41	8.344
40	0.01602	62.43	8.345
50	0.01602	62.41	8.343
60	0.01603	62.37	8.338
70	0.01605	62.31	8.329
80	0.01607	62.22	8.318
90	0.01610	62.12	8.304
100	0.01613	62.00	8.288
110	0.01617	61.86	8.270
120	0.01620	61.71	8.250
130	0.01625	61.55	8.228
140	0.01629	61.38	8.205
150	0.01634	61.19	8.180
160	0.01640	60.99	8.154
170	0.01645	60.79	8.126
180	0.01651	60.57	8.097
190	0.01657	60.34	8.067
200	0.01664	60.11	8.035
210	0.01670	59.86	8.002
212	0.01672	59.81	7.996
220	0.01678	59.61	7.969
240	0.01693	59.08	7.898
260	0.01709	58.52	7.823
280	0.01726	57.92	7.743
300	0.01745	57.31	7.661
350	0.01799	55.59	7.431
400	0.01864	53.65	7.172
450	0.01943	51.47	6.880
500	0.02043	48.95	6.543
550	0.02176	45.96	6.143
600	0.02364	42.30	5.655
650	0.02674	37.40	4.999
700	0.03662	27.30	3.651

Table A8 – Properties of Water: Kinematic Viscosity Centistokes

°F	(cSt)	°F	(cSt)
32	1.93	120	0.609
35	1.82	125	0.582
40	1.66	130	0.562
45	1.53	135	0.534
50	1.41	140	0.514
55	1.30	145	0.493
60	1.22	150	0.472
65	1.13	155	0.457
70	1.05	160	0.440
75	0.988	165	0.426
80	0.929	170	0.411
85	0.870	175	0.397
90	0.825	180	0.384
95	0.782	185	0.372
100	0.738	190	0.362
105	0.698	195	0.351
110	0.668	200	0.341
115	0.637	212	0.318

Table A9 – Properties of Saturated Steam

Pressure (lb/in²) Absolute P	Gauge P	Temperature (°F)	Heat of the Liquid (Btu/lb)	Latent Heat of Evaporation (Btu/lb)	Total Heat of Steam, Hg (Btu/lb)	Specific Volume (ft³/lb)
14.696	0.0	212.00	180.07	970.3	1150.4	26.80
15.0	0.3	213.03	181.11	969.7	1150.8	26.29
16.0	1.3	216.32	184.42	967.6	1152.0	24.75
17.0	2.3	219.44	187.56	965.5	1153.1	23.39
18.0	3.3	222.41	190.56	963.6	1154.2	22.17
19.0	4.3	225.24	193.42	961.9	1155.3	21.08
20.0	5.3	227.96	196.16	960.1	1156.3	20.089
21.0	6.3	230.57	198.79	958.4	1157.2	19.192
22.0	7.3	233.07	201.33	956.8	1158.1	18.375
23.0	8.3	235.49	203.78	955.2	1159.0	17.627
24.0	9.3	237.82	206.14	953.7	1159.8	16.938
25.0	10.3	240.07	208.42	952.1	1160.6	16.303
26.0	11.3	242.25	210.62	950.7	1161.3	15.715
27.0	12.3	244.36	212.75	949.3	1162.0	15.170
28.0	13.3	246.41	214.83	947.9	1162.7	14.663
29.0	14.3	248.40	216.86	946.5	1163.4	14.189
30.0	15.3	250.33	218.82	945.3	1164.1	13.746
31.0	16.3	252.22	220.73	944.0	1164.7	13.330
32.0	17.3	254.05	222.59	942.8	1165.4	12.940
33.0	18.3	255.84	224.41	941.6	1166.0	12.572
34.0	19.3	257.58	226.18	940.3	1166.5	12.226
35.0	20.3	259.28	227.91	939.2	1167.1	11.898
36.0	21.3	260.95	229.60	938.0	1167.6	11.588
37.0	22.3	262.57	231.26	936.9	1168.2	11.294
38.0	23.3	264.16	232.89	935.8	1168.7	11.015
39.0	24.3	265.72	234.48	934.7	1169.2	10.750
40.0	25.3	267.25	236.03	933.7	1169.7	10.498
41.0	26.3	268.74	237.55	932.6	1170.2	10.258
42.0	27.3	270.21	239.04	931.6	1170.7	10.029
43.0	28.3	271.64	240.51	930.6	1171.1	9.810
44.0	29.3	273.05	241.95	929.6	1171.6	9.601
45.0	30.3	274.44	243.36	928.6	1172.0	9.401
46.0	31.3	275.80	244.75	927.7	1172.4	9.209
47.0	32.3	277.13	246.12	926.7	1172.9	9.025
48.0	33.3	278.45	247.47	925.8	1173.3	8.848
49.0	34.3	279.74	248.79	924.9	1173.7	8.678
50.0	35.3	281.01	250.09	924.0	1174.1	8.515
51.0	36.3	282.26	251.37	923.0	1174.4	8.359
52.0	37.3	283.49	252.63	922.2	1174.8	8.208

Table A9 – Properties of Saturated Steam (continued)

Pressure (lb/in²)		Temperature (°F)	Heat of the Liquid (Btu/lb)	Latent Heat of Evaporation (Btu/lb)	Total Heat of Steam, Hg (Btu/lb)	Specific Volume (ft³/lb)
Absolute P	Gauge P					
53.0	38.3	284.70	253.87	921.3	1175.2	8.062
54.0	39.3	285.90	255.09	920.5	1175.6	7.922
55.0	40.3	287.07	256.30	919.6	1175.9	7.787
56.0	41.3	288.23	257.50	918.8	1176.3	7.656
57.0	42.3	289.37	258.67	917.9	1176.6	7.529
58.0	43.3	290.50	259.82	917.1	1176.9	7.407
59.0	44.3	291.61	260.96	916.3	1177.3	7.289
60.0	45.3	292.71	262.09	915.5	1177.6	7.175
61.0	46.3	293.79	263.20	914.7	1177.9	7.064
62.0	47.3	294.85	264.30	913.9	1178.2	6.957
63.0	48.3	295.90	265.38	913.1	1178.5	6.853
64.0	49.3	296.94	266.45	912.3	1178.8	6.752
65.0	50.3	297.97	267.50	911.6	1179.1	6.655
66.0	51.3	298.99	268.55	910.8	1179.4	6.560
67.0	52.3	299.99	269.58	910.1	1179.7	6.468
68.0	53.3	300.98	270.60	909.4	1180.0	6.378
69.0	54.3	301.96	291.61	908.7	1180.3	6.291
70.0	55.3	302.92	272.61	907.9	1180.6	6.206
71.0	56.3	303.88	273.60	907.2	1180.8	6.124
72.0	57.3	304.83	274.57	906.5	1181.1	6.044
73.0	58.3	305.76	275.54	905.8	1181.3	5.966
74.0	59.3	306.68	276.49	905.1	1181.6	5.890
75.0	60.3	307.60	277.43	904.5	1181.9	5.816
76.0	61.3	308.50	278.37	903.7	1182.1	5.743
77.0	62.3	309.40	279.30	903.1	1182.4	5.673
78.0	63.3	310.29	280.21	902.4	1182.6	5.604
79.0	64.3	311.16	281.12	901.7	1182.8	5.537
80.0	65.3	312.03	282.02	901.1	1183.1	5.472
81.0	66.3	312.89	282.91	900.4	1183.3	5.408
82.0	67.3	313.74	283.79	899.7	1183.5	5.346
83.0	68.3	314.59	284.66	899.1	1183.8	5.285
84.0	69.3	315.42	285.53	898.5	1184.0	5.226
85.0	70.3	316.25	286.39	897.8	1184.2	5.168
86.0	71.3	317.07	287.24	897.2	1184.4	5.111
87.0	72.3	317.88	288.08	896.5	1184.6	5.055
88.0	73.3	318.68	288.91	895.9	1184.8	5.001
89.0	74.3	319.48	289.74	895.3	1185.1	4.948
90.0	75.3	320.27	290.56	894.7	1185.3	4.896
91.0	76.3	321.06	291.38	894.1	1185.5	4.845

Table A9 – Properties of Saturated Steam (continued)

Pressure (lb/in²)		Temperature (°F)	Heat of the Liquid (Btu/lb)	Latent Heat of Evaporation (Btu/lb)	Total Heat of Steam, Hg (Btu/lb)	Specific Volume (ft³/lb)
Absolute P	Gauge P					
92.0	77.3	321.83	292.18	893.5	1185.7	4.796
93.0	78.3	322.60	292.98	892.9	1185.9	4.747
94.0	79.3	323.36	293.78	892.3	1186.1	4.699
95.0	80.3	324.12	294.56	891.7	1186.2	4.652
96.0	81.3	324.87	295.34	891.1	1186.4	4.606
97.0	82.3	325.61	296.12	890.5	1186.6	4.561
98.0	83.3	326.35	296.89	889.9	1186.8	4.517
99.0	84.3	327.08	297.65	889.4	1187.0	4.474
100.0	85.3	327.81	298.40	888.8	1187.2	4.432
101.0	86.3	328.53	299.15	888.2	1187.4	4.391
102.0	87.3	329.25	299.90	887.6	1187.5	4.350
103.0	88.3	329.96	300.64	887.1	1187.7	4.310
104.0	89.3	330.66	301.37	886.5	1187.9	4.271
105.0	90.3	331.36	302.10	886.0	1188.1	4.232
106.0	91.3	332.05	302.82	885.4	1188.2	4.194
107.0	92.3	332.74	303.54	884.9	1188.4	4.157
108.0	93.3	333.42	304.26	884.3	1188.6	4.120
109.0	94.3	334.10	304.97	883.7	1188.7	4.084
110.0	95.3	334.77	305.66	883.2	1188.9	4.049
111.0	96.3	335.44	306.37	882.6	1189.0	4.015
112.0	97.3	336.11	307.06	882.1	1189.2	3.981
113.0	98.3	336.77	307.75	881.6	1189.4	3.947
114.0	99.3	337.42	308.43	881.1	1189.5	3.914
115.0	100.3	338.07	309.11	880.6	1189.7	3.882
116.0	101.3	338.72	309.79	880.0	1189.8	3.850
117.0	102.3	339.36	310.46	879.5	1190.0	3.819
118.0	103.3	339.99	311.12	879.0	1190.1	3.788
119.0	104.3	340.62	311.78	878.4	1190.2	3.758
120.0	105.3	341.25	312.44	877.9	1190.4	3.728
121.0	106.3	341.88	313.10	877.4	1190.5	3.699
122.0	107.3	342.50	313.75	876.9	1190.7	3.670
123.0	108.3	343.11	314.40	876.4	1190.8	3.642
124.0	109.3	343.72	315.04	875.9	1190.9	3.614
125.0	110.3	344.33	315.68	875.4	1191.1	3.587
126.0	111.3	344.94	316.31	874.9	1191.2	3.560
127.0	112.3	345.54	316.94	874.4	1191.3	3.533
128.0	113.3	346.13	317.57	873.9	1191.5	3.507
129.0	114.3	346.73	318.19	873.4	1191.6	3.481
130.0	115.3	347.32	318.81	872.9	1191.7	3.455

Table A9 – Properties of Saturated Steam (continued)

Pressure (lb/in²) Absolute P	Gauge P	Temperature (°F)	Heat of the Liquid (Btu/lb)	Latent Heat of Evaporation (Btu/lb)	Total Heat of Steam, Hg (Btu/lb)	Specific Volume (ft³/lb)
131.0	116.3	347.90	319.43	872.5	1191.9	3.430
132.0	117.3	348.48	320.04	872.0	1192.0	3.405
133.0	118.3	349.06	320.65	871.5	1192.1	3.381
134.0	119.3	349.64	321.25	871.0	1192.2	3.357
135.0	120.3	350.21	321.85	870.6	1192.4	3.333
136.0	121.3	350.78	322.45	870.1	1192.5	3.310
137.0	122.3	351.35	323.05	869.6	1192.6	3.287
138.0	123.3	351.91	323.64	869.1	1192.7	3.264
139.0	124.3	352.47	324.23	868.7	1192.9	3.242
140.0	125.3	353.02	324.82	868.2	1193.0	3.220
141.0	126.3	353.57	325.40	867.7	1193.1	3.198
142.0	127.3	354.12	325.98	867.2	1193.2	3.177
143.0	128.3	354.67	326.56	866.7	1193.3	3.155
144.0	129.3	355.21	327.13	866.3	1193.4	3.134
145.0	130.3	355.76	327.70	865.8	1193.5	3.114
146.0	131.3	356.29	328.27	865.3	1193.6	3.094
147.0	132.3	356.83	328.83	864.9	1193.8	3.074
148.0	133.3	357.36	329.39	864.5	1193.9	3.054
149.0	134.3	357.89	329.95	864.0	1194.0	3.034
150.0	135.3	358.42	330.51	863.6	1194.1	3.015
152.0	137.3	359.46	331.61	862.7	1194.3	2.977
154.0	139.3	360.49	332.70	861.8	1194.5	2.940
156.0	141.3	361.52	333.79	860.9	1194.7	2.904
158.0	143.3	362.53	334.86	860.0	1194.9	2.869
160.0	145.3	363.53	335.93	859.2	1195.1	2.834
162.0	147.3	364.53	336.98	858.3	1195.3	2.801
164.0	149.3	365.51	338.02	857.5	1195.5	2.768
166.0	151.3	366.48	339.05	856.6	1195.7	2.736
168.0	153.3	367.45	340.07	855.7	1195.8	2.705
170.0	155.3	368.41	341.09	854.9	1196.0	2.675
172.0	157.3	369.35	342.10	854.1	1196.2	2.645
174.0	159.3	370.29	343.10	853.3	1196.4	2.616
176.0	161.3	371.22	344.09	852.4	1196.5	2.587
178.0	163.3	372.14	345.06	851.6	1196.7	2.559
180.0	165.3	373.06	346.03	850.8	1196.9	2.532
182.0	167.3	373.96	347.00	850.0	1197.0	2.505
184.0	169.3	374.86	347.96	849.2	1197.2	2.479
186.0	171.3	375.75	348.92	848.4	1197.3	2.454
188.0	173.3	376.64	349.86	847.6	1197.5	2.429

Table A9 – Properties of Saturated Steam (continued)

Pressure (lb/in²)		Temperature (°F)	Heat of the Liquid (Btu/lb)	Latent Heat of Evaporation (Btu/lb)	Total Heat of Steam, Hg (Btu/lb)	Specific Volume (ft³/lb)
Absolute P	Gauge P					
190.0	175.3	377.51	350.79	846.8	1197.6	2.404
192.0	177.3	378.38	351.72	846.1	1197.8	2.380
194.0	179.3	379.24	352.64	845.3	1197.9	2.356
196.0	181.3	380.10	353.55	844.5	1198.1	2.333
198.0	183.3	380.95	354.46	843.7	1198.2	2.310
200.0	185.3	381.79	355.36	843.0	1198.4	2.288
205.0	190.3	383.86	357.58	841.1	1198.7	2.234
210.0	195.3	385.90	359.77	839.2	1199.0	2.183
215.0	200.3	387.89	361.91	837.4	1199.3	2.134
220.0	205.3	389.86	364.02	835.6	1199.6	2.087
225.0	210.3	391.79	366.09	833.8	1199.9	2.0422
230.0	215.3	393.68	368.13	832.0	1200.1	1.9992
235.0	220.3	395.54	370.14	830.3	1200.4	1.9579
240.0	225.3	397.37	372.12	828.5	1200.6	1.9183
245.0	230.3	399.18	374.08	826.8	1200.9	1.8803
250.0	235.3	400.95	376.00	825.1	1201.1	1.8438
255.0	240.3	402.70	377.89	823.4	1201.3	1.8086
260.0	245.3	404.42	379.76	821.8	1201.5	1.7748
265.0	250.3	406.11	381.60	820.1	1201.7	1.7422
270.0	255.3	407.78	383.42	818.5	1201.9	1.7107
275.0	260.3	409.43	385.21	816.9	1202.1	1.6804
280.0	265.3	411.05	386.98	815.3	1202.3	1.6511
285.0	270.3	412.65	388.73	813.7	1202.4	1.6228
290.0	275.3	414.23	390.46	812.1	1202.6	1.5954
295.0	280.3	415.79	392.16	810.5	1202.7	1.5689
300.0	285.3	417.33	393.84	809.0	1202.8	1.5433
320.0	305.3	423.29	400.39	803.0	1203.4	1.4485
340.0	325.3	428.97	406.66	797.1	1203.7	1.3645
360.0	345.3	434.40	412.67	791.4	1204.1	1.2895
380.0	365.3	439.60	418.45	785.8	1204.3	1.2222
400.0	385.3	444.59	424.0	780.5	1204.5	1.1613
420.0	405.3	449.39	429.4	775.2	1204.6	1.1061
440.0	425.3	454.02	434.6	770.0	1204.6	1.0556
460.0	445.3	458.50	439.7	764.9	1204.6	1.0094
480.0	465.3	462.82	444.6	759.9	1204.5	0.9670
500.0	485.3	467.01	449.4	755.0	1204.4	0.9278
520.0	505.3	471.07	454.1	750.1	1204.2	0.8915
540.0	525.3	475.01	458.6	745.4	1204.0	0.8578
560.0	545.3	478.85	463.0	740.8	1203.8	0.8265

Table A9 – Properties of Saturated Steam (continued)

Pressure (lb/in²)		Temperature (°F)	Heat of the Liquid (Btu/lb)	Latent Heat of Evaporation (Btu/lb)	Total Heat of Steam, Hg (Btu/lb)	Specific Volume (ft³/lb)
Absolute P	Gauge P					
580.0	565.3	482.58	467.4	736.1	1203.5	0.7973
600.0	585.3	486.21	471.6	731.6	1203.2	0.7698
620.0	605.3	489.75	475.7	727.2	1202.9	0.7440
640.0	625.3	493.21	479.8	722.7	1202.5	0.7198
660.0	645.3	496.58	483.8	718.3	1202.1	0.6971
680.0	665.3	499.88	487.7	714.0	1201.7	0.6757
700.0	685.3	503.10	491.5	709.7	1201.2	0.6554
720.0	705.3	506.25	495.3	705.4	1200.7	0.6362
740.0	725.3	509.34	499.0	701.2	1200.2	0.6180
760.0	745.3	512.36	502.6	697.1	1199.7	0.6007
780.0	765.3	515.33	506.2	692.9	1199.1	0.5843
800.0	785.3	518.23	509.7	688.9	1198.6	0.5687
820.0	805.3	521.08	513.2	684.8	1198.0	0.5538
840.0	825.3	523.88	516.6	680.8	1197.4	0.5396
860.0	845.3	526.63	520.0	676.8	1196.8	0.5260
880.0	865.3	529.33	523.3	672.8	1196.1	0.5130
900.0	885.3	531.98	526.6	668.8	1195.4	0.5006
920.0	905.3	534.59	529.8	664.9	1194.7	0.4886
940.0	925.3	537.16	533.0	661.0	1194.0	0.4772
960.0	945.3	539.68	536.2	657.1	1193.3	0.4663
980.0	965.3	542.17	539.3	653.3	1192.6	0.4557
1000.0	985.3	544.61	542.4	649.4	1191.8	0.4456
1050.0	1035.3	550.57	550.0	639.9	1189.9	0.4218
1100.0	1085.3	556.31	557.4	630.4	1187.8	0.4001
1150.0	1135.3	561.86	564.6	621.0	1185.6	0.3802
1200.0	1185.3	567.22	571.7	611.7	1183.4	0.3619

Table A10 – Valve Selection: Materials and Applications

Valve Selection Guide for Application

Valve Terms

- **Breaking pressure** – The minimum pressure required to produce flow through a valve.
- **Duty cycle** – 100% duty cycle is defined as continuous operation without any damage occurring. For intermittent duty cycle (<100%), alternate energized and de-energized state at regular intervals to cause the valve to completely cool down to room temperature.
- **Flow pattern** – A diagram showing how flow can be directed using a particular valve.
- **Normally closed** – Valve stays closed in a de-energized state; opens when energized.
- **Normally open** – Valve stays open in a de-energized state; closes when energized.
- **Pressure differential or pressure drop** – The difference between the inlet and the outlet pressure through a valve. The outlet pressure is lower than the inlet pressure due to the restriction caused by the valve.
- **Three-way valve** – A valve with three ports. Depending on the particular valve, all three ports may be open, two ports may be open, or all ports may be closed.
- **Two-way valve** – A valve with a single inlet port and a single outlet port.

Selecting Your Valve

1. Choose a valve type depending on your application. This manual contains two valve application selection charts: Valve Selection Overview: Service Application Chart and Valve Selection Detailed: Service Application Chart.
2. Consider your fluid type (liquid, gas, or solid) and its characteristics to determine compatible valve materials. PTFE withstands many harsh or corrosive chemicals. For safety reasons, always use metal valves for pressurized gases.
3. Determine the temperature, pressure, and flow rate under which your valve will be operating. In general, metal valves withstand higher temperatures and pressures than plastic valves.
4. For solenoid valves, consider the response time and the length of time the valve will be energized. Continuous (100%) duty solenoid valves are best for frequent on/off cycling. Choose normally closed or normally open depending on the state the valve will be in most often.
5. Consider your maintenance requirements. Ball valves resist plugging and are easiest to service.

Valve Types and Descriptions

- **Angle-Seat Valves**
  Valves that utilize an ultracompact actuator to move a piston back and forth within the valve body. On/off control and continuous control designs are available for gases, steam, and liquids. The construction of the body causes extremely high flow rates, particularly in comparison to conventional globe valves.

- **Ball Valves**
  Designed primarily for on/off service. These valves contain a ball with a hole through it. A handle or electric actuator rotates the ball 90°, turning the flow on or off. Use plastic ball valves for liquid applications only.

Table A10 – Valve Selection: Materials and Applications (continued)

- **Butterfly Valves**
  Designed primarily for on/off service. These valves have low pressure drops and are self-sealing so they do not require an additional flange gasket for installation. For pipe sizes ranging from 1.5-inch to 12-inch in diameter.
- **Check Valves**
  Self-actuated valves designed to prevent fluid from flowing backward into your system. Flow forces a ball or disk in one direction to open the valve; when flow stops, the ball or disk seats to close the valve.
- **Diaphragm Valves**
  Valves that use a flexible diaphragm to shut off the flow; the center of the diaphragm is pushed down into a seat. Use these valves for fluids that are dirty or have a high particulate content.
- **Elliptic Valves**
  Similar to ball valves, except they use an elliptic O-ring to seal the cylinder during rotation. The seal design makes these valves ideal for vacuum applications.
- **Manifold Valves**
  Have one common port to two or more additional ports. For mixing, use the common port as the outlet; for distributing, use the common port as the inlet.
- **Metering Valves**
  Multi-turn valves designed to regulate the flow of fluid. These valves generally have low flow rates and high pressure differentials.
- **Needle Valves**
  Feature the most accurate flow control among the valves. They are an excellent choice for precise metering of liquids or gases.
- **Pinch Valves**
  Valves that use a solenoid to squeeze shut a piece of tubing. Fluid contacts only the tubing, making these valves ideal for high-purity fluid applications.
- **Plug Valves**
  Used primarily for on/off service. Controls flow using a plug with a hole through it. Plugs can be made of rigid materials such as PTFE, making them ideal for high-purity applications.
- **Pressure Valves**
  Designed to control or limit pressure, not flow, in a system. These self-actuating valves will either open a relief port or bypass the fluid when a preset limit is reached.
- **Proportioning Valves**
  Designed to produce variable flow rates. Valves open and close in proportion to the signal from a controller.
- **Sample Valves**
  Special three-way valves designed to tap off a portion of the main flow for sampling purposes.
- **Sanitary Valves**
  Feature Tri-Clamp connections and 316 stainless-steel body construction for sanitary and high-purity applications.
- **Solenoid Valves (Direct Lift)**
  Use a plunger that is actuated to open or close the fluid path. These valves usually have low flow rates and high differential pressure drops. They generally have a quicker response time than pilot-operated valves.
- **Solenoid Valves (Pilot-Operated)**
  Utilize pressure created by air or liquids for actuation purposes. These valves are either piston or diaphragm types and require differential pressure to keep the valve closed. Because they do not have a plunger, they are usually available in larger orifice sizes than direct lift valves.
- **Stopcocks**
  Similar to ball valves, but stopcocks are much smaller in size. Primarily used in the laboratory for on/off control of flow or as a crude regulation of flow rate.

Table A10 – Valve Selection: Materials and Applications (continued)

Valve Selection Overview: Service Application Chart

Valve Service Application Chart															
Valve Type	Media (Type of Fluid to Be Controlled)										Valve Trim Function		Mechanical Valve Stem Function		
	Liquid					Gas			Solids						
	Neutral (water, oil, etc.)	Corrosive (acid, alkaline, etc.)	Hygienic (food, beverages, drugs, etc.)	Slurry	Fibrous suspension	Neutral (air, steam, etc.)	Corrosive (acid vapors, chlorine, etc.)	Vacuum	Abrasive powder (silica, etc.)	Lubricating powder (graphite, talcum, etc.)	On/Off (blocking flow)	Control valve (modulating flow rate)	Linear	Rotary	Quarter turn
Ball valves	X	X		X		X	X	X			X			X	X
Butterfly valves	X	X	X	X		X	X	X			X	X		X	X
Diaphragm valves	X	X	X	X	X	X	X				X	X	X		
Pinch valves				X					X		X				
Squeeze valves				X					X		X				
Gate/knife valves	X	X		X	X	X		X		X	X	X			
Globe valves	X	X				X	X	X			X	X	X		
Needle valves	X					X	X					X			
Plug gate valves	X	X									X	X			
Plug valves	X	X				X	X				X			X	X

Table A10 – Valve Selection: Materials and Applications (continued)

Valve Selection Detailed: Service Application Chart

Valve Service Application Chart

Conveyed Fluid	Nature of Fluid	Valve Function	Type of Disc
Liquid	Neutral (water, oil, etc.)	On/Off	Gate
			Rotary ball
			Plug
			Diaphragm
			Butterfly
			Plug gate
		Control valve, modulating	Globe
			Butterfly
			Plug gate
			Diaphragm
			Needle
	Corrosive (acid, alkaline, etc.)	On/Off	Gate
			Plug gate
			Rotary ball
			Plug
			Diaphragm
			Butterfly
		Control valve, modulating	Globe
			Diaphragm
			Butterfly
			Plug gate
	Hygienic (food, beverages, drugs, etc.)	On/Off	Butterfly
			Diaphragm
		Control valve, modulating	Butterfly
			Diaphragm
			Squeeze
			Pinch
	Slurry	On/Off	Rotary ball
			Butterfly
			Diaphragm
			Plug
			Pinch
			Squeeze
		Control valve, modulating	Butterfly
			Diaphragm
			Squeeze
			Pinch
			Gate
	Fibrous Suspensions	On/Off; Control valve; modulating	Gate
			Diaphragm
			Squeeze
			Pinch

Valve Service Application Chart

Conveyed Fluid	Nature of Fluid	Valve Function	Type of Disc
Gas	Neutral (air, steam, etc.)	On/Off	Gate
			Globe
			Rotary ball
			Plug
			Diaphragm
		Control valve, modulating	Globe
			Needle
			Butterfly
			Diaphragm
			Gate
	Corrosive (Acid vapors, chlorine, etc.)	On/Off	Butterfly
			Rotary ball
			Diaphragm
			Plug
		Control valve, modulating	Butterfly
			Globe
			Needle
			Diaphragm
	Vacuum	On/Off	Gate
			Globe
			Rotary ball
			Butterfly
Solids	Abrasive Powder (Silica, etc.)	On/Off, Control valve, modulating	Pinch
			Squeeze
			Spiral sock
	Lubricating powder (graphite, talcum, etc.)	On/Off, Control valve, modulating	Pinch
			Gate
			Spiral sock
			Squeeze

Table A10 – Valve Selection: Materials and Applications (continued)

Valve Types: Advantages and Disadvantages

Valve	Advantages	Disadvantages
Butterfly	Quarter turn	Difficult to clean
	Open port	Difficulty with slurries
	Used with chemical or corrosive media	Throttling limited to low differential pressure
	Compact, lightweight design	Potential for cavitation and choke
	Available in large sizes	Unguided disc movement is affected by flow turbulence
	Low pressure drop and high-pressure recovery	
	High coefficient of flow	
Ball	Superior ease of operation	Poor throttling characteristics
	Maintains and regulates high volume, high pressure, and high temp flow	Difficult to clean, leads to contamination
	Rugged construction and long service life	
	Low purchase and maintenance cost	
	Able to function without side loads	
	Permits inspection and repair of seats and seals without removing the valves' body from the line	
	Does not require lubrication	
Diaphragm	Extremely clean	Can only be used in moderate temperature (–60°F to 450°F)
	Leak proof seal	Can only be used in moderate pressures (approximately 300 psi)
	Tight shutoff	Cannot be used in multi-turn operations
	Easy maintenance	No industry standard face-to-face dimensions
	Repairs can be made without interrupting pipeline	The body must be made of corrosive-resistant material
	Reduce leakage to the environment	
Gate	Available in large sizes	Low pressure limitations
	Can be used with slurries and viscous liquids	Slow open and close time
	Used as a shutoff valve	Erosion of the seat and disk can occur
	Easy to maintain and disassemble	Poor throttling characteristics
	Inexpensive	Difficult to repair
	Inherently fire-safe (when used with a metal sheet)	Should not be used in sanitary applications
	Bidirectional	
Globe	Can be fast-acting	High head loss
	Precise control	Large opening for disk assembly
	Can be used in high-pressure systems	Heavier than other valves
		Cantilevered mounting of the disk to the stem
		Low coefficient of flow
		Not good for clean or sterile applications
Pinch	Very clean	Cannot be used in high-temperature applications
	Excellent drainage	Cannot be used in high-pressure applications
	Minimal turbulence	Cannot be used with gas media
	Low airborne contaminants	
	Low maintenance	
	Low weight	
	Can be used in explosion-proof lines	
	Inexpensive	

Table A10 – Valve Selection: Materials and Applications (continued)

Standard Control Valve Body Materials

Cast Material	Standard	Comments
Carbon steel	ASTM A 216 Gr. WCB	Moderate services such as noncorrosive liquids. Higher pressures and temperatures than cast iron. Check codes for suitability at extended high temperatures.
Chrome-moly steel	ASTM A 217 Gr. CS	Used for mildly corrosive fluids such as sea water and oils. Resistant to erosion and creep at high temperatures. Can be used to 595°C (110°F).
Type 304 stainless steel	ASTM A 351 Gr. CSF	Used for oxidizing or very corrosive fluids. Can be used above 540°C (100°F).
Type 316 stainless steel	ASTM A 351 Gr. CSFM	Used for oxidizing or very corrosive fluids, resistant to corrosion pitting and creep. Provides greater strength than 304 stainless steel.
Monel	ASTM A 494 Gr. M35-1	Resistant to nonoxidizing acids. Used with seawater and other mildly corrosive fluids at high temperatures. Expensive.
Hastelloy-C	ASTM A 494 Gr. CW2N	Used particularly with chlorine and chloride compounds. Expensive.
Iron	ASTM A 126 Class B	Inexpensive and nonductile. Used for water and noncorrosive liquids.
Bronze	ASTM B 61 And B 62	ASTM B 61 typically used for trim. ASTM B 62 typically used for valve body. Can be used for water and dilute acid service (see Appendix B).

Note: Gr = grade: grade designation pursuant to the referenced standard. Source: Compiled by SAIC, 1998.

Valve Body Materials: Fisher Valves Conforming to ASME B16.34-2013 Standard

			ASME	
Category	**Specification**	**Grade**	**Material Group**	**Nominal Designation**
Carbon steel	SA-216	WCC(1)	1.2	S-Mn-Si
	SA-352	LCC(2)	1.2	S-Mn-Si
	SA-350	LF2	1.1	C-Mn-Si
Alloy steels	SA-217	WC6	1.9	1-1/4Cr-1/2 Mo
		WC9	1.10	2-1/4Cr-1 Mo
		C12A	1.15	9Cr-1 Mo-V
Stainless steel	SA-351	CF8M	2.2	16Cr-12 Ni-2Mo
		CF8	2.1	18Cr-8 Ni
		CF8C	2.11	18Cr-10 Ni-Cb
		CF3M	2.2	16Cr-12 Ni-2Mo
		CG8M	2.2	19Cr-10 Ni-3Mo
		CK3MCuN	2.8	20Cr-18 Ni-6Mo
	SA-995	CD3MN(3) (Grade 4A)	2.8	22Cr-5 Ni-3Mo-N
		CD3MWCuN (Grade 6A)	2.8	25Cr-7 Ni-3.5Mo-Cb

Notes:
1. WCC is a standard substitute for WCB material.
2. LCC is a standard substitute for LCB material.
3. Listed as A351-CD3MN in B16.34-2013.

Table A10 – Valve Selection: Materials and Applications (continued)

ASME Standard CL150 Valves[1] – 150# Temperature – Pressure Rating

Valve Body Temperature – Working Pressure (psig) – ASME Ratings															
Service Temp (°F)	LCC	LF2	WCC	WC6[2]	WC9[2]	C12A[2]	CF8[2,3] or 304[2,3]	CF8M[2,3] or 316[2,3]	CF3M	316L	CG8M	317[2,3]	CF8C[2,3]	347	CK3MCuN CD3MN CD3MWCuN
–20 to 100	290	285	290	290	290	290	275	275	275	230	275	275	275	275	290
200	260	260	260	260	260	260	230	235	235	195	235	235	255	255	260
300	230	230	230	230	230	230	205	215	215	175	215	215	230	230	230
400	200	200	200	200	200	200	190	195	195	160	195	195	200	200	200
500	170	170	170	170	170	170	170	170	170	150	170	170	170	170	170
600	140	140	140	140	140	140	140	140	140	140	140	140	140	140	140
650	125	125	125	125	125	125	125	125	125	125	125	125	125	125	125
700	–	110	110	110	110	110	110	110	110	110	110	110	110	110	110
750	–	95	95	95	95	95	95	95	95	110	95	95	95	95	95
800	–	80	80	80	80	80	80	80	80	80	80	80	80	80	–
850	–	–	–	65	65	65	65	65	65	65	65	65	65	65	–
900	–	–	–	50	50	50	50	50	–	–	50	50	50	50	–
950	–	–	–	35	35	35	35	35	–	–	35	35	35	35	–
1000	–	–	–	20	20	20	20	20	–	–	20	20	20	20	–
1050	–	–	–	20	20	20	20	20	–	–	–	20	20	–	–
1100	–	–	–	20	20	20	20	20	–	–	–	20	20	–	–
1150	–	–	–	–	–	20	20	20	–	–	–	20	20	–	–
1200	–	–	–	–	–	20	20	20	–	–	–	20	20	–	–
1250	–	–	–	–	–	–	20	20	–	–	–	20	20	–	–
1300	–	–	–	–	–	–	20	20	–	–	–	20	20	–	–
1350	–	–	–	–	–	–	20	20	–	–	–	20	20	–	–
1400	–	–	–	–	–	–	20	20	–	–	–	20	15	–	–
1450	–	–	–	–	–	–	20	20	–	–	–	20	10	–	–
1500	–	–	–	–	–	–	15	15	–	–	–	15	10	–	–

[1]Table information is extracted from the Valve–Flanged, Threaded, and Welding End, ASME Standard B16.34-2013. These tables must be used in accordance with the ASME standard. The user is advised that a valve used under the jurisdiction of the ASME Boiler and Pressure Vessel Code, ASME Code for Pressure Piping, or governmental regulations is subject to any limitation of that code or regulation. This includes any maximum temperature limitation for a material or rule governing the use of a material at a low temperature. Information copied with permission of the publisher; The American Society of Mechanical Engineers, 345 East 47th Street, New York, New York, 10017.

[2]Flanged end ratings terminate at 1000°F.

[3]At temperatures over 1000°F, use material only when carbon content is 0.04% or higher.

Table A10 – Valve Selection: Materials and Applications (continued)

Valve Trim Material Temperature Limits

Material	Application	Lower		Upper	
		OF	OC	OF	OC
304 SST, S30400, CF8	Uncoated plugs and seats	–450	–268	600	316
316 SST, S31600, CF8M	Uncoated plugs and seats	–450	–268	600	316
317 SST, S31700, CG8M	Uncoated plugs and seats	–450	–268	600	316
416 SST, S41600, 38 HRC min	Cages, plugs and seats	–20	–29	800	427
CA6NM, 32 HRC min	Cages, plugs and seats	–20	–29	900	482
Nitronic 50[1], S20910 high-strength condition	Shafts, stems and pins	–325	–198	1100	593
440 SST, S44004	Bushings, plugs and seats	–20	–29	800	427
17–4 PH, S17400, CB7Cu–1, H1075 condition	Cages, plugs and seats	–80	–62	800	427
Alloy 6, R30006, CoCr – A	Plugs and seats	–325	–198	1500	816
Electroless nickel coating	Trim coating	–325	–198	750	400
Hard chromium plating	Trim coating	–325	–198	600	316
Hard chromium plating on V – balls	Trim coating	–325	–198	800	427
Hard chromium coating	Trim coating	–325	–198	1100	593
Monel (2) K500, N05500	Uncoated plugs and seats	–325	–198	800	427
Monel (2) 400, N04400	Uncoated plugs and seats	–325	–198	800	427
Hastelloy[3] B2, N10665, N7M	Uncoated plugs and seats	–325	–198	800	427
Hastelloy[3] C276, N10276, CW2M	Uncoated plugs and seats	–325	–198	800	427
Titanium grades 2, 3, 4, C2, C3, C4	Uncoated plugs and seats	–75	–59	600	316
Nickel, N02200, CZ100	Uncoated plugs and seats	–325	–198	600	316
Alloy 20, N08020, CN7M	Uncoated plugs and seats	–325	–198	600	316
NBR, nitrile rubber	Seats	–20	–29	200	93
FKM Fluoroelastomer (Viton[4])	Seats	0	–18	400	204
PTFE, polytetrafluoroethylene	Seats	–450	–268	450	232
PA (nylon)	Seats	–60	–51	200	93
HDPE, high-density polyethylene	Seats	–65	–54	185	85
CR, chloroprene (neoprene[2])	Seats	–40	–40	180	82

[1]Trademark of Armco Steel Corp.
[2]Monel and Inconel are trade names of Special Metals Corp.
[3]Hastelloy is a trade name of Haynes International.
[4]Registered trademark of DuPont Performance Elastomers.

Table A10 – Valve Selection: Materials and Applications (continued)

Valve Service Temperature Limits for Nonmetallic Materials

ASTM Designations and Trade Names	Generic Description	Temperature Range
CR	Chloroprene	–40 to 180°F, –40 to 82°C
EPDM	Ethylene propylene terpolymer	–40 to 275°F, –40 to 135°C
FFKM, Kalrez,[1] Chemraz[2]	Perfluoroelastomer	0 to 500°F, –18 to 260°C
FKM, Viton[1]	Fluoroelastomer	0 to 400°F, –18 to 204°C
FVMQ	Fluorosilicone	–100 to 300°F, –73 to 149°C
NBR	Nitrile	–65 to 180°F, –54 to 82°C
NR	Natural rubber	–20 to 200°F, –29 to 93°C
PUR	Polyurethane	–20 to 200°F, –29 to 93°C
VMQ	Silicone	–80 to 450°F, –62 to 232°C
PEEK	Polyetheretherketone	–100 to 480°F, –73 to 250°C
PTFE	Polytetrafluoroethylene	–100 to 400°F, –73 to 204°C
PTFE, carbon filled	Polytetrafluoroethylene, carbon filled	–100 to 450°F, –73 to 232°C
PTFE, glass filled	Polytetrafluoroethylene, carbon filled	–100 to 450°F, –73 to 232°C
TCM Plus[3]	Mineral and MoS2 filled PTFE	–100 to 450°F, –73 to 232°C
TCM Ultra[3]	PEEK and MoS2 filled PTFE	–100 to 500°F, –73 to 260°C
Composition gasket		–60 to 300°F, –51 to 150°C
Flexible graphite, Grafoil[4]		–300 to 1000°F, –185 to 540°C

[1]Registered trademark of DuPont Performance Elastomers.
[2]Trademark of Greene, Tweed & Co.
[3]Trademark of Fisher Controls International LLC.
[4]Trademark of Union Carbide.

Valve Stem Packing Friction Values (Typical)

Stem Size (in)	PTFE Packing Class	Single	Double	Graphite Ribbon/ Filament
5/16	ALL	20	30	–
3/8	125 150 250 300	38	56	– 125 – 190
	600 900 1500			250 320 380
1/2	125 150 250 300	50	75	– 180 – 230
	600 900 1500 2500			320 410 500 590

Table A10 – Valve Selection: Materials and Applications (continued)

Valve Stem Packing Temperature – Pressure

Valve Sliding-Stem Environmental Packing Selection					
Packing System	Maximum Pressure and Temperature Limits for 500 ppm Service[1]		Seal Performance Index	Service Life Index	Packing Friction
	Customary US	Metric			
Single PTFE V-Ring	300 psi 0 to 200°F	20.7 bar −18 to 93°C	Better	Long	Very Low
ENVIRO-SEAL PTFE	See Fig. 3-25 −50 to 450°F	See Fig. 3-25 −46 to 232°C	Superior	Very Long	Low
ENVIRO-SEAL Duplex	750 psi −50 to 450°F	51.7 bar −46 to 232°C	Superior	Very Long	Low
ENVIRO-SEAL Graphite ULF	1500 psi 20 to 600°F	103 bar −7 to 315°C	Superior	Very Long	Moderate

[1]The values shown are only guidelines. These guidelines can be exceeded, but shortened packing life or increased leakage might result. The temperature ratings apply to the actual packing temperature, not to the process temperature.

Valve Seating Shutoff Pressure

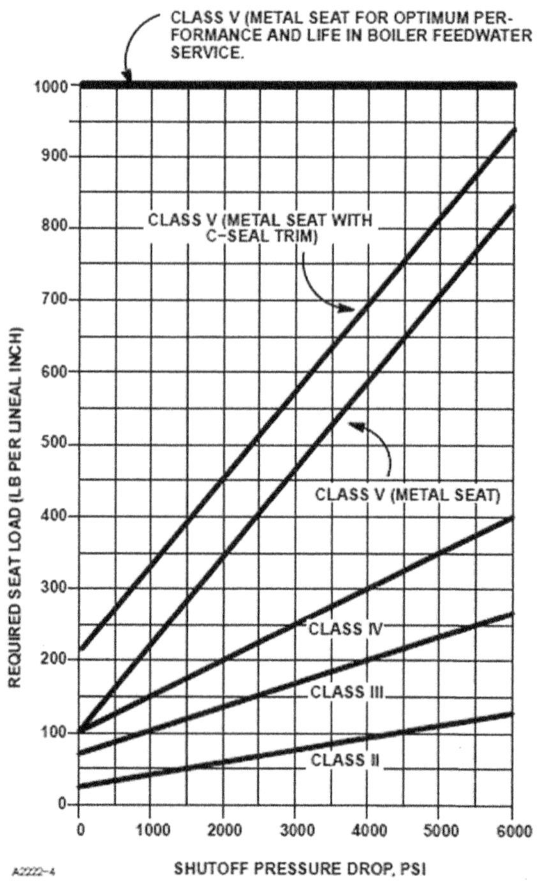

Figure Minimum Required Seat Load for Metal-Seated Valves for Improved Seat Life for Class II-V and Recommended Seat Load for Optimum Performance in Boiler Feedwater Service

Table A10 – Valve Selection: Materials and Applications (continued)

Leak Class	Recommended Seat Load
Class I	As required by user specification, no factory leak test required
Class II	20 lb per lineal inch of port circumference
Class III	40 lb per lineal inch of port circumference
Class IV	Standard (lower) seat only: 40 lb per lineal inch of port circumference
Class V	Metal seat: Determine pounds per lineal inch of port circumference from Table A-19
Class VI	Metal seat: 300 lb per lineal inch of port circumference

Abbreviations and Terminology

Symbol		Symbol	
C_v	Valve sizing coefficient	P_1	Upstream absolute static pressure
d	Nominal valve size	P_2	Downstream absolute static pressure
D	Internal diameter of the piping	P_c	Absolute thermodynamic critical pressure
F_d	Valve style modifier, dimensionless	P_v	Vapor pressure absolute of liquid at inlet temperature
F_F	Liquid critical pressure ratio factor, dimensionless	Δp	Pressure drop ($P_1 - P_2$) across the valve
F_k	Ratio of specific heats factor, dimensionless	$\Delta p_{max(L)}$	Maximum allowable liquid sizing pressure drop
F_L	Rated liquid pressure recovery factor, dimensionless	$\Delta p_{max(LP)}$	Maximum allowable sizing pressure drop with attached fittings
F_{LP}	Combined liquid pressure recovery factor and piping geometry factor of valve with attached fittings (when there are no attached fittings, FLP equals FL), dimensionless	q	Volume rate of flow
F_P	Piping geometry factor, dimensionless	q_{max}	Maximum flow rate (choked flow conditions) at given upstream conditions
G_f	Liquid specific gravity (ratio of density of liquid at flowing temperature to density of water at 60°F), dimensionless	T_1	Absolute upstream temperature (degree K or degree R)
G_g	Gas specific gravity (ratio of density of flowing gas to density of air with both at standard conditions,[1] that is, ratio of molecular weight of gas to molecular weight of air), dimensionless	w	Mass rate of flow
k	Ratio of specific heats, dimensionless	x	Ratio of pressure drop to upstream absolute static pressure (Δp/P1), dimensionless
K	Head loss coefficient of a device, dimensionless	x_T	Rated-pressure drop-ratio factor, dimensionless
M	Molecular weight, dimensionless	Y	Expansion factor (ratio of flow coefficient for a gas to that for a liquid at the same Reynolds number), dimensionless
N	Numerical constant	Z	Compressibility factor, dimensionless
		γ_1	Specific weight at inlet conditions
		ν	Kinematic viscosity, centistokes

[1]Standard conditions are defined as 60°F (15.5°C) and 14.7 psia (101.3 kPa).

Table A11 – Properties and Sizing C_v Coefficients for Fisher ED Globe Valves

Valve Size (in)	Valve Plug Style	Flow Characteristic	Port Diameter (in)	Rated Travel (in)	C_v	F_L	X_T	F_D
1/2	Post guided	Equal percentage	0.38	0.50	2.41	0.90	0.54	0.61
3/4	Post guided	Equal percentage	0.56	0.50	5.92	0.84	0.61	0.61
	Micro form M	Equal percentage	3/8	3/4	3.07	0.89	0.66	0.72
			1/2	3/4	4.91	0.93	0.80	0.67
1			3/4	3/4	8.84	0.97	0.92	0.62
	Cage guided	Linear	1–5/16	3/4	20.6	0.84	0.64	0.34
		Equal percentage	1–5/16	3/4	17.2	0.88	0.67	0.38
	Micro – form M	Equal percentage	3/8	3/4	3.20	0.84	0.65	0.72
			1/2	3/4	5.18	0.91	0.71	0.67
1 1/2			3/4	3/4	10.2	0.92	0.80	0.62
	Cage guided	Linear	1–7/8	3/4	39.2	0.82	0.66	0.34
		Equal percentage	1–7/8	3/4	35.8	0.84	0.68	0.38
2	Cage guided	Linear	2–5/16	1 1/8	72.9	0.77	0.64	0.33
		Equal percentage	2–5/16	1 1/8	59.7	0.85	0.69	0.31
3	Cage guided	Linear	3–7/16	1 1/2	148	0.82	0.62	0.30
		Equal percentage			136	0.82	0.68	0.32
4	Cage guided	Linear	4–3/8	2	236	0.82	0.69	0.28
		Equal percentage			224	0.82	0.72	0.28
6	Cage guided	Linear	7	2	433	0.84	0.74	0.28
		Equal percentage			394	0.85	0.78	0.26
8	Cage guided	Linear	8	3	846	0.87	0.81	0.31
		Equal percentage			818	0.86	0.81	0.26

Table A11 – Properties and Sizing C_v Coefficients for Fisher ED Globe Valves (continued)

Fisher ED CL125 – CL600						Linear Characteristic										
Valve Size, NPS	Port Diameter		Maximum Travel[2]		Flow Coefficient	Valve Opening – Percent of Total Travel										F_{LI}[1]
	mm	in	mm	in		10	20	30	40	50	60	70	80	90	100	
1 and 1–1/4	33.3	1.3125	19	0.75	C_v	3.21	5.50	8.18	10.9	13.2	15.0	16.9	18.6	19.9	20.6	0.84
					K_v	2.78	4.76	7.08	9.43	11.4	13.0	14.6	16.1	17.2	17.8	–
					X_T	0.340	0.644	0.494	0.509	0.532	0.580	0.610	0.629	0.628	0.636	–
1–1/2	47.6	1.875	19	0.75	C_v	4.23	7.84	11.8	15.8	20.4	25.3	30.3	34.7	37.2	39.2	0.82
					K_v	3.66	6.78	10.2	13.7	17.6	21.9	26.2	30.0	32.2	33.9	–
					X_T	0.656	0.709	0.758	0.799	0.738	0.729	0.708	0.686	0.683	0.656	–
					F_d	0.30	0.37	0.41	0.44	0.44	0.41	0.38	0.35	0.34	0.34	–
	33.3	1.3125	19	0.75	C_v	2.92	5.70	9.05	12.5	15.6	18.5	21.1	23.9	26.8	29.2	0.91
					K_v	2.53	4.93	7.83	10.8	13.5	16.0	18.3	20.7	23.2	25.3	–
					X_T	0.690	0.651	0.633	0.634	0.650	0.666	0.708	0.718	0.737	0.733	–
2	58.7	2.3125	29	1.125	C_v	7.87	16.0	24.9	33.4	42.1	51.8	62.0	68.1	70.6	72.9	0.77
					K_v	6.81	13.8	21.5	28.9	36.4	44.8	53.6	58.9	61.1	63.1	–
					X_T	0.641	0.720	0.728	0.767	0.793	0.754	0.683	0.658	0.652	0.638	–
					F_d	0.30	0.35	0.36	0.37	0.37	0.36	0.35	0.35	0.34	0.33	–
	33.3	1.3125	19	0.75	C_v	3.53	6.36	9.92	13.3	16.5	19.7	22.7	25.6	29.3	33.3	0.87
					K_v	3.05	5.50	8.58	11.5	14.3	17.0	19.6	22.1	25.3	28.8	–
					X_T	0.456	0.529	0.549	0.582	0.611	0.633	0.671	0.723	0.727	0.694	–
2-1/2	73.0	2.875	38	1.5	C_v	9.34	21.6	35.5	49.5	62.7	74.1	83.6	93.5	102	108	0.81
					K_v	8.08	18.7	30.7	42.8	54.2	64.1	72.3	80.9	88.2	93.4	–
					X_T	0.680	0.660	0.644	0.669	0.674	0.706	0.716	0.687	0.658	0.641	–
					F_d	0.27	0.33	0.35	0.36	0.35	0.34	0.32	0.29	0.27	0.27	–
	47.6	1.875	19	0.75	C_v	4.10	8.09	12.3	16.7	21.1	26.8	33.7	41.3	49.2	57.0	0.84
					K_v	3.55	7.00	10.6	14.4	18.3	23.2	29.2	35.7	42.6	49.3	–
					X_T	0.668	0.646	0.684	0.688	0.698	0.694	0.678	0.668	0.669	0.666	–
3	87.3	3.4375	38	1.5	C_v	14.5	32.9	52.1	70.4	88.5	105	118	133	142	148	0.82
					K_v	12.5	28.5	45.1	60.9	76.6	90.8	102	115	123	128	–
					X_T	0.671	0.699	0.697	0.720	0.733	0.718	0.707	0.650	0.630	0.620	–
					F_d	0.26	0.32	0.35	0.36	0.36	0.36	0.36	0.28	0.29	0.30	
	58.7	2.3125	29	1.125	C_v	8.06	16.9	26.7	37.5	49.0	61.4	73.8	85.3	94.7	102	0.85
					K_v	6.97	14.6	23.1	32.4	42.4	53.1	63.8	73.8	81.9	88.2	–
					X_T	0.592	0.614	0.662	0.672	0.674	0.676	0.694	0.722	0.736	0.732	–
4	111.1	4.375	51	2	C_v	23.3	50.3	78.1	105	127	152	181	203	223	236	0.82
					K_v	20.2	43.5	67.6	90.8	110	131	157	176	193	204	–
					X_T	0.691	0.714	0.720	0.731	0.764	0.757	0.748	0.762	0.732	0.688	–
					F_d	0.31	0.36	0.38	0.38	0.37	0.35	0.32	0.30	0.27	0.28	–
	73.0	2.875	38	1.5	C_v	9.77	22.6	37.2	51.8	65.7	77.5	87.5	97.9	107	113	0.84
					K_v	8.45	19.5	32.2	44.8	56.8	67.0	75.7	84.7	92.6	97.7	–
					X_T	0.926	0.899	0.873	0.904	0.919	0.962	0.972	0.937	0.891	0.872	–

[1]At 100% travel.

[2]If coefficients listed above for the NPS 8 linear cage with 51 mm (2-inch) travel are not sufficient for your application, consider using the quick-opening cage. The NPS 8 quick-opening cage with 51 mm (2-inch) travel has approximately a linear characteristic.

Table A11 – Properties and Sizing C_v Coefficients for Fisher ED Globe Valves (continued)

Fisher ED CL125 – CL600						Equal Percentage Characteristic										
Valve Size, NPS	Port Diameter		Maximum Travel		Flow Coeffi-cient	Valve Opening – Percent of Total Travel										
	mm	in	mm	in		10	20	30	40	50	60	70	80	90	100	F_L[1]
1 and 1–1/4	33.3	1.3125	19	0.75	C_v	0.783	1.54	2.20	2.89	4.21	5.76	7.83	10.9	14.1	17.2	0.88
					K_v	0.677	1.33	1.90	2.50	3.64	4.98	6.77	9.43	12.2	14.9	–
					X_T	0.766	0.614	0.587	0.667	0.672	0.687	0.743	0.760	0.733	0.667	–
1-1/2	47.6	1.875	19	0.75	C_v	1.52	2.63	3.87	5.41	7.45	11.2	17.4	24.5	30.8	35.8	0.84
					K_v	1.31	2.27	3.35	4.68	6.44	9.69	15.1	21.2	26.6	31.0	–
					X_T	0.780	0.735	0.716	0.715	0.738	0.727	0.690	0.685	0.685	0.679	–
					F_d	0.64	0.63	0.63	0.64	0.46	0.45	0.30	0.31	0.35	0.38	–
	33.3	1.3125	19	0.75	C_v	1.12	1.56	2.22	3.10	4.27	6.17	9.01	13.1	18.2	23.1	0.91
					K_v	0.969	1.35	1.92	2.68	3.69	5.34	7.79	11.3	15.7	20.0	–
					X_T	0.821	0.864	0.820	0.703	0.721	0.679	0.665	0.639	0.650	0.700	–
2	58.7	2.3125	29	1.125	C_v	1.66	2.93	4.66	6.98	10.8	16.5	25.4	37.3	50.7	59.7	0.85
					K_v	1.44	2.53	4.03	6.04	9.34	14.3	22.0	32.3	43.9	51.6	–
					X_T	0.827	0.834	0.774	0.727	0.687	0.684	0.702	0.736	0.686	0.687	–
					F_d	0.41	0.50	0.53	0.58	0.37	0.32	0.27	0.26	0.29	0.31	–
	33.3	1.3125	19	0.75	C_v	0.923	1.42	2.09	2.84	4.11	5.83	8.58	12.8	18.5	24.3	0.88
					K_v	0.798	1.23	1.81	2.46	3.56	5.04	7.42	11.1	16.0	21.0	–
					X_T	0.775	0.744	0.742	0.707	0.715	0.714	0.714	0.641	0.621	0.649	–
2-1/2	73.0	2.875	38	1.5	C_v	3.43	7.13	10.8	15.1	22.4	33.7	49.2	71.1	89.5	99.4	0.84
					K_v	2.97	6.17	9.34	13.1	19.4	29.2	42.6	61.5	77.4	86.0	–
					X_T	0.778	0.702	0.678	0.677	0.658	0.654	0.661	0.665	0.661	0.660	–
					F_d	0.45	0.49	0.49	0.47	0.35	0.32	0.30	0.24	0.25	0.27	–
	47.6	1.875	19	0.75	C_v	1.57	2.57	3.82	5.44	7.64	11.5	18.2	26.7	35.1	43.9	0.89
					K_v	1.36	2.22	3.30	4.71	6.61	9.95	15.7	23.1	30.4	38.0	–
					X_T	0.801	0.756	0.713	0.677	0.648	0.672	0.628	0.635	0.706	0.710	–
3	87.3	3.4375	38	1.5	C_v	4.32	7.53	10.9	17.1	27.2	43.5	66.0	97.0	120	136	0.82
					K_v	3.74	6.51	9.43	14.8	23.5	37.6	57.1	83.9	104	118	–
					X_T	0.774	0.706	0.682	0.635	0.616	0.602	0.663	0.693	0.670	0.675	–
					F_d	0.52	0.63	0.68	0.39	0.36	0.29	0.26	0.28	0.30	0.32	–
	58.7	2.3125	29	1.125	C_v	1.75	3.11	4.77	7.07	10.7	17.0	27.9	41.5	58.0	70.7	0.87
					K_v	1.51	2.69	4.13	6.12	9.26	14.7	24.1	35.9	50.2	61.2	–
					X_T	0.944	0.840	0.803	0.757	0.735	0.642	0.531	0.613	0.629	0.702	–
4	111.1	4.375	51	2	C_v	5.85	11.6	18.3	30.2	49.7	79.7	125	171	205	224	0.82
					K_v	5.06	10.0	15.8	26.1	43.0	68.9	108	148	177	194	–
					X_T	0.731	0.650	0.643	0.645	0.632	0.625	0.672	0.742	0.737	0.716	–
					F_d	0.45	0.42	0.40	0.33	0.30	0.28	0.23	0.24	0.26	0.28	–
	73.0	2.875	38	1.5	C_v	3.82	7.65	11.4	16.9	25.5	38.2	60.5	85.7	105	112	0.89
					K_v	3.30	6.62	9.86	14.6	22.1	33.0	52.3	74.1	90.8	96.9	–
					X_T	0.746	0.700	0.694	0.669	0.640	0.627	0.591	0.644	0.735	0.813	–

[1]At 100% travel. Restricted trim.

Table A12 – Properties and Sizing C_v Coefficients for Fisher Rotary Valves

Valve Size (in)	Valve Style	Degrees of Valve Opening	C_v	F_L	X_T	F_D
1	V–notch ball valve	60	15.6	0.86	0.53	
		90	34	0.86	0.42	
1–1/2	V–notch ball valve	60	28.5	0.85	0.5	
		90	77.3	0.74	0.27	
2	V–notch ball valve	60	59.2	0.81	0.53	
		90	132	0.77	0.41	
	High-performance butterfly Valve	60	58.9	0.76	0.5	0.49
		90	80.2	0.71	0.44	0.7
3	V–notch ball valve	60	120	0.8	0.5	0.92
		90	321	0.74	0.3	0.99
	High-performance butterfly valve	60	115	0.81	0.46	0.49
		90	237	0.64	0.28	0.7
4	V–notch ball valve	60	195	0.8	0.52	0.92
		90	596	0.62	0.22	0.99
	High-performance butterfly valve	60	270	0.69	0.32	0.49
		90	499	0.53	0.19	0.7
6	V–notch ball valve	60	340	0.8	0.52	0.91
		90	1100	0.58	0.2	0.99
	High-performance butterfly valve	60	664	0.66	0.33	0.49
		90	1260	0.55	0.2	0.7
8	V–notch ball valve	60	518	0.82	0.54	0.91
		90	1820	0.54	0.18	0.99
	High-performance butterfly valve	60	1160	0.66	0.31	0.49
		90	2180	0.48	0.19	0.7
10	V–notch ball valve	60	1000	0.8	0.47	0.91
		90	3000	0.56	0.19	0.99
	High-performance butterfly valve	60	1670	0.66	0.38	0.49
		90	3600	0.48	0.17	0.7
12	V–notch ball valve	60	1530	0.78	0.49	0.92
		90	3980	0.63	0.25	0.99
	High-performance butterfly valve	60	2500			0.49
		90	5400			0.7
16	V–notch ball valve	60	2380	0.8	0.45	0.92
		90	8270	0.37	0.13	1
	High-performance butterfly valve	60	3870	0.69	0.4	
		90	8600	0.52	0.23	

Table A13 – Numerical Constants for Control Valve Sizing Formulas

		N	w	q	p^{2}	*y*	T	d, D
N_1		0.0865	–	m^3/h	kPa bar	–	–	–
		0.865	–	m^3/h	psia	–	–	–
		1.00	–	gpm		–	–	–
N_2		0.00214	–	–	–	–	–	mm in
		890	–	–	–	–	–	
N_5		0.00241	–	–	–	–	–	mm in
		1000	–	–	–	–	–	
N_6		2.73	kg/h	–	kPa bar	kg/m^3	–	–
		27.3	kg/h	–	psia	kg/m^3	–	–
		63.3	lb/h	–		lb/ft^3	–	–
N_7[1]	Normal conditions $T_N = 0°C$	3.94	–	m^3/h	kPa bar	–	deg K	–
		394	–	m^3/h		–	deg K	–
	Standard conditions $T_s = 15.5°C$	4.17	–	m^3/h	kPa bar	–	deg K	
		417	–	m^3/h		–	deg K	
	Standard conditions $T_s = 60°F$	1360	–	Scfh	psia	–	deg R	
N_8		0.948	kg/h	–	kPa bar	–	deg K	–
		94.8	kg/h	–	psia	–	deg K	–
		19.3	lb/h	–		–	deg R	–
N_9[1]	Normal conditions $T_N = 0°C$	21.2	–	m^3/h	kPa bar	–	deg K	–
		2120	–	m^3/h		–	deg K	–
	Standard conditions Ts = 15.5°C	22.4	–	m^3/h	kPa bar	–	deg K	–
		2240	–	m^3/h		–	deg K	–
	Standard conditions TS = 60°F	7320	–	scfh	psia	–	deg R	–

[1]Pressure base is 101.3 kPa (1.013 bar)(14.7 psia).

Table A14 – Critical Pressure and Temperature of Elements

Element or Compound	Critical Pressure, p_c psia	Critical Pressure, p_c bar (abs)	Critical Temperature, T_c °F	Critical Temperature, T_c °C	k C_p/C_v
Acetic acid, CH_3-CO-OH	841	58.0	612	322	322
Acetone, CH_3-CO-CH_3	691	47.6	455	235	–
Acetylene, C_2H_2	911	62.9	97	36	1.26
Air, $O_2 + N_2$	547	37.8	–222	–141	1.40
Ammonia, NH_3	1638	113.0	270	132	1.33
Argon, A	705	48.6	–188	–122	1.67
Benzene, C_6H_6	701	48.4	552	289	1.12
Butane, C_4H_{10}	529	36.5	307	153	1.09
Carbon dioxide, CO_2	1072	74.0	88	31	1.30
Carbon monoxide, CO	514	35.5	–218	–139	1.40
Carbon tetrachloride, CCl_4	661	45.6	541	283	–
Chlorine, Cl_2	1118	77.0	291	144	1.36
Ethane, C_2H_6	717	49.5	90	32	1.22
Ethyl alcohol, C_2H_5OH	927	64.0	469	243	1.13
Ethylene, CH_2=CH_2	742	51.2	50	10	1.26
Ethyl ether, C_2H_5-O-C_2H_5	522	36.0	383	195	–
Fluorine, F_2	367	25.3	–247	–155	1.36
Helium, He	33.2	2.29	–450	–268	1.66
Heptane, C_7H_{16}	394	27.2	513	267	–
Hydrogen, H_2	188	13.0	–400	–240	1.41
Hydrogen chloride, HCl	1199	82.6	124	51	1.41
Isobutane, (CH_3) CH-CH_3	544	37.5	273	134	1.10
Isopropyl alcohol, CH_3-CHOH-CH_3	779	53.7	455	235	–
Methane, CH_4	673	46.4	–117	–83	1.31
Methyl alcohol, H-CH_2OH	1156	79.6	464	240	1.20
Nitrogen, N_2	492	34.0	–233	–147	1.40
Nitrous oxide, N_2O	1054	72.7	99	37	1.30
Octane, CH_3-(CH_2)6-CH_3	362	25.0	565	296	1.05
Oxygen, O_2	730	50.4	–182	–119	1.40
Pentane, C_5H_{12}	485	33.5	387	197	1.07
Phenol, C_6H_5OH	889	61.3	786	419	–
Phosgene, $COCl_2$	823	56.7	360	182	–
Propane, C_3H_8	617	42.6	207	97	1.13
Propylene, CH_2=CH-CH_3	661	45.6	198	92	1.15
Refrigerant 12, CCl_2F_2	582	40.1	234	112	1.14
Refrigerant 22, $CHClF_2$	713	49.2	207	97	1.18
Sulfur Dioxide, SO_2	1142	78.8	315	157	1.29
Water, H_2O	3206	221.0	705	374	1.32

Table A15 – Pipe Standard Dimensions and Data

Carbon Steel Pipes – Working Pressure

Nominal Size (in)	Pipe Outside Diameter OD (in)	Schedule Number or Weight	Wall Thickness, *t* (in)	Inside Diameter, *d* (in)	Working Pressure ASTM A53 B to 400°F		
					Manufacturing Process	Joint Type	psig
1/4	0.540	40ST	0.088	0.364	CW	T	188
		80XS	0.119	0.302	CW	T	871
3/8	0.675	40ST	0.091	0.493	CW	T	203
		80XS	0.126	0.423	CW	T	820
1/2	0.840	40ST	0.109	0.622	CW	T	214
		80XS	0.147	0.546	CW	T	753
3/4	1.050	40ST	0.113	0.824	CW	T	217
		80XS	0.154	0.742	CW	T	681
1	1.315	40ST	0.133	1.049	CW	T	226
		80XS	0.179	0.957	CW	T	642
1-1/4	1.660	40ST	0.140	1.380	CW	T	229
		80XS	0.191	1.278	CW	T	594
1-1/2	1.900	40ST	0.145	1.610	CW	T	231
		80XS	0.200	1.500	CW	T	576
2	2.375	40ST	0.154	2.067	CW	T	230
		80XS	0.218	1.939	CW	T	551
2-1/2	2.875	40ST	0.203	2.469	CW	W	533
		80XS	0.276	2.323	CW	W	835
3	3.500	40ST	0.216	3.068	CW	W	482
		80XS	0.300	2.900	CW	W	767
4	4.500	40ST	0.237	4.026	CW	W	430
		80XS	0.337	3.826	CW	W	695
6	6.625	40ST	0.280	6.065	ERW	W	696
		80XS	0.432	5.761	ERW	W	1209
8	8.625	30	0.277	8.071	ERW	W	526
		40ST	0.322	7.981	ERW	W	643
		80XS	0.500	7.625	ERW	W	1106
10	10.75	30	0.307	10.136	ERW	W	485
		40ST	0.365	10.020	ERW	W	606
		XS	0.500	9.750	ERW	W	887
		80	0.593	9.564	ERW	W	1081

Table A15 – Pipe Standard Dimensions and Data (continued)

Carbon Steel Pipes – Working Pressure							
Nominal Size (in)	Pipe Outside Diameter OD (in)	Schedule Number or Weight	Wall Thickness, *t* (in)	Inside Diameter, *d* (in)	Working Pressure ASTM A53 B to 400°F		
					Manufacturing Process	Joint Type	psig
12	12.75	30	0.330	12.090	ERW	W	449
		ST	0.375	12.000	ERW	W	528
		40	0.406	11.938	ERW	W	583
		XS	0.500	11.750	ERW	W	748
		80	0.687	11.376	ERW	W	1076
14	14.00	30ST	0.375	13.250	ERW	W	481
		40	0.437	13.126	ERW	W	580
		XS	0.500	13.000	ERW	W	681
		80	0.750	12.500	ERW	W	1081
16	16.00	30ST	0.375	15.250	ERW	W	421
		40XS	0.500	15.000	ERW	W	596
18	18.00	ST	0.375	17.250	ERW	W	374
		30	0.437	17.126	ERW	W	451
		XS	0.500	17.000	ERW	W	530
		40	0.562	16.876	ERW	W	607
20	20.00	20ST	0.375	19.250	ERW	W	337
		30XS	0.500	19.000	ERW	W	477
		40	0.593	18.814	ERW	W	581

Table A16 – NEC Wire Ampacity Table 310.16

Three Conductors in Raceway, Cable, or Earth

Table 310.16 Allowable Ampacities of Insulated Conductors Rated 0 Through 2000 V. 60°C Through 90°C (140°K Through 194°F)
No More Than Three Current-Carrying Conductors in Raceway, Cable, or Earth (Directly Buried). Based on Ambient Temperature of 30°C <86°F)

Size AWG or kcmil	Temperature Rating of Conductor (See Table 310.11)					
	60°C (140°F)	75°C (167°F)	90°C (194°F)	60°C (140°F)	75°C (167°F)	90°C (194°F)
	Types TW, UF	Types RHW. THHW, THW, THWN, XHHW, USE. ZW	Types TBS. SA. SIS. EEP, EEPB, MI, RHH, RHW-2, THHN, THHW, THW-2, THWN-2, USE-2, XHH, XHHW, XHHW-2, ZW-2	Types TW, UK	Types RHW, THHW, THW, THWN, XHHW, USE	Types TBS, SA, SIS, THHN, THHW, THW-2, THWN-2, RHH, RHW-2, USE-2, XHH, XHHW, XHHW-2, ZW-2
	Copper			Aluminum or Copper-Clad Aluminum		
18			14			
16			18			
14*	20	10	25			
12*	25	25	30	20	20	25
10*	30	35	40	25	30	35
8	40	50	55	30	40	45
6	55	65	75	40	50	60
4	70	85	95	55	65	75
3	85	100	110	65	75	85
2	95	115	130	75	90	100
1	110	130	150	85	100	115
1/0	125	150	170	100	120	135
2/0	145	175	195	115	135	150
3/0	165	200	225	130	155	175
4/0	195	230	260	150	180	205
250	215	255	290	170	205	250
300	240	285	320	190	230	300
350	260	310	350	210	250	350
400	280	335	380	225	270	400
500	320	380	430	260	310	500
600	355	420	475	285	340	385
700	385	460	520	310	375	420
750	400	475	535	320	385	435
800	410	490	555	330	395	450
900	435	520	585	355	425	480
1000	455	545	615	375	445	1000
1250	495	590	665	405	485	1250
1500	520	625	705	435	520	1500
1750	545	650	735	455	545	1750
2000	650	665	750	470	560	2000

Table A17 – NEC Conductor Properties and Impedance

NEC Table 8 Conductor Properties (DC Resistance)

	Table 8 Conductor Properties					Direct Current Resistance at 75°C (167°F)		
		Stranding		Overall		Copper		Aluminum
Wire Size AWG	Wire Area Circular Mills (cm)	Quantity	Diam. in	Diam. in	Area in^2	Uncoated Ohms per 1000 ft	Coated Ohms per 1000 ft	Ohms per 1000 ft
18	1620	1	–	0.040	0.001	7.77	8.08	12.8
18	1620	7	0.015	0.046	0.002	7.95	8.45	13.1
16	2580	1	–	0.051	0.002	4.89	5.08	8.05
16	2580	7	1.019	0.058	0.003	4.99	5.29	8.21
14	4110	1	–	0.064	0.003	3.07	3.19	5.06
14	4110	7	0.024	0.073	0.004	3.14	3.26	5.17
12	6530	1	–	0.081	0.005	1.93	2.01	3.18
12	6530	7	0.030	0.092	0.006	1.98	2.05	3.25
10	10380	1	–	0.102	0.008	1.21	1.26	2.00
10	10380	7	0.038	0.116	0.011	1.24	1.29	2.04
8	16510	1	–	0.128	0.013	0.764	0.786	1.26
8	16510	7	0.049	0.146	0.017	0.778	0.809	1.28
6	26240	7	0.061	0.184	0.027	0.491	0.510	0.808
4	41740	7	0.077	0.232	0.042	0.308	0.321	0.508
3	52620	7	0.087	0.260	0.053	0.245	0.254	0.403
2	66360	7	0.097	0.292	0.067	0.194	0.201	0.319
1	83690	19	0.066	0.332	0.087	0.154	0.160	0.253
1/0	105600	19	0.074	0.372	0.109	0.122	0.127	0.201
2/0	133100	19	0.084	0.418	0.137	0.0967	0.101	0.159
3/0	167800	19	0.094	0.470	0.173	0.0766	0.0797	0.126
4/0	211600	19	0.106	0.528	0.219	0.0608	0.0626	0.100
250	–	37	0.082	0.575	0.260	0.0515	0.0535	0.0847
300	–	37	0.090	0.630	0.312	0.0429	0.0446	0.0707
350	–	37	0.097	0.681	0.364	0.0367	0.0382	0.0605
400	–	37	0.104	0.728	0.416	0.0321	0.0331	0.0529
500	–	37	0.116	0.813	0.519	0.0258	0.0265	0.0424
600	–	61	0.099	0.893	0.626	0.0214	0.0223	0.0353
700	–	61	0.107	0.964	0.730	0.0184	0.0189	0.0303
750	–	61	0.111	0.998	0.782	0.0171	0.0176	0.0282
800	–	61	0.114	1.030	0.834	0.0161	0.0166	0.0265
900	–	61	0.122	1.094	0.940	0.0143	0.0147	0.0235
1000	–	61	0.128	1.152	1.042	0.0129	0.0132	0.0212
1250	–	91	0.117	1.289	1.305	0.0103	0.0106	0.0169
1500	–	91	0.128	1.412	1.566	0.00858	0.00883	0.0141
1750	–	127	0.117	1.526	1.829	0.00735	0.00756	0.0121
2000	–	127	0.126	1.632	2.092	0.00643	0.00662	0.0106

Notes:

1. These resistance values are valid only for the parameters given. Using conductors having coated strands, different stranding type, and, especially, other temperatures changes the resistance.
2. Formula for temperature change: R2 = R1 [1 + a(T2-75)]; where: aCU = 0.00323, aAL = 0.00330. Conductors with compact and compressed stranding have about 9% and 3%, respectively, smaller bare conductor diameters than those shown. See Table 5A for actual compact cable dimensions.
3. The IACS conductivities used: bare copper = 100%, aluminum = 61%.
4. Class B stranding is listed as well as solid for some sizes. Its overall diameter and area are that of its circumscribing circle.
5. (FPN): The construction information is per NEMA WC8-1988. The resistance is calculated per National Bureau of Standards Handbook 100, dated 1966, and Handbook 109, dated 1972.

Table A17 – NEC Conductor Properties and Impedance (continued)

NEC Table 9 Conductor Properties (AC Resistance and Reactance)

AC Inductance and AC Resistance at 60 Hz Only

AC Resistance and Reactance for 600 V Cables, 3-Ø, 60 Hz, 75°C (167°F)
(Note: Three Single Conductors in Conduit, Cable, or Air)

Size AWG/ kcmil	XL (Reactance) for all Wires		AC Resistance for Uncoated Copper Wires			AC Resistance for Aluminum Wires		
	PVC, AL Conduit	Steel Conduit	PVC Conduit	AL Conduit	Steel Conduit	PVC Conduit	AL Conduit	Steel Conduit
14	0.058	0.073	3.1	3.1	3.1	–	–	–
12	0.054	0.068	2	2	2	3.2	3.2	3.2
10	0.05	0.063	1.2	1.2	1.2	2	2	2
8	0.052	0.065	0.78	0.78	0.78	1.3	1.3	1.3
6	0.051	0.064	0.49	0.49	0.49	0.81	0.81	0.81
4	0.048	0.06	0.31	0.31	0.31	0.51	0.51	0.51
3	0.047	0.059	0.25	0.25	0.25	0.4	0.41	0.4
2	0.045	0.057	0.19	0.2	0.2	0.32	0.32	0.32
1	0.046	0.057	0.15	0.16	0.16	0.25	0.26	0.25
1/0	0.044	0.055	0.12	0.13	0.12	0.2	0.21	0.2
2/0	0.043	0.054	0.1	0.1	0.1	0.16	0.16	0.16
3/0	0.042	0.052	0.077	0.082	0.079	0.13	0.13	0.13
4/0	0.041	0.051	0.062	0.067	0.063	0.1	0.11	0.1
250	0.041	0.052	0.052	0.057	0.054	0.085	0.09	0.086
300	0.041	0.051	0.044	0.049	0.045	0.071	0.076	0.072
350	0.04	0.05	0.038	0.043	0.039	0.061	0.066	0.063
400	0.04	0.049	0.033	0.038	0.035	0.054	0.059	0.055
500	0.039	0.048	0.027	0.032	0.029	0.043	0.048	0.045
600	0.039	0.048	0.023	0.028	0.025	0.036	0.041	0.038
750	0.038	0.048	0.019	0.024	0.021	0.029	0.034	0.031
1000	0.037	0.046	0.015	0.019	0.018	0.023	0.027	0.025

Table A17 – NEC Conductor Properties and Impedance (continued)

Table Continued – Effective Impedance at 85% Power Factor at 60 Hz Only

Size AWG/ kcmil	Effective Z AT 0.85 PF for Uncoated Copper Wires			Effective Z AT 0.85 for Aluminum Wires		
	PVC Conduit	AL Conduit	Steel Conduit	PVC Conduit	AL Conduit	Steel Conduit
14	2.7	2.7	2.7	–	–	–
12	1.7	1.7	1.7	2.8	2.8	2.8
10	1.1	1.1	1.1	1.8	1.8	1.8
8	0.69	0.69	0.7	1.1	1.1	1.1
6	0.44	0.45	0.45	0.71	0.72	0.72
4	0.29	0.29	0.3	0.46	0.46	0.46
3	0.23	0.24	0.24	0.37	0.37	0.37
2	0.19	0.19	0.2	0.3	0.3	0.3
1	0.16	0.16	0.16	0.24	0.24	0.25
1/0	0.13	0.13	0.13	0.19	0.2	0.2
2/0	0.11	0.11	0.11	0.16	0.16	0.16
3/0	0.088	0.092	0.094	0.13	0.13	0.14
4/0	0.074	0.078	0.08	0.11	0.11	0.11
250	0.066	0.07	0.073	0.094	0.098	0.1
300	0.059	0.063	0.065	0.082	0.086	0.088
350	0.053	0.058	0.06	0.073	0.077	0.08
400	0.049	0.053	0.056	0.066	0.071	0.073
500	0.043	0.048	0.05	0.057	0.061	0.064
600	0.04	0.044	0.047	0.051	0.055	0.058
750	0.036	0.04	0.043	0.045	0.049	0.052
1000	0.032	0.036	0.04	0.039	0.042	0.046

Notes:

1. These values are based on the following constants: UL-type RHH wires with Class B stranding, in cradled configuration. Wire conductivities are 100% IACS copper and 61% AICS aluminum, and aluminum conduit is 45%.
2. Capacitive reactance is ignored because it is negligible at these voltages. These resistance values are valid only at 75°C (167°F) and for the parameters as given but are representative for 600 V wire types operating at 60 Hz.
3. Effective Z is defined as R cos (Ø) plus X sin (Ø), where Ø is the power factor angle of the circuit. Multiplying current by effective impedance gives a good approximation for line-to-neutral voltage drop.
4. Effective impedance values shown in this table are valid only at 0.85 power factor (PF). For another circuit PF, effective impedance (Zc) can be calculated from R and XL values given in this table as follows: Ze = R • PF + XL sin [arccos (PF)].

Table A18 – NEC Full-Load Motor Currents

	60 Hz AC Induction Motor						
	Single-Phase		Three-Phase				
Horsepower	115 V	230 V	200 V	230 V	380–415 V	460 V	575 V
1/6	4.4	2.2	–	–		–	–
1/4	5.8	2.9	–	–		–	–
1/3	7.2	3.6	–	–		–	–
1/2	9.8	4.9	2.5	2.2	1.3	1.1	0.9
3/4	13.8	6.9	3.7	3.2	1.8	1.6	1.3
1	16.0	8.0	4.8	4.2	2.3	2.1	1.7
1-1/2	20.0	10.0	6.9	6.0	3.3	3.0	2.4
2	24.0	12.0	7.8	6.8	4.3	3.4	2.7
3	34.0	17.0	11.0	9.6	6.1	4.8	3.9
5	56.0	28.0	17.5	15.2	9.7	7.6	6.1
7-1/2	80.0	40.0	25.0	22.0	14.0	11.0	9.0
10	100	50.0	32.0	28.0	18.0	14.0	11.0
15	135	68.0	48.0	42.0	27.0	21.0	17.0
20	–	88.0	62.0	54.0	34.0	27.0	22.0
25	–	110	78.0	68.0	43.0	34.0	27.0
30	–	136	92.0	80.0	51.0	40.0	32.0
40	–	176	120	104	66.0	52.0	41.0
50	–	216	150	130	83.0	65.0	52.0
60	–	–	177	154	103	77.0	62.0
75	–	–	221	192	128	96.0	77.0
100	–	–	285	248	165	124	99.0
125	–	–	359	312	208	156	125
150	–	–	414	360	240	180	144
175	–	–	475	413	275	207	168
200	–	–	552	480	320	240	192
250	–	–	692	602	403	302	242
300	–	–	–	–	482	361	289
350	–	–	–	–	560	414	336
400	–	–	–	–	636	477	382
450	–	–	–	–	711	515	412
500	–	–	–	–	786	590	472

Notes:

1. The information in this chart was derived from Table 430-148 and 430-150 of the NEC and Table 50.1 of UL standard 508A. The voltages listed are rated motor voltages. The currents listed shall be permitted for system voltage ranges of 110–120, 220–240, 380–415, 440–480, and 550–600 V.
2. The full-load current values are for motors running at usual speeds and motors with normal torque characteristics. Motors built for especially low speeds or high torques may have higher full-load currents, and multi-speed motors will have full-load currents varying with speed. In these cases, the nameplate current ratings shall be used.

Caution: The actual motor amps may be higher or lower than the average values listed above. For more reliable motor protection, use the actual motor current as listed on the motor nameplate. Use this table as a guide only.

Table A19 – NEC Grounding and Bonding Conductors

Table 250.66 Grounding-Electrode Conductor for Alternating Current Systems

Size of Largest Ungrounded Service-Entrance Conductor or Equivalent Area for Parallel Conductors (AWG/kcmil)		Size of Grounding-Electrode Conductors (AWG/kcmil)	
Copper	Aluminum or Copper-Clad Aluminum	Copper	Aluminum or Copper-Clad Aluminum
2 or smaller	1/0 or smaller	8	6
1 or 1/0	2/0 or 3/0	6	4
2/0 or 3/0	4/0 or 250	4	2
Over 3/0 through 350	Over 250 through 500	2	1/0
Over 350 through 600	Over 500 through 900	1/0	2/0
Over 600 through 1100	Over 900 through 1750	2/0	3/0
Over 1100	Over 1750	3/0	250

Notes:

1. Where multiple sets of service-entrance conductors are used as permitted in 230.40. Exception No. 2: The equivalent size of the largest service-entrance conductor shall be determined by the largest sum of the areas of the corresponding conductors of each set.
2. Where there are no service-entrance conductors, the grounding-electrode conductor size shall be determined by the equivalent size of the largest service-entrance conductor required for the load to be served.

Table 250.122 Minimum Size Equipment Grounding Conductors for Grounding Raceway and Equipment

Rating or Setting of Automatic Overcurrent Device in Circuit Ahead of Equipment, Conduit, etc., Not Exceeding (Amperes)	Size (AWG or kcmil)	
	Copper	Aluminum or Copper-Clad Aluminum*
15	14	12
20	12	10
30	10	8
40	10	8
60	10	8
100	8	6
200	6	4
300	4	9
400	3	1
500	2	1/0
600	1	2/0
800	1/0	3/0
1000	2/0	4.0
1200	3/0	250
1600	4/0	350
2000	250	400
2500	350	600
3000	400	600
4000	500	750
5000	700	1200
6000	800	1200

Note: Where necessary to comply with 250.4(A)(5) or (BX4), the equipment grounding conductor shall be sized larger than given in this table. See installation restrictions in 250.120.

Table A20 – Specific Gravity and Gas Constants for Some Common Gases

Gas	Specific Gravity SG or G_f	Molecular Weight M	Ratio of Specific Heat k
Acetylene (ethyne) – C_2H_2	0.907	26.038	1.234
Air[1)]	1.000	28.967	**1.399**
Ammonia – NH_3	0.588	17.032	1.304
Argon – Ar	1.379	39.944	1.668
Arsine	2.69		
Benzene	2.559	78.114	1.113
Blast furnace gas	1.02		
Butadiene	1.869		
n-Butane – C_4H_{10}	2.007	58.124	1.093
I-Butene – C_4H_8	1.937	56.108	1.111
Carbon dioxide – CO_2	1.519	44.011	1.288
Carbon monoxide – CO	0.967	28.011	1.399
Carbureted water gas	0.63		
Chlorine – Cl_2	2.486	70.910	
Coke oven gas	0.44		
Cyclobutane	1.938		
Cyclohextane	2.905	84.161	1.07
Cyclopentane	2.422	70.135	1.08
Cyclopropane	1.451		
Dodecane – $C_{12}H_{26}$	5.88	170.340	1.031
Digestive gas (sewage or biogas)	0.8		
Ethane – C_2H_6	1.038	30.070	1.188
Ethylene (ethene) – C_2H_4	0.9685	28.054	1.236
Fluorine	1.31	38.000	
Freon, F-12		120.925	1.136
Helium - He	0.138	4.003	1.667
n-Heptane – C_7H_{16}	3.459	100.205	1.053
n-Hexane – C_6H_{14}	2.9753	86.178	1.062
Hydrogen	0.069	2.016	1.405
Hydrogen chloride – HCl	1.268	36.470	
Hydrogen sulfide – H_2S	1.177	34.082	
Isobutane – C_4H_1	2.007	58.124	1.094
Isopentane – C_5H_{12}	2.4911	72.151	1.074

(Continued)

Table A20 – Specific Gravity and Gas Constants for Some Common Gases (continued)

Gas	Specific Gravity *SG or G_f*	Molecular Weight *M*	Ratio of Specific Heat *k*
Krypton	2.89		
Methane – CH_4	0.554	16.043	1.304
Methyl chloride	1.74	50.490	
Natural gas (typical)	0.60 – 0.70 (0.65)	(18.829)	(1.32)
Neon	0.696	20.183	1.667
Nitric oxide–- NO	1.0359	30.008	1.386
Nitrogen – N_2	0.967	28.016	1.40
Nitrous oxide - N_2O	1.530	44.020	
Nonane	4.428	128.258	1.04
Octane	3.944	114.232	1.046
Oxygen – O_2	1.105	32.000	1.396
Ozone	1.660		
n-Pentane – C_5H_{12}	2.4908	72.151	1.074
Phosgene	1.39		
Propane – C_3H_8	1.522	44.097	1.128
Propene (propylene) – C_3H_6	1.4527	42.081	1.187
Sasol	0.42		
Silane	1.11		
Sulfur dioxide – SO_2	2.2117	64.066	1.264
Toluene-methylbenzene	3.176		
Water gas (bituminous)	0.71		
Water vapor	0.622	18.016	1.329
Xenon	4.533	131.300	1.667

Table A21 – Specific Gravity: Common Fluids

Product	Temperature °F	Temperature °C	Specific Gravity *SG or G_f*
Acetaldehyde CH_3CHO	61 68	16.1 20	0.79 0.76
Acetic acid 5% – vinegar	59	15	1.006
Acetic acid – 10%	59	15	1.014
Acetic acid – 50%	59	15	1.061
Acetic acid – 80%	59	15	1.075
Acetic acid – concentrated	59	15	1.055
Acetic acid anhydride $(CH_3COO)_2O$	59	15	1.087
Acetone CH_3COCH_3	68	20	0.792
Alcohol – allyl	68	20	0.855
Alcohol – butyl-n	68 158	20 70	0.81 0.78
Alcohol – ethyl (grain) C_2H_5OH	68 104	20 40	0.789 0.772
Alcohol – methyl (wood) CH_3OH	68	20	0.79
Alcohol – propyl	68 32	20 0	0.804 0.817
Aluminum sulfate 36% solution	60	15.6	1.055
Ammonia	0	–17.8	0.662
Aniline	68 32	20 0	1.022 1.035
Automotive crankcase oils SAE-5W/10W/20W/30W/40W/50W	60	15.6	0.88–0.94
Automotive gear oils SAE-75W/80W/85W/90W/140W/150W	60	15.6	0.88–0.94
Beer	60	15.6	1.01
Benzene (benzol) C6H6	32 60	0 15.6	0.899 0.885
Bone oil	60	15.6	0.918
Boric acid H3BO3	46.4 59	8 15	1.014 1.025
Bromine	32	0	2.9
Butane-n	60	15.6	0.584
Butyric acid	68	20	0.959
Calcium chloride 5%	65	18.3	1.040
Calcium chloride 25%	60	15.6	1.23
Carbolic acid (phenol)	65	18.3	1.08
Carbon tetrachloride CCl4	68	20	1.594
Carbon disulfide CS_2	32 68	0 20	1.293 1.263
Castor oil	68 104	20 40	0.96 0.95

Table A21 – Specific Gravity: Common Fluids (continued)

Product	Temperature °F	Temperature °C	Specific Gravity *SG or G_f*
China wood oil	60	15.6	0.943
Chloroform	68 140	20 60	1.489 1.413
Coconut oil	60	15.6	0.925
Cod liver oil	59	15	0.920–0.925
Corn oil	60	15.6	0.924
Cotton seed oil	60	15.6	0.88–0.93
Creosote	60	15.6	1.04–1.10
Crude oil 48° API	60 130	15.6 54.4	0.79 0.76
Crude oil 40° API	60 130	15.6 54.4	0.825 0.805
Crude oil 35.6° API	60 130	15.6 54.4	0.847 0.824
Crude oil 32.6° API	60 130	15.6 54.4	0.832 0.84
Crude oil Salt creek	60 130	15.6 54.4	0.843 0.82
Decane-n	68	20	0.73
Diethylene glycol	60	15.6	1.12
Diethyl ether	68	20	0.714
Diphenylamine			1.16
Diesel fuel oil 2D 13D 14D 15D	60	15.6	0.81–0.96
Dowtherm	77	25	1.056
Ethyl acetate $CH_3COOC_2H_3$	59 68	15 20	0.907 0.90
Ethyl bromide C_2H_3Br	59	15	1.45
Ethylene bromide	68	20	2.18
Ethylene chloride	68	20	1.246
Ethylene glycol	60	15.6	1.125
Formic acid – 10%	68	20	1.025
Formic acid – 50%	68	20	1.121
Formic acid – 80%	68	20	1.186
Formic acid – concentrated	68	20	1.221
Freon – 11	70	21.1	1.49
Freon – 12	70	21.1	1.33
Freon – 21	70	21.1	1.37
Furfurol	68	20	1.159
Fuel oils 12, 13, 15A, 15B, 16	60	15.6	0.82–0.95
Gas oils	60	15.6	0.89
Gasoline a	60	15.6	0.74
Gasoline b	60	15.6	0.72

Table A22 – Kinematic Viscosity: Common Fluids (continued)

	Temperature		Kinematic Viscosity	
Fluid	°F	°C	Centistokes (cSt)	Seconds Saybolt Universal (SSU)
Automotive gear oil SAE 80W	210	98.9	7.0 min	49 min
Automotive gear oil SAE 85W	210	98.9	11.0 min	63 min
Automotive gear oil SAE 90W	210	98.9	14–25	74–120
Automotive gear oil SAE 140	210	98.9	25–43	120–200
Automotive gear oil SAE150	210	98.9	43–min	200 min
Beer	68	20	1.8	32
Benzene (Benzol) $C6H_6$	32 68	0 20	1.0 0.74	31
Bone oil	130 212	54.4 100	47.5 11.6	220 65
Bromine	68	20	0.34	
Butane-n	–50 30	–1.1	0.52 0.35	
Butyric acid-n	68 32	20 0	1.61 2.3 cp	31.6
Calcium chloride 5%	65	18.3	1.156	
Calcium chloride 25%	60	15.6	4.0	39
Carbolic acid (phenol)	65 194	18.3 90	11.83 1.26 cp	65
Carbon tetrachloride CCl_4	68 100	20 37.8	0.612 0.53	
Carbon disulfide CS2	32 68	0 20	0.33 0.298	
Castor oil	100 130	37.8 54.4	259–325 98–130	1200–1500 450–600
China wood oil	69 100	20.6 37.8	308.5 125.5	1425 580
Chloroform	68 140	20 60	0.38 0.35	
Coconut oil	100 130	37.8 54.4	29.8–31.6 14.7–15.7	140–148 76–80
Cod oil (fish oil)	100 130	37.8 54.4	32.1 19.4	150 95
Corn oil	130 212	54.4 100	28.7 8.6	135 54
Corn starch solution 22 Baume	70 100	21.1 37.8	32.1 27.5	150 130
Corn starch solution 24 Baume	70 100	21.1 37.8	129.8 95.2	600 440

Table A22 – Kinematic Viscosity: Common Fluids (continued)

Fluid	Temperature		Kinematic Viscosity	
	°F	°C	Centistokes (cSt)	Seconds Saybolt Universal (SSU)
Corn starch solution	70	21.1	303	1400
25 Baume	100	37.8	173.2	800
Cotton seed oil	100	37.8	37.9	176
	130	54.4	20.6	100
Crude oil 48° API	60	15.6	3.8	39
	130	54.4	1.6	31.8
Crude oil 40o API	60	15.6	9.7	55.7
	130	54.4	3.5	38
Crude oil 35.6o API	60	15.6	17.8	88.4
	130	54.4	4.9	42.3
Crude oil 32.6o API	60	15.6	23.2	110
	130	54.4	7.1	46.8
Decane-n	0	17.8	2.36	34
	100	37.8	1.001	31
Diethyl glycol	70	21.1	32	149.7
Diethyl ether	68	20	0.32	
Diesel fuel 20	100	37.8	2–6	32.6–45.5
	130	54.4	1.–3.97	–39
Diesel fuel 30	100	37.8	6–11.75	45.5–65
	130	54.4	3.97–6.78	39–48
Diesel fuel 40	100	37.8	29.8 max	140 max
	130	54.4	13.1 max	70 max
Diesel fuel 60	122	50	86.6 max	400 max
	160	71.1	35.2 max	165 max
Ethyl acetate $CH_3COOC_2H_3$	59	15	0.4	
	68	20	0.49	
Ethyl bromide C_2H_5Br	68	20	0.27	
Ethylene bromide	68	20	0.787	
Ethylene chloride	68	20	0.668	
Ethylene glycol	70	21.1	17.8	88.4
Formic acid 10%	68	20	1.04	31
Formic acid 50%	68	20	1.2	31.5
Formic acid 80%	68	20	1.4	31.7
Formic acid concentrated	68	20	1.48	31.7
	77	25	1.57cp	
Freon 11	70	21.1	0.21	
Freon 12	70	21.1	0.27	
Freon 21	70	21.1	1.45	
Furfurol	68	20	1.45	31.7
	77	25	1.49cp	
Fuel Oils No. 2–No.6				

Table A22 – Kinematic Viscosity: Common Fluids (continued)

Viscosity - Saybolt Universal Seconds (SSU) versus temperature for typical fuel oils is indicated in the diagram below:

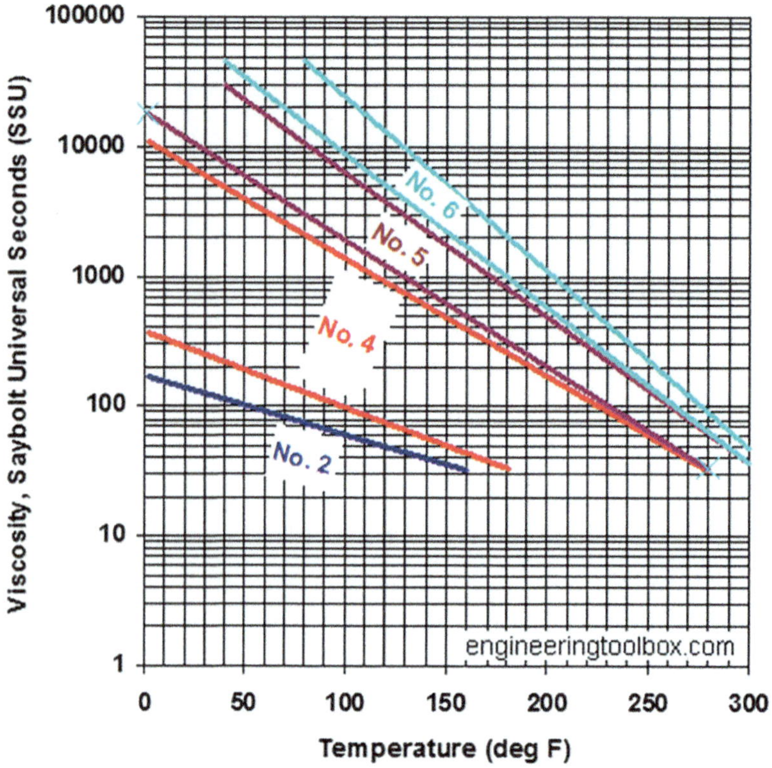

Fluid	Temperature °F	Temperature °C	Kinematic Viscosity Centistokes (cSt)	Kinematic Viscosity Seconds Saybolt Universal (SSU)
Fuel oil 1	70 100	21.1 37.8	2.39–4.28 –2.69	34–40 32–35
Fuel oil 2	70 100	21.1 37.8	3.0–7.4 2.11–4.28	36–50 33–40
Fuel oil 3	70 100	21.1 37.8	2.69–5.84 2.06–3.97	35–45 32.8–39
Fuel oil 5A	70 100	21.1 37.8	7.4–26.4 4.91–13.7	50–125 42–72
Fuel oil 5B	70 100	21.1 37.8	26.4 13.6–67.1	125– 72–310
Fuel oil 6	122 160	50 71.1	97.4–660 37.5–172	450–3M 175–780
Gas oils	70 100	21.1 37.8	13.9 7.4	73 50
Gasoline a	60 100	15.6 37.8	0.88 0.71	
Gasoline b	60 100	15.6 37.8	0.64	
Gasoline c	60 100	15.6 37.8	0.46 0.40	
Glycerin 100%	68.6 100	20.3 37.8	648 176	2950 813

Table A22 – Kinematic Viscosity: Common Fluids (continued)

Fluid	Temperature		Kinematic Viscosity	
	°F	°C	Centistokes (cSt)	Seconds Saybolt Universal (SSU)
Glycerin 50% water	68	20	5.29	43
	140	60	1.85 cp	
Glucose	100	37.8	7.7M–22M	35M–100M
	150	65.6	880–2420	4M–11M
Heptanes-n	0	–17.8	0.928	
	100	37.8	0.511	
Hexane-n	0	–17.8	0.683	
	100	37.8	0.401	
Honey	100	37.8	73.6	349
Ink, printers	100	37.8	550–2200	2500–10M
	130	54.4	238–660	1100–3M
Insulating oil	70	21.1	24.1 max	115 max
	100	37.8	11.75 max	65 max
Kerosene	68	20	2.71	35
Jet fuel	–30	–34.4	7.9	52
Lard	100	37.8	62.1	287
	130	54.4	34.3	160
Kinematic Viscosity: 1 cSt (centistoke) = 10^{-6} m²/s				
Milk	68	20	1.13	31.5
Molasses A, first	100	37.8	281–5070	1300–23500
	130	54.4	151–1760	700–8160
B, second	100	37.8	1410–13.2M	6535–61180
	130	54.4	660–3.3M	3058–15294
C, blackstrap	100	37.8	2630–55M	12190–255M
	130	54.4	1320–16.5M	6120–76.5M
Naphthalene	176	80	0.9	
	212	100	0.78 cp	
Neatstool oil	100	37.8	49.7	230
	130	54.4	27.5	130
Nitrobenzene	68	20	1.67	31.8
Nonane-n	0	–17.8	1.728	32
	100	37.8	0.807	
Octane-n	0	–17.8	1.266	31.7
	100	37.8	0.645	
Olive oil	100	37.8	43.2	200
	130	54.4	24.1	
Palms oil	100	37.8	47.8	
	130	54.4	26.4	
Peanut oil	100	37.8	42	200
	130	54.4	23.4	
Pentane-n	0	17.8	0.508	
	80	26.7	0.342	
Petrolatum	130	54.4	20.5	100
	160	71.1	15	77
Petroleum ether	60	15.6	31(est)	1.1
Propionic acid	32	0	1.52 cp	31.5
	68	20	1.13	

Table A22 – Kinematic Viscosity: Common Fluids (continued)

Fluid	Temperature °F	Temperature °C	Kinematic Viscosity Centistokes (cSt)	Kinematic Viscosity Seconds Saybolt Universal (SSU)
Propylene glycol	70	21.1	52	241
Quenching oil (typical)			100–120	20.5–25
Rapeseed oil	100 130	37.8 54.4	54.1 31	250 145
Rosin oil	100 130	37.8 54.4	324.7 129.9	1500 600
Rosin (wood)	100 200	37.8 93.3	216–11M 108–4400	1M–50M 500–20M
Sesame seed oil	100 130	37.8 54.4	39.6 23	184 110
Sodium chloride 5%	68	20	1.097	31.1
Sodium chloride 25%	60	15.6	2.4	34
Sodium hydroxide (caustic soda) 20%	65	18.3	4.0	39.4
Sodium hydroxide (caustic soda) 30%	65	18.3	10.0	58.1
Sodium hydroxide (caustic soda) 40%	65	18.3		
Soya bean oil	100 130	37.8 5.4	35.4 19.64	165 96
Sperm oil	100 130	37.5 54.4	21–23 15.2	110 78
Sulfuric acid 100%	68 140	20 60	14.56 7.2 cp	76
Sulfuric acid 95%	68	20	14.5	75
Sulfuric acid 60%	68	20	4.4	41
Sulfuric acid 20%				3M–8M 650–1400
Tar, coke oven	70 100	21.1 37.8	600–1760 141–308	15M–300M 2M–20M
Kinematic Viscosity: 1 cSt (centistoke) = 10^{-6} m²/s				
Tar, gas house	70 100	21.1 37.8	3300–66M 440–4400	2500 500
Tar, pine	100 132	37.8 55.6	559 108.2	200–300 55–60
Toluene	68 140	20 60	0.68 0.38 cp	185.7
Triethylene glycol	70	21.1	40	400–440 185–205
Turpentine	100 130	37.8 54.4	86.5–95.2 39.9–44.3	1425 650
Varnish, spar	68 100	20 37.8	313 143	
Water, distilled	68	20	1.0038	31
Water, fresh	60 130	15.6 54.4	1.13 0.55	31.5
Water, sea			1.15	31.5
Whale oil	100 130	37.8 54.4	35–39.6 19.9–23.4	163–184 97–112
Xylene-o	68 104	20 40	0.93 0.623 cp	

Table A23 – Absolute Viscosity: Common Liquids

Temperature °F	Viscosity Centipoise	Temperature °F	Viscosity Centipoise	Temperature °F	Viscosity Centipoise
Acetic acid		Ethylene oxide		Isohexane	
59	1.31	−57	0.577	32	0.376
64	1.30	−37	0.488	68	0.306
77	1.155	−5.8	0.394	104	0.254
86	1.04	32	0.320	Isopentane	
106	1.00	Fluorobenzene		32	0.273
212	0.43	68	0.598	68	0.223
Acetic anhydride		140	0.389	Kerosene	
32	1.24	212	0.275	68	2.69
59	0.971	Fuel oil, #2		100	2.0
64	0.90	70	3.0–7.4	Methyl alcohol	
86	0.783	100	2.11–4.28	−48	1.98
212	0.490	Fuel oil, #6		32	0.82
Acetone		122	97.4–660	59	0.623
14	0.450	160	37.5–172	68	0.597
32	0.399	Gasoline		77	0.546
59	0.337	60	0.46–0.88	86	0.510
77	0.316	100	0.40–0.71	Methyl chloride	
Ammonia		Glycerin		0	0.25
−92	0.475	32	12,110	20	0.23
−58	0.317	43	6,260	40	0.21
−40	0.276	59	2,330	60	0.19
−28	0.255	68	1,490	100	0.16
Benzene		77	954	Naphthalene	
32	0.912	86	629	176	0.967
50	0.758	Heptane		212	0.776
68	0.652	32	0.524	Nitric acid	
86	0.564	63	0.461	32	2.275
104	0.503	68	0.409	Nitrobenzene	
122	0.542	77	0.386	37	2.91
Carbon tetrachloride		104	0.341	42	2.71
32	1.329	Hexane		50	2.48
59	1.038	32	0.401	68	2.03
68	0.969	68	0.326	Nitromethane	
86	0.843	77	0.386	32	0.853
104	0.739	104	0.341	77	0.620
140	0.585	Hydrochloric acid, 31.5%		n-Octane	

(*continued*)

Table A23 – Absolute Viscosity: Common Liquids (continued)

Temperature °F	Viscosity Centipoise	Temperature °F	Viscosity Centipoise	Temperature °F	Viscosity Centipoise
Chlorine liquid		0	3.4	32	0.706
–40	0.505	20	2.9	68	0.240
–20	0.462	40	2.5	104	0.433
20	0.400	60	2.0	Pentane	
60	0.350	80	1.8	32	0.289
100	0.313	100	1.6	68	0.524
Ethylbenzene		140	1.2	Phenol	
63	0.691	Iodine Liquid		65	12.7
Ethylene glycol		241	2.27	122	3.49
68	19.9	32	0.481	158	2.03
104	9.13	68	0.384	194	1.26
140	4.95	104	0.315		
176	3.02				
Phosphorous liquid		Sodium liquid		Sulfuric acid	
71	2.34	0	2.4	32	48.4
88	2.01	26	1.3	59	32.8
110	1.73	40	1.2	68	25.4
123	1.60	60	1.1	86	15.7
140	1.45	100	1.0	104	11.5
158	1.32	140	0.85	122	8.82
Sodium hydroxide		Sulfur (gas free)		140	7.22
0	100	253	10.9	Turpentine	
100	40	276	8.7	60	2.11
120	25			100	2.0
140	15	301	7.1		
160	9.5	314	7.2	Water	
200	3.7	317	7.6	60	1.13
220	2.4	319	14.5	130	0.55
250	1.4	Sulfur dioxide			
		–28	0.5508		
		13	0.4285		
		32	0.3936		

Table A24 – Absolute Viscosity: Common Gases

Temperature °F	Viscosity Centipoise	Temperature °F	Viscosity Centipoise	Temperature °F	Viscosity Centipoise
Acetylene		Carbon monoxide		Nitrogen	
32	0.00935	32	0.0166	–6.7	0.0156
Air		59	0.0172	51.6	0.0171
32	0.0171	260.8	0.0218	81	0.0178
104	0.0190	Chlorine		261	0.0219
444	0.0264	122	0.0147	440	0.0256
633	0.0305	212	0.0168	Oxygen	
674	0.0312	302	0.0187	32	0.0189
768	0.0341	392	0.0208	67	0.0202
Ammonia		Ethane		262	0.0257
32	0.0092	32	0.0085	440	
68	0.0098	63	0.0090		0.0302
212	0.0128	Ethylene		n-Pentane	
302	0.0146	32	0.0091	77	0.0068
482	0.0181	68	0.0101	212	0.0084
Argon		122	0.0110	Propane	
32	0.0210	212	0.0126	64.2	0.0079
68	0.0222	Heliumb		213	0.0101
212	0.0269	32	0.0186	Propylene	
392	0.0322	68	0.0194	62	0.0083
Benzene		Hydrogen		122	0.0093
0	0.0065	–172	0.0057	32	0.0116
40	0.0070	–143.5	0.0062	64.4	0.0124
70	0.0075	–25	0.0077	68.9	0.0125
100	0.0080	32	0.0084	213	0.0161
200	0.0091	69	0.0088		
Butene		264	0.0108		
0	0.0075	Hydrogen chloride			
40	0.0080	54.4	0.0139		
70	0.0085	61.8	0.0141		
100	0.0090	212	0.0182		
200	0.0104	Hydrogen sulfide			
Butylene		32	0.0117		
66	0.0074	62.6	0.0124		
212	0.0095	212	0.0159		
Carbon dioxide		Methane			
–144	0.009	32	0.0102		
–76	0.0106	68	0.0109		
32	0.0139	212	0.0133		
68	0.0148				
86	0.0153				

Table A25 – Density of Elements in English and Metric Units

Element or Compound	Density – lb/ft³ 14.7 psia and 60°F		Density – kg/m³ 1013 mbar and 15.6°C		Mol. Wt.
	Liquid	Gas	Liquid	Gas	M_r
Acetic acid, CH_3-CO-OH	65.7		1052.4		66.1
Acetone, CH_3-CO-CH_3	49.4		791.3		58.1
Acetylene, C_2H_2		0.069		1.11	26.0
Air, $O_2 + N_2$		0.0764		1.223	29.0
Ammonia, NH_3		0.045		0.72	17.0
Argon, A		0.105		1.68	39.9
Benzene, C_6H_6	54.6		874.6		78.1
Butane, C_4H_{10}		0.154		2.47	58.1
Carbon dioxide, CO_2		0.117		1.87	44.0
Carbon monoxide, CO		0.074		1.19	28.0
Carbon tetrachloride, CCl_4	99.5		1593.9		153.8
Chlorine, Cl_2		0.190		3.04	70.9
Ethane, C_2H_6		0.080		1.28	30.1
Ethyl alcohol, C_2H_5OH	49.52		793.3		46.1
Ethylene, CH_2=CH_2		0.074		1.19	28.1
Ethyl ether, C_2H_5-O-C_2H_5	44.9		719.3		74.1
Fluorine, F_2		0.097		1.55	38.0
Helium, He		0.011		0.18	4.00
Heptane, C_7H_{16}	42.6		682.4		100.2
Hydrogen, H_2		0.005		0.08	2.02
Hydrogen chloride, HCl		0.097		1.55	36.5
Isobutane, $(CH_3)_2$ CH-CH_3		0.154		2.47	58.1
Isopropyl alcohol, CH_3-CHOH-CH_3	49.23		788.6		60.1
Methane, CH_4		0.042		0.67	16.0
Methyl alcohol, H-CH_2OH	49.66		795.5		32.0
Nitrogen, N_2		0.074		1.19	28.0
Nitrous oxide, N_2O		0.117		1.87	44.0
Octane, CH_3-$(CH_2)_6$-CH_3	43.8		701.6		114.2
Oxygen, O_2		0.084		1.35	32.0
Pentane, C_5H_{12}	38.9		623.1		72.2
Phenol, C_6H_5OH	66.5		1065.3		94.1
Phosgene, $COCl_2$		0.108		1.73	98.9
Propane, C_3H_8		0.117		1.87	44.1
Propylene, CH_2=CH-CH_3		0.111		1.78	42.1
Refrigerant 12, CCl_2F_2		0.320		5.13	120.9
Refrigerant 22, $CHClF_2$		0.228		3.65	86.5
Sulfur dioxide, SO_2		0.173		2.77	64.1
Water, H_2O	62.34		998.6		18.0

Table A26 – Metric Conversion Tables

Multiply	By	To Obtain
	Length	
millimeters	0.10	centimeters
millimeters	0.001	meters
millimeters	0.039	inches
millimeters	0.00328	feet
centimeters	10.0	millimeters
centimeters	0.010	meters
centimeters	0.394	inches
centimeters	0.0328	feet
inches	25.40	millimeters
inches	2.54	centimeters
inches	0.0254	meters
inches	0.0833	feet
feet	304.8	millimeters
feet	30.48	centimeters
feet	0.304	meters
feet	12.0	inches
	Area	
sq. millimeters	0.010	sq. centimeters
sq. millimeters	$10.^{-6}$	sq. meters
sq. millimeters	0.00155	sq. inches
sq. millimeters	$1.076 \cdot 10^{-5}$	sq. feet
sq. centimeters	100	sq. millimeters
sq. centimeters	0.0001	sq. meters
sq. centimeters	0.155	sq. inches
sq. centimeters	0.001076	sq. feet
sq. inches	645.2	sq. millimeters
sq. inches	6.452	sq. centimeters
sq. inches	0.000645	sq. meters
sq. inches	0.00694	sq. feet
sq. feet	$9.29 \cdot 10^{4}$	sq. millimeters
sq. feet	929	sq. centimeters
sq. feet	0.0929	sq. meters
sq. feet	144	sq. inches
	Flow Rates	
gallons US/minute		
gpm	3.785	L/min
gallons US/minute	0.1337	ft^3/min
gallons US/minute	8.0208	ft^3/h
gallons US/minute	0.227	m^3/h
gallons US/minute	34.2857	barrels/day (42 US gal)
cubic feet/minute	7.4805	gpm
cubic feet/minute	28.32	liters/minute

Multiply	By	To Obtain
	Flow Rates	
cubic feet/minute	60.0	Ft^3/h
cubic feet/minute	1.699	m^3/h
cubic feet/minute	256.5	barrels/ day
cubic feet/hour	0.1247	gpm
cubic feet/hour	0.472	L/min
cubic feet/hour	0.01667	ft^3/min
cubic feet/hour	0.0283	m^3/h
cubic meters/hour	4.403	gpm
cubic meters/hour	16.67	L/min
cubic meters/hour	0.5886	ft^3/min
cubic meters/hour	35.31	ft^3/h
cubic meters/hour	150.9	barrels/ day
	Velocity	
feet/second	60	ft/min
feet/second	0.3048	m/s
feet/second	1.097	km/h
feet/second	0.6818	mi/h
meters/second	3.28084	ft/s
meters/second	196.85	ft/min
meters/second	3.6	km/h
meters/second	2.237	miles/h
	Weight (Mass)	
pounds	0.0005	short ton
pounds	0.000446	long ton
pounds	0.453	kilogram
pounds	0.000453	metric ton
short ton	2000.0	pounds
short ton	0.8929	long ton
short ton	907.2	kilogram
short ton	0.9072	metric ton
long ton	2240	pounds
long ton	1.120	short ton
long ton	1016	kilogram
long ton	1.016	metric ton
kilogram	2.205	pounds
kilogram	0.0011	short ton
kilogram	0.00098	long ton
kilogram	0.001	metric ton
metric ton	2205	pounds
metric ton	1.102	short ton
metric ton	0.984	long ton
metric ton	1000	kilogram

Table A26 – Metric Conversion Tables (continued)

Some units shown on this page are not recommended by SI, for example, kilogram/sq. cm should be read as kilogram (force)/sq. cm.

Multiply	By	To Obtain
Volume and Capacity		
cubic cm	0.06102	cubic inches
cubic cm	3.531×10^{-5}	cubic feet
cubic cm	10^{-6}	cubic meters
cubic cm	0.0001	liters
cubic cm	2.642×10^{-4}	gallons (US)
cubic meters	10^{6}	cubic centimeters
cubic meters	61,023	cubic inches
cubic meters	35.31	cubic feet
cubic meters	1000	liters
cubic meters	264.2	gallons
cubic feet	28,320	cubic centimeters
cubic feet	1728	inches
cubic feet	0.0283	cubic meters
cubic feet	28.32	liters
cubic feet	7.4805	gallons
liters	1000	cubic centimeters
liters	61.02	cubic inches
liters	0.03531	cubic feet
liters	0.001	cubic meters
liters	0.264	gallons
gallons	0.00379	cubic meters
gallons	231	cubic inches
gallons	0.1337	cubic feet
gallons	3785	milliliters
gallons	3.785	L
Pressure and Head		
pounds/sq. inch	0.06895	bar atmosphere
pounds/sq. inch	0.06804	kg/cm^2
pounds/sq. inch	0.0703	kPa
pounds/sq. inch	0.433	ftH_2O (4°C)
pounds/sq. inch	2.307	mH_2O (4°C)
pounds/sq. inch	0.703	cm of Hg (0°C)
pounds/sq. inch	5.171	torr (mm of Hg)
pounds/sq. inch	51.71	(0°C)
pounds/sq. inch	2.036	in of Hg (0°C)

Multiply	By	To Obtain
Pressure and Head		
atmosphere	14.69	psi
atmosphere	1.013	bar
atmosphere	1.033	Kg/cm^2
atmosphere	101.3	kPa
atmosphere	33.9	ft of H_2O
atmosphere	10.33	m of H_2O
atmosphere	76	cm of Hg
atmosphere	760	torr (mm of Hg)
atmosphere	29.92	in of Hg
bar	14.5	psi
bar	0.9869	atmosphere
bar	1.02	Kg/cm^2
bar	100	kPa
bar	33.45	ft of H_2O
bar	10.2	m of H_2O
bar	75.01	cm of Hg
bar	750.1	torr (mm of Hg)
bar	29.53	in of Hg
kilogram/sq. cm	14.22	psi bar
kilogram/sq. cm	0.9807	atmosphere
kilogram/sq. cm	0.9678	kPa
kilogram/sq. cm	98.07	ft of H_2O (4°C)
kilogram/sq. cm	32.81	m of H_2O (4°C)
kilogram/sq. cm	10	cm of Hg
kilogram/sq. cm	73.56	torr (mm pf Hg)
kilogram/sq. cm	735.6	in of Hg
kilogram/sq. cm	28.96	psi
kilopascal	0.145	bar
kilopascal	0.01	atmosphere
kilopascal	0.00986	kg/cm^2
kilopascal	0.0102	ft of H_2O
kilopascal	0.334	m of H_2O
kilopascal	0.102	cm of Hg
kilopascal	0.7501	torr (mm of Hg)
kilopascal	7.501	in of Hg
kilopascal	0.295	bar
millibar	0.001	

Table A27 – Standard Conditions and Gas Laws

Standards and Conditions	Laws of Gases
Standard conditions (US customary gases) are: 14.69 psia and 68°F	Universal gas equation
Standard conditions (US customary liquids) are: 14.69 psia and 60°F	$P_V = mRTZ$
Normal conditions (metric) are: 1.013 bar and 0°C and 4°C water	where *English units* P = press in pounds per foot squared
Note: In the *Masoneilan Control Valve Sizing Handbook*, the metric factors are: 1.013 bar and 15.6°C.	v = volume in cubic feet m = mass in pounds
Specific gravity of air SG = 1 @ standard condition for gases	R = gas constant = $\frac{1545}{Mr}$
Specific gravity of water = 1 @ std. cond. for liquids	*Metric units*
1 US gallon of water = 8.3378 lb @ standard. condition 1 cubic foot of water = 62.34 lb @ standard condition (= density) 1 cubic foot of water = 7.48 gal 1 cubic foot of air = 0.076 lb @ standard condition (= air density)	P = press in pascals v = volume in cubic meters m = mass in kilograms R = gas constant = $\frac{8314}{Mr}$
Air specific volume = 1/density = 13.1 cubic ft /lb Air molecular weight $M_r = 29$	Gas expansion (perfect gas)
SG of any gas = density of gas/0.076	$\frac{P_1V_1}{T_1} = \frac{P_2V_2}{T_2}$
SG of any gas = molecular weight M of gas/29	Velocity of sound, C, in feet per second $C = 223\sqrt{\frac{k(T+460)}{M_r}}$
	Velocity of sound, C, in meters per second $C = 91.2\sqrt{\frac{k(T+273)}{M_r}}$

Table A28 – Head Loss in Piping Systems

Friction Losses in Pipe Fittings: Resistance Coefficient K

Head loss in fittings, h_f: $h_f = [K \cdot v^2/2g]$

Head loss in pipe, h_p: $h_p = f \cdot [L \text{ (in ft)}/D \text{ (in ft)}] \cdot [v^2/2g]$ *Note*: f = Moody friction factor for pipe

Head loss flowmeter, h_o: $h_o = [\text{inH}_2\text{O}/12]$ in feet

Head loss in piping and fittings: $h_L = h_p + h_f + h_O + h_V$

Head loss for valve $h_V = h_L \cdot (0.15 \text{ to } 0.4)$ *Note*: Verify valve works for percentage pressure drop.

Head required for pump in feet = $h_p + h_f + h_o + h_V$ + any working vessel pressure in feet

			Nominal Pipe Size in Inches											
			1/2	3/4	1	1-1/4	1-1/2	2	2-1/2-3	4	6	8-10	12-16	18-24
Fitting		LD	K Value											
Angle valve		55	1.48	1.38	1.27	1.21	1.16	1.05	0.99	0.94	0.83	0.77	0.72	0.66
Angle valve		150	4.05	3.75	3.45	3.30	3.15	2.85	2.70	2.55	2.25	2.10	1.95	1.80
Ball valve		3	0.08	0.08	0.07	0.07	0.06	0.06	0.05	0.05	0.05	0.04	0.04	0.04
Butterfly valve								0.86	0.81	0.77	0.68	0.63	0.35	0.30
Gate valve		8	0.22	0.20	0.18	0.18	0.15	0.15	0.14	0.14	0.12	0.11	0.10	0.10
Globe valve		340	9.2	8.5	7.8	7.5	7.1	6.5	6.1	5.8	5.1	4.8	4.4	4.1
Plug valve branch flow		90	2.43	2.25	2.07	1.98	1.89	1.71	1.62	1.53	1.35	1.26	1.17	1.08
Plug valve straightaway		18	0.48	0.45	0.41	0.40	0.38	0.34	0.32	0.31	0.27	0.25	0.23	0.22
Plug valve 3-way Thru-Flow		30	0.81	0.75	0.69	0.66	0.63	0.57	0.54	0.51	0.45	0.42	0.39	0.36
Standard elbow	90°	30	0.81	0.75	0.69	0.66	0.63	0.57	0.54	0.51	0.45	0.42	0.39	0.36
	45°	16	0.43	0.40	0.37	0.35	0.34	0.30	0.29	0.27	0.24	0.22	0.21	0.19
	long radius 90°	16	0.43	0.40	0.37	0.35	0.34	0.30	0.29	0.27	0.24	0.22	0.21	0.19
Close return bend		50	1.35	1.25	1.15	1.10	1.05	0.95	0.90	0.85	0.75	0.70	0.65	0.60
Standard tee	Thru-Flow	20	0.54	0.50	0.46	0.44	0.42	0.38	0.36	0.34	0.30	0.28	0.26	0.24
	Thru-Branch	60	1.62	1.50	1.38	1.32	1.26	1.14	1.08	1.02	0.90	0.84	0.78	0.72
90 bends, Pipe bends, Flanged elbows, Butt-welded elbows	r/d = 1	20	0.54	0.50	0.46	0.44	0.42	0.38	0.36	0.34	0.30	0.28	0.26	0.24
	r/d = 2	12	0.32	0.30	0.28	0.26	0.25	0.23	0.22	0.20	0.18	0.17	0.16	0.14
	r/d = 3	12	0.32	0.30	0.28	0.26	0.25	0.23	0.22	0.20	0.18	0.17	0.16	0.14
	r/d = 4	14	0.38	0.35	0.32	0.31	0.29	0.27	0.25	0.24	0.21	0.20	0.18	0.17
	r/d = 6	17	0.46	0.43	0.39	0.37	0.36	0.32	0.31	0.29	0.26	0.24	0.22	0.20
	r/d = 8	24	0.65	0.60	0.55	0.53	0.50	0.46	0.43	0.41	0.36	0.34	0.31	0.29
	r/d = 10	30	0.81	0.75	0.69	0.66	0.63	0.57	0.54	0.51	0.45	0.42	0.39	0.36
	r/d = 12	34	0.92	0.85	0.78	0.75	0.71	0.65	0.61	0.58	0.51	0.48	0.44	0.41
	r/d = 14	38	1.03	0.95	0.87	0.84	0.80	0.72	0.68	0.65	0.57	0.53	0.49	0.46
	r/d = 16	42	1.13	1.05	0.97	0.92	0.88	0.80	0.76	0.71	0.63	0.59	0.55	0.50
	r/d = 18	45	1.24	1.15	1.06	1.01	0.97	0.87	0.83	0.78	0.69	0.64	0.60	0.55
	r/d = 20	50	1.35	1.25	1.15	1.10	1.05	0.95	0.90	0.85	0.75	0.70	0.65	0.60
Miter bends	a = 0°	2	0.05	0.05	0.05	0.04	0.04	0.04	0.04	0.03	0.03	0.03	0.03	0.02
	a = 15°	4	0.11	0.10	0.09	0.09	0.08	0.08	0.07	0.07	0.06	0.06	0.05	0.05
	a = 30°	8	0.22	0.20	0.18	0.18	0.17	0.15	0.14	0.14	0.12	0.11	0.10	0.10
	a = 45°	15	0.41	0.38	0.35	0.33	0.32	0.29	0.27	0.26	0.23	0.21	0.20	0.18
	a = 60°	25	0.68	0.63	0.58	0.55	0.53	0.48	0.45	0.43	0.38	0.35	0.33	0.30
	a = 75°	40	1.09	1.00	0.92	0.88	0.84	0.76	0.72	0.68	0.60	0.56	0.52	0.48
	a = 90°	60	1.62	1.50	1.38	1.32	1.26	1.14	1.08	1.02	0.90	0.84	0.78	0.72

Note: Fittings are standard with full openings.

Table A29 – Maximal Flow Velocity in Pipes

Maximal Flow Velocity in Pipes		
Water	**m/s**	**ft/s**
Tap water (low noise)	0.5–0.7	1.6–8.2
Tap water	1.0–2.5	3.3–8.2
Cooling water	1.5–2.5	4.9–8.2
Boiler feedwater - suction	0.5–1.0	1.6–3.3
Boiler feedwater - discharge	1.5–2.5	4.9–8.2
Condensate	1.0–2.0	3.3–6.5
Heating circulation	1.0–3.0	3.3–9.8
Lower velocity in thin lines. Higher in thick lines. The pressure drop normally calculates separately for longer lines with several separate bend-resistances.		
Steam	**m/s**	**ft/s**
Saturated steam. high pressure	25–40	82–131
Saturated steam. in special cases	0–60	0–197
Saturated steam. medium and low pressure	30–40	99–131
Saturated steam. at peak load	0–50	0–164
Steam/water emulsion	0–25	0–82
The pressure drop normally calculates separately for longer lines with several separate bend-resistances.		
Oil	**m/s**	**ft/s**
Suction lines for pumps	0–0.5	0–1.6
Suction lines for pump (low pressure)	0.1–0.2	0.3–0.65
Discharge line for booster pump	1.0–2.0	3.3–6.5
Discharge line for burner pump	0–1.0	0–3.3
Air	**m/s**	**ft/s**
Combustion air ducts	12–20	40–66
Air inlet to boiler room	1–3	3.3–9.8
Warm air for house heating	0.8–1.0	2.6–3.3
Vacuum cleaning pipe	8–15	26–49
Compressed air pipe	20–30	66–98
Ventilation ducts (hospitals)	1.8–4	5.9–13
Ventilation ducts (office buildings)	2.0–4.5	6.5–15
Exhaust Gas	**m/s**	**ft/s**
Ducts at minimum load	0–4.0	0–13
Stack at minimum load	0–5.0	0–16
Boiler with one-step burner (on-off)	5.0–8.0	16–26
Boiler with two-step burner (high-low)	10–15	31–49
Boiler with modulating burner (3:1)	15–25	49–82
To keep the surface free from soot, the velocity should always exceed	3.0–4.0	9.8–13

Table A30 – Pressure Vapor Chart of Common Liquids

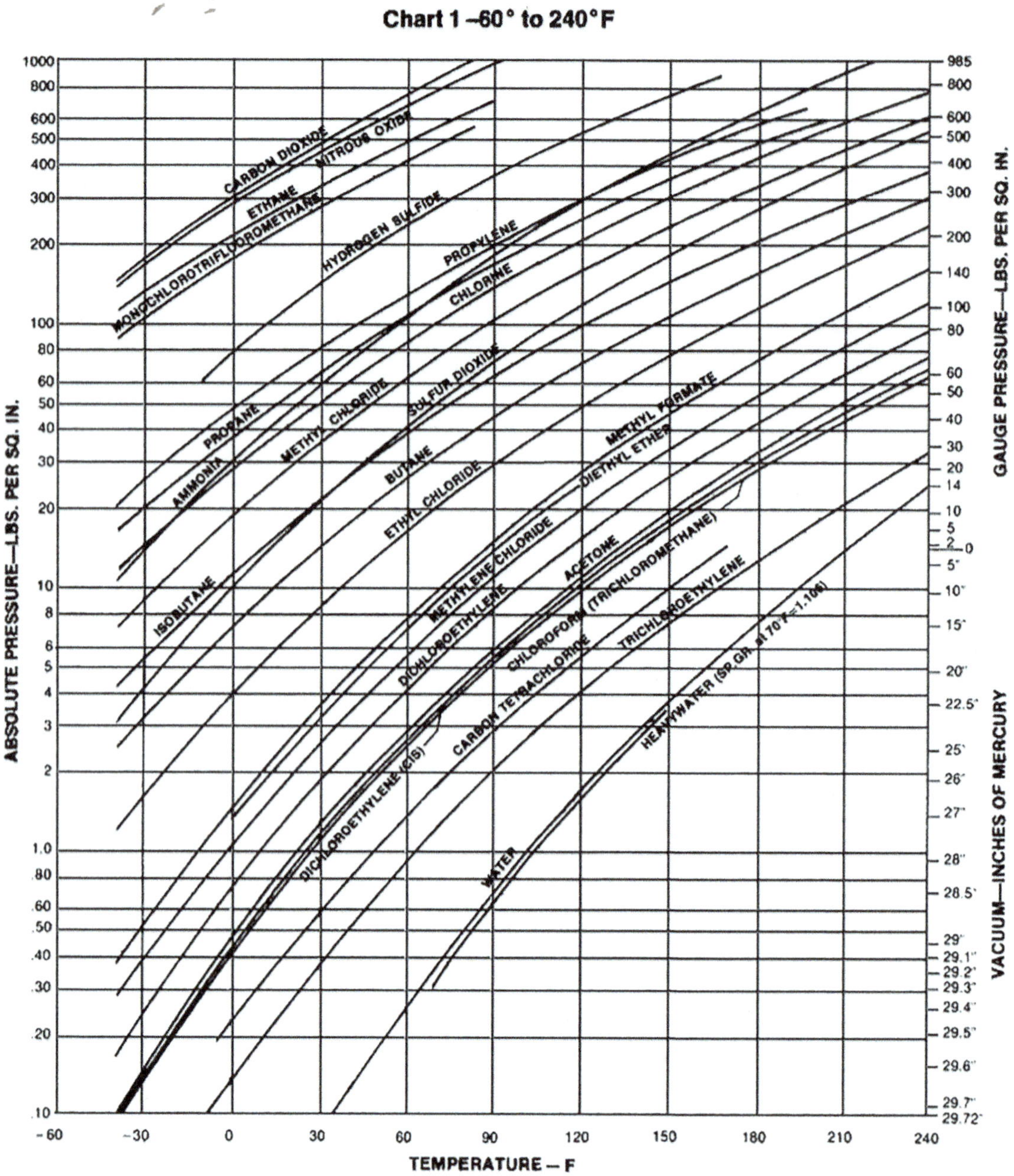

References

1. Fisher Controls, *Fisher Control Valve Handbook*, 4th ed
2. Norman A. Anderson, *Instrumentation for Process Measurement and Control*, 3rd ed
3. Dale Seborg, Thomas Edgar, Mellichamp, *Process Dynamics and Control*, 1st ed

4. Bela Liptak, *Instrument Engineer's Handbook*
5. Dr. Angela Summers - President, SIS-TECH
6. Gene F. Franklin, et al., *Feedback Control of Dynamic Systems*, 3rd ed
7. Tony R. Kuphaldt, Series on electronics and instrumentation
8. Jacques Smuts, PhD, PE, *Process Control for Practitioners*
9. Myke King, *Process Control: A Practical Approach*
10. Michael Brown, *Control Engineering* - http://www.controlloop.co.za
11. Emerson Process
12. Cisco Networks
13. Yokogawa Corporation
14. Endress - Hauser
15. Smar Corporation
16. Samson
17. Masoneilan
18. ISA, Control Systems Engineer Study Guide, 5th ed
19. ANSI/ISA-51.1-1979(R1993) Process Instrumentation Terminology
20. ANSI/ISA-75.01.01-2002 Flow Equations for Sizing Control Valves
21. ANSI/ISA-5.1-1984(R1992) Instrumentation Symbols and Identification
22. NFPA 30 Flammable and Combustible Liquids Code
23. NFPA 70E Standard for Electrical Safety in the Workplace
24. NFPA 70 National Electrical Code
25. NFPA 77 Static Electricity

26. NFPA 78 Lightning Protection

27. NFPA 79 Industrial Machinery

28. NFPA 496 Purged and Pressurized Systems

29. ASME Boiler and Pressure Vessel Code Section VIII, Div. 1, UG-125 through UG-136

30. http://www.atlascopco.us

31. http://www.instrumentationportal.com/

32. http://www.chemengonline.com/

33. http://www.automation.com/

34. http://www.pumpsandsystems.com/

35. http://www.EngineeringToolbox.com

36. http://arachnoid.com/calculus/volume2.html

37. https://www.osha.gov/doc/outreachtraining/htmlfiles/hazloc.html

38. http://www.specificsystems.com/

39. http://www.spiraxsarco.com/resources/steam-engineering-tutorials.asp

40. http://engineeronadisk.com/

41. http://pcuserinfo.com/open-system-interconnect-osi-and-tcpip-network-layer-model/

42. http://www.google.com (Images)

43. http://www. wikipedia.org/

44. https://instrumentationtools.com/

Notes